Achieving sustainable cultivation of maize

Volume 2: Cultivation techniques, pest and disease control

It is widely recognised that agriculture is a significant contributor to global warming and climate change. Agriculture needs to reduce its environmental impact and adapt to current climate change whilst still feeding a growing population, i.e. become more 'climate-smart'. Burleigh Dodds Science Publishing is playing its part in achieving this by bringing together key research on making the production of the world's most important crops and livestock products more sustainable.

Based on extensive research, our publications specifically target the challenge of climate-smart agriculture. In this way we are using 'smart publishing' to help achieve climate-smart agriculture.

Burleigh Dodds Science Publishing is an independent and innovative publisher delivering high quality customer-focused agricultural science content in both print and online formats for the academic and research communities. Our aim is to build a foundation of knowledge on which researchers can build to meet the challenge of climate-smart agriculture.

For more information about Burleigh Dodds Science Publishing simply call us on +44 (0) 1223 839365, email info@bdspublishing.com or alternatively please visit our website at www.bdspublishing.com.

Related titles:

Achieving sustainable cultivation of maize Volume 1: From improved varieties to local applications
Print (ISBN 978-1-78676-008-1); Online (ISBN 978-1-78676-010-4, 978-1-78676-011-1)

Achieving sustainable cultivation of rice Volume 1: Breeding for higher yield and quality
Print (ISBN 978-1-78676-024-1); Online (ISBN 978-1-78676-026-5, 978-1-78676-027-2)

Achieving sustainable cultivation of rice Volume 2: Cultivation, pest and disease management
Print (ISBN 978-1-78676-028-9); Online (ISBN 978-1-78676-030-2, 978-1-78676-031-9)

Achieving sustainable cultivation of wheat Volume 1: Breeding, quality traits, pests and diseases
Print (ISBN 978-1-78676-016-6); Online (ISBN 978-1-78676-018-0, 978-1-78676-019-7)

Achieving sustainable cultivation of wheat Volume 2: Cultivation techniques
Print (ISBN 978-1-78676-020-3); Online (ISBN 978-1-78676-022-7, 978-1-78676-023-4)

Chapters are available individually from our online bookshop: https://shop.bdspublishing.com

climate•SMART•publishing
IN AGRICULTURAL SCIENCE

BURLEIGH DODDS SERIES IN AGRICULTURAL SCIENCE

NUMBER 02

Achieving sustainable cultivation of maize

Volume 2: Cultivation techniques, pest and disease control

Edited by Dr Dave Watson, CGIAR Maize Research Program Manager, CIMMYT, Mexico

burleigh dodds SCIENCE PUBLISHING

Published by Burleigh Dodds Science Publishing Limited
82 High Street, Sawston, Cambridge CB22 3HJ, UK
www.bdspublishing.com

Burleigh Dodds Science Publishing, 1518 Walnut Street, Suite 900, Philadelphia, PA 19102-3406, USA

First published 2017 by Burleigh Dodds Science Publishing Limited

Library of Congress Control Number: 2017932196

British Library Cataloguing in Publication Data
A catalogue record for this book is available from the British Library

ISBN 978-1-78676-012-8 (print)
ISBN 978-1-78676-014-2 (online)
ISBN 978-1-78676-015-9 (online)
ISSN 2059-6936 (print)
ISSN 2059-6944 (online)

Typeset by Deanta Global Publishing Services, Chennai, India
Printed by Lightning Source

Contents

Series list

Title	Series number
Achieving sustainable cultivation of maize - Vol 1 From improved varieties to local applications Edited by: Dr Dave Watson, CGIAR Maize Research Program Manager, CIMMYT, Mexico	001
Achieving sustainable cultivation of maize - Vol 2 Cultivation techniques, pest and disease control Edited by: Dr Dave Watson, CGIAR Maize Research Program Manager, CIMMYT, Mexico	002
Achieving sustainable cultivation of rice - Vol 1 Breeding for higher yield and quality Edited by: Prof. Takuji Sasaki, Tokyo University of Agriculture, Japan	003
Achieving sustainable cultivation of rice - Vol 2 Cultivation, pest and disease management Edited by: Prof. Takuji Sasaki, Tokyo University of Agriculture, Japan	004
Achieving sustainable cultivation of wheat - Vol 1 Breeding, quality traits, pests and diseases Edited by: Prof. Peter Langridge, The University of Adelaide, Australia	005
Achieving sustainable cultivation of wheat - Vol 2 Cultivation techniques Edited by: Prof. Peter Langridge, The University of Adelaide, Australia	006
Achieving sustainable cultivation of tomatoes Edited by: Dr Autar Mattoo, USDA-ARS, USA & Prof. Avtar Handa, Purdue University, USA	007
Achieving sustainable production of milk - Vol 1 Milk composition, genetics and breeding Edited by: Dr Nico van Belzen, International Dairy Federation (IDF), Belgium	008
Achieving sustainable production of milk - Vol 2 Safety, quality and sustainability Edited by: Dr Nico van Belzen, International Dairy Federation (IDF), Belgium	009
Achieving sustainable production of milk - Vol 3 Dairy herd management and welfare Edited by: Prof. John Webster, University of Bristol, UK	010
Ensuring safety and quality in the production of beef - Vol 1 Safety Edited by: Prof. Gary Acuff, Texas A&M University, USA & Prof.James Dickson, Iowa State University, USA	011
Ensuring safety and quality in the production of beef - Vol 2 Quality Edited by: Prof. Michael Dikeman, Kansas State University, USA	012
Achieving sustainable production of poultry meat - Vol 1 Safety, quality and sustainability Edited by: Prof. Steven C. Ricke, University of Arkansas, USA	013
Achieving sustainable production of poultry meat - Vol 2 Breeding and nutrition Edited by: Prof. Todd Applegate, University of Georgia, USA	014
Achieving sustainable production of poultry meat - Vol 3 Health and welfare Edited by: Prof. Todd Applegate, University of Georgia, USA	015

Introduction

Maize is one of the most widely-grown crops in the world, both for food and livestock feed, biofuels and other uses. Cultivated on 184 million hectares globally, maize is cultivated in 160 countries throughout the continents of America, Europe, Oceania, Africa, and Asia. Maize is fundamental to global food and economic security, providing up to 30% of total calorie intake in some developing countries. It has been estimated that maize yields need to increase by 60% by 2050 in the context of increasing competition for land, water, energy and labour. Maize production is currently held back by factors such as: lack of available improved cultivars or failure to take up new improved varieties, inadequate crop management and storage, poor soil quality, the impact of pests and diseases, and more extreme weather related to climate change. Research needs to focus on supporting more productive, sustainable and nutritionally-valuable maize cultivation, particularly for smallholders in the developing world.

The two volumes of *Achieving sustainable cultivation of maize* summarise key research addressing these challenges. This volume looks at ways of improving maize cultivation as well as the management of pests and diseases. Volume 1 focuses on breeding, improving nutritional quality and ways of supporting smallholders.

Part 1 Maize cultivation techniques

Crop models both help to identify opportunities for yield improvement and to predict the effects of factors such as changing weather patterns. Chapter 1 reviews the principles and development of existing crop models for maize. It then focuses on the problem of drought stress which is likely to increase in range and severity with climate change. Robust crop models are needed to predict likely drought impacts on maize production. The chapter reviews current research on predicting plant responses under drought and heat stress, particularly biomass production and grain yield formation. These models provide the foundation for assessing what interventions, whether in terms of breeding or cultivation practices, may be needed to maintain yields and grain quality.

To be sustainable maize production must meet a number of objectives. These range from producing a good quality product, optimising yields and ensuring a livelihood for the farmer to protecting the natural resource base in such areas as soil health and the broader environment. Good agricultural practices (GAP) allow farmers to achieve these objectives. However, the gap between potential and actual yields can make it complex to set realistic and achievable yield goals. In addition implementation of GAP will vary according to the demands of different production conditions. Chapter 2 addresses ways setting targets, establishing the right mix of cultivation practices and measuring performance. It includes case studies of practical GAP implementation in three different situations: irrigated production in Nebraska, smallholder rainfed production in the semi-arid Central Rift Valley of Ethiopia (CRVE), and rainfed maize in the Pampas of Argentina.

New maize varieties will not be effective if they are not properly disseminated and adopted by local farmers. Chapter 3 reviews the development and effectiveness of the collaborative drought and low nitrogen maize breeding program initiated by the International Maize and Wheat Improvement Centre (CIMMYT) to increase yields in low input and drought prone environments in southern Africa. Reflecting themes discussed in Volume 1, it describes initiatives such as CIMMYT's use of mother-baby trial (MBTs), an innovative farmer participatory evaluation scheme that evaluates farmer responses to new maize varieties in their own environmental conditions. This approach has accelerated the development and dissemination of new stress tolerant and high yielding open pollinated varieties (OPVs) and hybrids in southern Africa. The chapter also highlights the development and selection of maize germplasm resistant to a range of biotic and abiotic stresses, and the role of the seed industry in disseminating improved maize germplasm.

Maize is the most important staple food crop for over 300 million people in Sub-Saharan Africa but poor management practices and problems related to climate and soil quality mean that yields regularly fall below what is needed to feed the population. Chapter 3 begins by describing the effect on maize yields of poor soil fertility, drought and weeds (especially Striga hermonthica). Building in particular on Chapter 2, it then identifies good agricultural practices (GAP) to address these problems. These techniques include targeted nutrient management to improve soil health and techniques for weed management. The chapter also examines ways of dealing with the challenge of drought, including improved irrigation methods and development of drought-resistant maize varieties.

Chapter 5 builds on Chapter 4 by focussing on the role of zero-till cultivation of maize. This technique aims to reduce erosion, improve soil structure and nutrient composition, decrease costs, and improve long-term sustainability of maize cultivation. However, it is important to understand what factors determine whether the technique is effective or not. The chapter reviews current research on the effects of zero till cultivation in such areas as reduced soil erosion and nutrient losses, water infiltration and soil organic matter. This chapter discusses suggestions for best management of zero-till maize systems.

Zero-till cultivation is one element in a broader set of practices known as 'conservation agriculture' (CA). CA emphasises practices designed to improve soil health such as: zero or reduced tillage; retention of adequate levels of crop residues and soil surface cover to improve water productivity and to enhance soil physical, chemical and biological properties; and use of crop rotations to help control weeds and pests, improve soil health and provide more diverse sources of income for farmers. Based on long-term experimental sites managed by CIMMYT, Chapter 6 provides an authoritative review of the effectiveness of CA in such areas as physical soil quality (including reducing erosion), soil chemical and biological quality, and weed management. As it shows, forty years of research show that conservation agriculture has proved the better choice in providing continuously higher and more stable yields for both wheat and maize when compared to traditional practices in rainfed conditions. The chapter discusses the future potential of CA in maize cultivation and, specifically, its further development in Latin America.

As Chapter 7 points out, information and communication technologies are transforming the way maize is cultivated around the world by providing farmers with a suite of novel tools and techniques for improving crop production, enhancing input use efficiency, and increasing profitability while achieving environmental sustainability. Chapter 7 provides an overview of the current state of precision maize cultivation techniques at different stages of maize cultivation. A section on pre-planting assesses current knowledge on the spatial variability of soil properties and how it influences precision maize cultivation techniques. Subsequent sections address planting and the current state of variable-rate seeding, nutrition and various approaches for site-specific fertilizer management. The chapter also reviews developments in precision irrigation, different aspects of precision pest control and weed management, as well as the opportunities enabled by digital yield maps and big data management.

Sustainable cultivation requires more efficient resource use, including in the use of fertilisers. Chapter 8 looks at the various ways of improving nutrient management in maize cultivation. It includes a range of case studies from South Asia, Sub-Saharan Africa, Latin America as well as the use of the Nutrient Expert® tool in China. As an example, in the case of Sub-Saharan Africa, the chapter reviews ways of targeting fertilisers according to different soil conditions, establishing optimum nitrogen application rates and more effective timing of nitrogen application (depending on prevailing rainfall) as well as more targeted, site-specific fertilizer application. It shows for example that, when combined with mineral nitrogen fertilizers, manure resulted in better crop yields than fertilizer treatments alone. The chapter also demonstrates the success of more flexible systems of fertilization, in which optimum rates of P, K and S fertilizers are applied based on yield potential in an average rainfall season, while N is applied as a series of split applications, which are adjusted according to the evolving rainfall pattern in any one season.

Modern agriculture has led to a simplification of maize-based cropping systems and the extensive practice of monoculture maize. This has led to problems such as yield reduction and soil quality degradation. Meanwhile, as pointed out in Chapter 8, the overuse of inorganic fertilizer to increase yields comes with a high environmental cost. As suggested in Chapter 6, maize grown in rotation with grain or forage legume crops often yields more and requires less application of synthetic chemicals (fertilizers, pesticides, herbicides, etc.) than continuous monoculture maize. Rotations in tandem with the application of farmyard or dairy manure or other organic wastes has proved an environmentally-friendly strategy for sustainable agriculture development. Chapter 9 summarizes recent research on yield enhancement in maize-based rotation systems, as well as the impact of crop rotation on soil quality, fertility and nutrient use efficiency.

Chapter 10 reviews the existing evidence on the productivity and resource use efficiency of maize–legume intercropping systems compared to the use of single crops. Findings indicate that system productivity in intercropping systems is more resource-use efficient and productive, particularly in low-yielding environments and production systems where soil quality is poor and rainfall low. As the level of resource availability increases, the differences between intercropping and sole cropping are reduced, though intercropping still presents advantages over sole cropping in some cases.

Climate change, and the likelihood of hotter, drier conditions in many parts of the world, presents a particular threat to rainfed maize cultivation. Chapter 11 assesses the vulnerability of rainfed maize cultivation to changing conditions and the risk of rainfall variability. Supporting farmers to identify and manage risks and opportunities in these challenging conditions is essential. The chapter summarises recent research on risk management strategies for smallholder maize cropping. It concludes with a case study exploring practical options for the sustainable intensification of rainfed maize cropping in Mozambique.

Part 2 Maize pests, diseases and weeds

Maize yields in many countries are negatively affected by damage from insect pests. Chapter 13 describes the main pests of maize. These include below-ground pests such as various types of corn rootworm, above-ground pests such as the corn leaf aphid, as well as pests of stored maize products such as the larger grain borer. The chapter also includes a detailed case study of the European corn borer in the USA, assessing its life cycle and what this means for control using integrated pest management (IPM) techniques. Finally, the chapter assesses potential future directions for research in this area.

Plant-feeding, or plant-parasitic, nematodes are ubiquitous in soils that support plant growth. The effects of phytophagous nematodes are, however, notoriously difficult to test and measure, and depend upon many factors associated with the plant, its environment, and the nematodes themselves. This chapter begins with an introduction to soil and plant nematodes, covering their general characteristics, identification and an overview of our current knowledge of their effects on maize. Subsequent sections review various types of nematodes, with particular focus on those which are highly likely to cause yield losses and other damage to maize crops such as lesion, root-knot and vermiform nematodes.

Rodents cause significant damage at the sowing, seedling and maturity stage of maize. Rodent management programs in regions such as Sub-Saharan Africa have traditionally been reactive and most suited to managing low-density local rodent populations. More effective rodent management requires a more proactive and integrated approach. Chapter 15 describes the range of cultural, chemical and biological methods of control that make up an integrated rodent management programme. It also looks at improvements in surveillance and early warning systems.

Emerging plant diseases (EPDs) can cause significant losses to maize and other crops. Chapter 16 reviews research on factors influencing the origins and spread of EPDs affecting maize. Effective control of EPDs relies on rapid detection, accurate diagnosis, as well as speedy deployment of preventive and containment measures to prevent the spread of disease. The efficiency of a response also depends on networking and collaboration among all the stakeholders, including intergovernmental and nongovernmental organizations and specialized agencies in developed and developing countries. The chapter reviews strategies for mitigating risks from emerging maize diseases, components of an effective rapid response system and strategies for managing maize disease outbreaks. These include effective models to predict the likely spread of disease, quarantine procedures, effective seed systems, integrated disease management and pre-emptive breeding measures. The chapter includes a case study of how procedures worked in practice with the emergence of maize lethal necrosis (MLN) in eastern Africa, reviewing how effective various agencies were in responding to the MLN outbreak and minimizing its impact.

Maize is subject to pre- and post-harvest contamination with aflatoxins, which are acutely toxic and carcinogenic compounds. This chapter describes in detail mechanisms of aflatoxin contamination in maize. It also reviews the current strategies employed for aflatoxin control and the challenges associated with them, including pre- and post-harvest methods of control and prevention. The chapter also addresses the detoxification of aflatoxin-affected maize as well as suggesting lines of future research in this area.

Weeds are one of the major constraints on maize productivity. Chapter 18 discusses the impact of weeds on maize cultivation and methods of control. It focuses on integrated weed management (IWM) techniques which reduce the need for herbicides. IWM techniques include methods of cultural control such as: land preparation, choice of cultivars, planting schemes and methods of cultivation. They also include cropping systems such as rotations, intercropping and cover crops, as well as physical methods of control. The development of herbicide-tolerant maize cultivars and image-based site specific herbicide applications are expected to contribute significantly in improving weed management in maize.

Chapter 18 is complemented by Chapter 19 which reviews the types of maize cropping systems used to grow maize under temperate conditions, with a particular focus on the situation in the US and Europe. After reviewing weed flora in these regions, it discusses current weed management systems and the herbicide resistance issues that have developed through the continuous use of herbicides. Echoing Chapter 18, it emphasises the importance of integrated weed management (IWM) techniques in achieving more sustainable maize production. The chapter concludes with IWM case studies in three European maize production regions.

Part 1

Maize cultivation techniques

Chapter 1

Modelling crop growth and grain yield in maize cultivation

Alam Sher, Xiaoli Liu and Jincai Li, Anhui Agricultural University, China; and Youhong Song, Anhui Agricultural University, China and The University of Queensland, Australia

1 Introduction

Maize is one of three most important grain crops in addressing global food security. It is a C4 crop with high productivity, but requires considerable water during the growing season. Maize can be grown over a wide range of climate scenarios, but performs especially well in warmer temperate regions and humid subtropics. However, maize cultivation is constrained by water availability worldwide. There have been continuous efforts in breeding new maize varieties adapted to diminished water use (Gong et al., 2015). It is noted that drought stress, often accompanied with heat stress in summer, is projected to be more frequent and severe due to global climate change (Feller and Vaseva, 2014). Therefore, it is essential to predict likely drought impacts on maize production and then take action, which largely relies on robust crop models. An emerging challenge is to enhance the ability of crop models in predicting crop growth, development and grain yield under a wide range of environmental conditions (Jin et al., 2016).

Crop physiological models have been developed since the 1960s and are used to simulate crop growth and development, grain yield in response to various environments and managements. The framework of crop models is composed of various modules including phenological development, leaf area development, biomass estimation, grain yield formation, and water and nitrogen balance. The prediction of maize growth, development and grain yield under drought stress has been implemented in existing crop models such as CERES-Maize (Panda et al., 2004; DeJonge et al., 2012) and APSIM-Maize

http://dx.doi.org/10.19103/AS.2016.0002.01

(Song et al., 2010) by calibrating the model parameters including hybrids and environments. These models have been widely applied in arid and semi-arid areas worldwide (Cavero et al., 2000; López-Cedrón et al., 2008; Song et al., 2010; DeJonge et al., 2012). Drought stress during the vegetative stage is less detrimental to grain yield loss compared to drought stress during reproductive stage. Bracketing silking drought stress causes fatal impacts on final grain yield. However, when facing changing climate, how these models perform in capturing the impacts of drought and heat remain to be determined (Barnabás et al., 2008). This review will cover the development of crop models in accounting for drought stress and will examine the scope of the existing model in predicting impacts of drought and heat stress under climate change.

2 Crop modelling principles

Crop models are reliable tools in integrating information from different crop subsystems to better understand how a given crop system responds to climatic and management practices. Crop modelling provides a framework for how the interaction of crop and environments can be used to improve the prediction of maize growth and yield under water-limited conditions (Hammer et al., 2002). For example, APSIM modelling principles consist of different modules: i) a set of biophysical modules that predict biological and physical processes in different cropping patterns, ii) a set of management modules that are used to specify the intended management rules that improve the scenario being simulated and that control the conduct of the simulation, iii) various modules that are available to facilitate data input and output for the simulation and iv) a simulation engine that uses the simulation and manages all messages passing between the independent modules (Keating et al., 2003). According to modelling principles, some major maize models (APSIM-Maize, CERES-Maize, Hybrid-Maize and MAIZSIM) with their modules (phenology, photosynthesis, apical development, biomass growth and partitioning, transpiration, soil nitrogen balance, drought and heat stress, pest damage, residue management, planting date and plant density), are presented in Table 1. This table also include how these models predict drought and heat stress impact on different developmental stages of maize crop.

Among all maize crop models, APSIM's capability is well developed for different farming systems, especially in temperate and tropical areas. Finally, improvement and progress in terms of specific aspects of modelling required for use under drought and heat stress conditions for future prospect have been identified.

3 Predicting crop development under drought and heat stress

Crop development is governed by phenological events, sensitive to drought and heat stress (Birch et al., 1998; Parent et al., 2010), so the precise prediction of phenology is critical in modelling crop growth and yield under drought/heat stress. Modelling phenological development expresses the rate of change in life stage as a function of environmental variables, mainly temperature stress (Wilson et al., 1995; Yin et al., 1995; Yan and Hunt 1999). Phyllochron is defined as the thermal intervals of two consecutive

Table 1 Modules of different maize models to predict drought/heat stress impact on maize crop

Model name	Model modules	Reference	Predict impacts of drought/heat stress
APSIM-Maize	Phenology, photosynthesis, biomass growth and partitioning, transpiration, soil nitrogen and water deficit, soil temperature, pest damage, residue management	Wilson et al. (1995)	Predicted drought impacts on maize growth and development in southeast Queensland, Australia (Song et al., 2010)
		Jones et al. (2001)	Modelling impacts of heat and drought stresses on yield (Jin et al., 2016)
			Extreme heat stress for maize production in the United States (Lobell et al., 2013)
CERES-Maize	Phenology, apical development, biomass accumulation and partitioning, soil water and nitrogen balance	Jones and Kiniry (1986)	Predicted drought impacts on maize growth and development and yield with a modified version of CERES-Maize (Nouna et al., 2003)
		Castrignano et al. (1998)	New modelling of drought stress impact on leaf area functioning and maize growth (Nouna et al., 2003; Mastrorilli et al., 2003)
		Mastrolli et al. (2003)	Predicted heat stress impact on maize production in the United States and its use on food security (Chung et al., 2014)
			Modelling heat stress impact on maize growth and yield (Ban et al., 2015)
Hybrid-Maize	Phenology, soil moisture status and management, yield prediction based on planting date and plant density	Yang et al. (2004, 2006)	Prediction of corn yield under heat stress by University of Nebraska (Lincoln 2014). Modelling drought stress effect on yield under different planting density and date in maize crop (Yang et al., 2006)
MAIZSIM	Phenology, range of environmental, soil, and management conditions	Yan and Hunt (1999)	Predicted heat stress effect on leaf growth, development and biomass in maize (Kim et al., 2012)
		Kim and Reddy (2004)	Prediction of transpiration and photosynthesis under drought stress in maize (Yang et al., 2009)
		Kim et al. (2012)	Drought stress effects on stomatal conductance (Tuzet et al., 2003)

leaf tip appearances, which is often used to characterise crop development and growth stages. Different thermal functions and phyllochron values are available (Birch et al., 1998) and crop models that use the relationship between maize development and thermal time

Table 2 List of different models that predict maize phenology under drought/heat stress

Model name	Details to predict phenology	References	Predict impacts of drought/ heat stress
APSIM-Maize	Emergence, leaf initiation, floral initiation, grain fill and maturity according to thermal time	Wilson et al. (1995)	Predicted drought stress effect on leaf area index (Song et al., 2010)
			Predicted heat stress effect on anthesis, silking and physiological maturity in maize crop (Kumudini et al., 2014)
CERES-Maize	Thermal time (GDD 8,34)	Jones and Kiniry (1986)	Modelling drought stress impact on leaf growth and senescence in Australia (Carberry et al., 1989)
			Modelling soil temperature impact on leaf area index (Sandor and Fodor, 2012)
			Predicted heat stress effect on anthesis, silking and physiological maturity (Kumudini et al., 2014)
MAIZSIM	This model simulates leaf phenology by four processes: initiation, appearance, expansion and senescence of individual leaves	Yan and Hunt (1999)	Predicted heat stress effect on leaf area index (Kim et al., 2012)
		Kim and Reddy (2004)	Predicted heat stress effect on anthesis, silking, and physiological maturity (Kumudini et al., 2014)
		Kim et al. (2012)	
Hybrid-Maize	Phenological stage is determined by GDD and photoperiod	Yang et al. (2004, 2006)	Prediction of emergence and silking by GDD (Yang et al., 2004)

but these functions vary in a number of ways (Kumudini et al., 2014). Vegetative stages are simply the repetition of vegetative organs while tassel initiation marks the cessation of vegetative organ occurrence and onset of reproductive organs. It is widely reported that maize tassel initiation, silking and anthesis are delayed by drought stress (Abrecht and Carberry, 1993; Edmeades et al., 2000; Campos et al., 2004; Barnabas et al., 2008) and heat stress (Basetti and Westgate, 1993; Rattalino Edreira et al., 2011; Cicchino et al., 2010). For maize under drought stress, flowering is slightly delayed, but seed filling and physiological maturity is accelerated.

The capacity of different maize phenological models in APSIM-Maize, CERES-Maize, MAIZSIM and Hybrid-Maize to predict phenology and drought and heat stress impact on different phenological stages in maize is presented in Table 2. The APSIM model uses multi-linear functions derived from experimental observation to predict emergence, leaf initiation, floral initiation, grain filling and maturity. MAIZSIM predicts leaf phenology composed of four processes: initiation, appearance, expansion and senescence of individual leaves using *Beta* function; CERES-Maize and Hybrid-Maize phenology model integrating growing degree days (GDD) and photoperiod, for example, prediction of silking stage (Wilson et al., 1995; Jones and Kiniry, 1986; Kim et al., 2012; Yang et al., 2004; 2006).

4 Predicting biomass production under drought and heat stress

In the absence of stress, crop dry matter accumulation is determined by the intercepted quantity of radiation by the canopy (Monteith, 1977; Sinclair and Muchow, 1999) and radiation use efficiency (RUE). The simple relationships are often used to predict biomass production in many crop models (Ko et al., 2010; also see review of White et al., 2011). Light absorbed by the canopy depends on canopy architecture, which can be affected by drought stress (Song et al., 2010) and heat stress (Stone 2001), and the RUE parameter varies with varieties, location and environment (Gallagher and Biscoe, 1978; Gosse et al., 1986). RUE is reduced by drought stress (Monteith, 1977) and heat stress (Rattalino et al., 2012). Heat stress has a negative effect on biomass production due to the amount of light captured by the crop. Heat stress during vegetative stage reduces RUE in the maize crop (Andrade et al., 1992; Oyun et al., 2011). In crop modelling, biomass accumulation is predicted in different ways: (1) estimation of RUE, which relates biomass to intercepted light in CERES, STICS, APSIM and CropSyst; (2) CERES-PR and CSM-IXIM at leaf level predicted photosynthesis and respiration; and (3) a more mechanistic approach with the integration of a model of leaf photosynthesis (Farquhar, 1980) in GECROS. Mostly, a reduction factor is used to predict the impact of drought stress on RUE or its components, or indirectly, as in APSIM-Maize, where biomass accumulation is simulated as the minimum of 'the light-limited biomass accumulation' and 'the water deficit biomass accumulation', which uses available water and transpiration efficiency (Hammer et al., 2010).

5 Predicting grain yield formation under drought and heat stress

Grain yield is determined by kernel number and kernel weight. Grain yield is closely associated with kernel number at harvest, and this component is a function of the physiological condition of the crop (Jacobs and Pearson, 1991; Otegui and Andrade, 2000; Tollenaar et al., 2000). Kernel number is very sensitive to environmental stresses including drought and heat stress (Nesmith and Ritchie, 1992). Anthesis-silking interval (ASI) under drought becomes shorter in modern hybrids, the growth of spikelets and ears increases, and the final kernel number and yield decreases (Banziger et al., 2000). Grain yield was reduced by 47% due to lower kernel weight when the water deficit occurred 22–26 days after silking (NeSmith a and Ritchie, 1992). Drought stress at tasselling and grain-filling stages decreased grain yield by 66.4% and 44.9%, respectively, and it reduced grain number/ear by 59.3% and 30.1%, respectively (Moosavi et al., 2012). Maize yield trials in Africa over an eight-year period showed that for every degree above 30°C grain yield was reduced by 1–1.7% under drought conditions (Lobell et al., 2011). Direct heating of kernels above 30°C and 35°C lead to reduced kernel number and yield (Jones et al., 1984; Lobell and Burke 2010; Carcova and Otegui, 2001). The effect of heat stress from flowering to grain filling often limits final crop yield due to reduction in kernel growth (Guilioni et al., 2003).

Maize yield reduction is more sensitive to drought stress than to heat stress in the US Midwest. There are different models that predict the impacts of heat and drought stress on maize biomass and yield production (Jin et al., 2016). For example, some models like CSIRO-Mk3.0 and MIROC 3.2 also give the same results and predict that, due to heat stress, average grain yields would decrease by 38% in 2050s (Chung et al., 2014). CERES-Maize, Crop-Syst and the EPIC model predict biomass and yield of maize under drought conditions (DeJonge et al., 2012; Ko et al., 2010; López-Cedrón et al., 2008). Some models are unreliable in predicting yield because they have relatively large errors in simulating kernel number and kernel weight under heat stress (Ban et al., 2015).

6 Conclusions

Climate change causes more frequent drought and heat events, which limits the productivity of maize in arid and semi-arid areas, therefore threatening future food security. Crop models have evolved to embrace the interaction of crops by environments and precisely predict climate change impacts on maize production. By calibrating and revising specific modules, the existing crop models have been successfully applied to arid and semi-arid areas accounting for drought impacts. Predicting heat impact was seldom considered, but has received more attention recently. The effects of drought and heat stress on grain yield are hypo-additive, which makes it difficult to predict the impact of drought and heat stress. The reproductive process during which kernel number is determined is most sensitive to drought and heat stress. The quantification of maize grain yield formation, including determining kernel number in relation to drought and heat stress, is critical in enabling the models to be used under climate change.

7 Where to look for further information

The emerging challenge is to improve the ability in predicting drought and heat stress impact on crop productivity. This chapter overviewed the scope of existing crop models in considering drought and heat stress. Further information can be found in the key references, for example, Barnabás et al. (2008), Feller et al. (2014) and Kumudini et al. (2014).

8 References

Abrecht, D. G. and Carberry, P. S. 1993. The influence of water deficit prior to tassel initiation on maize growth, development, and yield. *Field Crops Res.* 31, 55–69.

Andrade, F. H., Uhart, S. A. and Frugone, M. I. 1992. Intercepted radiation at flowering and kernel number in maize: shade versus plant density effects. *Crop Sci.* 33, 482–5.
Ban, H., Sim, D., Lee, K. et al. 2015. Evaluating maize growth models 'CERES-Maize' and 'IXIM-Maize' under elevated temperature conditions. *J. Crop Sci. Biotechnol.* 18, 265. doi:10.1007/s12892-015-0071-3.
Barnabás, B., Katalin, J. and Attila, F. 2008. The effect of drought and heat stress on reproductive processes in cereals. *Plant Cell Environ.* 31, 11–38.
Basetti, P. and Westgate, M. E. 1993. Water deficit affects receptivity of maize silks. *Crop Sci.* 33, 279–82.
Birch, C. J., Vos, J., Kiniry, J., Bos, H. J. and Elings, A. 1998. Phyllochron responds to acclimation to temperature and irradiance in maize. *Field Crops Res.* 59, 187–200.
Cavero, J., Farré, I., Debaeke, P. and Faci, J. M. 2000. Simulation of maize yield under water stress with the EPIC phase and CROPWAT models. *Agron. J.* 92, 679–90.
Chung, U., Sika, G., Bekele, S., Richard, R., Jin, I. Y., Kindie, T., Gerrit, H. and Kai, S. 2014. Modeling the effect of a heat wave on maize production in the USA and its implications on food security in the developing world. *Weather Clim. Extremes* (5–6), 67–77.
Cicchino, M., Rattalino Edreira, J. I., Uribelarrea, M. and Otegui, M. E. 2010. Heat stress in field grown maize: response of physiological determinants of grain yield. *Crop Sci.* 50, 1438–48.
Craufurd, P. Q., Prasad, P. V. and Summerfield, R. J. 2002. Dry matter production and rate of change of harvest index a high temperature in peanut. *Crop Sci.* 42, 146–51.
DeJonge, K. C., Ascough, I. I., Andales, J. C., Hansen, A. A., Garcia, N. C. and Arabi, L. A. M. 2012. Improving evapotranspiration simulations in the CERES-Maize model under limited irrigation. *Agric. Water Manage.* 115, 92–103.
Demba, T. N. A., Heidi, W., Wilson, A. A., Mathias, F., Jesse, N. and Thomas, G. 2015. Modelling heat stress effect on two maize varieties in Northern Region of Ghana. *Glo. Adv. Res. J. Agric. Sci.* 4 (3), 145–55.
Edmeades, G. O., Bolanos, J., Elings, A., Ribaut, J. M., Banziger, M. and Westgate, M. E. 2000. The role and regulation of the anthesis silking interval in maize. In *Physiology and Modeling Kernel Set in Maize* (eds M. E. Westgate and K. Boote), pp. 43–73. Crop Science Society of America, Madison, WI, USA.
Farquhar, G. D., Caemmerer, S. V. and Berry, J. A. 1980. A biochemical model of phtosynthetic CO^2 assimilation in leaves of C-3 species. *Planta* 149, 78–90.
Feller, U. and Vaseva, I. I. 2014. Extreme climatic events: impacts of drought and high temperature on physiological processes in agronomically important plants. *Front. Environ. Sci.* 2, 1–16.
Gong, F., Wu, X., Zhang, H., Chen, Y. and Wang, W. 2015. Making better maize plants for sustainable grain production in a changing climate. *Front Plant Sci.* 6, 835.
Guilioni, L., Wery, J. and Lecoeur, J. 2003. High temperature and water deficit may reduce seed number in field pea purely by decreasing plant growth rate. *Funct. Plant Biol.* 30, 1151–64.
Hammer, G. L., van Oosterom, E., McLean, G., Chapman, S. C., Broad, I., Harland, P. and Muchow, R. C. 2010. Adapting APSIM to model the physiology and genetics of complex adaptive traits in field crops. *J. Exp. Bot.* 61, 2185–2202.
Hammer, G. L., Kropff, M. J., Sinclair, T. R. and Porter, J. R. 2002. Future contributions of crop modelling – from heuristics and supporting decision making to understanding genetic regulation and aiding crop improvement. *Eur. J. Agron.* 18, 15–31.
Jacobs, B. C. and Pearson, C. J.1991. Potential yield of maize, determined by rates of growth and development of ears. *Field Crops Res.* 27, 281–98.
Jin, Z., Zhuang, Q., Tan, Z., Dukes, J. S., Zheng, B. and Melillo, J. M. 2016. Do maize models capture the impacts of heat and drought stresses on yield? Using algorithm ensembles to identify successful approaches. *Glob. Change Biol.* 22 (9), 3112–26.
Jones, C. A. and Kiniry, J. R. 1986. *CERES-Maize: A Simulation Model of Maize Growth and Development.* Texas A&M University Press, College Station, TX.

Keating, B. A., Carberry, P., Hammer, G., Probert, M. E., Robertson, M., Holzworth, D., Huth, N., Hargreaves, J., Meinke, H. and Hochman, Z. 2003. An overview of APSIM, a model designed for farming systems simulation. *Eur. J. Agron.* 18, 267–88.

Kim, S. H., Yang, Y., Timlin, D. J., Fleisher, D. H., Dathe, A., Reddy, V. R. and Staver, K. 2012. Modeling temperature responses of leaf growth, development, and biomass in maize with MAIZSIM. *Agron. J.* 104, 1523–37.

Ko, J., Piccinni, G. and Steglich, E. 2010. Using EPIC model to manage irrigated cotton and maize. *Agric. Water Manage.* 96, 1323–31.

Kumudini, S., Andrade, F. H., Boote, K. J., Brown, G. A., Dzotsi, K. A., Edmeades, G. O., et al. 2014. Predicting maize phenology: intercomparison of function for developmental response to temperature. *Agron. J.* 106, 1–11.

Lindquist, J. L., Arkebauer, T. J., Walters, D. T., Cassman, K. G. and Dobermann, A. 2004. Maize radiation use efficiency under optimal growth conditions. *Agron. J.* 97 (1), 72–8.

Lobell, D. B. and Burke, M. B. (2010). On the use of statistical models to predict crop yield responses to climate change. *Agric. Forest Metero.* 150, 1443–52.

Lobell, D. B., Bänziger, M., Magorokosho. C. and Vivek, B. 2011. Nonlinear heat effects on African maize as evidenced by historical yield trials. *Nat. Clim. Change* 1, 42–5.

Lobell, D. B., Hammer, G., Mclean, G., Messina, C. Roberts, M. J. and Sclenker, W. 2013. The critical role of extreme heat for maize production in the United States. *Nat. Climate Change* 3, 497–501, doi:10.1038/nclimate1832.

López-Cedrón, F. X., Boote, K. J., Pineiro, J. and Sau, F. 2008. Improving the CERES-Maize model ability to simulate water deficit impact on maize production and yield components. *Agron. J.* 100, 296–307.

Mastrorilli, M. N., Katerije, B. Bnouna. 2003. Using Ceres-Maize model in a semi- arid Mediterranean environment. Validation of three revised versions. *Europ. J. Agron.* 19, 125–34.

Monteith, J. L. and Moss, C. J. 1977. *Climate and the efficiency of crop production in Britain.* Philosophical Transactions Royal Society London. Series B, Biological Sciences Vol. 281, No. 980, The management of inputs for yet greater agricultural yield and efficiency (Nov. 25, 1977), pp. 277-94.

Moosavi, S. G. 2012. The effect of water deficit stress and nitrogen fertilizer levels on morphology traits, yield and leaf area index in maize. *Pak. J. Bot.* 44 (4), 1351–5.

NeSmith, D. S. and Ritchie, J. T. 1992. Maize (Zea mays L.) response to a severe soil water-deficit during grain-filling. *Field Crops Res.* 29, 23–35.

Nouna, B. B., Katerji, N. and Mastrorilli, M. 2003. Using the CERES-Maize model in a semi-arid Mediterranean environment. New modelling of leaf area and water stress functions. *Eur. J. Agron.* 19 (2), 115–23.

Otegui, M. E. and Andrade, F. R. 2000. New relationships between light interception, ear growth, and kernel set in maize. In M. E. Westgate and K. J. Boote (eds), *Physiology and Modeling. Kernel set in Maize*, CSSA, Madison, WI, pp. 89–102.

Oyun, T. B., Masato, S. and Mitsuru, T. 2011. Effect of water and temperature stresses on radiation use efficiencies in a semi- arid grassland. *J. Plant Interac.* 7 (3), 214–24.

Panda, R. K., Behera, S. K. and Kashyap, P. S. 2004. Effective management of irrigation water for maize under stressed conditions. *Agric. Water Manage.* 66, 181–203.

Parent, B., Turc, O., Gibon, M., Stitt, M. and Tardieu, F. 2010. Modelling temperature-compensated physiological rates, based on the co-ordination of responses to temperature of developmental processes. *J. Exp. Bot.* 61, 2057–69. doi:10.1093/jxb/erq003.

Rattalino Edreira, J. I., Budakli Carpici, E., Sammarro, D. and Otegui, M. E. 2011. Heat stress effects around flowering on kernel set of temperate and tropical maize hybrids. *Field Crops Res.* 123, 62–73.

Sandor, R. and Fodor, N. 2012. Simulation of soil temperature dynamics with models using different concepts. *Sci. World J.* 8. Article ID 590287.

Shaw, R. H. and Thom, H. C. S. 1951. On the phenology of field corn, silking to maturity. *Agron. J.* 43, 541–6. doi:10.2134/agronj1951.00021962004300110004x.

Song, Y., Birch, C., Shanshan, Q., Doherty, A. and Hanan, J. 2010. Analysis and modelling of the effects of water stress on maize growth and yield in dryland conditions. *Plant Prod. Sci.* 13 (2), 199–208.

Stewart, D. W., Dwyer, L. M. and Carrigan, L. L. 1998. Phenological temperature response of maize. *Agron. J.* 90, 73–9.

Stone, P. 2001. The effects of heat stress on cereal yield and quality. In A. S. Basra (ed.), *Crop Responses and Adaptations to Temperature Stress*. Food Products Press, Binghamton, NY, USA, pp. 243–91.

Tollenaar, M., Dwyer, L. M., Stewart, D. W. and Ma, B. L. 2000. Physiological parameters associated with differences in kernel set among maize hybrids. In M. E. Westgate and K. Boote (eds), *Physiology and Modeling Kernel Set in Maize*. CSSA Special Publication Number 29. Madison, WI, pp. 115–30.

Wilson, D. R., Muchow, R. C. and Murgatroyd, C. J. 1995. Model analysis of temperature and solar radiation limitations to maize potential productivity in a cool climate. *Field Crops Res.* 43, 1–18.

Yan, W. and Hunt, L. A. 1999. An equation for modelling the temperature response of plants using only the cardinal temperatures. *Ann. Bot.* 84, 607–14.

Yang, H., Dobermann, A., Lindquist, J. L., Walters, D. T., Arkebauer, T. J. and Cassman, K. G. 2004. Hybrid-maize a maize simulation model that combines two crop modeling approaches. *Field Crop Res.* 87 (2–3), 131–54.

Yang, H., Dobermann, A., Walters, D. T., Arkebauer, T. J. and Cassman, K. G. 2006. Hybrid-Maize: A simulation model for corn growth and yield. Department of Agronomy and Horticulture, University of Nebraska, Lincoln.

Yin, X., Kropff, M. J., MacLaren, G. and Visperas, R. M. 1995. A nonlinear model for crop development as a function of temperature. *Agric. For. Meteorol.* 77, 1–16.

Chapter 2

Optimizing maize-based cropping systems: sustainability, good agricultural practices (GAP) and yield goals

Charles Wortmann, Patricio Grassini and Roger W. Elmore, University of Nebraska-Lincoln, USA

1 Introduction

This chapter addresses sustainable crop production and how good agricultural practices (GAP) can optimize sustainable production. We outline here a range of metrics for setting and monitoring targets such as yield potential, yield gaps and yield goals and review metrics for sustainability targets and considerations in selecting between competing priorities. These are discussed in consideration of three contrasting maize cultivation systems: irrigated production in Nebraska, smallholder rain-fed production in the semi-arid Central Rift Valley of Ethiopia (CRVE) and rain-fed maize in the Pampas of Argentina.

2 Defining sustainable crop production

Definitions of sustainable production and sustainable intensification abound but commonly include meeting current needs without compromising the ability to meet future needs (Petersen and Snapp, 2015). The essential goal of farming is to provide food, feed, fibre and fuel while protecting the environment and ensuring long-term sustainability (Loomis, 1984; Sadras et al., 2009; Connor et al., 2011). The failure to take proper account of long-term sustainability has led to the collapse of some agrarian societies (Diamond, 2005).

http://dx.doi.org/10.19103/AS.2016.0002.02

A sustainable crop production system not only produces adequate food supplies but also generates profit for the farmer; protects the integrity of the resource base; improves the quality of life of individuals and communities and ensures the nutritional value and safety of food (FAO Council, 1989; Sadras et al., 2009).

The long-term temporal scale associated with the concept of sustainability is crucial: a production system can be considered sustainable only if production levels can be maintained over years at the same site (Marten, 1988). Spatial scale is also important when assessing sustainability. An indicator that may be useful to assess field-scale sustainability may have little value at larger spatial levels and *vice versa*. For example, a high water productivity may be desirable for an irrigated field, but not so relevant at a watershed/region level if excess water is reused in adjacent fields or used to recharge streams or aquifers.

3 Good agricultural practices

GAP are a key tool in achieving sustainable agriculture. GAP for a specific crop need to balance the numerous dimensions of sustainable crop production. Through additive and interactive effects, as part of a set of practices, GAP are expected to contribute to maintenance or improvement of the following characteristics of a crop production system (e.g. FAO Council, 1989):

1. Productivity of land, water and human resource
2. Farmer profitability
3. Farm family/community livelihood and operational security
4. The natural resource base of the farm, particularly soil and water
5. The surrounding environment
6. Provision of quality and safe produce
7. Worker safety
8. Capacity for adaptive management and resilience to extreme stresses
9. Resilience to extreme stresses
10. Economy of the broader community
11. Social values

Finally, GAP also need to be achievable and to demonstrate benefits for farmers compared to existing cultivation practices. GAP include the following operations (www.fao.org/prods/gap):

- Soil and water management
- Crop production (including harvesting, on-farm storage and processing)
- Crop protection
- Energy and waste management
- Health and safety requirements
- Wildlife and landscape conservation

The importance of GAP can be seen even for relatively advanced, highly productive agricultural systems, such as those in Nebraska and other states in the Great Plains of the United States. An example is soil management. Many vulnerable fields in states such as

Nebraska were rendered less productive or unsuitable for crop production by wind erosion, most notably during the American Dust Bowl years of the 1930s (Hornbeck, 2009). Soil loss from highly erodible lands by water erosion has also caused much land degradation, compared to soil conditions at the time of conversion of perennial grass prairies to annual crop production (Anderson, 2012). Most of the damaged fields are currently components, albeit often with reduced land productivity, of successful farm enterprises.

One of the greatest threats to fields and the surrounding environment in the semi-arid CRVE is soil erosion mostly by water but also by wind, due to inadequate ground cover and lack of incentives to implement land conservation measures. Berry (2003) estimated the direct cost of soil erosion in Ethiopia to equal 3% of the agricultural gross domestic product. Soil erosion control and soil organic matter (SOM) maintenance are difficult for farmers of the CRVE in part because of weather conditions which make soil erosion more likely, notably the heavy rains of July and August and winds during the long, hot, dry season. Lack of secure land ownership or tenure is a disincentive in making investments for delayed returns. The tradition and prevalence of transhumance or the movement of livestock to distant grasslands during the crop season and to cropland areas with unrestricted grazing during the dry season restricts GAP options for leaving crop residues in the field or integrating perennial grass or trees into the cropping system. The value of manure is recognized but is often perceived to be greater as a fuel compared with its use for maintaining and improving land productivity.

Aquifer protection is also a major sustainability concern in Nebraska, which has more irrigated land than any other U.S. state (Census of Agricultural Highlights, 2012). The formation and enablement of the Natural Resource Districts (NRD) in the 1970s was an effective response to this challenge, with NRDs having local watershed responsibilities to monitor aquifer levels and regulate pumping (Ganzel, 2006; NRD, 2014). Farmers also have learnt to substantially reduce water use per unit of grain production (Grassini et al., 2014). These gains in water use efficiency have buffered the increasingly higher regional water withdrawals that resulted from the increase in U.S. maize irrigated area from 0.9 million ha in 1970 to 3.2 million ha in 2015. Indeed, the estimated decline in the water level of the Great Plains Aquifer underlying Nebraska since pre-development to 2013 has been negligible relative to the drastic depletion observed in southern states (Scanlon et al., 2012; USGS Groundwater Resources Program, 2014).

A further example of the importance of GAP is the wider impact of crop production on the environment. A revolutionary call for environment protection came with the 1962 publication of *Silent Spring* by Rachel Carson (Carson, 1962), especially with regard to pesticide use. Alternatives for pest management were revisited and developed; the phrase Integrated Pest Management (IPM) came into common usage. Pesticide use remains an important component of IPM and the overall levels of pesticide use in Nebraska in 1992 and 2009 were similar (Nebraska Department of Agriculture, not dated). However, some older pesticides have been replaced with pesticides with less residual activity. For example, in Nebraska, atrazine use, a common maize herbicide, dropped from a high of 3770 tons in 1999 to 2500 tons in 2010 (USDA-NASS, 2012). Meanwhile, nearly 3600 tons of glyphosate, a post-emergence herbicide with much lower residual activity, were used in 2010. Glyphosate was little used in maize production systems prior to the late 1990s (Nebraska Department of Agriculture, not dated).

Nutrient losses from agricultural land to water resources gained attention in the 1960s and remain an on-going challenge, including nitrate-N leaching to groundwater and nutrient losses to surface waters. Partly in response to this, U.S. maize farmers have greatly reduced the amount of applied N per unit of production in recent decades (Cassman et al., 2002; Ferguson, 2014; Grassini et al., 2014) and manure has regained recognition as an agronomically valuable resource. Improvement in N fertilizer efficiency at the field level has helped mitigate the impact of an increasingly higher regional N fertilizer use over time due to expansion of U.S. maize area, from 22 million ha in 1965 to 32 million ha in 2015 (USDA-NASS, 2015).

The impacts of nitrous oxide and ammonia emissions on the atmosphere are a more recent sustainability concern with wide environmental impact. The above-mentioned gains in N use efficiency imply reduced field-level greenhouse gas (GHG) emission intensities per unit of production, with no apparent yield or economic penalty. The full impact of climate change is yet to be realized, but farmers have already adapted to longer-term trends of climate change, partly accounting for average corn planting being advanced to about two weeks now compared to 1975 (Abendroth and Elmore, 2010; Grassini et al., 2014) even as awareness of the need for greater adaptability to weather extremes is growing.

Worker safety is associated with implementation of GAP, given that carelessness in the use of pesticides can pose a major health risk. Farm accidents have declined in the United States over the decades because of far fewer farm workers and also safer equipment and better procedures (Myers and Hendricks, 2010; NCCRAHS, 2014), but the impact of pesticide-use-related illness is not well established. Worker safety and farm family safety have been and remain a concern with maize production in CRVE as there is more human input per unit of production, employment of child labour, disregard for safety in pesticide use and accidents related to mechanization, including manual, animal-draft or machine operations (Mekonnen and Agonafir, 2002; USDL, 2012).

If GAP benefits are easily observed or confirmed in the field, adoption is more likely than for GAP of less obvious benefits. In the intensively managed Nebraska crop system, changed practices often result in maize yield increases of only 1–3% that can be confirmed only through well-conducted on-farm trials (Nebraska On-farm Research Network, 2014). In CRVE, GAP may result in a very substantial yield increase in some years but not in other years, making it difficult for the farmer to justify change. Reasons for inconsistency in GAP implementation include insufficient knowledge and training, the level of available external support and advice (e.g. extension services) and insufficient finance and resources. This supporting infrastructure for GAP is much stronger in Nebraska than either in CRVE or in Argentina. In addition, it may be difficult to adapt GAP to fit diverse types and scales of cropping systems, which is a greater challenge for the diverse, finance limited systems of CRVE than for the less diverse cropping systems of Nebraska. Barriers to GAP adoption by smallholders are addressed in more detail by Mwangi and Kariuki (2015).

4 Setting targets and measuring performance

A sustainable future can be achieved only by greater productivity underpinned by an improved understanding of cropping systems and a careful and comprehensive analysis of yield goals and environmental impact (Sadras et al., 2009).

4.1 Yield potential, yield gaps and yield goals

Yield potential (Yp) is defined as the yield of an adapted crop cultivar as determined by solar radiation, temperature, CO_2 and genetic traits that govern the length of the growing

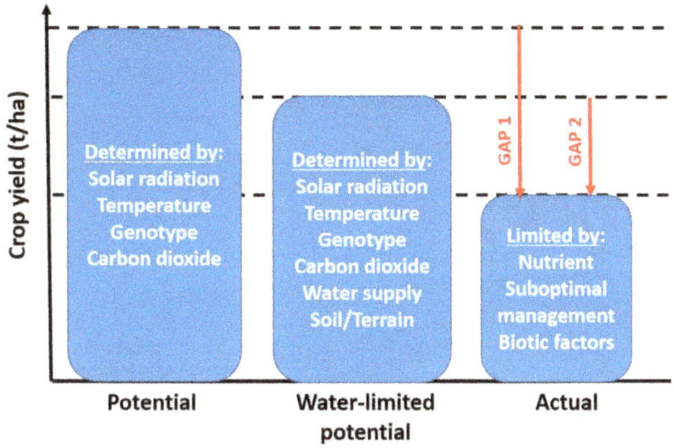

Figure 1 Schematic figure showing yield potential (Yp), water limited yield potential (Yw) and yield gaps of actual mean yield compared with Yp and Yw.

Source: Adapted from van Ittersum and Rabbinge (1997) and Lobell et al. (2009).

period, light interception by the crop canopy and its conversion to biomass, and partition of biomass to the harvestable organs (Evans, 1996; van Ittersum and Rabbinge, 1997) (Fig. 1). Water-limited yield potential (Yw) is determined by the amount of water supply and distribution during the crop growth period and by field and soil properties that affect soil water availability, such as slope, plant-available soil water holding capacity and depth of the root zone (Lobell et al., 2009; van Ittersum and Rabbinge, 1997; Van Ittersum et al., 2013).

For a specific location year, the crop yield gap (Yg) is defined as the difference between Yp (irrigated systems) or Yw (rain-fed systems) and average actual farm yield (Gap 1 and Gap 2 in Fig. 1). The magnitude of Yg provides a benchmark for current productivity, in relation to the biophysical yield ceiling, and an estimate of the additional crop production that could potentially be achieved, on existing cropland area, through improved management that alleviates all limiting factors other than weather factors. The Global Yield Gap Atlas project (www.yieldgap.org) is an on-going project to map yield gaps of major food crops for crop-producing countries by using a consistent methodology and a robust spatial framework (Grassini et al., 2015a; van Bussel et al., 2015).

Both Yp and Yw are defined by crop species, cultivar, climate, soil type (Yp) and water supply (Yw), and thus both Yp and Yw are highly variable across and within regions. However, it is impossible for a large population of farmers to achieve the perfection in crop and soil management required for Yp or Yw, and generally it is not cost-effective to do so because yield response to applied inputs follows diminishing returns when farm yields approach ceiling yields. Because average farm yields tend to plateau when they reach 75–85% of Yp or Yw, the exploitable yield gap is smaller than Yg (Cassman et al., 2003).

Estimating current Yg is crucial in determining how much extra food can be produced on existing land area and/or how much extra land (or exports/food aid) will be needed. Yield gap analysis is a powerful tool to assess the degree to which crop intensification can help ensure food security and alleviate pressure on cropland expansion. There is an inverse relationship between the land area required to produce a certain amount of grain and crop yield (Fig. 2). For example, less than 1 ha is required in the United States and

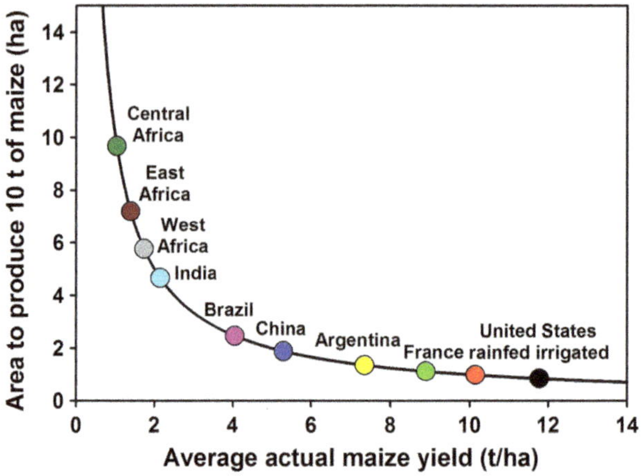

Figure 2 Relationship between the required land to produce 10 tons of maize grain and average actual yield (solid line). Data points indicate the area/yield for selected maize-producing countries or regions. Source: FAOSTAT (http://faostat.fao.org/).

southern Europe to produce 10 t of maize grain due to high productivity per unit of land. In contrast, more than 5 ha are needed to produce the same amount of maize grain in sub-Saharan Africa (FAOSTAT, 2015). A modest increase in actual average yield by 1 t ha^{-1} in sub-Saharan Africa, equivalent to bringing current Yg from 70% to 50% of Yw, would result in an approximate 33% reduction of land area requirement to produce 10 tons of maize yield.

Farmers need to set a realistic yield goal in order to make sound decisions on variety choice, seeding rate and fertilizer and irrigation application. In commercial agriculture, the goal should be the most profitable yield that can be expected for the particular set of soil, climate and management practices. When such practices overcome biotic and abiotic constraints to yield, the yield goal is close to Yp and is relatively stable from year to year. Management must be focused on minimizing yield-reducing factors that determine the exploitable Yg. In rain-fed crops, the yield goal is typically lower than that for irrigated crops, and actual yield fluctuates more widely across years and fields, depending on the initial soil water status at planting, soil water holding capacity and amount and distribution of rainfall during the growing season. Therefore, setting a realistic yield goal is more difficult in rain-fed farming than in irrigated agriculture because Yw can vary greatly spatially and temporally.

How can farmers set realistic yield goals? Maximum yields obtained in yield contests are reasonable estimates for the best genotype × environment × management interaction across a large population of farms (Duvick and Cassman, 1999). Crops are typically grown in these plots at a high density with abundant water and nutrient supply, and complete weed and pest control. Although these 'record yields' illustrate the upper yield limits that could be achieved, it is neither possible nor economical to strive for such yields. To set a realistic yield goal, estimates of the attainable yield must be obtained for a specific location and management and combined with information about the yield history of the field. The yield goal should reflect what can be grown if normal weather prevails and if GAP are achieved. Yield goals cannot exceed Yp (irrigated systems) or Yw (rain-fed systems). Unrealistically, high yield goals typically lead to a lower net profit, low resource-use efficiency and potentially higher negative environmental impact.

4.2 Sustainability metrics

Sustainability metrics and indices have been developed (FAO Council, 1989; Moore, 2009; https://www.fieldtomarket.org/fieldprint-calculator/; Petersen and Snapp, 2015). A combination of metrics should be used to account for both environmental and production objectives and needs to be measurable in a cost-effective way and contextualized using a robust spatial framework that allows trade-off analysis at different spatial scales (Grassini et al., 2015b). For example, Grassini and Cassman (2012) provide a set of metrics for productivity and environmental performance of irrigated and rain-fed maize fields in Nebraska (USA) on the basis of producer-reported data (Table 2). Interestingly, these authors found that productivity and environmental goals were not conflicting objectives in high-yield intensive irrigated systems. Indeed, high-yield irrigated maize fields achieved higher, and more stable, yields and higher efficiencies when compared to low-input rain-fed fields, despite greater input use in the former. The Field to Market Fieldprint Calculator estimates field-level performance by taking into consideration land use, conservation, soil C, irrigation water use, water quality, energy use and GHG emissions (https://www.fieldtomarket.org/fieldprint-calculator/).

5 Case study: improving maize cultivation in the United States, Ethiopia and Argentina

In this case study, sustainable production, GAPs, yield gaps and yield goals are discussed for maize production with examples for irrigated production in Nebraska, smallholder rain-fed production in the semi-arid CRVE, and rain-fed maize in the Pampas of Argentina (Table 1). These also show some of the competing priorities and trade-offs in each system, which need to be addressed.

5.1 Estimating yield gaps

Level of crop-system intensification affects Yg, as illustrated for the three contrasting production situations (Table 1). Yp (irrigated maize) and Yw (rain-fed maize) for each of these systems were estimated by using crop simulation models for average sowing date, dominant cultivar maturity and plant population density at each location (Fig. 3). Average farm yields are shown as percentage of the simulated Yp or Yw. The difference between actual yield and the 100% dashed line represents the Yg.

In the case of irrigated maize in Nebraska, Yp is high with little yearly variation due to irrigation and Yg is only 10% of the simulated Yp. In Argentina, Yw is lower and coefficient of variation (CV) is higher, relative to Yp of irrigated maize in Nebraska, due to inadequate water supply to fully satisfy crop water requirements. Farmers in Argentina have access to inputs, technology and markets, but a risk component is associated with unpredictable rainfall. Therefore, management practices are more conservative, resulting in a larger Yg,

Table 1 Approximate characterization of three maize-production situations

Production area characteristics	Irrigated maize in Nebraska	Rain-fed maize for the Argentina Corn Belt	Rain-fed maize for the CRVE
Mean farm size (ha)	~400	~300	<2.5
Mean maize yield (Mg/ha)	12.7	8.0	<2.5
Cropping system: maize relative to total farm crop value	Continuous maize and maize–soybean rotation; >65%	100% no till. 5-year rotation with 3 years of soybean, 1 year of maize and 1 year of wheat/soybean double crop; 25%	Diverse; maybe 25%
Commodity sales per farm	>$250 000 for 29% of the operations	~$195 000	Average of <$500; much consumed in the household
Finance availability	Very good	Good, but costly	Very poor
Input use	High and efficient	Intermediate and efficient	Little
Risk management	Relatively stable yield; revenue protected insurance	Relatively stable yield; crop insurance for hail	Highly variable yield; no insurance; crop diversity
Market system	Very efficient	Regulated by government	Inefficient
Water productivity (kg/m^3 of water)	Very high	High	Very low
Yield gap (per cent of potential yield)	10	41	60

as per cent of Yp, compared with irrigated maize in Nebraska. Finally, Yw of maize in CRVE is smaller and more variable than in Argentina as a result of frequent occurrence of soil water deficits and Yg, as per cent of Yp, is largely due to low input use to counter other edaphic and biotic constraints.

It is clear, therefore, that water supply is the major determinant of Yw and its variability across years but may not be a major determinant of actual yield in sub-Saharan Africa (Fig. 4). Yw increases with increasing water availability and a boundary function, with a slope (water productivity) of 32 kg ha^{-1} mm^{-1} and an x-intercept of 160 mm, delimits the maximum Yw along the range of water supply. Variation in Yw, at the same level of water availability, is explained by the erratic distribution of water supply during the crop growing season and/or by the differences in vapour pressure deficit across climate zones. However, actual yield exhibits a nearly flat response to increasing water supply. Indeed, there is always a huge difference between Yw and actual yields across the entire range of water availabilities. This finding suggests that other constraints, in addition to soil water deficits, are limiting maize yields in sub-Saharan Africa. Among these, inadequate nutrient supply and control of biotic stresses are likely to be very important.

The historical yield trends for maize in the three regions since the onset of the Green Revolution in 1960s are informative (Fig. 5). Yield gain rate in irrigated maize in Nebraska has been remarkably high and stable since 1965 (124 kg ha^{-1} y^{-1}), though an incipient yield

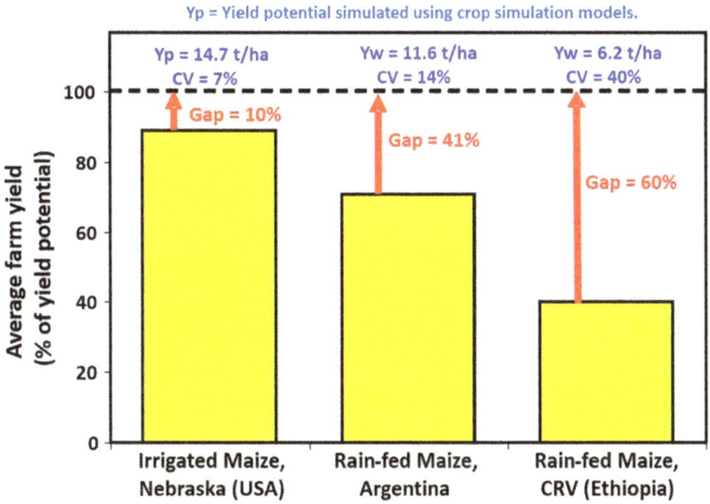

Figure 3 Actual yield and yield gap in three cropping systems: irrigated maize in Nebraska (western U.S. Corn Belt), rain-fed maize in Argentina and CRVE. Actual yield and yield gaps are expressed as percentage of the simulated Yp (potential yield of irrigated maize) or Yw (potential yield of rain-fed maize) for each crop system. Year-to-year variability in Yp and Yw was calculated with the CV.

Source: Grassini et al. (2011), van Ittersum et al. (2013), Aramburu et al. (2015) and Global Yield Gap Atlas (www.yieldgap.org).

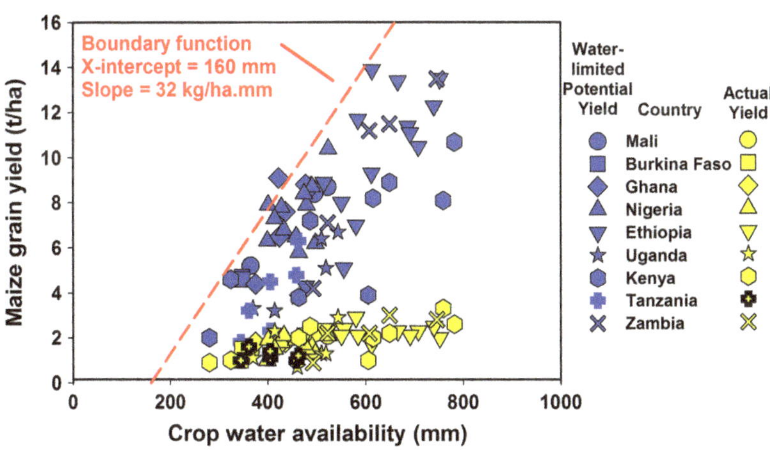

Figure 4 Boundary function for the relationship between water-limited yield potential and water supply during the crop growing season in sub-Saharan Africa (SSA). Black symbols represent simulated water-limited yields, while the gray symbols represent the actual fields attained by farmers. Each point represents a long-term average for rain-fed maize in SSA for a given climate zone. Crop water availability was calculated as the total water supply (stored soil water at sowing plus rainfall) minus soil water left at maturity and water losses through runoff and deep drainage.

Source: Global Yield Gap Atlas (www.yieldgap.org).

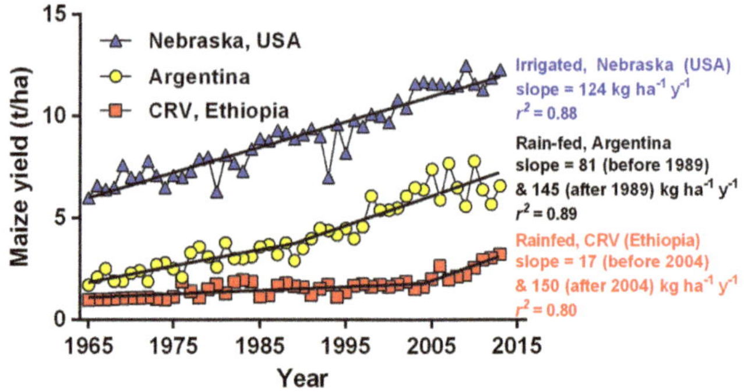

Figure 5 Historical yield trends for irrigated maize in Nebraska (western U.S. Corn Belt) and rain-fed maize in Argentina and Ethiopia. Slopes and coefficients of determination (r^2) of the fitted linear (Nebraska) or piecewise (Argentina and Ethiopia) models are shown.

Source: USDA-NASS and FAOSTAT.

plateau is evident during the last ten years. The small Yg estimated for irrigated maize in Nebraska highlights the size of the challenge for maintaining the historical yield gain rate in the coming decades (Fig. 5; Grassini et al., 2011). Some opportunities to increase yields by fine-tuning the current crop management may be limited by economic or environmental costs, but there may be significant opportunities for improving the fertilizer N and irrigation water use efficiency (Grassini et al., 2012; Farmaha et al., 2016). Remarkably, average yields have increased steadily during the last 50 years with no detectable increase in N and water inputs (Ferguson, 2014; Grassini et al., 2014).

Yield gain rate was relatively small in Argentina (81 kg ha^{-1} y^{-1}) but has been 145 kg ha^{-1} y^{-1} since 1990 (Fig. 5). Since the 1990s, simple and specialized cropping systems were developed and the Pampas have experienced profound technological changes, including adoption of no-till methods, increased fertilizer use and improved insect control as a result of widespread use of Bt maize (Satorre, 2011). Current average maize yield is *ca.* 40% below the Yw, which indicates that there is still room to exploit the Yg. Although fertilizer use in Argentina has increased over time, it is likely that a large portion of current Yg is still related to inadequate nutrient supply (Aramburu et al., 2016).

In Ethiopia, the maize yield trend was relatively flat until 2005 (only 17 kg ha^{-1} y^{-1}), with an incipient increasing trend in most recent years (150 kg ha^{-1} y^{-1}) (Fig. 5). The recent abrupt yield increase in Ethiopia has been attributed to the increasing use of mineral fertilizer and improved varieties (Abate et al., 2015). From a perspective of Yg, there is a very good scope for improving yields. To maintain current high yield gain and close the Yg, it is necessary to overcome the chronic problems related to diverse biotic and abiotic yield constraints, lack of markets, infrastructure and extension services, together with the constrained financial abilities of smallholder farmers and high input costs.

5.2 Setting targets for GAP: resource productivity

Yield gaps can identify the potential scope for improvement. However, farmers need specific, measurable targets to effect improvements in practice. Optimizing productivity per unit of the most limiting resource (land, water, labour or other inputs) is a commonly desired objective throughout the world (Herdt and Mandac, 1981; Hopper, 1965; Sheriff, 2005). Yield per hectare may be the most common concern. However, water productivity is often the most important constraint with semi-arid production in the CRVE and with irrigated production for Nebraska in cases of water use exceeding aquifer recharge or surface water supply.

Labour productivity may be most important, even in CRVE or in other areas of high un- or underemployment, because of shortages during labour peaks or a family health crisis. Often, increased productivity to land and labour can occur with a GAP, but increasing productivity to labour may in some cases decrease land productivity. For example, while care is taken in Nebraska to achieve optimal maize plant density with uniform spacing within rows, planting in CRVE is often done in groups of two or more plants with relatively wide row spacing to save labour required for planting and weeding, even though mean yield may be reduced.

Management is complicated in highly variable production environments because the most limiting constraint varies unpredictably by year as in semi-arid CRVE. Soil water deficits may be the most commonly limiting constraint for maize in CRVE, as was found for sorghum (Wortmann et al., 2009). However, in other years, N deficiency or one of several biotic constraints may be most limiting.

When the most limiting constraint to a crop is not addressed, the effectiveness of other investments such as N application may be reduced. For example, the mean yield increase due to N application was > 4 t ha^{-1} for irrigated maize production in Nebraska but only 2 t ha^{-1} for rain-fed maize in Uganda, where other factors constrained production and where there was little response to > 50 kg N ha^{-1} (Fig. 6).

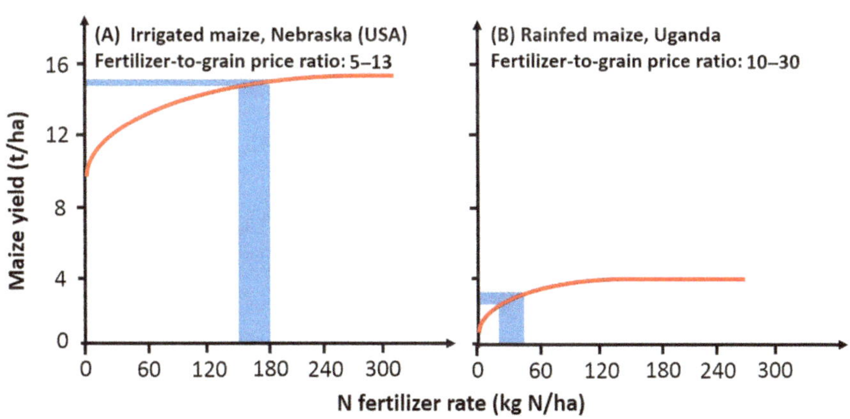

Figure 6 Nitrogen response curves for irrigated maize in Nebraska (a) and rain-fed maize in Uganda (b). The red line shows the relationship between maize yield and applied N fertilizer. Blue-shaded areas indicate the range of economically optimum N fertilizer rates (EONR) and the maize yields at EONR for a range of fertilizer-to-grain price ratios that are representative of each country (Nebraska: 5–13 and Uganda: 10–30).

Source: Adapted from Dobermann et al. (2011) and Kaizzi et al. (2012).

5.3 Setting targets for GAP: farmer profitability

Farmer profitability is commonly considered as net returns per hectare and is often associated with increased land productivity, at least when access to finance for input use allows for maximizing profit per hectare, such as is typical for Nebraska. Smallholder farmers, such as in CRVE, however, commonly cannot access enough finance to maximize net returns per hectare and need to maximize returns on their constrained investment.

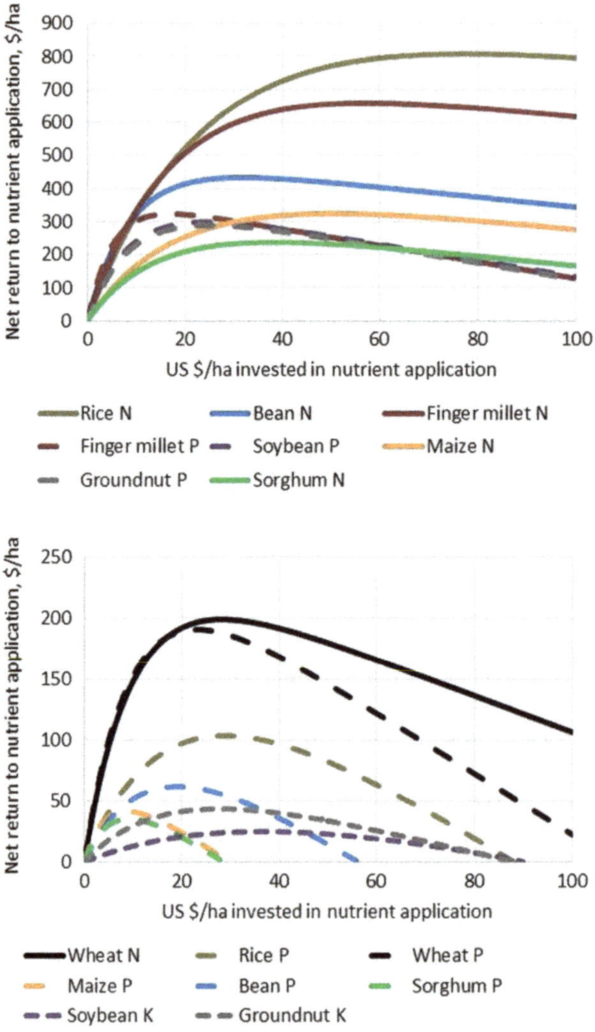

Figure 7 Marginal rates of net return for investment in crop-nutrient combinations with relatively high (top) and low (bottom) net returns. These results depend on the nature of the crop-nutrient response function, grain values and fertilizer nutrient costs.

Source: Jansen et al. (2013).

When management is finance constrained, net returns may or may not be greatest by addressing the most yield limiting constraint. Optimization across several investment options can offer the farmer the greatest profit potential. Decisions on fertilizer use for financially constrained situations, therefore, need to consider which crop-nutrient-rate combinations have the greatest potential of net returns on investment with the expectation that the improved profit will gradually improve investment ability (Fig. 6, 7; Jansen et al., 2013).

For example, in a study of six crops in Uganda, Jansen et al. (2013) showed that potential net returns to investment in nutrient application varied with crop-nutrient choices, as in the following sequence: upland rice N = finger millet N > bean N > finger millet P = soybean P = groundnut P > maize N > sorghum N > wheat N and P > upland rice P > bean P > maize P = groundnut K = sorghum P = soybean K and > K applied to cereal crops (Fig. 7).

5.4 Setting targets for GAP: livelihood security for farmers and local communities

Livelihood security includes consistent meeting of financial and basic needs (e.g. food, shelter, health care, education). Low or failed productivity and profitability are concerns of all maize producers, but the GAP implications vary with farmer context. For Nebraska, simple cropping systems such as continuous maize and maize–soybean rotation managed for high yield can be highly productive and profitable than alternative cropping systems. Income-protected crop insurance, which is currently federally subsidized, and efficient financial institutions for provision of credit provide a secure foundation which allows farmers to focus on high yields. Earnings from farm enterprises can be efficiently used to meet operational and livelihood needs because of efficient marketing of goods and services.

For CRVE, where insurance and credit are less accessible or not available and are less cost-effective in providing a security net compared with Nebraska, the choice of the cropping system and GAP are important for security. Cropping systems of CRVE are, therefore, more diverse than those of Nebraska for reduced risk of total failure and for providing for the diverse needs of the family and the local market, even though the crop choices and sequences may not be potentially the most productive or profitable (Lin, 2011; Porter et al., 2003). Marketing and using much of the production locally is currently likely to result in more economic and security benefits for the community than might occur with a more commercial and market-oriented cropping system with less diversity. Failed investments threaten operational and livelihood security, and financial recovery can take years. Investment in a full package of GAPs, which results in high maize yield, assuming that there is adequate financial capacity, is not often a preferred option for CRVE if the risk of occasional poor crop or market performance jeopardizes financial security and capacity.

The adoption of GAP which have allowed many individuals to farm large tracts of land and achieve high productivity in Nebraska has resulted in fewer people being involved directly or indirectly in agricultural production (such as through input supply and marketing), and a population decline in most small rural communities, while the economy of Nebraska has grown (USDA, 2014). At the same time, the increased production has enabled growth in livestock feeding, bio-fuel production, food processing and agricultural support sectors.

While many migrate from rural communities of CRVE, the communities continue to grow with a heavy dependence on local crop production for essential commodities and economic growth. Adoption of GAP with more financial investment for increased crop productivity is likely to favour rural communities, but an extreme shift to adopt Nebraska-type agro-ecosystems is likely to be harmful to rural communities, with economic growth largely by-passing the rural community. Consideration of social values is generally of little consequence in the selection of GAP for Nebraska. Social values and considerations are very important for adopting GAP in CRVE where dependence on extended families and neighbourhoods is crucial to family livelihood and security.

5.5 Setting targets for GAP: conserving natural resources and the environment

Maintaining or improving the natural resource base, especially land productivity and water availability, should be a concern for maize producers globally. Soil erosion is a major soil degradation concern for Nebraska and CRVE with loss of SOM, nutrients, available water holding capacity and depth. Farmers strive to maintain or increase SOM recognizing its importance to nutrient supply, microbial activity and soil physical properties. Some Nebraska farmers' practices on highly erodible lands are inappropriate for soil erosion control, for example, continuous maize–soybean production with tillage and no conservation practices on steeply rolling deep silt loam soils derived from loess. Most Nebraska farmers, however, operate with little erosion and maintain or increase SOM with the result that most of their surface soil is in water-stable aggregates.

Off-field effects on the surrounding environment (including the atmosphere and ground water) are considerations in the selection of GAP. Some of the very important factors for Nebraska are leaching to aquifers of nitrate and pesticide-related compounds, ammonia and nitrous oxide emissions to the atmosphere, and erosion transport of sediment, nutrients, organic materials and pesticide-related compounds to surface waters. Contaminant transport through tile drainage is a major concern in some U.S. maize production areas. Input application rates affect losses, especially for N, but the timing and method of application, in relation to crop requirements, is also important.

Ultimately, N losses depend on the synchrony between N inputs and crop N requirements (Broadbent and Carlton, 1978). For example, reducing nitrate–N loss through tile drainage for maize production in the U.S. Corn Belt by economic optimization of N fertilizer rates was estimated to be more cost-effective than alternative practices with cost-effectiveness decreased in the order of constructed wetlands, drainage water management, winter cover crops and avoiding fall application of N (Wortmann et al., 2008).

The amount of environmental contamination relative to the amount of production is an important consideration; for example, the quantity of NO_2 emission per ton of maize produced at least needs to be reduced even if an absolute reduction in emissions is not achievable. Situations of highly productive maize commonly have potential for decreased environmental contamination per unit of production compared with lower potential situations such as lower CO_2 intensity with irrigated corn than with rain-fed corn (Table 2). A crop with only minor constraints to growth should be able to produce more per unit of N (Wortmann et al., 2011) or pesticide applied and provide better ground cover for protection against erosion than a crop that encounters major constraints to growth.

Table 2 Several metrics related to production and environmental aspects of maize crop production in rain-fed and irrigated fields in Nebraska (USA)

Variable	Rain-fed maize	Irrigated maize	Irrigated-to-rain-fed ratio
Yield (t/ha)	5.9	13.2	2.2
Year-to-year yield variation (CV, %)	23	3	0.1
N fertilizer rate (kg N/ha)	110	183	1.7
Partial N productivity factor (kg grain/kg N)	54	71	1.3
Total water supply (mm)*	670	943	1.4
Water productivity (kg grain/m^3 water supply)	0.88	1.40	1.6
Fossil-fuel energy input (GJ/ha)[†]	10.8	30.0	2.8
Net energy yield (GJ/ha)[§]	74	159	2.2
GHG emission (kg CO_2e/ha) [‡]	2289	3001	1.3
GHG intensity (kg CO_2e/t grain)[‡]	388	231	0.6

*Based on a soil water at planting and in-season rainfall plus irrigation.
†Includes embodied energy associated with manufacturing, transportation and application of agricultural inputs.
§Calculated as harvested grain energy output minus fossil-fuel energy inputs.
‡Includes emissions of CO_2, N_2O and CH_4 adjusted for CO_2 warming equivalent.
Source: Adapted from Grassini and Cassman (2012).

5.6 Setting targets for GAP: adapting to climate change

Changes in commodity values, climate and input costs often imply some adjustment in management for profit maximization. The adaptive potential of a package of GAPs affects the ability to respond to changing opportunities, constraints, perspectives and values (Moore, 2009; Wortmann et al., 2008). Adaptation to gradual climate change, especially to increased growing season length (Shulski et al., 2015), has occurred in Nebraska, for example, early planting of maize and soybean (Grassini et al., 2014). The low crop diversity of Nebraska cropping systems may be restrictive to adaptation, while the availability of finance and information enables adaptation.

Conversely, having flexible crop diversity as a part of a cropping system GAP package with relatively low crop-specific capital investment in the CRVE allows for much adaptive management, while low financial capability and information availability are major constraints to adaptation. Most maize is sown in a three-week period in Nebraska, while maize sowing occurs from April through July in semi-arid CRVE, with farmers making decisions on crop species and varieties, sole crop and intercrop and fertilizer use in response to how the season progresses and on the basis of the outcomes of earlier plantings (Fujisaka et al., 1997). Sowing of late maturing and often photoperiod-sensitive maize and sorghum in CRVE on soils of good water holding capacity may occur in April, with sowing decisions continuing into August when some teff varieties might still be sown, depending on the success of earlier sowings. Dry soil sowing in anticipation of the onset of heavy seasonal rainfall adds to the flexibility (Liben et al., 2015a,b). These response farming practices of CRVE are not highly productive, but such adaptive management allows for improved security.

Attention to the resilience of GAP has increased with a more frequent occurrence of weather extremes in recent years. The most important factors that contribute to resilience are crop or revenue protection insurance and financial security, both relatively strong in Nebraska but weak in CRVE. Irrigation adds tremendously to resilience in Nebraska (Table 2). While the adaptive management ability in CRVE discussed above contributes to resilience, resource-poor farmers are very susceptible to extreme crop stresses, and therefore, financial and security recovery is difficult to achieve.

It is notable, however, that the global trends in crop agriculture in the past 50 years are towards the adoption of modern, improved cultivars grown in pure stands and fewer and simpler crop sequences because of higher yields, greater responsiveness to fertilizers, reduced labour and easier management (e.g. weed control, sowing and harvesting) once farmers have access to inputs and markets.

6 Conclusions

The major challenge to modern agriculture is to narrow the gap between potential yield and actual yield while maximizing profit, protecting the natural resource base, improving the quality of life of individuals and communities and ensuring the nutritional value and safety of food. Such sustainable intensification requires appropriate choices of GAP for each cropping system context.

The Yp of maize growing areas differs greatly and setting realistic yield goals is important for GAP choice. The socio-economic context is also critical. Farmers operating under severe financial constraints, producing to directly meet much of the family's material needs and lacking control over numerous and diverse biotic and abiotic constraints to production, need to consider different criteria compared to farmers operating in a high productivity situation. Secure land ownership and access to finance and credit, markets and technical support contribute to increased potential of GAP, but GAP need to be relevant to the farmers' context. However, the farmer's context is generally not factored in in these attributes. Choice of GAP to approach Yp needs to consider the challenges faced by farmers and the complex and often conflicting priorities faced in meeting these challenges. Consultations with farmers to determine their priorities can lead to the making of informed decisions for effecting sustainable improvements in the field.

7 References

Abate, T., Shiferaw, B., Menkir, A., Wegary, D., Kebede, Y., Tesfaye, K., Kassie, M., Bogale, G., Tadesse, B. and Keno, T. (2015), Factors that transformed maize productivity in Ethiopia. *Food Sec.* doi.1007/s12571-015-0488-z.

Abendroth, L. and Elmore, R. (2010), Planting date trends. Iowa State University Corn Production. http://www.agronext.iastate.edu/corn/production/management/planting/earlier.html (accessed 29 December 2015).

Anderson, D. (2012), Organic matter in prairie soils. *Crops and Soils,* January to February. https://www.agronomy.org/files/publications/crops-and-soils/organic-matter-in-prairie-soils.pdf (accessed 29 December 2015).

Aramburu Merlos, F., Monzon, J. P., Mercau, J. L., Taboada, M., Andrade, F. H., Hall, A. J., Jobbagy, E., Cassman, K. G. and Grassini, P. (2015), Potential for crop production increase in Argentina through closure of existing yield gaps. *Field Crops Res.*, 184, 145–54

Berry, L. (2003), Land degradation in Ethiopia: its extent and impact. https://www.google.com/#q=soil+erosion+in+ethiopia (accessed 29 December 2015).

Broadbent, F. E. and Carlton, A. B. (1978), Field trials with isotopically labeled nitrogen fertilizer. In Nielsen, D. R. and MacDonald, J.G. (Eds), *Nitrogen in the Environ.* Academic Press, San Diego, CA. pp. 1–41.

Carson, R. (1962), *Silent Spring.* Houghton Mifflin Co. Boston.

Cassman, K. G., Dobermann, A. and Walters, D. T. (2002), Agroecosystems, nitrogen-use efficiency, and nitrogen management. *Ambio*, 31, 132–40.

Cassman, K. G., Dobermann, A., Walters, D. T. and Yang, H. (2003), Meeting cereal demand while protecting natural resources and improving environmental quality. *Annu. Rev. Environ. Resour.*, 28, 315–58.

Census of Agricultural Highlights (2012), http://www.agcensus.usda.gov/Publications/2012/Online_Resources/Highlights/Irrigation/Irrigation_Highlights.pdf (accessed 30 January 2016).

Connor, D. J., Loomis, R. S. and Cassman, K. G. (2011), *Crop Ecology: Productivity and Management in Agricultural Systems.* Cambridge University Press, Cambridge, UK.

Diamond, J. (2005), *Collapse: How Societies Choose to Fail or Succeed.* Penguin books, Ltd. London, England.

Dobermann, A., Wortmann, C. S., Ferguson, R. B., Hergert, G. W., Shapiro, C. A., Tarkalson, D. D. and Walters, D. (2011), Nitrogen response and economics for irrigated corn in Nebraska. *Agron. J.*, 103, 67–75.

Duvick, D. N. and Cassman, K. G. (1999), Post-green revolution trends in yield potential of temperate maize in the North-Central United States. *Crop Sci.*, 39, 1622–30.

Evans, L. T. (1996), *Crop Evolution, Adaptation and Yield.* Cambridge University Press, Cambridge. p. 500

FAO Council (1989), Sustainability assessment of food and agriculture systems (SAFA). http://www.fao.org/nr/sustainability/sustainability-assessments-safa/en/ (accessed 29 December 2015).

FAOSTAT (2015), http://faostat.fao.org/site/567/default.aspx#ancor (accessed 29 December 2015).

Farmaha, B. S., Lobell, D. B., Boone, K., Cassman, K. G., Yang, S. H. and Grassini, P. (2016), Contribution of persistent factors to yield gaps in high-yield irrigated maize. *Field Crops Res.*, 186, 124–32.

Ferguson, R. B. (2014), Groundwater quality and nitrogen use efficiency in Nebraska's Central Platte River Valley. *J. Enviorn. Qual.* doi:10.2134/jeq2014.02.0085.

Fujisaka, S., Wortmann, C. S. and Adamassu, H. (1997), Resource poor farmers with complex technical knowledge in a high risk system in Ethiopia. Can research help? *J. Farm. Syst. Res. Extens.*, 6, 1–14.

Ganzel, B. (2006), Nebraska's unique natural resource districts. http://www.livinghistoryfarm.org/farminginthe50s/water_11.html (accessed 29 December 2015).

Global Yield Gap Atlas (GYGA). www.yieldgap.org (accessed October 2015).

Grassini, P., Thornburn, J., Burr, C. and Cassman, K. G. (2011), High-yield irrigated maize in the Western U.S. Corn Belt: I. On-farm yield, yield potential, and impact of agronomic practices. *Field Crops Res.*, 120, 144–52.

Grassini, P. and Cassman, K. G. (2012), High yield maize with large net energy yield and low global warming potential. *Proc. Natl. Acad. Sci. (PNAS)*, 109, 1074–9.

Grassini, P., Specht, J., Tollenaar, T., Ciampitti, I. and Cassman, K. G. (2014), High-yield maize-soybean cropping systems in the U.S. Corn Belt. In V. O. Sadras and D. F. Calderini (Eds), *Crop Physiology – Applications for Genetic Improvement and Agronomy*, 2nd ed., Elsevier, The Netherlands.

Grassini, P., Van Bussel, L. G. J., Van Wart, J., Wolf, J., Claessens, L., Yang, H., Boogaard, H., de Groot, H., Van Ittersum, M. K. and Cassman, K. G. (2015a), How good is good enough? Data requirements for reliable crop yield simulations and yield-gap analysis. *Field Crops Res.*, 177, 49–63.

Grassini, P., Cassman, K. G., Yang, H. S., Archontoulis, S., Licht, M., Lamkey, K. R., Ciampitti, I.A., Coulter, J. A., Pittelkow, C., Broder, S. M. and Volenec, J. V. (2015b), Improved spatial framework to better leverage research on the food-energy-water nexus. ASA/CSSA/SSSA White Paper. https://www.agronomy.org/files/science-policy/white-papers/5.pdf.

Herdt, R. W. and Mandac, A. M. (1981), Modern technology and economic-efficiency of Philippine rice farmers. *Econ. Dev. Cult. Change*, 29, 375–99.

Hopper, W. D. (1965), Allocation efficiency in a traditional Indian agriculture. *J. Farm. Econ.*, 47, 611–24.

Hornbeck, R. (2009), The enduring impact of the American Dust Bowl: short and long-run adjustments to environmental catastrophe. *Working Paper 15605, National Bureau of Economic Research.* http://www.nber.org/papers/w15605 (accessed 29 December 2015).

Jansen, J. A., Wortmann, C. S., Stockton, M. A. and Kaizzi, C. K. (2013), Maximizing net returns to financially constrained fertilizer use. *Agron. J.*, 105, 573–8.

Kaizzi, C. K., Byalebeka, J., Semalulu, O., Alou, I., Zimwanguyizza, W., Nansamba, A., Musinguzi, P., Ebanyat, P., Hyuha, T. and Wortmann, C. S. (2012), Maize response to fertilizer and nitrogen use efficiency in Uganda. *Agron. J.*, 104, 73–82.

Liben, F., Tesfaye, K. and Wortmann, C. S. (2015), Dry soil planting of maize for variable onset of rains in Ethiopia. *Agron. J.*, 107, 1618–25.

Liben, F., Wortmann, C. S. and Fantaye, K. T. (2015), Risks associated with dry soil planting time in Ethiopia. *Intern. J. Agric. Agron. Res.*, 6, 14–23.

Lin, B. B. (2011), Resilience in agriculture through crop diversification: adaptive management for environmental change. *BioSci.*, 61(3), 183–93.

Lobell, D. B., Cassman, K. G. and Field, C. B. (2009), Crop yield gaps: their importance, magnitude, and causes. *Annu. Rev. Environ. Resour.*, 34, 1–26.

Loomis, R. S. (1984), Traditional agriculture in America. *Ann. Rev. Ecol. Syst.*, 15, 449–78.

Marten, G. G. (1988), Productivity, stability, sustainability, equitability and autonomy as properties for agroecosystem assessment. *Agric. Syst.*, 26, 291–316.

Mekonnen, Y. and Agonafir, T. (2002), Pesticide sprayers' knowledge, attitude and practice of pesticide use on agricultural farms of Ethiopia. *Occup. Med.*, 52, 311–15.

Moore, K. M. (2009), Landscape systems framework for adaptive management. In K. M. Moore (Ed.), *The Sciences and Art of Adaptive Management: Innovating For Sustainable Agriculture and Natural Resource Management.* http://www.swcs.org/en/publications/the_sciences_and_art_of_adaptive_management/ (accessed July 2015).

Mwangi, M. and Kariuki, S. (2015), Factors determining adoption of new agricultural technology by smallholder farmers in developing countries. *J. Econ. Sustain. Develop.*, 5, 208–16.

Myers, J. R. and Hendricks, K. J. (2010), Agricultural tractor overturn deaths: assessment of trends and risk factors. *Am. J. Ind. Med.*, 53, 662–72. doi: 10.1002/ajim.20775.

NCCRAHS (2014), National Children's Center for Rural and Agricultural Health and Safety. 2014 fact sheet childhood agricultural injuries in the U. S. https://www3.marshfieldclinic.org/proxy/MCRF-Centers-NFMC-NCCRAHS-2014_Child_Ag_Injury_FactSheet.1.pdf (accessed 3 January 2016).

Nebraska Department of Agriculture (n.d.), Pesticide use trends. http://www.nda.nebraska.gov/pesticide/trends.html (accessed 29 December 2015).

Nebraska On-farm Research Network (2014), 2014 growing season results. http://cropwatch.unl.edu/OnFarmResearch/Post%20Conference%20Report%202014%20Color%20Final%20Updated2.pdf (accessed 29 December 2015).

NRD (2014), NRD groundwater regulations across Nebraska. https://www.nrdnet.org/sites/default/files/state_map_water_management_status_14feb2014.pdf (accessed 29 December 2015).

Petersen, B. and Snapp, S. (2015), What is sustainable intensification? Views from experts. *Land Use Policy*, 46, 1–10.

Porter, P., Huggins, D. R., Perillo, C. A., Quiring, S. R. and Crookston, R. K. (2003), Organic and other management strategies with two- and four-year crop rotations in Minnesota. *Agron. J.*, 95, 233–44.

Sadras, V., Calederini, D. and Connor, D. (2009), Sustainable agriculture and crop production. In V. O. Sadras and D. F. Calderini (Eds), *Crop Physiology: Applications for Genetic Improvement and Agronomy*, Academic Press, Elsevier.

Satorre, E. H. (2011), Recent changes in pampean agriculture: possible new avenues in coping with global change challenges. In J. L. Araus and G. A. Slater (Eds), *Crop Stress Management and Global Climate Change*, pp. 47–57.

Scanlon, B. R., Faunt, C. C., Longuevergne, L., Reedy, R. C., Alley, W. M., McGuire, V. L. and McMahon, P. B. (2012), Groundwater depletion and sustainability of irrigation in the U.S. high plains and central valley. *Proc. Natl. Acad. Sci.*, 109, 9320–5.

Sheriff, G. (2005), Efficient waste? Why farmers over-apply nutrients and the implications for policy design. *Rev. Agric. Econ.*, 27, 542–57.

Shulski, M. D., Baule, W. J., Stiles, C. and Umphlett, N. (2015), A historical perspective on Nebraska's variable and changing climate. *Great Plains Res.*, 25, 109–20.

USDA (2014), USDA 2012 Census of Agriculture. http://www.neded.org/files/research/stathand/esecc1.pdf and http://www.agcensus.usda.gov/Publications/2012/Preliminary_Report/Highlights.pdf (accessed 29 December 2015).

USDA-NASS (2012), http://www.nda.nebraska.gov/pesticide/1992_2012_usda_nass_field_crop_ag_chem_%20usage.pdf (accessed 29 January 2016).

USDA-NASS (2015), National statistics for corn. http://quickstats.nass.usda.gov/results/08638011-B478-3942-B44F-FA130ADDE283?pivot=short_desc (accessed 29 December 2015).

USDL United States Department of Labor's Bureau of International Labor Affairs (2012), 2012 Findings of Worst forms of child labor – Ethiopia. https://www.dol.gov/ilab/reports/child-labor/findings/2012TDA/ethiopia.pdf (accessed 29 December 2015).

USGS Groundwater Resources Program (2014), Water-Level Changes and Change in Water in Storage in the High Plains Aquifer, Predevelopment to 2013 and 2011–13. http://pubs.usgs.gov/sir/2014/5218/pdf/sir2014_5218.pdf (accessed 29 December 2015).

Van Bussel, L. G. J., Grassini, P., Van Wart, J., Wolf, J., Claessens, L., Yang, H., Boogaard, H., de Groot, H., Saito, K., Cassman, K. G. and Van Ittersum, M. K. (2015), From field to atlas: Upscaling of location-specific yield gap estimates. *Field Crops Res.*, 177, 98–108.

Van Ittersum, M. K. and Rabbinge, R. (1997), Concepts in production ecology for analysis and quantification of agricultural input–output combinations. *Field Crops Res.*, 52, 197–208.

Van Ittersum, M. K., Cassman, K. C., Grassini, P., Wolf, J., Tittonell, P. and Hochman, Z. (2013), Yield gap analysis with local to global relevance – a review. *Field Crops Res.*, 143, 4–17.

Wortmann, C. S., Helmers, M., Gelder, B., Wright Morton, L., Devlin, D., Barden, C., Anderson, S., Broz, R., Franti, T., Regassa, T., Shea, P., Tomer, M., Frees, L. and Griffith, D. (2008), Targeting of watershed management practices for water quality protection. University of Nebraska-Lincoln Extension, RP195.

Wortmann, C. S., Mamo, M., Mburu, C., Letayo, E., Abebe, G., Kayuki, K. C., Chisi, M., Mativavarira, M., Xerinda, S. and Ndacyayisenga, T. (2009), Atlas of sorghum (*Sorghum bicolor* (L.) Moench) production in eastern and southern Africa. University of Nebraska-Lincoln, Lincoln NE. http://agronomy.unl.edu/documents/Sorghum%20Atlas%20ESA.pdf (accessed 29 December 2015).

Wortmann, C. S., Tarkalson, D. D., Shapiro, C. A., Dobermann, A. R., Ferguson, R. B., Hergert, G. W. and Walters, D. (2011), Nitrogen use efficiency for three cropping systems in Nebraska. *Agron. J.*, 103, 76–84.

Chapter 3

Maize seed variety selection and seed system development: the case of southern Africa

Peter S. Setimela, Global Maize Program, International Maize and Wheat Improvement Center (CIMMYT), Zimbabwe

1 Introduction

1.1 Overview

Maize is an important lifeline in southern Africa (SA) where over 150 million people consume close to 13 million metric tons annually (FAOSTAT, 2014). Maize production in Africa extends from SA through eastern Africa (EA) and across the savanna of western Africa. In SA, the maize area is estimated around 5.2 million hectares. Modern maize varieties are estimated to represent more than three-quarters of maize area in Kenya, Zimbabwe, Swaziland, Lesotho and South Africa. With total production of 17 million tons per annum, SA is a net surplus producer in most years (SADC, 2011). The surplus is, however, accounted for by only three countries, Malawi, South Africa and Zambia, out of the eleven countries in the region, with the majority countries (Mozambique, Namibia, Zimbabwe, Angola, Lesotho, Swaziland and Democratic Republic of Congo and Botswana) perpetually registering deficits. South Africa alone accounts for 55% of the production depending on rainfall distribution (BMI USDA, 2012). Despite the skewed regional production, SA has the highest average per capita consumption of maize ranging from 52 to 328 g/person/day (FAO, 2012). In some countries such as Lesotho, Malawi, South Africa, Zambia and Zimbabwe the average consumption exceeds 100 kg/capita/year, representing more than 50% of total calorie intake (FAOSTAT, 2010).

http://dx.doi.org/10.19103/AS.2016.0002.03

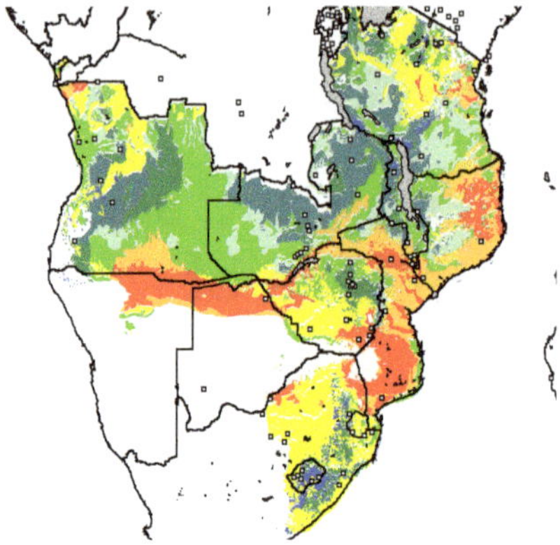

Key		
Maize mega-environments	Area in southern Africa (%)	Typical environment
A	29%	Mid Altitude Humid Warm
B	27%	Mid Altitude Humid Hot
C	19%	Mid Altitude Dry
D	7%	Low Tropical Humid
E	15%	Low Tropical Dry
H	3%	Highlands

Figure 1 Map of agro-ecological zones of SADC, with colour definitions described in Table 1. Source: Setimela et al., 2005

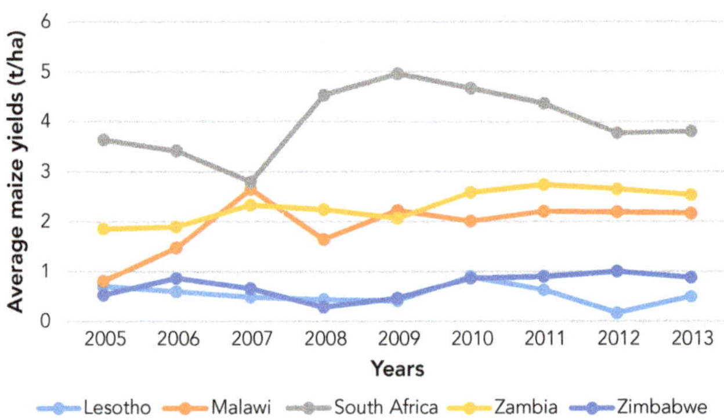

Figure 2 Average maize yields in selected countries in southern Africa (2005–2013). FAOSTAT 2013.

Maize has a unique strategic importance in the food security and socio-economic stability of the region. It is also the dominant driver in support of agriculture as a whole as it provides the volume to make services necessary for efficient grain trade and other agro industries (BMI USDA, 2012). Maize is mostly grown under rain-fed conditions by smallholder farmers (Smale, 1995; Kassie et al., 2012). Smallholder farmers cultivate 1–3 ha of land using hand hoe and animal draft power (Thierfelder et al., 2015b). The crop is mono-cropped (about 30%) and the rest is intercropped with legumes (70%) – mostly pigeon peas, cowpeas and groundnuts. For smallholders, the ability to produce sufficient amounts of maize is core to family food and income security: it determines whether children attend school, enables families to take care of their basic needs and to develop assets as a reserve for harsher times (Eicher, 1995). Despite being one of the most important crops in the region, yields remain low 1.5–2 t ha^{-1} compared to the United States and other regions (Shiferaw et al., 2011; Ray et al., 2013).

Average production is predicted to decrease further due to changes in weather patterns. Climate predictions suggest increases in maximum temperatures and a reduction in the rainfall season duration, which will affect maize production (Cairns et al., 2013). Although

Table 1 Six zones delineated by combinations of maximum air temperature and risk of drought (i.e. average seasonal precipitation)

Zone	Description	Maximum air temperature (°C)	Risk of drought	Seasonal precipitation (9 mm)	Area in SADC (%)
A	Mid-altitude, humid, warm	24–27	low	>600	29.6
B	Mid-altitude, humid, hot	28–30	Medium	>600	26.4
C	Mid-altitude dry	24–30	High	350–600	19.0
D	Lowland humid	>30	Medium	>800	6.8
E	Lowland dry	>30	High	350–800	15.1
F	Highland	18–24	Medium	>350	3.1

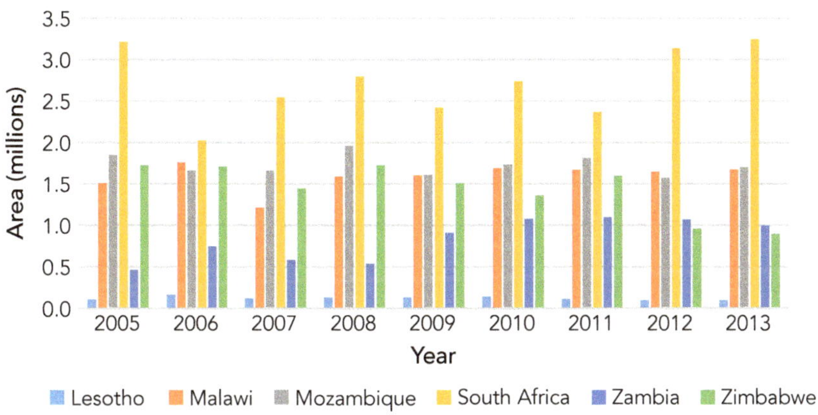

Figure 3 Trends in average maize production, Lesotho, Malawi Mozambique, South Africa, Zambia and Zimbabwe (2005–2013) (FAOSTAT, 2014).

the total rainfall is expected to increase slightly during the growing season, increased temperatures are likely to offset the increase due to increased evapotranspiration. Spatial analyses and crop modelling studies also predict a decline in maize yields by up to 10% by 2050, associated with a decline in yields rather than a reduction in area (Tesfaye et al., 2015). Therefore, the development and deployment of drought-tolerant maize seed is an important intervention in combating climate change. This chapter reviews key activities and the roles of key stakeholders in improving maize productivity for smallholder farmers in SA. It also highlights the development and selection of first-generation maize germplasm and the development of the seed industry and its role in disseminating improved maize germplasm.

1.2 Maize production distribution by environment

The mid-altitude humid and hot zones account for the largest area under maize in SA while the highlands account for the smallest proportion (Table 1). The mid-altitude zones have a low to medium risk of drought and therefore ideal for growing maize. Due to its popularity, maize is now widely grown in the semi-arid environments were sorghum and millets used to be the dormant crops. These zones are highly drought prone, leading to farmers harvesting maize once every five seasons (Hodgson et al., 2002; Kontandini et al., 2013). The extreme year-to-year variability in maize yields has profound effects on national and regional food security (Fig. 1; Smale et al., 2011). In Lesotho and Zimbabwe the coefficient of variation in maize production over the last five years has been above 50% (FAOSTAT, 2015). Maize is also photosensitive but it is grown over a wider range of altitudes and latitudes than any other food crop, ranging from cool to very hot temperatures, on wet to semi-arid lands, and in many different types of soil. Some of the maize germplasm is highly adapted to specific locations, for example, highland and temperate germplasm cannot easily be transferred to tropical environments in developing countries, and progress achieved in one tropical environment cannot be easily replicated in another. In sub-Saharan Africa, preference is for white maize which means germplasm from temperate areas have to be converted first before being used in the region (Hassan et al., 2001).

2 Maize variety development in southern Africa (SA)

Despite maize being the staple, it is relatively a new world crop that arrived in Africa only a half millennium ago and became SA's primary food after Second World War (MacCann, 2005). Maize has moved from being a garden crop to a large-scale industrial agriculture crop in South Africa, Zambia, Malawi and Zimbabwe (Smale, 1995; MacCann, 2005). The early SA maize varieties were largely flint and early maturing varieties adapted to the mid- and low altitudes of the region. The American white dent types (Hickory, Iowa Silver Mine, Hickory King and Horsetooth King) became the most popular and were commercialized by the white farmers in South Africa and Zimbabwe (MacCann, 2005). In 1932 Rhodesia's Salisbury Agricultural Experimental Station (SAES) began efforts to improve open-pollinated varieties (OPVs), leading to the first most popular single cross-hybrid SR52, developed from two inbred lines. Inbred lines are small in size, have small cobs and reduced yields but when two of them are crossed, the vigour is restored in the resulting increased yields were the offspring is better than the two parents (Muir-Leresche, 1984; Robbach, 1988). From the SAES new hybrids were introduced into Zambia, following independence in 1964 (Howard, 1994).

In contrast Malawian farmers resisted adopting new hybrids developed from Rhodesia and other parts of the region preferring flint mostly OPVs. Compared to hybrids, OPVs are low yielding and highly variable, which delayed Malawi's maize revolution compared to Zimbabwe, South Africa and Zambia, where farmers adopted high-yielding hybrids (Smale, 1995). The National Maize Breeding Program of Malawi developed its first hybrid in 1958. The first hybrid (Mthenga (LH7)) was, however, dent, while farmers preferred flinty maize (which has a higher proportion of hard starch granules in the kernel) for making *nsima* (the stiff porridge), leading to its rejection. The breakthrough came when MH17 and MH18, flinty varieties of intermediate maturity with yield potential of 6–7 t/ha, were released in 1991 (Heisey and Smale, 1995). Their poundability was good, since the hard endosperm 'mphale' extraction was 68% which was close to the unimproved local maize (72%) (Smale, 1995).

According to surveys undertaken by the National Statistical Office in 2006, MH17 and MH18 were still planted in over half of the area of improved maize in this densely populated, maize-dependent, nation (Sauer and Tchale, 2009). The development of MH17 and MH18 in Malawi shows the importance of incorporating farmers' preferences in variety development. Since the development of the of SR52 by Rhodesia's SAES, many hybrids and OPVs have been developed and made available to smallholder farmers in SA, mostly driven by national maize breeding programmes and private seed companies.

The highest adoption rates of improved varieties has been by large commercial farmers (Setimela et al., 2005; Langyintuo et al., 2008). Smallholder farmers on the other hand continue to experience chronic difficulties in accessing seed from the formal sector due to limited finances, inaccessible input and product markets, and lack of options and information in regard to varieties (Bänziger et al., 2006).

3 Maize seed industries

After First World War maize became a major crop first in Kenya, where farmers were encouraged to grow maize by the colonial government and maize slowly replaced sorghum and millets (Gerhart, 1975). In SA, due to mining, maize slowly became more important to feed the urban population (Miracle, 1966). With increased demand for maize, research to develop high-yielding varieties also increased leading to the development of more than 300 varieties between 1966 and 1990 (Morris, 2002). These varieties were mostly developed and released in Kenya, Malawi, South Africa, Tanzania, Zambia and Zimbabwe, where colonial governments invested heavily in supporting farmers of European descent compared to West Africa. During the 1990s, the private sector effectively took over the seed supply function throughout EA and SA from the governments with World Bank structural adjustments. The private seed companies now control more than 97% of all commercial maize seed sales in SA (Rashid et al., 2000). The maize seed sales are estimated around US$500 million followed by horticulture crops at US$250 million (African Centre for Biodiversity, 2015).

During the colonial period the public sector led most maize research but has now been replaced by private sector companies. A recent World Bank report named just five African seed companies with the capacity to develop new types of maize germplasm: Zamseed Research Institute (MRI) of Zambia, Pannar Seed of South Africa (now part of DuPont Pioneer), Zimbabwe's Seed Co Limited and Kenya's National Seed Research Institute (World Bank, 2012). Of the African seed companies, Seed Co has the largest geographical footprint, with a presence in 15 countries and research centres (African Center for Biodiversity, 2015).

By region, EA accounts for the largest sales of hybrid seed; Kenya accounting for the largest hybrid seed sales in EA. While, in SA, Zimbabwe accounted for the largest sales (Langyintuo et al., 2008). The seed need ranges from 64 000 t in Tanzania to 14 000 t in Zambia. Compared to the adoption rates recorded in 2001 by Hassan et al. (2001) Zambia and Malawi had the highest adoption rates of improved maize seed due to government seed and fertilizer subsides that encourages the use of improved seed and also a large number of seed companies compared to Mozambique and Angola were there are only few seed companies (Langyintuo et al., 2008; Kassie et al., 2013).

During the 2006/07 season, an estimated 103 600 t of improved maize seed was marketed in EA and SA. Improved OPVs of maize account for about 23 000 t while hybrid maize seed is estimated at 80 000 t and the rest of the seed is sourced from the informal seed sector through seed exchanges and recycling of OPVs and hybrids (Table 2).

Table 2 Estimated maize seed supply and need in eastern and southern Africa

Region/ country	Maize area (x mil ha)	Estimated seed need (x 1000 MT)[1]	Improved OPV maize seed sales (x 1000 MT)			Hybrid maize seed sales in 2006/07 (x 1000 MT)	Adoption rate 2006/07 (as % of maize area)[2]	Adjusted adoption rate in 2006/07 (as % of maize area)[3]
			2004/05	2005/06	2006/07			
Eastern Africa	**6.6**	**161.8**	**4.0**	**3.5**	**11.1**	**42**	**33 (23)**	**37**
Ethiopia	1.7	42.4	0.4	0.4	2	6.2	19 (8)	21
Kenya	1.6	38.9	0.6	0.1	1.7	26.3	72 (71)	74
Tanzania	2.6	64	0.6	2	3.9	7.3	18 (4)	22
Uganda	0.7	16.5	2.3	1	3.5	2.2	35 (9)	54
Southern Africa	**5.4**	**133.4**	**9.3**	**9.8**	**12**	**38.5**	**38 (28)**	**52**
Angola	0.8	19.3	0.8	0.1	0.8	0.2	5 (12)	10
Malawi	1.4	35.3	5.2	4.5	5.4	2.5	22 (14)	50
Mozambique	1.2	30.3	1.2	2.2	3.1	0.2	11 (9)	22
Zambia	0.6	14.1	0.3	1	0.5	9.7	73 (23)	81
Zimbabwe	1.4	34.4	1.8	2.1	2.2	25.9	80 (82)	93
Total/ average	**12.0**	**295.1**	**13.3**	**13.3**	**23.1**	**80.5**	**35(26)**	**44**

[1] Estimate based on area and planting rate of 25 kg/ha.
[2] In parentheses are figures observed in 1997 by Hassan et al. (2001). Only seed sales in 2006/07 were used in the estimation.
[3] Adjusted for OPV sales in 2004/05, 2005/06 and 2006/07 assuming that similar quantities purchased in the first two years were recycled in 2006/07. That is, total improved OPV seed planted is aggregated over 2004/05, 2005/06 and 2006/07. Note that total area under improved maize varieties is 4.2 million ha (0.92 million ha under OPV) before and 5.3 million ha (2 million ha under OPV) after adjustment with previously purchased OPV seed.
Source: DTMA seed sector survey, 2007/2008.

In a review of varietal releases between 2002 and 2006 under the Drought Tolerant Maize for Africa (DTMA) Project in 13 countries (excluding South Africa), 600 maize varieties were released averaging about four varieties per year per country. More than 60% of the varieties were from the private seed sector, mostly in Kenya, Malawi, Zambia and Zimbabwe. Many of the hybrids were based on CIMMYT and IITA germplasm. The releases in SA were mainly hybrids indicating strong presence of the private sector (Setimela et al., 2009).

The development and selection of stress-tolerant varieties in SA first started around 1900 in Zimbabwe, South Africa, Malawi and Zambia but significant improved cultivar development efforts started in the 1950s and 1960s (Magorokosho, 2006; Mason, 2012) The following hybrids and OPVs are widely grown and occupying the largest land area DK8053, PAN53, SC513 PAN67, SC627, ZM521, ZM309 and ZM523. Most of these hybrids and OPVs are early to intermediate maturing (De Groote et al., 2012).

To sustain varietal development in the late 1990s, CIMMYT and seed companies introduced new maize germplasm, resistant to different biotic and abiotic stresses. The new germplasm was resistant to a) southern corn leaf blight (*Bipolaris maydis*), b) southern rust (*Puccinia polysora*), c) northern corn leaf blight (*Exserohilum turcicum*), d) common rust (*P. sorghi*), e) grey leaf spot (*Cercospora* spp.), f) stalk and ear rots caused by *Diplodia* spp. and *Fusarium* spp. and g) kernel rot caused by several *Fusarium* and *Aspergillus* species. From the introduction of disease-resistant germplasm, several hybrids and OPVs from CIMMYT were taken by seed companies for release or to extract genes for further improvement (Banziger et al., 2006).

To address the gap in maize genetic improvement for smallholder farmers, CIMMYT initiated a collaborative drought and low N maize breeding programme in 1996 dubbed Southern Africa Drought and Low Fertility (SADLF) programme. The project aimed to increase yields in low-input and drought-prone environments that characterize most of the region (Bänziger et al., 2006; Table 3). The programme focused on developing maize hybrids and OPVs for drought, low N and major leaf and ear diseases. A regional network was established in collaboration with National Agricultural Research Systems (NARS) and private seed companies and initial testing sites established across SA (Bänziger and de Meyer, 2002). Elite maize hybrids and OPVs were tested and validated through the regional testing network under optimal conditions, managed random drought stress, low N stress and maize streak virus (MSV) infection. The breeding strategy was to simultaneously

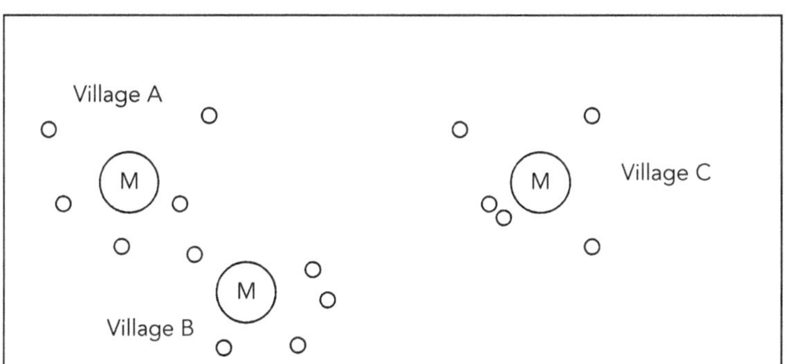

Figure 4 Mother-baby trial design to evaluate maize cultivars by researchers and farmers. 'M' indicates a mother trial.

Table 3 Mean grain yield of 16 maize varieties evaluated in 27 on-farm experiments grown under recommended (green trial) and farmer-representative (yellow trial) fertilizer levels and farmer practice during the 2006/2007 season

		Mother trials		Green		Yellow trials			Farmer fields (Babies)		Farmer perception rank
Variety	Origin	t/ha	Rank	t/ha	Rank	t/ha	Rank	t/ha	t/ha	Rank	Rank
Early maturing hybrids											
PHB30G19	Pioneer	2.51	4	2.8	4	2.23	4	1.89	1.38	8	10
SC525	Seed Co	2.38	6	2.69	6	2.07	5	1.96	1.32	6	7
SC513	Seed Co	2.27	9	2.6	10	1.94	9	1.89	1.16	7	1
PAN413	Pannar	2.1	14	2.49	12	1.7	16	1.3	1.4	16	11
Intermediate-maturing hybrids											
R201	ARDA	2.06	15	2.31	15	1.81	11	2.09	1.41	3	12
PHB3253	Pioneer	2.6	2	2.93	1	2.26	3	2.21	1.55	2	14
SC637	Seed Co	2.66	1	2.92	2	2.39	1	1.72	1.47	14	13
PAN5503	Pannar	2.22	10	2.67	8	1.78	15	1.88	1.38	9	9
ZS108	AGPY	2.35	7	2.62	9	2.07	6	2.09	1.29	4	6
ZS261	CIMMYT	2.53	3	2.68	7	2.38	2	1.96	1.2	5	3
VP05181	CIMMYT	2.03	16	2.25	16	1.81	12	1.82	1.2	13	3
Late-maturing hybrids											
PAN67	Pannar	2.39	5	2.82	3	1.96	8	1.83	1.2	11	3
Early maturing OPVs											
ZM521	ARDA	2.13	13	2.46	13	1.8	13	1.55	1.69	15	15
ZM423	AGPY	2.17	11	2.54	11	1.79	14	1.84	1.69	10	15
ZM309	CIMMYT	2.15	12	2.34	14	1.97	7	2.47	1.18	1	2
ZM401	CIMMYT	2.32	8	2.79	5	1.86	10	1.82	1.33	12	8
Mean		2.3		2.62		1.99		1.89	1.36		
Min		2.03		2.25		1.7		1.3	1.16		
Max		2.66		2.93		2.39		2.47	1.69		
LSD (0.05)		1.8		2.4		1.1		1.6	1.2		

select maize varieties under optimal (recommended agronomic management/high rainfall conditions), low N, managed and random drought stress conditions.

Although the main focus was on hybrid breeding, OPVs were developed as a by-product, mainly as synthetic varieties derived from elite inbred lines. OPVs were suitable for farmers with either limited adoption rates for hybrids or limited capital investment (Langyintuo and Setimela et al., 2010). Improved OPVs contributed to farmers' own seed security, as they can be recycled without yield penalty. Effective regional seed and information exchanges were essential for improving seed security at the national and regional level. In 1997, over 50% of all maize varieties, public and private, released in sub-Saharan Africa were based on CIMMYT and IITA germplasm (Hassan et al., 2001; Manyong et al., 2000).

In 1999, under the SADLF project CIMMYT in Zimbabwe initiated a new variety testing system that evaluates the performance and acceptance of new maize varieties under farmers' conditions, and creates a cost-effective and simple flow of information between breeders, extension staff and farmers. The SADLF began to breed a new generation of seed types as a response to declining maize production in SA and the increasing role of smallholders. The project started to screen OPVs under low potential conditions as compared to seed companies that were concentrating on high potential areas and also the project began to evaluate maize varieties under decentralized research conditions, in collaboration with smallholder farmers using the mother-baby trial approach. The resulting testing scheme was based on two concepts: (i) an innovative experimental design – the 'mother-baby trial (MBT) design' (Snapp, 1999) – that was adapted for on-farm variety evaluation, and (ii) wide-ranging and organized partnerships with farmers and farmer organizations interested in variety testing (Bänziger et al., 1998).

The MBT design is an innovative farmer-participatory scheme in which a set of experiments are undertaken by farming communities in collaboration with a local partner, such as an extension agent, NGO, secondary school or a research station. The design has

Table 4 Criteria importance, farmer perception and general perception

	Farmer assessment										
Variety	Husk cover	Ear size	Ear aspect	Taste	Kernel size	Resistance weevil	Time to maturity	Grain colour	Ear rot	Grain yield	Lodging
ZM 421	0.71	0.71	0.86	0.92	0.14	0.86	0.79	0.86	0.57	0.83	0.79
ZM 423	0.71	0.71	0.86	0.9	0.5	1	0.93	0.93	0.93	0.5	0.86
ZM 521	1	0.64	0.93	0.8	0.5	1	0.86	1	1	0.63	0.71
ZM 523	0.88	0.5	0.5	0.75	0.63	0.88	0.94	0.81	0.81	0.75	0.56
ZM 623	0.63	0.63	0.88	0.81	0.63	0.88	0.75	0.88	0.63	0.38	0.56
ZMS 606	0.63	0.88	0.88	0.7	0.69	0.56	0.63	0.75	0.5	0.88	0.81
MRI EP	0.75	0.69	0.88	0.86	0.5	0.75	0.81	0.5	0.75	0.7	0.56
MRI 514	0.88	0.88	0.69	0.67	0.63	1	0.81	0.94	0.81	0.75	0.81
SC 513	0.88	0.69	0.81	0.75	0.81	0.69	0.81	0.88	0.69	0.38	0.69
PAN 6M-55	0.25	0.5	0.56	0.43	0.25	0.56	0.75	0.63	0.69	0.6	0.88
PAN 67	0.69	0.63	0.69	0.4	0.44	0.31	0.88	0.81	0.56	0.88	0.25
PRA-78-32-44 C5	0.71	0.64	0.29	0.67	0.29	0.5	0.93	0.93	0.36	0.5	0.57

Table 5 Open-pollinated varieties released across SADC countries

Variety	Country of release	Year of release and registration	Agro-ecological zone of adaptation
ZM309	Malawi	2009	Dry lowlands, mid-altitude dry
	Swaziland	2007	
	Mozambique	2010	
	Zimbabwe	2009	
	Botswana	2012	
	Zambia	2011	
ZM421	Malawi	2001	Dry lowlands, mid-altitude dry
	Zimbabwe	2002	
	Tanzania	2002	
	Mozambique	2003	
	Zambia	2003	
	Zimbabwe	2003	
	South Africa	2002	
ZM521	Malawi	2001	Dry lowlands, mid-altitude dry, warm humid, mid-altitude
	Tanzania	2002	
	South Africa	2001	
	Angola	2004	
	Ethiopia	2006	
	Lesotho	2004	
	Mozambique	2003	
	Swaziland	2006	
	Zambia	2004	
ZM523	Malawi	2009	Dry lowlands, mid-altitude dry, warm humid, mid-altitude
	South Africa	2006	
	Angola	2010	
	Zambia	2008	
	Zimbabwe	2008	
	Mozambique	2010	
ZM623	DRC	2008	Warm humid mid-altitude, cool humid altitude
	Lesotho	2008	
	South Africa	2006	
	Malawi	2006	

proved useful for farmer-participatory research because it permits farmers to choose from a range of new varieties they evaluate in baby trials on their own farms in subsequent seasons (Snapp, 1999). The mother trial: A replicated researcher-managed trial usually planted by partners at the centre of a farming community. It has two different levels of fertilizer: The 'green' experiment was grown under recommended fertilizer levels. The 'yellow' experiment was grown under farmer-representative fertilizer levels. The two experiments of one trial were planted on the same date. Each experiment had an alpha (0,1) lattice design with three replicates (Patterson and Williams, 1976).

The MBTs concentrate on the collection of qualitative data (Table 4). The objective of these trials was not to simulate conditions experienced in farmers' fields but to simulate stress that occurs in smallholder farmer's fields where the grain yields are between 1 and 2 t ha^{-1}). Thus, low N and managed drought trials were considered to represent the breeding target. Each baby trial evaluates four of the 12 varieties under completely farmer-managed conditions. The baby trials are planted by farmers in the area around the mother trial. If there are 12 varieties in the mother trial, there will be three baby trials, each with four varieties. The size of the plot planted to each of the four varieties in the baby trials is determined by the amount of seed available for planting.

The participating farmers plant the baby trials when they plant the rest of their own maize. A sufficient number of MBTs needs to be set up to obtain data that is representative of the target environment. As an ideal, there should be a minimum of 15 MBTs. From the baby trials, farmers were asked to rank the importance of 12–16 characteristics when deciding on the relative merit of a maize cultivar. Options were: 'very important', 'regular' or 'not important'. A value of 1 was allocated to 'very important', a value of 0.5 was allocated to 'regular' and a value of −1 was allocated to 'not important'. Criteria importance was the average score given to a characteristic (Table 4). The MBTs provided a forum for farmers to raise their voice on the choice of varieties they want and has resulted in fast tracking of varieties for release. As a result of the scheme, several OPVs were released across the region (Table 5). One of the OPVs, ZM521, became very popular with farmers in several countries because of its drought tolerance and flintiness (Table 5). The OPVs were multiplied by multinational seed companies and several community-based seed production schemes (Setimela et al., 2015). Besides breeding for low soil fertility the project also improved maize varieties for MSV, turcicum leaf blight, grey leaf spot, southern leaf rust, blight, stalk borers (*Busseola fusca, Chilo partellus, Eldana saccharina*), the parasitic weed *Striga hermonthica* and socio-economic factors including inadequate access and affordability of improved seed and fertilizers.

4 The development of maize seed systems in SA

Seed companies in the region also participated along with other partners to implement the MBTs in farmers' fields and contributed their new entries to be tested along with the new varieties developed by the SADLF project. The level of contribution varied from country to country depending on the strength of the seed industry. Zimbabwe, Zambia and Malawi have number of multinational and national seed companies developing and marketing hybrids, whereas Angola and Mozambique have very few seed companies, most of which do not have any research capacity and depend on CIMMYT and IITA for new products (Kassie et al., 2013; Rusike, 1998). In Zimbabwe and South Africa the average seed company is about 25.4 years old, five times the average age of those in Angola and Mozambique (Kassie et al., 2013). Hybrid maize seed contributes about 95% to seed companies' revenues (Kassie et al., 2013; Langyintuo et al., 2008). In Angola, Mozambique and Democratic Republic of Congo, where there are few seed companies, CIMMYT, in collaboration with the NARS partners, contributed to a number of variety releases which were taken by small seed companies for multiplication and marketing.

CIMMYT has over the years licensed more than 80 hybrids and OPVs to seed companies in SA. For hybrids, CIMMYT gives exclusive rights on condition that a seed company produces enough quantities of certified seed to make a difference in the seed market. For OPVs there are no exclusive rights; any seed company can market and also provide breeders' seed to other companies who are interested in marketing the same OPVs. OPVs are licensed on non-exclusive rights because of slow turnover in the seed market compared to hybrids that have high turnover in some countries. Seed companies select hybrids from the regional trials and send their request to CIMMYT for approval. Once a seed company identifies a hybrid, CIMMYT will avail the parental lines for the seed company to start production and provide support to the company to register the selected product.

The seed development protocol in the region requires that varieties undergo performance testing and approval by National Seed Authorities (Setimela et al., 2009). The new varieties are subjected to standardized tests for value for cultivation and use (VCU) and distinctness, uniformity and stability (DUS). Depending on the country, VCU and DUS tests may take 2–3 years, regardless of whether the variety has been tested and released in another country with similar growing conditions or mega-environment (Table 1). The purposes of these tests are to ensure that new varieties have minimum VCU and use and are novel from the existing registered varieties.

The use of these tests is deemed appropriate in the Southern African Development Community (SADC), as many smallholder farmers are not able to absorb or claim losses resulting from poor performing varieties because the seed markets are insufficiently transparent. The variety release and registration requirements in SADC are overlapping and rigid, making it difficult to commercialize and market new improved varieties quickly (Setimela et al., 2005; Zulu et al., 2003). The SADC member states have recently developed procedures and guidelines to harmonize seed systems. They address crop variety testing, registration and release, seed certification, quality assurance, quarantine and phytosanitary measures for seed. These agreements allow a variety to be marketed regionally once it has been entered into the SADC variety catalogue in order to improve the seed trade across the region. The delay in variety release has also been apparent in maize OPVs developed by the project where a number of OPVs have been released in different years across the various countries in the SADC region (Table 2). The multi-country releases for maize OPVs shows regional adaptation and the evidence for harmonized variety release system. Although there have been multi-country variety releases, each member state continue to have its own independent variety release and registration procedures, delaying farmers accesses to new improved germplasm.

5 The adoption of modern maize varieties

Zambia and Malawi has had the highest adoption rates of improved maize seed (Langyintuo et al., 2008). The high adoption rate in Malawi is attributed to the large number of seed companies and the government's Farm Input Subsidy Program (Langyintuo et al., 2008; Table 2). Maize hybrid seed dominates the seed market compared to OPVs (Kassie et al.,

2013; Langyintuo et al., 2008). The high sales may be attributed to seed companies as they can derive more profit from hybrid seed. Additionally, the yield penalty for recycling hybrids is higher in comparison to OPVs, so hybrid users have to purchase new seed every season. In Zimbabwe more than 80% of the farmers are using hybrid seed, the highest outside South Africa. In Angola and Mozambique, 90% and 99% of maize land, respectively, is sown to recycled or traditional maize varieties (Kassie et al., 2013).

Fisher et al. (2015) found that the major barriers to adoption were lack of availability of improved seed, inadequate information, lack of resources, high seed price and perceived attributes of different varieties. New maize varieties with improved tolerance to the conditions typically found under smallholder farming (drought, low fertility, acidic soils and diseases) have been developed and deployed. Through innovative farmer-participatory trials, farmers have been given a voice in deciding the kind of varieties that make a difference to their livelihoods. It is estimated that more than 1.5 million hectares are now planted with more drought-tolerant farmer-selected maize varieties in SA, mainly the result of the last eight years of investment (Fisher et al., 2015). These maize varieties have become the most frequently distributed within the framework of government- and NGO-led seed distribution projects. They have resulted in emergence of new local seed companies and farmers attaining improved food security and increased incomes from better maize harvests and seed production. Other projects from CIMMYT and partners are providing complementary soil fertility-enhancing and conservation agriculture-based technologies.

6 Conclusions and future trends

CIMMYT's maize breeding programme has had significant impacts in eastern and southern Africa a source of germplasm for the various traits especially to combat fall armyworm, (MLN) and climate change. Emerging seed companies will continue to depend on CIMMYT's finished products such as hybrids and OPVs since they do not have their own breeding programmes to develop new products. The recent policy reforms in many countries on seed production and support by Alliance for Green Revolution (AGRA) will increase private investment in maize breeding to support new challenges in maize production research. Maize remains crucial for food security in southern Africa but questions still remain about the viability of the small seed companies marketing maize seed, with multinational companies coming on board.

6.1 Conclusions

It is clear that Africa's seed systems are in the midst of transition and transformation from the public researchers as primary agents of breeding, selection, and distribution of improved maize seed to a system that is shaped more by the private sector. These changes vary from country to country in which some countries are ahead and others are lagging behind. Seed companies are also at different stages of consolidating their seed business in SSA; for example, DuPont Pioneer have been on the continent for many years, while others which are new like Limagrain are moving into the continent to consolidate their position. On the other hand international organisations like CIMMYT and IITA will continue to influence NARS and private seed companies to incorporate stress breeding especial for drought and low N.

6.2 Emerging challenges

Besides drought and soil fertility constraints, maize faces challenges from new diseases that appeared in the farmers' fields in EA in 2011. MLN disease is caused by synergistic interaction of *maize chlorotic mottle virus* and *sugarcane mosaic virus*, and has emerged as a serious threat to maize production in EA. The grain yield losses were estimated at 90% in farmer's fields in 2012 (Mahuku et al., 2015). Seed companies in the region are under significant pressure to develop resistant varieties that can replace highly susceptible germplasm. CIMMYT with the Kenyan Agricultural Research Institute has established a screening site to accelerate development and deployment of improved maize varieties with resistance to MLN. A new lepidopteran invasive pest of American origin, the fall armyworm *S. frugiperda* (J. E. Smith), was reported for the first time in January 2016 in Nigeria (Goergen et al. 2016), and it has since assumed epidemic proportions in several southern African countries, including Zambia, Zimbabwe, Malawi, Mozambique, Namibia and South Africa. Fall armyworm larvae burrow into the growing point of maize plants and destroy and clip up the leaves, affecting photosynthesis.

7 Where to look for further information

For further information consult the CIMMYT website for recent publications on stress breeding.

8 References

Ariga, J. and Jayne, T. S. 2010. Unlocking the Market: Fertilizer and Maize in Kenya. In Spielman, D. and Pandya-Lorch, R. (Eds), *Millions Fed: Proven Successes in Agricultural Development, Chapter 14*. International Food Policy Research Institute: Washington, DC.

Bänziger, M. and de Meyer, J. 2002. 'Collaborative maize variety development for stress-prone environments in southern Africa.' *Farmers, scientists, and plant breeding: Integrating knowledge and practice*. CABI, Oxon, UK, pp. 269–96.

Bänziger, M. 1998. *Annual Report the Southern African Drought and Low Soil Fertility Project*. CIMMYT: Harare.

Bänziger, M., Setimela, P. S., Hodson, D. and Vivek, B. (2006). Breeding for improved abiotic stress tolerance in maize adapted to southern Africa. *Agr. Water Manag.* 80(1–3), 212–24.

Business Monitor International (BMI). (2011). *Southern Africa Agribusiness Report for 1st Quarter 2012*.

Byerlee, D. and Heisey, P. W. 1997. Evolution of the African Maize Economy. In Byerlee, D. and Eicher, C. K. (Eds), *Africa's Emerging Maize Revolution*. Lynne Rienner Publishers: Boulder, CO, 9–22.

De Groote, H., Gitonga, Z., Smale, M., Kasuta, M., Asare-Marfo, D., Sonder, K. and Birol, E. 2012. Adoption and diversification of modern maize varieties in Zambia in 2011. A baseline report prepared for HarvestPlus. CIMMYT: Mexico, D.F.

FAO. 2012. FAOSTAT, Food Supply. Cited 10 October 2013. http://faostat.fao.org/site/345/default.aspx.

FAOSTAT. (2010). Statistical databases and data-sets of the Food and Agriculture Organization of the United Nations. http://faostat.fao.org/default.aspx. Accessed April 2010.

Fisher, M., Abate, T., Lunduka, R., Asnake, W., Alemayehu, Y. and Madulu, R. 2015. Drought tolerant maize for farmer adaptation to drought in sub-Saharan Africa: Determinants of adoption in eastern and southern Africa. *Clim. Change* 133(2), 238–99. doi: 10.1007/s10584-015-1459-2.

Gerhart, J. 1975. The diffusion of hybrid maize in western Kenya. Ph.D. thesis, Princeton University. Abridged by CIMMYT. CIMMYT, Mexico.

Goergen, G., Kumar, P. L., Sankung, S. B. et al. (2016) First report of outbreaks of the Fall Armyworm *Spodoptera frugiperda* (J.E. Smith) (Lepidoptera, Noctuidae), a new alien invasive pest in West and Central Africa. *PLoS ONE* 11(10): e0165632.

Hodson, D. P., Martinez-Romero, E., White, J. W., Corbett, J. D. and Bänziger, M., 2002. *Africa Maize Research Atlas* (v. 3.0), CD-ROM Publication. CIMMYT: Mexico, D.F., Mexico.

Holden, S. and Lunduka, R.. 2010a. *Too Poor to be Efficient? Impacts of the Targeted Fertilizer Subsidy Program in Malawi on Farm Plot Level Input Use, Crop Choice* and *Land Productivity*. Mimeo. April 2010.

Kassie, G. T., Erenstein, O., Mwangi, W., La Rovere, R., Setimela, P. and Langyintuo, A. 2012. Characterization of maize production in southern Africa: Synthesis of CIMMYT/DTMA household level farming system surveys in Angola, Malawi, Mozambique, Zambia and Zimbabwe. Socioeconomics Working Paper 4, CIMMYT. Mexico, DF (Mexico).

Kassie, G. T., Erenstein, O., Mwangi, W., MacRobert, J., Setimela, P. and Shiferaw, B. 2013. Political and economic features of the maize seed industry in southern Africa. *Agrekon* 52(2), 104–27. ISSN Print 0303-1853/Online 2078-0400 © Agricultural Economics Association of South Africa. doi: 10.1080/03031853.2013.798067.

Kostandini, G., La Rovere, R. and Abdoulaye, T. 2013. Potential impacts of increasing average yields and reducing maize yield variability in Africa. *Food Policy* 43, 213–26.

Langyintuo, A. L., Mwangi, W., Diallo, A. O., MacRobert, J., Dixon, J. and Bänziger, M. 2008. *An Analysis of the Bottlenecks Affecting the Production and Deployment of Maize Seed in Eastern and Southern Africa*. CIMMYT: Harare, Zimbabwe.

Langyintuo, A. L. and Setimela, P. 2010. Assessing the effectiveness of a technical assistance program: The case of maize seed relief to vulnerable households in Zimbabwe. *Food Policy* 34 (2009), 377–87.

Magorokosho, C. 2006. The Impact of farmers' selections on adapting maize landraces to diverse agro-ecological conditions in Zimbabwe, Zambia, and Malawi, and genetic gain and diversity trends in maize varieties released and grown in Zimbabwe from 1900 to 2004. Texas A&M University.

Mahuku, G., Lockhart, B. E., Wanjala, B., Jones, M. W., Kimunye J. N., Stewart L.R., Cassone, B. J., Subramanian, S., Nyasani, J. O., Kusia, E., Lava Kumar, P., Niblett, C. L., Kiggundu, A., Asea, G., Pappu, H. R., Wangai, A., Prasanna, B. M. and Redinbaugh, M. G. 2015. Maize lethal necrosis (MLN), an emerging threat to maize-based food security in sub-Saharan Africa. *Phytopathology* 105 (7), 956–65.

Mason, N. M. and Ricker-Gilbert, J. 2012. Disrupting demand for commercial seed: Input subsidies in Malawi and Zambia. *World Dev.* 45, 75–91.

McCann, J. C. (2005). *Maize and Grace: Africa's Encounter with a New World Crop, 1500–2000.* Harvard University Press: Cambridge Cambridge.

Miracle, J. P. 1966. *Maize in tropical Africa*. University of Wisconsin Press: Madison, WI.

Morris, M. L. 2002. *Impact of International Maize Breeding Research in Developing Countries, 1966–98.* CIMMYT: Mexico.

Muir-Leresche, K. 1984. Crop price and wage policy in the light of Zimbabwe's development goals, D. Phil. Thesis (Harare: University of Zimbabwe).

Patterson, H. D. and Williams, E. R. 1976. A new class of resolvable incomplete block designs. *Biometrika* 63, 83–9.

Rashid, M. H., Mekuria, M. and Mwangi, W. 2001. *Maize Breeding Research in Eastern and Southern Africa: Current Status and Impacts of Past Investments Made by the Public and Private Sectors 1966–97*. CIMMY: Mexico, D.F.

Ray, D. K., Mueller, N. D., West, P. C. and Foley, J. A. 2013. Yield trends are insufficient to double global crop production by 2050. *PLoS One* 8(6), e66428. doi: 10.1371/journal.pone.0066428.

Rohrbach, D. D. 1988. The Economics of Smallholder Maize Production in Zimbabwe: Implications for Food Security. MSU International Development Paper No. 11. Michigan State University: East Lansing

SADC (2011). *SADC Statistics Yearbook 2011*. Southern African Development Community (SADC): Gaborone, Botswana.

Setimela, P. S, MacRobert, J., Atlin, G. N., Magorokosho, C., Tarekegne, A., Makumbi, D. and Taye, G. 2012. Performance of elite maize varieties tested on-farm trials in eastern and southern Africa. Presented at the ASA, CSSA, and SSSA International Annual Meetings, 21–24 October 2012. Cincinnati, Ohio, USA.

Setimela, P. S., Badu-Apraku, B. and Mwangi, W. 2009. *Variety Testing and Release Approaches in DTMA Project Countries in sub-Saharan Africa*. CIMMYT: Harare.

Setimela, P. S., Chitalu, Z., Jonazi, J., Mambo, J. A., Hodson, D. and Bänziger, M. 2005. Environmental classification of maize testing sites in the SADC region and its implication for collaborative maize breeding strategies in the subcontinent. *Euphytica* 145, 123–32.

Shiferaw, B., Tesfaye, K., Kassie, M., Abate, T., Prasanna, B. M. and Menkir, A. 2014. Managing vulnerability to drought and enhancing livelihood resilience in sub-Saharan Africa: Technological, institutional and policy options. *Weather Clim. Extremes* 3, 67–79.

Smale, M. and Mason, N. 2013. Hybrid seed, income, and inequality among smallholder maize farmers in Zambia. *IAPRI working paper 72*. Indaba Agricultural Policy Research Institute (IAPRI): Lusaka.

Smale, M. 1995. Maize is Life: Malawi's Delayed Green Revolution. *World Dev.* 235, 819–31.

Smale, M. and Jayne, T. S. 2003. *Maize in eastern and southern Africa: Seeds of success in retrospect.* EPTD Discussion Paper 97. International Food Policy Research Institute: Washington, DC.Smale, M., Byerlee, D. and Jayne, T. 2011. *Maize revolutions in Sub-Saharan Africa*. Forthcoming, World Bank, Washington, DC, and Tegemeo Institute, Kenya.

Thierfelder, C., Matemba-Mutasa, R. and Rusinamhodzi, L. (2015b). Yield response of maize (Zea mays L.) to conservation agriculture cropping system in Southern Africa. *Soil Tillage Res.* 146, 230–42.

World Bank. 2013. *Agribusiness Indicators: Kenya*. The World Bank: Washington, D.C.

Zulu, E. 2003. Experiences with variety testing, registration, and release, in SADC region. Paper presented at the Regional Technical Meeting for Formulation of SADC Regional Variety Release System, 7–9 April 2003, Gaborone, Botswana.

Chapter 4

Good agricultural practices for maize cultivation: the case of West Africa

Alpha Kamara, International Institute of Tropical Agriculture, Nigeria

1 Introduction

Approximately 80% of the population of sub-Saharan Africa (SSA) depends on agriculture for food, income and employment (WDR, 2008). However, the agricultural sector is characterized by very low productivity, leading to very high poverty levels and undernourishment. Approximately 300 million people live below the poverty line in SSA (WDR, 2008). The incidence and severity of poverty is highest in rural areas, where over 70% of the households are poor (WDR, 2008). Approximately 51% of the population is food insecure, and malnutrition is pervasive in many households due to low intake of carbohydrates, proteins and micronutrients. The World Development Report (2008) concludes that accelerating agricultural growth through increased smallholder production, integrated with the development of the rural non-farming sector and markets will lead to a significant reduction of poverty and improvement in the living conditions of poor households in rural and urban areas.

Maize is the most important staple food crop for over 300 million people in SSA (http://dtma.cimmyt.org/index.php/background). It is an important source of calories, protein, vitamins and minerals for resource-poor rural and urban communities. Africa produces approximately 70 million tons of maize annually from about 34 million ha (FAO database, 2012). Smallholder farmers in 46 countries with a combined agricultural population of about 553 million produce maize under diverse and varied agro-ecologies and socio-economic

http://dx.doi.org/10.19103/AS.2016.0002.05

conditions (http://dtma.cimmyt.org/index.php/about/background). Maize (*Zea mays* L.) is an especially important cereal crop in the countries of West and Central Africa (WCA). The importance of maize in human diet, livestock feed and as raw material for some industries increased rapidly in the past two or three decades of the twentieth century. Although maize was introduced to Africa, the crop rapidly gained popularity as a major food item as well as an intra- and international trade commodity in the WCA sub-region (Fakorede et al., 2003). In the past three decades, maize has spread rapidly into the moist savannas of West Africa, replacing traditional cereal crops such as sorghum and millet, particularly in areas with good access to fertilizer inputs and markets (Manyong et al., 1996). In the West African moist savannas, higher radiation levels, lower night temperatures and a reduced incidence of diseases and insect pests have helped to increase maize yield potentials compared with traditional areas for maize cultivation (Kassam et al., 1975). Owing to the availability of short-season early maturing varieties, cultivation has also gradually spread to the Sudan savanna where the growing period is 90–100 days. Despite the expansion in these production areas, maize yields in farmers' fields average from 1 t/ha to 2 t/ha, in contrast to the higher yields of about 5 to 7 t/ha reported on breeding stations in the region (Kamara et al., 2013).

Maize production in WCA has been greatly constrained by many abiotic and biotic stresses. Most of the constraints also apply to other countries in ecologically similar zones. The most important constraints on maize yields are infestation by the stem borer complex, the parasitic weed *Striga hermonthica*, drought and poor soil fertility.

2 Poor soil fertility and effects of drought on maize in West Africa

2.1 Soil fertility

Lack of nitrogen (N) fertilizer is the greatest limitation on maize production in the savannas of WCA (Carsky and Iwuafor, 1995). In most cases, less than 20 kg ha^{-1} of N fertilizer is applied to maize crops. The high price of fertilizer and poor access to it (Smith et al., 1997), as well as low recovery due to losses through denitrification, ammonia volatization, run off and leaching, and inefficient utilization by maize crops (Akintoye et al., 1999) are all constraints on N fertilizer use among farmers in the sub-region. One important characteristic of maize is its high and relatively rapid nutrient requirements. Maize grain generally contains up to 2% N, that is, 100 kg of harvested grain yield contains 2 kg N. Total exports are 2.6 kg N per 100 kg of grain produced if above-ground residues are removed (Cretenet et al., 1994; Sanchez, 1976). Soil N can be quickly depleted by maize, especially when yields are high and stover is exported. Yields on farmers' fields are low. It is estimated that annual loss of maize yields due to low N stress varies from 10% to 50% (Wolfe et al., 1988). But even at these low yields, soil nutrients are being mined beyond the soil's ability to replenish them. Van der Pol (1991) estimated that average depletion of soil N by maize in south Mali was approximately 25 kg ha^{-1}. In contrast, phosphorus inputs and outputs were more or less in balance (Van der Pol, 1991), thereby justifying an emphasis on increasing nitrogen supply for maize.

The amount and distribution of rainfall are highly important factors in successful maize production. A minimum range of 480–880 mm of well-distributed rainfall is adequate for maize depending on the variety (IITA, 1982). The moisture requirement of maize is small during the early stages of development but increases rapidly as the flowering stage approaches, before decreasing again as the crop matures (Edmeades et al., 1997). The maize crop is especially sensitive to moisture stress during flowering, when a short spell of stress can reduce the crop yield by up to 30–35% (Azeez, et al., 2005). The ecological zones for maize production in the West African savannas have been demarcated on the basis of rainfall and vegetation cover and reflect diversity in cropping systems and production constraints. Three ecologies have been identified: southern and northern Guinea savannas and Sudan savanna. In the southern Guinea savanna, annual rainfall is about 1000 mm spread over 170 rainy days, between late May and early October. In the northern Guinea savanna, rainfall is about 800–900 mm spread over 150–60 rainy days, while in the Sudan savanna, annual rainfall is rarely up to 700 mm spread over about 120 rainy days (Ogungbile et al., 1998). The potential for maize production in these zones as well as the varieties adapted to them vary remarkably. The potential for sustainable maize yields increases gradually from the northern to the southern parts of the sub-region.

The amount of rainfall determines the growth and yield of maize especially where irrigation is not feasible. In most parts of the West African savannas, rainfall is unreliable and frequently less than that required for a good maize crop. Droughts occur in three forms in the region (Eckebil, 1991). In the southern Guinea savanna, where rainfalls are higher than in other savanna zones, instability in the establishment of the rains frequently occurs and often compels farmers to replant maize. In the northern Guinea savanna, owing to the instability of late establishment of the rains, mid-season and terminal drought conditions regularly occur. In the Sudan savanna, early season and terminal drought conditions are almost an annual event. In addition to the parasitic weed *S. hermonthica*, maize production in the West African savannas is limited by recurrent drought. The risk of drought is high in the savanna zone because rainfall is unreliable and its distribution is erratic (Eckebil, 1991). Annual maize yield loss due to recurrent drought is about 15% in SSA (Edmeades et al., 1995). However, localized losses in the savanna may be much higher (Edmeades et al., 1995). Grain yield losses can be even greater if the stress on maize plants coincides with the flowering and grain-filling period. NeSmith and Ritchie (1992) reported that maize yield can be reduced by as much as 90% if drought stress occurs between a few days before tassel emergence and the beginning of grain filling. Under induced moisture stress from the tassel-emergence stage to the end of the crop cycle of maize, Badu-Apraku et al. (2005) observed yield reduction of 62% compared with well-watered plants. Maize is sensitive to drought at the seedling, flowering and grain-filling stages. Drought, which coincides with the flowering and grain-filling periods, can cause serious yield instability at the farm level, as it allows no opportunities for farmers to replant or otherwise compensate for loss of yield (Kamara et al., 2003).

3 Weed competition with maize

Weeds are a major source of yield loss in maize in West Africa. Among other biotic and abiotic stresses, maize production is limited by low soil fertility and weed competition (Chikoye et al., 2004; Sanginga et al., 2003). Owing to reduced soil fertility and increasing weed pressure, resource-poor farmers have often abandoned land to prolonged fallow periods of up to ten years (Hobbs and Bellinder, 2004). Currently, long fallows are no longer possible in many farming systems in West Africa because of the high human population density on limited arable land and the consequent demand for food (Chikoye and Ekeleme, 2003). Low soil fertility, declining crop yields and high weed pressure are among the major problems that have been associated with the intensification of cropping systems (Tian et al., 1995).

In many parts of the West African savannas, *S. hermonthica* is a very notorious weed parasite of maize that has continuously posed a great threat to maize production. Under severe infestation, the maize seedling is usually seriously attacked even before emergence of the *Striga*, with the maize seedlings looking yellowish, stunted and wilted (Berner et al., 1995; Lagoke et al., 1991). Surveys in the northern Guinea savanna of Nigeria (NGS) showed that *Striga* has remained a serious problem, attacking millet, sorghum [*Sorghum bicolor* L. Moench], maize [*Zea mays* L.] and upland rice (Weber et al., 1995; Kim et al., 1997; Showemimo et al., 2002). In north-east Nigeria, over 85% of fields planted with maize and sorghum were infested with *Striga* (Dugje et al., 2006). *S. hermonthica* is estimated to have infested approximately 2.4 to 4 million ha of land under maize production, causing yield losses of 30 to 80% valued at between US$383 and US$1 billion and negatively affecting the livelihoods of about 100 million people in SSA, who are unable to grow enough food (AATF, 2006; Ejeta, 2007; Gressel et al., 2004; Woomer et al., 2008). *Striga* infestation is often so severe that maize can suffer total yield loss and farmers may abandon their fields. In a study to compare the effects of drought stress and *S. hermonthica* on maize under field conditions, Badu-Apraku et al. (2004) reported that grain yield was reduced by 53% under drought stress and by 42% under *Striga* infestation. The increasing incidence of *Striga* has been attributed to poor soil fertility and structure, intensification of land use through continuous cultivation and an expansion of cereal production (Rodenburg et al. 2005).

4 Maize nutrient management

To maintain a good crop and increase grain yield/unit area, agronomic best practices should be followed to address and mitigate these constraints, allowing farmers to close the yield gap for maize in the West African savannas. In order to achieve a high yield, adequate and balanced nutrition is important for maize. In the 1960s and 1970s, attention was focused worldwide on chemical fertilizers as a way to provide sufficient nitrogen for crop production (Carksy and Iwuafor, 1995). With better crop management practices, increased cultivation intensity and introduction of new varieties, maize has responded strikingly to the addition of nitrogen fertilizers in the West African savannas. Several researchers have recommended 100 to 120 kg N ha^{-1} (Balasubramanian et al., 1978; Chude et al., 1994). Mughogho et al. (1986) also suggested 100 kg N ha^{-1} as the optimum in the sub-humid savanna in all of West Africa. Although maize responds dramatically to nitrogen application, the low levels of P and K in the savanna soils (Kwari et al., 2011) limit the response of maize to nitrogen. The wider recommendation in the savannas for enhanced maize growth is the application of a compound fertilizer in the form of N:P:K as a basal dose. The first dose is applied at planting or within the first week of planting using N:P:K at a rate of 50 kg N ha^{-1}, 50 kg P_2O_5 ha^{-1} and 50 kg K_2O ha^{-1}. The second dose of 50 kg N in the form of urea is applied between four and five weeks after planting in the southern and northern Guinea savanna zones to medium- and late-maturing maize varieties, and between three and four weeks after planting in the Sudan savanna zone to extra-early maturing maize varieties. All fertilizers are buried during application. Farmers are advised to ensure that the second weeding has been conducted before the application of the second fertilizer dose.

4.1 Integrated soil fertility management approach

The use of inorganic fertilizers on their own, as per standard recommendations, is not always profitable to many small farmers in West Africa, particularly where rainfall is erratic. By following an Integrated soil fertility management (ISFM) approach, the efficiency and hence profitability of inorganic fertilizers can be improved. ISFM has been defined as 'a set of soil fertility management practices that necessarily include the use of fertilizer, organic inputs, and improved germplasm combined with the knowledge on how to adapt these practices to local conditions, aiming at maximizing agronomic use efficiency of the applied nutrients and improving crop productivity' (Vanlauwe et al., 2010). Increased use of organic and inorganic fertilizers, together with diversification of cropping to include legumes are important tools in restoring or sustaining the soil fertility for the intensive cropping systems of the savannas of West Africa (Vanlauwe et al., 2001; Sanginga et al., 2003;). These so-called 'balanced nutrient management systems' can be further enhanced through the use of improved cultivars that are tolerant to drought, resistant to *Striga* and use available nutrients efficiently, such as maize cultivars developed at the International Institute of Tropical Agriculture (IITA), Nigeria (Kamara et al., 2005).

Although an adequate supply of soil nitrogen (N) is critically important to maximize maize yield, especially for modern cultivars, which are specifically designed to respond to high N fertility (Celik et al., 2010; Liu et al., 2012), N fertilization rates are frequently chosen to maximize grain yield without considering potential negative impacts. Synthetic fertilizer influences crop productivity, which directly or indirectly induces changes in the physiochemical and biological environment of the soil (Jung et al., 2011; Li et al., 2012), especially in the case of long-term fertilizer application. Synthetic fertilizers can, however, increase soil acidity (Divito et al., 2011), and they can promote N losses via NO_3 leaching and gaseous N emissions (Pelster et al., 2011; Perego et al., 2012). Long-term experiments are usually used to determine how crop production systems affect profitability and soil, water and air resources (Zhao et al., 2010; Liu et al., 2011). Long-term field studies can be prohibitively expensive (Liu et al., 2013) and, as a result, detailed crop production and soil process models are increasingly being applied as time and cost-saving measures (Liu et al., 2011).

4.2 Extrapolating nitrogen recommendations beyond West Africa

Most of the nitrogen recommendations that have been made for maize production in West Africa are site-specific and cannot be extrapolated to other areas. There is no information on how the maize varieties respond to nitrogen more widely, considering the differences in soils and climate. Agronomic field experiments conducted at similar locations provide a reliable and valuable information source on crop management (Kassie et al., 2014), but collecting exhaustive data at multiple locations to support management decisions on a larger scale is limited by time and logistical costs (Kassie et al., 2014; Kisaka et al., 2015).

Also, experimental data are expensive, as they require several years of experimental data gathering. Moreover, experimental trials do not report consistent results from year-to-year and from site-to-site because nutrient management of maize is site-specific. Thus, most N recommendations cannot be extrapolated to wider areas (Kasika et al., 2015). This is simply because soils vary considerably in terms of their properties (Tremblay et al., 2015). The CERES-Maize model was used to determine the nitrogen fertilizer requirements of early maturing maize varieties in the Sudan savanna of northern Nigeria. The model predicted grain yield and harvest index reasonably well for the two varieties. Increasing N application from 0 kg to 30 kg N ha^{-1}, increased grain yield by 105%. When the N rate was increased to 60 kg N ha^{-1}, grain yield increased by 226%. Yield increases of 364%, 451% and 461% were observed when the N rate increased from 0 to 90, 120 and 150 kg ha^{-1}, respectively (Adnan et al., 2016). In Ghana, simulations with the CERES-Maize model showed that the addition of N fertilizer generally increased maize yields for all the years across sites. The median yield of maize increased with increasing N fertilizer application until 60 kg N ha^{-1}. Subsequently, the rate of increase either levelled off or diminished (4) across sites. Even though generally maize yield response occurs at 90 kg ha^{-1}, the additional increase in yield compared to that obtained at 60 kg ha^{-1} is small, hence the potential additional returns in monetary terms favour the use of 60 kg N ha^{-1}, which is currently the recommendation for maize production in the study region. Sites with deeper soils may, however, respond to mineral N additions beyond the 90 kg ha^{-1} rate of fertilization (McCarthy et al., 2016).

5 Weed management for maize cultivation

Good control of weeds in maize is a prerequisite for high yields. It is clear that weeds usually compete with the crops for nutrients, light and moisture. In order to permit the maize crop to express its full potential on the field, it is necessary to weed regularly. Timely weeding is essential, especially at the early stages of growth. Two manual weedings are recommended, which should be conducted at two weeks after planting and at 4–5 weeks after planting. Whenever it is economically feasible to do so, herbicides can be used to control weeds on a maize farm. Although some have been recommended for sole-crop maize, it is important to note that the effectiveness of the herbicides is enhanced by very good land preparation (Onyibe et al., 2006). The growing problem of lack of appropriate labour has increased the adoption of chemical weed control. Several pre- and post-emergent herbicides are recommended for effective control of non-parasitic weeds in the West African savannas (Onyibe et al, 2006).

Research has resulted in a range of component technologies that have been found to be effective in combating *Striga*. For instance, the use of resistant maize varieties (leading to reduced parasitism by *Striga*) or tolerant maize varieties (leading to reasonable crop yields even with high parasitism), maize–legume rotation using trap crops that stimulate suicidal germination and the application of nitrogen fertilizer can all be effective in reducing infestation and damage (Ellis-Jones et al., 2004; Franke et al., 2006). At the same time, farmers cope with low soil fertility and *Striga* through the use of crop rotation, organic manure and inorganic fertilizer (Dugje et al., 2006). It has been generally accepted that *Striga* control is more likely to be effective when a range of individual technologies are combined in a programme of integrated *Striga* control (ISC) that can provide sustainable control over a wide range of biophysical and socio-economic environments (Berner and Kling, 1997; Ellis-Jones et al., 2004). In fact, researcher-developed *Striga* control options have already shown that ISC is effective in reducing infestation as well as increasing maize grain yield on farmers' fields (Franke et al., 2006; Schulz et al., 2003). Ellis-Jones et al. (2004) showed that *Striga*-resistant maize grown after a soya bean trap crop increased the net benefit over two cropping seasons by over 100% compared with continuous maize cultivation. Franke et al. (2006) found that ISC which combined all the above-mentioned components reduced the *Striga* seed bank by 46% and improved crop productivity by 88%.

6 Management of soil moisture stress in maize

6.1 Drought-tolerant maize varieties

The use of drought-tolerant maize has been recommended for maize production in West Africa. A major outcome of the Drought-Tolerant Maize Project (DTMA), an agricultural initiative, is the release of several drought-tolerant maize varieties in West Africa. These varieties are bred to tolerate mid-season drought. Maize breeders at IITA have incorporated a *Striga*-resistant gene in these varieties. There are now several extra-early, early, medium and late-maturing maize varieties that combine drought tolerance with *Striga* resistance. These varieties have been made available to seed companies in the region for multiplication and marketing.

6.2 Irrigation for maize cultivation

In addition to the use of drought-tolerant maize varieties, soil and water management (mulching, water harvesting, trenches/terraces, irrigation and conservation tillage) can increase maize productivity in the West African savannas. Through adapted tillage techniques, moisture loss in the soil can be reduced and soil quality and agricultural production can be improved. Thus, water use for irrigation can be decreased and the local water cycle can be optimized. Soil moisture loss (and also soil degradation) can be minimized through different techniques of soil moisture conservation: soil cover and reforestation, mulching, soil amendments and different tillage techniques. Soil tillage comprises all the physical, mechanical, chemical or biological actions conducted to prepare the seedbed for seed germination, emergence and establishment, root development and crop growth (Biovision, 2010). Conservation tillage limits the devastating consequences of drought, conserves soil moisture (thus decreasing water use for irrigation) and some techniques even enhance the local nutrient cycle (e.g. mulch tillage) (Franzen, 1997). There are some limited irrigation schemes in maize-producing countries in West Africa. Investment in irrigation infrastructure constitutes an important poverty reduction strategy since it would boost agricultural productivity by reducing the risks associated with rainfall unreliability in arid regions. The Governments of Ghana, Mali and Nigeria have invested in large-scale irrigation schemes for production of cereal and vegetable crops. Maize is grown in the dry season under irrigation. Farmers also grow maize in these irrigated areas during the main cropping season. The farmers have access to supplementary irrigation when mid-season drought occurs.

7 Future trends and conclusion

Increasing agricultural production in sub-Saharan Africa is likely going to be the result of intensification and area expansion (Vanlauwe et al., 2014). Maize production in Africa is intensifying because of demand and perhaps increase in population. Yields of maize is however, quite low despite the availability of improved varieties that are tolerant to biotic and abiotic stresses. For maize, nutrients, especially nitrogen fertilization, should be the focus in increasing the productivity in Africa. Response to fertilizer is however dependent on good weed control and availability of water. Supplementary irrigation and deployment of drought-tolerant maize varieties would help in improving maize productivity.

8 Where to look for further information

- Badu-Apraku, B., M. A. B. Fakorede, A. F. Lum, A. Menkir and M. Ouedraogo (Eds), (2007). Demand-driven technologies for sustainable maize production in West and Central Africa. *Proceedings of the Fifth Biennial Regional Maize Workshop*, IITA-Contonou, Benin, 3–6 May 2005. WECAMAN/IITA, Ibadan, Nigeria, p. 515.
- Badu-Apraku, B., A. Menkir, M. A. B. Fakorede, S. Ajala and J. Ellis-Jones (2012). *Building partnerships and encouraging innovation for sustainable maize production: the West and Central Africa Collaborative Maize Research Network, Achievements and Impact*, IITA, Ibadan, Nigeria, p. 82.
- Hershey, C. H. and P. Neate (Eds) (2013). *Eco-efficiency: From vision to reality* (Issues in Tropical Agriculture series)/Clair H. Hershey and Paul Neate. Cali, CO: Centro Internacional de Agricultura Tropical (CIAT), p. 252 (CIAT Publication No. 381).

9 References

AATF. 2006. Empowering African Farmers to Eradicate Striga from Maize Crop lands. ISBN 9966-775-02-01.

Adnan, A. Adnan, Jibrin, M. Jibrin, Alpha, Y. Kamara, Bassam, L. Abdulrahman, Abdulwahab, S. Shaibu (2016). Using CERES-Maize model to determine the nitrogen fertilization requirements of early maturing maize in the Sudan Savanna of Nigeria. *Journal of Plant Nutrition*. In Press.

Akintoye, H. A., J. G. Kling and E. O. Lucas (1999). N-use efficiency of single, double and synthetic maize lines grown at four levels in three ecological zones of west Africa. *Field Crops Research* 60: 189–99.

Azeez, J. O., D. Chikoye, A. Y. Kamara, A. Menkir and M. T. Adetunji (2005). Effect of drought and weed management on maize genotypes and the tensiometric soil water content of an eutric nitisol in south western Nigeria. *Plant and Soil* 276:61–8.

Badu-Aparaku, B. M., A. Fakorede, A. Menkir, A. Y Kamara, S. Dapaah (2005). Screening maize for drought tolerance in the Guinea Savanna. *West Central African Cereal Research Communication* 33:2–3.

Badu-Apraku, B., M. A. B. Fakorede, A. Menkir, A. Y. Kamara,, A. Adam (2004). Effects of drought-screening methodology on genetic variances and co-variances in Pool 16 DT maize population. *Journal of Agricultural Science* 142:445–52.

Balasubramanian, V., L. A. Nnadi and A. U. Mokwunye (1978). Fertilizing sole crop maize for high yields, *Samara Miscellaneous Paper No. 76*. Institute of Agricultural Research, Ahmadu Bello University, Zaria, Nigeria. 14p.

Berner, D. K., and J. G. Kling (1997). Sustainable control of *S. hermonthica* spp. through a focused integrated pest management programme. In B. Badu-Apraku, M. O. Akoroda, M. Ouderaoogo and F. M. Quin (eds), *Contributing to Food Self Sufficiency: Maize Research and Development in West and Central Africa*. Ibadan, Nigeria: International Institute of Tropical Agriculture, pp. 1–11.

Berner, D. K., J. G. Kling and B. B. Singh (1995). Striga research and control: a perspective from Africa. *Plant Disease* 79:652–60.

Biovision –Foundation for ecological development, Annual Report 2010.

Carsky, R. J. and E. N. O. Iwuafor (1999). Contribution of soil fertility research/maintenance to improved maize production and productivity in sub-Saharan Africa. In B. Badu-Apraku, M. A. B. Fakorede, M. Ouedraago and F. M. Quin (eds), *Strategy for Sustainable Maize Production in West and Central Africa*. Proceedings of a Regional Maize Workshop, 21–25 April 1997 – IITA Cotonou, Benin Republic, IITA Ibadan, pp. 3–20.

Celik, I., H. Gunal, M. Budak, and C. Akpinar (2010). Effects of long-term organic and mineral fertilizers on bulk density and penetration resistance in semi-arid Mediterranean soil conditions. *Geoderma* 160:236–43.

Chikoye, D. and F. Ekeleme (2003). Cover crops for cogon grass (Imperata cylindrica) management and effects on subsequent corn yield. *Weed Science*, 51(5):792–7.

Chikoye, D., S. Schultz and F. Ekeleme (2004). Evaluation of integrated weed management practices for maize in the northern Guinea savanna of Nigeria. *Crop Protection* 23:895–900.

Chude, V. O., E. N. O. Iwuafor and G. Lombin (1994). Field evaluation of Nigeria made granular, urea on maize production in the semi-arid zone of Nigeria. In J. M. Menyonga, T Bezuneh, J. Y. Yayock and I. Soumana (eds), *Progress in Food Grain Research and Production in Semi-arid Africa*. Proceedings of a conference at Niamey, Niger, 7–14 March 1991, OAU/STRC-SAFGRAD Ouagadougou, Burkina Faso.

Cretenet, M., D. Dureau, B. Traore and D. Ballo (1994). Fertilit ´e et fertilisation´ dans la region sud du Mali: du diagnostic au pronostic. *Agriculture et ´ Developpement* 2:4–13.

Dugje, I. Y., A. Y. Kamara and L. O. O moigui (2006). Infestation of crop fields by Striga species in the savanna zones of northeast Nigeria. *Agriculture, Ecosystems, and Environment* 116:251–2.

Divito, G. A., H. R. Sainz Rozas, H. E. Echeverrı'a, G. A. Studdert and N. Wyngaard (2011). Long term nitrogen fertilization: soil property changes in an Argentinean pampas soil under no tillage. *Soil Tillage Research* 114:117–26

Eckebil, J. P. (1991). New Frontiers of Food Grains Research for the 1990s. In J. M. Menyonga, T. Benzumeh, J. Y. Yayock and J. Soumana (eds), *Progress in Food Grain Research and Production in Semi-arid Africa*, OAU/STRC-SAFGRAD, pp. 3–20.

Edmeades, G. O., M. Banziger, S. C. Chapman, J. M. Ribaut and J. Bolaños (1995). Recent Advances in Breeding for Drought Tolerance in Maize. Paper presented at the West and Central Africa Regional Maize and Cassava Workshop, May 28 to June 2 1995, Cotonou, Benin Republic.

Edmeades, G. O., Bolaños, J. and S. C. Chapman (1997). Value of secondary traits in selecting for drought tolerance in tropical maize. In *Developing Drought and Low N-Tolerant Maize*. Edmeades, G. O., Banziger, M., Mickelson, H. R. and Peña-Valdivia, C. B. (Eds), *Proceedings of a Symposium*, 25–29 March, CIMMYT, EL Batan, Mexico D. F. Mexico.

Ejeta, G. (2007). The Striga scourge in Africa: a growing pandemic. In Gebisa Ejeta and Jonathan Gressel (ed.). *Integrating New Technologies for Striga Control: Towards Ending the Witch Hunt*. Singapore: World Scientific Publishers, pp. 3–16.

Ellis-Jones, J., S. Schulz, B. Douthwaite, M. A. Hussaini, B. D. Oyewole, A. S. Olanrewaju and R. White (2004). An assessment of integrated *Striga hermonthica* control and early adoption by farmers in northern Nigeria. *Experimental Agriculture* 40:353–68.

Fakorede, M. A. B., B. Badu-Apraku, A. Y. Camara, A. Menkir and S. O. Ajala (2003). In B. Badu-Apraku, M. A. B. Fakorede, M. Ouedraogo, R. J. Carsky and A. Menkir (eds), *Maize Revolution in West Central Africa*. Proceedings of a Regional Workshop, 14–18 May 2001, IITA-Cotonou, Benin Republic. WECAMAN/IITA, Ibadan, Nigeria, pp. 3–15.

FAO. (2008). FAO Methodology for the Measurement of Food Deprivation: Updating the minimum dietary energy requirements. FAO Statistics Division.

Franke, A. C., J. Ellis-Jones, G. Tarawali, S. Schulz S, M. A. Hussaini, I. Kureh, R. White, D. Chikoye, B. Douthwaite, B. D. Oyewole and A. S. Olanrewaju (2006). Evaluating and scaling up integrated *Striga hermonthica* control technologies among farmers in northern Nigeria. *Crop Protection* 25(8):868–78.

Franzen, D. (1997). *Tillage Techniques that Can Save Moisture*. Fargo: North Dakota State University.

Gressel, J., A. Hanafi, G. Head, W. Marasas, B. A. Obilana, J. Ochanda, T. Souissi and G. Tzotzos (2004). Major heretofore intractable biotic constraints to African food security that may be amenable to novel biotechnological solutions. *Crop Protection* 23:661–89.

Hobbs, P. R., and R. R. Bellinder (2004). Weed Management in Less Developed Countries, pp: 1295–8. Encyclopedia of plant and crop science, Marcel Dekker, Inc.

IITA, Maize Production Manual, Volume 1. Manual Series No. 8.

INFONET BIOVISION (Editor) (2010). Conservation Tillage Systems. Zuerich: Biovision. URL [Accessed 19 June 2012].

Jung, M. Y., S. J. Park and D. Min (2011). Enrichment and characterization of an autotrophic ammonia-oxidizing archaeon of mesophilic crenarchaeal group I.1a from an agricultural soil. *Appl. Environ. Microbiol.* 77:8635–864.

Kamara, A. Y., S. U. Ewansiha and A. Menkir (2013). Assessment of nitrogen uptake and utilization in drought tolerant and Striga resistant tropical maize varieties. *Archives of Agronomy and Soil Science*, doi:10.1080/03650340.2013.783204.

Kamara, A. Y., J. G. Kling, A. Menkir and O. Ibikunle (2003). Agronomic performance of maize lines derived from a low nitrogen maize population. *Journal of Agriculture Science* 141:221–30.

Kamara, A. Y., A Menkir, S. O. Ajala and I. Kureh (2005). Performance of diverse maize genotypes under nitrogen deficiency in the northern Guinea savanna of Nigeria. *Experimental Agriculture* 41:1999–1212.

Kamara, A. Y., F. Ekeleme, A. Menkir, D. Chikoye and L. O. Omoigui (2009). Influence of nitrogen fertilization on the performance of early and late maturing maize cultivars under natural infestation with *Striga hermonthica*. *Archives of Agronomy and Soil Science* 55(2):125–45.

Kassam, A. H., J. Kowal, M. Dagg and M. N. Harrison (1975). Maize in West Africa and its potential in the savanna area. *World Crops* 27(1):73–8.

Kassie, B. T., M. K. Van Ittersum, H. Hengsdijk, S. Assengc, J. Wolfa, R. P. Rötter (2014). Climate-induced yield variability and yield gaps of maize (*Zea mays* L.) in the Central Rift Valley of Ethiopia. *Field Crops Research* 160:41–53.

Kisaka, M. O., M. Mucheru-Muna, F. K. Ngetich, J. N. Mugwe, D. N. Mugendi, F. Mairura and J. Muriuki (2015). Using APSIM-Model as a decision-support-tool for long-term integrated nitrogen-management and maize productivity under semi-arid conditions in Kenya. *Experimental Agriculture* page 1 of 21 C. Cambridge *University Press 2015*, doi:10.1017/S0014479715000095.

Kim, S. K. (1997). Achievements, challenges and future direction of hybrid maize research and production in West and Central Africa. In B. Badu-Apraku, M. O. Akoroda, M. Ouedrago and F. M. Quin (Eds), *Contributing to Food Self Sufficiency in Maize Research and Development in West and Central Africa*. Proceedings of a regional maize workshop, 29 May to 2 June, 1995, IITA, Cotonou, Benin Republic, pp. 42–82.

Kwari, J. D., A. Y. Kamara, F Ekeleme and L. O. Omoigui (2011). Soil fertility variability in relation to the yields of maize and soybean under intensifying cropping systems in the tropical savannas of north-eastern Nigeria. In A. Bationo, B. Waswa, J. M. Okeyo, F. Maina, J. Job Kihara (eds), *Innovations as Key to the Green Revolution in Africa*, Vol 1. Exploring the scientific facts. London: Springer Science + Business Media, doi:10.100071978-90-481-2543-2-47.

Lafitte, H. R., and G. O. Edmeades (1994). Improvement for tolerance to low soil nitrogen in tropical maize. II. Grain yield, biomass production, and N accumulation. *Field Crops Research* 39:15–25.

Lagoke, S. T. O., V. Parkinson and R. M. Agunbiade (1991). Parasitic weeds and controlmethods in Africa. In S. K. Kim (ed.), *Combating Striga in Africa*. Proc. Int. Workshop org. IITA, ICRISAT, and IDRC. IITA, Ibadan, Nigeria 61-274, pp. 3–14.

Li, X. H., X. Z. Han, H. B. Li, C. Song, J. Yan and Y. Liang (2012). Soil chemical and biological properties affected by 21-year application of composted manure with chemical fertilizers in a Chinese Mollisol. *Canadian Journal of Soil Science* 92:419–28.

Liu, H. L., J. Y. Yang, C. F. Drury, W. D. Reynolds, C. S. Tan, Y. L. Bai, P. He, J. Jin and G. Hoogenboom (2011a). Using the DSSAT-CERES-Maize model to simulate crop yield and nitrogen cycling in fields under long-term continuous maize production. *Nutrient Cycling in Agroecosystems* 89:313–28.

MacCarthy, D. S., G. S. K. Adiku, B. S. Freduah and A. Y. Kamara (2016). Using Decision Support Tools for climate-sensitive and soil management decisions for sustainable maize production in the northern regions of Ghana. *Archives of Agronomy and Soil Science*, under review.

Manyong, V. M., J. Smith, G. K. Weber, S. S. Jagtap and B. Oyewole (1996). Macro characterization of agricultural systems in west Africa: an overview. Resource and Crop Management Research Division Monograph. No 21. IITA. Ibadan.

Mughogho, S. K., A. Bationo, B. Christianson and P. L. G. Vlek (1986). Management of N fertilizer for tropical African soils. In A. U. Mokwunye and P. L. G. Vlek (eds), *Management of N and P Fertilizers in sub-Saharan Africa*. Dordrecht, The Netherlands: Martinus Nijhoff Publishers.

NeSmith, D. S., and J. T. Ritchie (1992). Short- and long-term responses of corn to a pre-anthesis soil water deficit. *Agronomy Journal* 84:107–13.

Ogungbile, A. O., R. Tabo, N. Van, Duivenbooden, and S. K. Debrah (1998). Analysis of constraints to agricultural production in the Sudan Savanna zone of Nigeria using multi-scale characterization. *Netherlands Journal of Agricultural Science* 46:27–38.

Onyibe, J. E., A. Y. Kamara and L. O. Omoigui (2006). Guide to maize production in Borno State Nigeria. Published by IITA press Nigeria.

Pelster, D. E., F. Larouche, P. Rochette, M. H. Chantigny, S. Allaire and D. A. Angers (2011). Nitrogen fertilization but not soil tillage affects nitrous oxide emissions from a clay loam soil under a maize-soybean rotation. *Soil Tillage Research* 115–16:16–26.

Perego, A., A. Basile, A. Bonfante, R. De Mascellis, F. Terribile, S. Brenna and M. Acutis (2012). Nitrate leaching under maize cropping systems in Po Valley (Italy). *Agriculture, Ecosystems and Environment* 147:57–65.

Rodenburg, J., L. Bastiaans, E. Weltzien and D. E. Hess (2005). How can selection for Striga resistance and tolerance in sorghum be improved? *Field Crops Research* 93:34–50.

Sanchez, P. A. (1976). *Properties and Management of Soils of the Tropics*. New York: John Wiley & Sons.

Sanchez, P. A. (1976). *Properties and Management of the Soils in the Tropics*. Chichester, U.K: John Wiley.

Sanginga,N., K. E. Dashiell, J. Diels, B. Vanlauwe, O. Lyasse, R. J. Carsky, S. Tarawali, B. Asafo-Adjei, A. Menkir, S. Schulz, B. B. Singh, D. Chikoye, D. Keatinge and R. Ortiz (2003). Sustainable resource management coupled to resilient germplasm to provide new intensive cereal-grain-legume-livestock systems in the dry savanna. *Agriculture, Ecosystems and Environment* 100:305–14.

Schulz, S., A. N. Honlonkou, R. J. Carsky, V. M. Manyong and B. D. Oyewole (2003). Alternatives to Mucuna for soil fertility management in southern Bénin: farmer perception and use of traditional and exotic grain legumes. *Experimental Agriculture* 39(3):267–78.

Showemimo, F. A., C. A. Kimbeng and S. O. Alabi (2002). Genotype response of sorghum cultivars to nitrogen fertilization in the control of *Striga hermonthica*. *Crop Protection* 21:867–70.

Smith, J., J Weber, M. V. Mangong and M. A. B. Fakorede (1997). Fostering sustainable increases in maize productivity in Nigeria. In D. Byerlee and C. K. Eicher (eds), Chapter 8 in *Africa's Emerging Maize Revolution*, London, UK: Lynne Rienner Publishers, pp. 107–15.

Tian, G., L. Brussaard and B. T. Kang (1995). Biological effects of plant residues with contrasting chemical compositions under humid tropical conditions: effects on soil fauna. *Soil Biology & Biochemistry* 25:731–7.

Tremblay, N., Y. M. Bouroubi, C. Bélec, R.W. Mullen, N. R. Kitchen, W. E . Thomason, S. Ebelhar, D. B. Mengel, W. R. Raun, D. D. Francis, E. D. Vories and I. Ortiz-Monasterio (2012). Corn response to nitrogen is influenced by soil texture and weather. *Agron. J.* 104:1658–71. doi:10.2134/agronj2012.0184.

Van der Pol, F. (1991). L'epuisement des terres, une source de revenus pour les paysans au mali-Sud. In *Savanes d'Afrique, terres fertiles*? Proceedings of and International Meeting at Montpellier, 10–14 December 1990. Ministere de la Cooperation et du Developpement (France), Paris, pp. 403–18.

van der Pol, F., and B. Traore (1993). Soil Nutrient depletion by agricultural production in Southern Mali. *Fertilizer Research* 36:79–90.

van Dijk, M., G. W. Meijerink, M. L. Rau and K. Shutes (2012). 'Mapping maize yield gaps in Africa: can a leopard change its spots?' LEI, Wageningen.

Vanlauwe, B., A. Bationo, J. Chianu, K. E. Giller, R. Merckx, U. Mokwunye, O. Ohiokpehai, P. Pypers, R. Tabo, K. Shepherd, E. M. A. Smaling and P. L. Woomer (2010). Integrated soil fertility management: Operational definition and consequences for implementation and dissemination. *Outlook Agriculture* 39:17–24.

Vanlauwe, B., J. Wendt and J. Diels (2001). Combined application of organic matter and fertilizer. In G. Tian, F. Ishida and J. D. H. Keatinge (eds), *Sustaining Soil Fertility in WestAfrica*, Madison, USA: SSSA Special Publication Number 58, pp. 247–80.

WDR (World Development Report) (2008). Agriculture for Development. The Word Bank, Washington, DC. p. 32.

Weber, G. K., P. S. Chindo, K. A. Elemo and S. Oikeh (1995). Nematodes as Production Constraints in Intensifying Cereal-Based Cropping Systems of the Northern Guinea Savanna. *Resource and Crop Management Research Monograph* 17:1–37.

Woomer, P. L., M. Bokanga and G. D. Odhiambo (2008). Striga management and the African farmer. *Outlook on Agriculture* 37:247–52.

Wolfe, D. W., D. W. Henderson, T. C. Hsiao and A. Alvino (1988). Interactive water and nitrogen effects on senescence of maize. II. Photosynthetic decline and longevity of individual leaves. *Agron. J.* 80:865–70.

Zhang, T. Q., T. O. Oloya, N. B. McLaughlin and J. D. Gaynor (2013). Reducing nitrate loss in tile drainage water with cover crops and water table management systems. *Journal of Environmental Quality* 43:587–98.

Zhao, B. Q., X. Y. Li, X. P. Li, X. J. Shi, S. M. Huang, B. R. Wang, P. Zhu, X. Y. Yang, H. Liu and Y. Chen (2010). Long-term fertilizer experiments network in China: crop yields and soil nutrient trends. *Agron. J.* 102:216–30.

Chapter 5

Zero-tillage cultivation of maize

Wade E. Thomason, Bee Khim Chim and Mark S. Reiter, Virginia Tech University, USA

1 Introduction

Zero-tillage cultivation of maize (*Zea mays* L.) is a relatively widely used production system whereby the only soil disturbance occurs with the seeding operation. The system aims to reduce erosion, improve soil physical attributes, decrease costs and improve long-term sustainability of maize cultivation. However, positive impacts from zero-tillage maize cultivation do not always occur and it is important to understand which factors favour this system and which do not. This chapter will also discuss suggestions for best management of zero-tillage maize systems.

2 History of zero-tillage maize cultivation

While zero-tillage crop production has been adopted in many areas of the world, this review will focus on adoption only in countries with significant maize production. Figure 1 illustrates the adoption of Conservation Agriculture (CA) systems for cropland in Argentina, Brazil, China, India, Ukraine and the United States. Conservation agriculture is an agricultural production system that is characterized by (1) continuous minimum mechanical soil disturbance, (2) permanent organic soil cover and (3) diversification of crop species grown

http://dx.doi.org/10.19103/AS.2016.0002.06

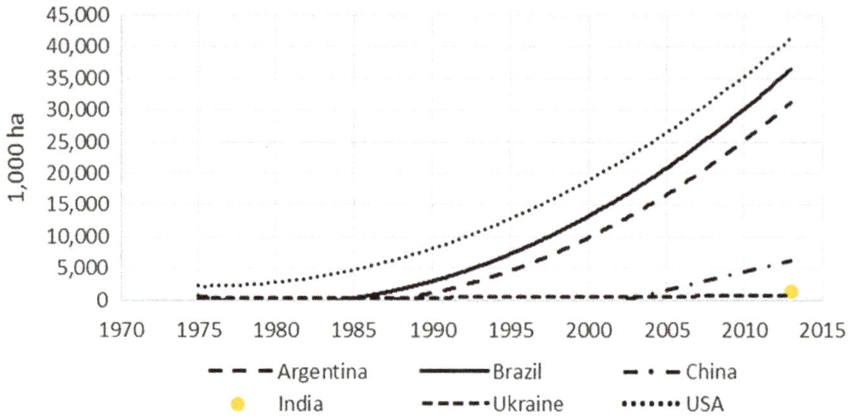

Figure 1 Conservation agriculture adoption in major maize-producing countries. Source: FAO, 2016.

in sequences and/or associations (FAO, 2011). All countries have shown an increasing trend in adoption of CA practices on cropland and one of the previously mentioned core tenets of CA is reduced or zero-tillage. This increase represents improvements in knowledge dissemination, equipment and management. The implementation of zero-tillage maize in South America began in the early 1970s and has become common on many hectares (FAO, 2016). Compared to North and South America, the proportion of maize cropland in the rest of the world farmed using zero-tillage practices is much less (FAO, 2016). There are likely many reasons for these differences. In some areas, particularly sub-Saharan Africa, it is likely the result of inadequate information and equipment. In areas of very small landholdings, the culprit is often the lack of affordable, effective, small-scale equipment (Johansen et al., 2012). However, there are many other factors including farmer attitudes (Wall, 2007) and labour changes (Knowler and Bradshaw, 2007) that influence adoption locally. In others, such as most of Europe, zero-tillage was tested and found not to provide significant benefits over conventional tillage systems; as constraints such as a build-up of previous crop residues and increased crop disease (Soane et al., 2012) outweighed benefits.

The majority of the increase in zero-tillage maize area has occurred since the early 1990s and has been fuelled by the use of improved equipment, herbicide availability and awareness of erosion prevention. Concurrently, the development of herbicide tolerant crops has improved in-crop weed control allowing greater benefits to accrue for zero-tillage systems (Dill et al., 2008). The current trend is for an increase in the maize area managed with zero-tillage, especially in Asia (Stevenson et al., 2014) and potentially in Africa (Thierfelder et al., 2013). However, it should be noted that the actual rate of expansion of zero-tillage management in the USA has fallen short compared to earlier predictions (USDA, 1985). There are many reasons for this, not the least of which is that, in the early years of adoption of zero-tillage, some believed that the system could be successfully applied under almost all soil and environmental conditions. With time, it has become apparent that zero-tillage is less practical on fine-textured soils and in areas where cold soil temperatures limit early-season germination and growth.

2.1 USA

Edward Faulkner's 1943 publication of 'Plowman's Folly' likely marks the first text to scientifically challenge the need for ploughing. His statement, 'No one has ever advanced a scientific reason for plowing', was a highly disruptive idea and one that would cause many scientists and farmers the world over to rethink the use of the plough and the need for tillage in maize production. In the late 1950s and early 1960s the invention and availability of new herbicides provided the means to implement zero-tillage seeding in maize and other crops. Some of the earliest published reports on the topic summarized the effect of reduced tillage and residue management practices on maize emergence and yield in Ohio beginning in 1937 (Page et al., 1947), and these authors suggested that for maize, tillage should be kept to a minimum consistent with meeting acceptable seedbed requirements. Zero-tillage experiments were conducted in North Carolina in 1948 (Klingman and Spain, 1965) but failed due to inadequate weed control (Strohm, 1952). The first on-farm demonstration of zero-tillage planting and production was conducted in Kentucky in 1962 (Phillips and Young, 1973) and thousands of farmers visited the fields to learn from this pioneering effort. In Virginia, zero-tillage experiments were begun in 1960 and found that zero-tillage planting into killed sod resulted in greater soil moisture throughout the season, faster maize growth and similar yields compared to conventional tillage (Moody et al., 1961). Jones et al. (1968) reported increased maize yield, reduced runoff and greater soil moisture with zero-tillage. These early research highlighted the positive impacts that could be captured with zero-tillage but adoption lagged due to difficulties in scaling up the techniques and in overcoming negative perceptions of 'trashy' fields.

Planter technology was a major impediment to zero-tillage maize production as almost all early units required some degree of tillage in order to accurately place seeds. In 1966

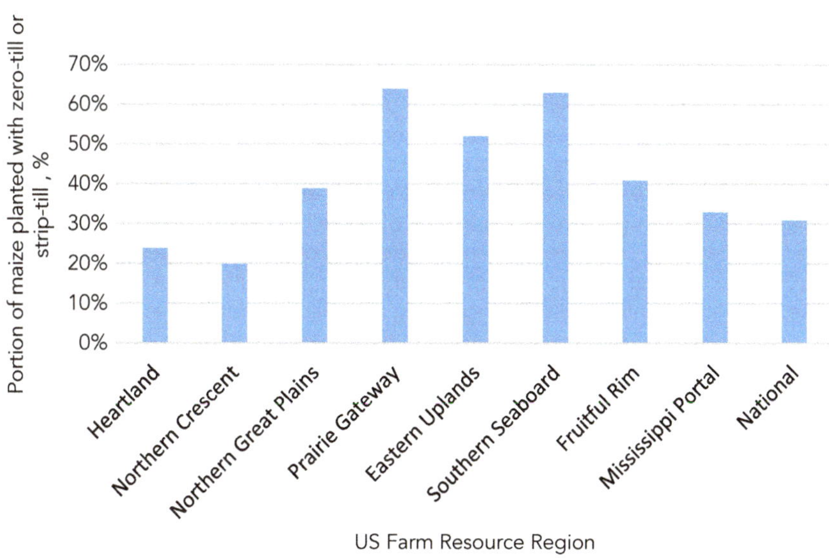

Figure 2 Portion of maize area planted with zero- or strip-tillage, by US farm resource region, 2010–11. Source: Wade et al. (2015).

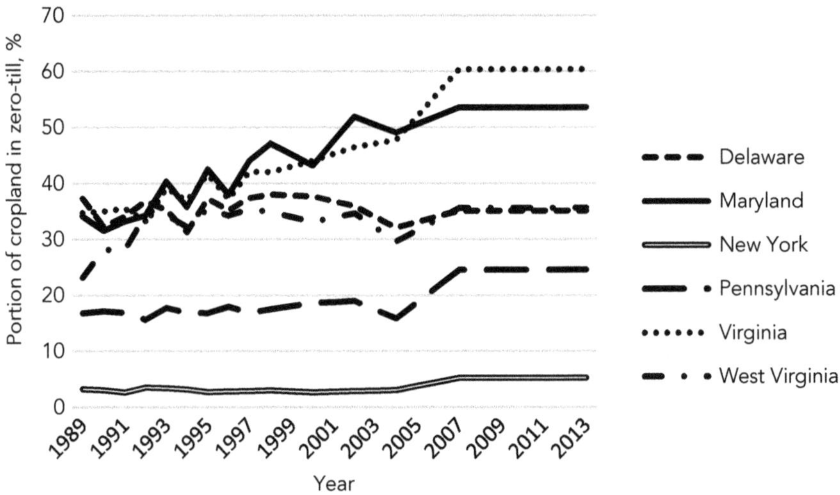

Figure 3 Zero-till implementation, portion of cropland by state in the Chesapeake Bay watershed, 1989–2013. Source: Conservation Technology Information Center tillage survey. http://www.ctic.purdue.edu/.

Allis Chalmers introduced a new planter with a leading fluted coulter to cut residue and open the soil, slanted opening discs to open the seed furrow and presswheels to close the furrow. This was only the first of many equipment innovations that allowed greater success and greater adoption of zero-tillage farming.

Since that time, adoption of zero-tillage/strip-till maize production in the USA has increased to approximately 30% of all planted acres (Wade et al., 2015). However, implementation varies widely among agricultural regions. High rates of adoption (>60% of area) exist in the Prairie Gateway, which includes portions of Nebraska, Kansas, Colorado, Oklahoma, Texas and New Mexico (Fig. 2) and in the Southern Seaboard. It is interesting to note that these higher rates are observed in areas where cold, wet soils in spring are not typically an impediment to maize germination and early growth. These areas often also experience acute drought stress due to lack of rainfall in-season.

At a finer scale, Fig. 3 shows the trend in adoption of zero-tillage adoption since 1989 in states that make up the Chesapeake Bay watershed in the eastern USA. This area represents diverse cropping systems and soils. Increasing zero-tillage adoption as a means to improve water quality has been a policy goal throughout this period and those effects are seen in the high levels of adoption in certain states, particularly Maryland and Virginia. Similar to the nationwide trend, adoption is lower in the more northern areas of the watershed; however, since many farms in Pennsylvania and New York include intensive animal production, this trend is likely confounded with land application of manure and the need for incorporation.

2.2 South America

Farmers and farmer associations have been pioneers in the adoption of zero-tillage in many areas in Latin America. This is especially the case in parts of Brazil and Argentina (Fig. 4). This expansion since the early 1990s has been driven by the lower cost of production for

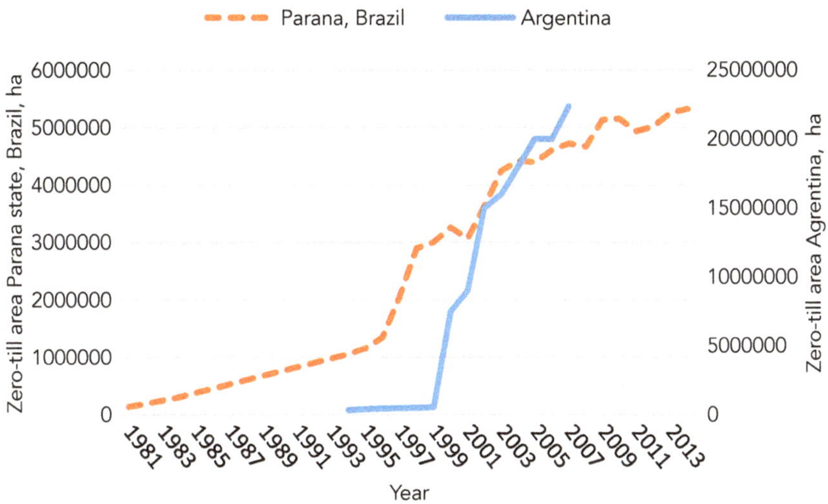

Figure 4 Area of zero-till cropland in Argentina, 1994–2008 and in Parana state, Brazil, 1981–2013. Source: Trigo et al. (2009) and Bublitz (2014).

zero-tillage compared to conventional tillage and by the time savings associated with zero-tillage practices. Farmers can manage larger areas of cropland because less labour per ha is required.

3 Advantages of zero-tillage maize production: reduced soil erosion

One of the most widely accepted benefits of zero-tillage maize production is reduced soil erosion, especially on sloping land (Fig. 5). The presence of plants and plant residues on the surface of the soil reduces erosion by preventing the direct impact of raindrops on soil which prevents the initial displacement of soil particles or soil splash. The presence of residue on the surface also reduces sheet and rill erosion by reducing the flow energy of water. Redirecting and slowing the flow of water moving downhill results in more shallow flow depth of water and less erosive force. Similarly, the presence of residue generates a rough surface that reduces wind erosion by reducing wind velocity at the soil surface and disrupting the energy moving soil particles impart to one another. Residue also captures moving soil particles preventing avalanching (Weil and Brady, 2017). Standing residue is more effective at reducing wind erosion compared to flat residue and several small residue pieces are typically more effective than very large pieces. Coarse-textured soils typically require greater residue amounts to prevent wind erosion because they typically have less soil OM and fewer stable aggregates compared to finer soils.

In a recent literature review of the effect of residue cover on erosion from studies pertinent to the eastern and Mid-Atlantic areas of the USA, the mean sediment loss reduction from two small watershed studies was 64.5% less erosion for zero-tillage than for conventional,

Table 1 Reported sediment reduction values of high residue systems over conservation tillage from applicable peer-reviewed studies and RUSLE2 model runs

Brief citation	Sediment loss reduction, %
Small Watershed-scale studies	
Shipitalo and Edwards, 1998	−61.5%
Staver, 2004	−67.5%
AVG	**−64.5%**
Small plot studies	
Verbree et al., 2010	−85.2%
Truman et al., 2005	−91.5%
Benham et al., 2007	−77.2%
Eghball and Gilley, 2001	−79.6%
Kleinman et al., 2009	−38.0%
AVG	**−74.3%**
RUSLE2 model runs	
[†]Coastal Plain, 1% slope	−49%
Coastal Plain, 2% slope	−80%
Coastal Plain, 4% slope	−78%
[‡]Piedmont, 3–4% slope	−65%
Piedmont, 5–6% slope	−68%
Piedmont, 9–10% slope	−58%
[§]Ridge & Valley, 3–4% slope	−66%
Ridge & Valley, 5–6% slope	−71%
Ridge & Valley, 9–10% slope	−70%
[††]Appalachian Plateau, 4% slope	−75%
Appalachian Plateau, 6% slope	−77%
Appalachian Plateau, 10% slope	−76%
AVG	**−69.4%**

[†]Coastal Plain values represent mean estimates over the following soil series at the indicated slopes: Emporia loamy fine sand, Pamunkey fine sandy loam, Norfolk loamy sand, Bojac loamy fine sand, State silt loam.
[‡]Piedmont values represent mean estimates for the following soil series at the indicated slopes: Cecil sandy loam, Hagerstown silt loam.
[§]Ridge and Valley values represent mean estimates over the following soil series at the indicated slopes: Frederick silt loam, Pope fine sandy loam.
[††]Appalachian Plateau values represent mean estimates over the following soil series at the indicated slopes: Ernest silt loam, Laidig loam.

full inversion tillage (Table 1). A summary of five small plot research studies, mostly using simulated rainfall techniques, produced a mean sediment reduction of 74% with a range in reduction from 38 to 92% (Table 1). Similarly, RUSLE2 erosion estimates from computer

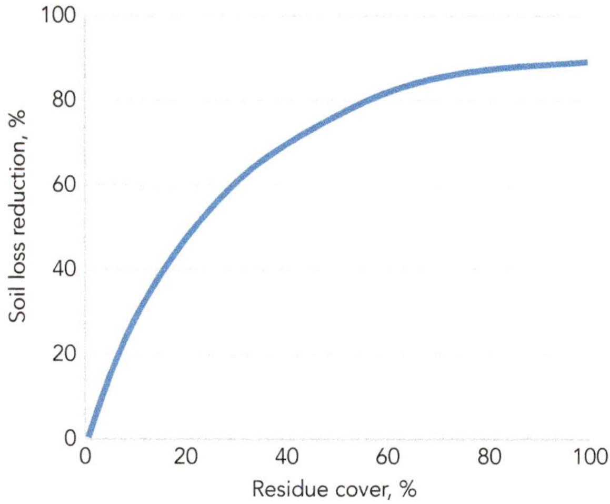

Figure 5 Generalized relationship of soil residue cover and soil loss reduction.

simulation runs indicate a mean reduction of 69% with a range of 49–80% depending on soil series and slope. Not only does soil erosion damage soil productivity directly (Pimentel et al., 1995; Pimentel and Kounang, 1998), but it also often results in losses of sediment-bound pesticides (Clark, 1987) and nutrients to water (Carpenter et al., 1998).

4 Advantages of zero-tillage maize production: reduced nutrient losses

4.1 Nitrogen

For nitrogen (N), zero-tillage can reduce losses through various mechanisms including: reduced surface losses of enriched sediment; reduced soluble surface N losses through increased infiltration; increased soil OM resulting in N being held in labile but not water-soluble forms; reduced soil mineralization rates; and decreased overall leaching due to preferential flow pathways. Dinnes (2004) summarizes the effect of N loss reduction for zero-tillage compared to moderate (conservation) tillage to −90 to +95%, highlighting the impact of site-specific conditions on N losses.

Similar to the reduction in soil erosion, the greater infiltration and reduced runoff of water also result in reductions in soluble N losses in runoff. Figure 6 shows the relationship between runoff N losses and surface residue, the extent of which was achieved with tillage practices of various intensity. In most rotations that include grass crops such as maize, zero-tillage systems typically have at least 50% surface residue cover, which would relate to approximately a 50% reduction in runoff N.

The impact of zero-tillage on N leaching losses is variable. Some authors have reported an increase in soil N balance (Constantin et al., 2010) or in short-term immobilization of N with zero-tillage (Rice and Smith, 1984), which could result in less leaching. Randall and Iragavarapu (1995) reported on an 11-year study conducted on a poorly drained silt loam

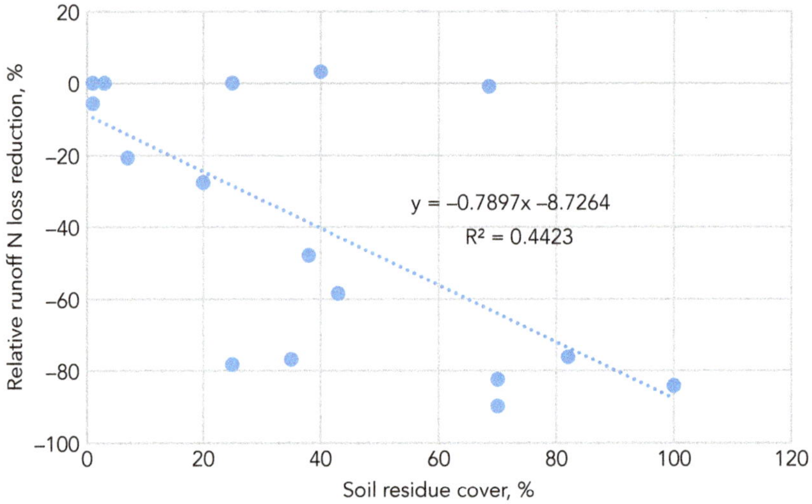

Figure 6 Runoff N loss as a function of soil surface cover. Data from: Chichester, 1977; McDowell and McGregor, 1984; Owens and Edwards, 1993; Romkens et al., 1973; Shipitalo et al., 2013.

soil and found that leaching losses were 5% greater for conventional tillage compared to zero-tillage. Others (Kanwar et al., 1997) have reported similar leaching losses between zero-tillage and conventional tillage systems. And still others have reported greater nitrate leaching with zero-tillage due to greater infiltration rates and preferential flow paths (Tyler and Thomas, 1977). It appears that site-specific conditions are the overriding factor that determines the effect tillage on N leaching, making it inappropriate to generalize.

4.2 Phosphorus

Phosphorus (P) losses from agricultural fields also occur via surface runoff as water-soluble P, as sediment-bound P in runoff, and as water-soluble P via leaching (Carpenter et al., 1998). Which of the loss pathways predominates depends on many factors, but is heavily influenced by slope (soil loss), soil texture and soil P concentration. In systems where the main mechanism for loss is sediment-bound P zero-tillage can dramatically reduce P losses. Dinnes (2004) summarizes the effect of converting from intensive tillage to zero-tillage and concludes that a 50–90% reduction in P losses would occur at any given point in time. It should be noted that in zero-tillage systems, fertilizer and manures are typically applied to the soil surface and that enrichment of runoff may occur if runoff is greater than infiltration. This is of particular concern when large rainfall events occur soon after application. Because of this risk, some advocate for the low-disturbance incorporation of P when possible.

In a recent review of literature relevant to the Chesapeake Bay Watershed of Virginia differential effects of tillage practices on P losses were found between studies conducted on soils defined as agriculturally well and moderately well drained and those that were conducted fully or partially on poorly drained soils (Table 2). On well-drained soils, reducing the intensity of tillage resulted in a decline in P losses (68% reduction compared

Table 2 Effect of change in tillage practice on P runoff loss by agricultural drain class

	Intensive tillage to conservation tillage	
Drainage category	Citation	Change in runoff P loss, %
Well	Angle et al., 1984	−90%
	Mostaghimi et al., 1988	−71%
	Butler and Coale, 2005	0%
	Benham et al., 2007	−64%
	Ross et al., 2001	−93%
Poor	Kleinman et al., 2002	149%
	Conservation tillage to zero-tillage	
Drainage category	Citation	Change in runoff P loss, %
Well	Benham et al., 2007	−29%
	Ross et al., 2001	−4%
	Kleinman et al., 2009	−49%
	Sharpley and Kleinman, 2003	−39%
	Verbree et al., 2010	−49
Poor	Verbree et al., 2010	51%
	Staver, 2004	238%
	Kibet et al., 2011	184%
	Kleinman et al., 2009	26%

to conventional tillage). This is likely due to greater infiltration rates associated with zero-tillage and a resulting reduction in runoff volume in these systems. On poorly drained soils, P losses were typically increased in zero-tillage compared to conventional tillage. On these soils, it is probable that zero-tillage practices do not result in increased infiltration. In the absence of this benefit, P losses were higher. This could be due to greater losses of soluble P from surface-applied fertilizer and manure. It is also possible that the stratification of P within zero-tillage systems resulting in higher P concentration in the surface few cm of soil also results in greater soluble P losses in runoff water through enrichment.

5 Advantages of zero-tillage maize production: water infiltration and use, economics and soil organic matter

5.1 Infiltration

Infiltration rates are typically higher for tilled soils immediately after tillage operations, but because of the destruction of soil aggregates, surface crusts often form after initial rainfall events, thus restricting infiltration and increasing runoff. The major drivers for increased infiltration are: plant residues themselves, improved soil aggregation and intact macropores (Blevins and Frye, 1993). Plant residues physically slow the movement of water over the soil surface allowing greater time for infiltration to occur. Reduced tillage often results in the maintenance of aggregates created via the combining of organic substances from plants and microorganisms with soil particles while tillage destroys these aggregates (Unger and McCalla, 1980). On a Cecil sandy clay loam soil, Golabi et al. (1995) used blue dye to track the flow of water to a depth of 25 cm in soils that had been managed with 10 yr of either conventional till or zero-tillage. Figure 7 shows the percentage of soil area affected by dye at various depths when added with water at a rate of 3–5 cm hr^{-1} for one hour. At the end of the period, the zero-tillage soil had 26% more area affected than conventional tillage and while the conventional till soil had retained 99% of the dye in the top 5 cm, the distribution of dye in the zero-tillage treatments were: 34, 34 and 22% of the total in the 5, 10 and 15 cm depths, respectively.

5.2 Economics

In mechanized agriculture systems, equipment and operating cost of equipment make up a major portion of crop production input costs (Pflueger, 2005). Tillage requires large, powerful tractors which consume significant amounts of fuel. Ground speed is also limited by the need to limit slippage and maximize the effectiveness of tillage operations so the hours $acre^{-1}$ to complete these operations are greater as well. Zero-tillage systems typically require less equipment and fewer field operations, compared to full-tillage crop production. As the amount of tillage decreases, the cost for machinery and labour decrease, especially if the machinery complement used is close to the optimum for the size of the overall operation. An example of this in shown in Table 3 which lists the relative differences of estimated costs for relevant expenses between conventional and zero-tillage maize based on enterprise budgets available from Virginia Cooperative Extension. Adoption of zero-tillage systems does typically require the purchase of new or additional equipment, especially planters and seeders specifically designed to operate in zero-tillage conditions. These seeders are typically heavier and more costly than those suitable for convention tillage (Dickey and Jasa, 1983). Many producers have made the shift to reduced- or

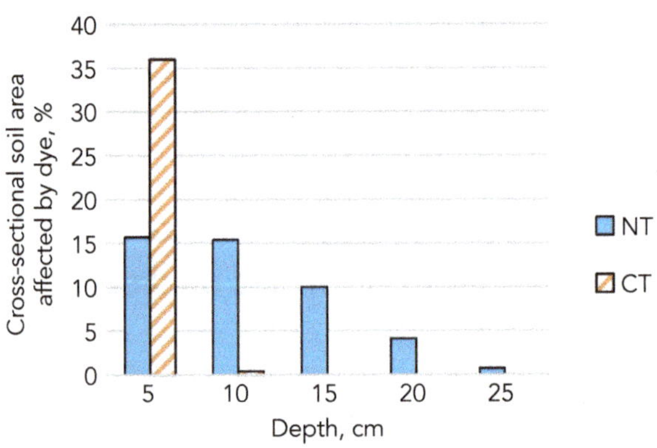

Figure 7 Percentage of soil area affected by dye at various depths. Adapted from Golabi et al., 1995.

Table 3 Variable and machinery cost for conventional and zero-till maize production. Based on baseline Virginia Cooperative Extension enterprise budget for maize

	Conventional tillage	Zero-tillage
Pre-harvest variable costs	$ ha^{-1}	
Herbicides	$55.17	$79.63
Fuel, oil, lube	$34.06	$14.94
Repairs	$31.34	$23.09
Pre-harvest labour	$25.43	$13.26
Machinery fixed costs		
Tractor and machinery	$161.42	$144.27
Total	$307.42	$275.18

Conventional tillage equipment complement includes:	Zero-tillage equipment complement includes:
140 hp MFWD tractor	120 hp MFWD tractor
180 hp tractor	20 m boom sprayer
120 hp tractor	8-row planter
6 m chisel plot	
5 m offset disk	Zero-till includes burn-down herbicide and surfactant cost
7m tandem disk	
8-row planter	

zero-tillage systems in tandem with replacing seeding equipment. Similarly, because herbicide applications are the main mechanism for weed control in zero-tillage, many users will purchase larger sprayers with greater precision for use in these systems. So while some individual equipment components in the zero-tillage system are more expensive, the overall result is less total machinery overhead in the farming operation and savings in labour and fuel (Table 3).

5.3 Water use efficiency

Maintaining soil cover of 30% or more generally reduces evaporation and increases water infiltration leading to greater available soil water. This increased soil moisture is extremely important in regions with low rainfall and high evapotranspiration, in all regions when rainfall is below normal, and on soils with low water-holding capacity. Papendick et al. (1973) reported that tillage increased soil water evaporation compared to zero-tillage during fallow. Tillage increased evaporation by moving moist soil to the surface and exposing that moisture to air. Fabrizzi et al. (2005) similarly reported greater water availability (0–80cm) for zero-tillage compared to minimum tillage during critical growth stages of maize and wheat in the Pampas of Argentina. Greater infiltration due to zero-tillage was the major driver for this effect, but lower evaporation from the soil surface is also important.

5.4 Carbon sequestration and soil organic matter

West and Post (2002) summarized the effect of tillage practice on soil organic carbon (SOC) from 67 published studies. They report that compared to conventional tillage, adoption of zero-tillage resulted in mean sequestration of 48 g C $m^{-2} yr^{-1}$ and that the greatest rate of C sequestration increase occurred when adopting zero-tillage in a maize:soybean rotation with a mean sequestration rate of 90 g C $m^{-2} yr^{-1}$. Similar comparisons were compiled for 62 studies in Canada (VandenBygaart et al., 2003). These authors reported a SOC increase of 0.38 t ha^{-1} yr^{-1} in studies samples to 30 cm or less. In the eastern USA, Spargo et al. (2008) collected samples (0–15 cm) from 63 fields under various duration of zero-tillage management and reported a mean rate of SOC sequestration of 0.308 Mg C ha^{-1} yr^{-1}. Blanco Canqui and Lal (2008) compared long-term, zero-tillage and ploughed plots to a depth of 60 cm on 11 sites in Ohio, Kentucky and Pennsylvania, USA. They reported greater SOC for zero-tillage in five of 11 sites, but only within the top 10 cm. Over the entire measured soil profile, no differences in SOC were observed between the two systems and the authors concluded that while zero-tillage increases SOC concentration in the upper layers of some soils, it does not store more SOC over the 0–60 cm depth. Similar conclusions were reached by VandenBygaart et al. (2003) for their studies that were sampled to depths greater than 30 cm (5 experiments) and by Baker et al. (2007) in their summary paper on the topic. It is likely that the increase in SOC in the top 30 cm of soil due to adoption of zero-tillage is due to a combination of greater root growth and proliferation in this zone with zero-tillage as well as reduced rates of organic matter decomposition when tillage is not used.

5.5 Yield under water stress

When zero-tillage results in greater available soil moisture in seasons where rainfall is limited, maize yield is often increased (Blevins et al., 1971; Norwood, 2000; Wagger and Cassel, 1993). When seasonal rainfall is generally adequate to meet crop demand no advantage is generally seen with zero-tillage systems (Licht and Al-Kaisi, 2005). As early as 1968, the complementary advantages of surface residue and zero-tillage for maize were recognized (Triplett et al., 1968). When double residue was applied to zero-tillage plots, available soil moisture (0–46 cm) was increased by 38% over ploughed plots, and over the three years of the study, zero-tillage plot with double residue yields were 19% higher than those from ploughed plots. A recent meta-analysis summarized the overall effects of zero-tillage on maize yields from 38 countries (Pittelkow et al., 2015). Overall they found a maize yield reduction of over 7%, but higher yields on average, in studies conducted under rainfed conditions in dry areas. However, they also highlight the relative importance of agronomic practices on yields and list duration of zero-tillage, residue management and aridity as the main influences.

6 Potential disadvantages of zero-tillage maize production

6.1 Weeds

Weed control is often a challenge in zero-tillage maize systems, especially in the initial years of adoption. Herbicides are typically required for weed control unless adequate surface mulch exists to suppress weed seedlings. This typically dictates that herbicide programmes in zero-tillage require greater planning and management. However, zero-tillage systems typically see a reduction in annual weed populations with a few years of adoption. Surface tillage, such as disking creates a favourable environment for germination of many annual weeds. Inversion tillage buries weed seeds, and seed buried deeper than 5 cm typically do not emerge. However, some weed seed can remain viable in soil for years so inversion tillage in subsequent years returns those seeds to the soil surface, maintaining the population. So that, over time, maintaining good weed control in zero-tillage results in a reduction in the weed seed bank and soil surface conditions that are unfavourable for annual weed emergence. Control of deep-rooted perennial weeds with herbicides is often only marginally effective as the dose required for control may be higher than those in typical application rates. Over time, weed populations in zero-tillage systems often shift away from annuals and toward perennials. This highlights the need for managers to understand establishment and growth patterns of different weed species in order to develop effective control strategies.

6.2 Cold soils

Residue cover on soils tends to keep soils cooler through greater reflectance of incident light, less heat accumulation, and an insulating effect on the soil (Kaspar et al., 1990). There is also typically an interaction with soil moisture since drier soils warm and cool faster than wet soils because soil particles have a greater conductivity and lower heat capacity than water. This may be an advantage during hot periods, but can be very detrimental in the early season, especially in areas with cooler climates (DeFelice et al., 2006). These authors reviewed 61 maize experiments comparing zero- and conventional tillage in the USA and Canada. They found that in the southern and western areas (warmer zones) zero-tillage yielded 12% higher than conventional tillage, while in the northern (coolest) zone, zero-tillage maize yielded 6% less and much of this is likely due to slow emergence and early growth in zero-tillage due to cooler soils early in the season. This reduction in corn yield with zero-tillage caused by cooler soils has been documented by numerous researchers working in the central and northern USA (Allmaras et al., 1964; Kaspar et al., 1990).

6.3 Decreased emergence

Decreased seedling emergence has been reported for zero-tillage maize in some instances (Gupta et al., 1988). Surface residue may impair the ability of the seeder to consistently place seed at the proper depth. This is especially problematic with non-uniform distribution of previous crop residue after harvest. Inconsistent seeding depth can result in differential emergence of plants in the stand and can cause undue stress through interplant competition (Liu et al., 2004). Seeds planted too shallow are exposed to greater variation in diurnal temperatures resulting in greater early-season stress. Seed that is placed too deep must use more energy to reach the soil surface and seed resources may be exhausted before the seedling can emerge. In addition to the direct stresses associated with inaccurate or imprecise seeding, seeds that take longer to emerge are at greater risk and exposure to early-season diseases and other pests.

6.4 Compaction

Soil compaction is frequently a major concern in zero-tillage systems, especially in the initial years after conversion from conventional tillage (Thomason et al., 2006). The elimination of deep or inversion tillage can limit the ability of managers to solve compaction problems. Soil bulk density in the surface layers is typically greater for zero-tillage compared to tilled systems. In many cases, however, the rate of water movement into the soil and soil root exploration are greater for zero-tillage. Many researchers report positive changes in soil physical properties with increased duration of zero-tillage management (Derpsch, 2003; Lal, 1976).

Controlled traffic management systems are one way to help alleviate the concern of compaction issues with zero-tillage (Tullberg et al., 2007). These systems confine equipment traffic to the same areas (rows) of the field for all equipment and all crops. If the cause of soil compaction is equipment induced, this is a very effective scheme to limit the negative effects of field traffic.

6.5 Plant disease

Many authorities have recognized the contribution of previous crop residue to higher root and foliar disease. In maize this effect of greater disease is most common with grassy residue, especially previous maize-crop residue. The most common foliar diseases that are exacerbated by the presence of maize residue include:

- Grey leaf spot (*Cercospora zeae-maydis* L.) is the most prevalent foliar disease in humid maize production areas (Munkvold and White, 2016). The disease begins with small pinpoint lesions but matures into rectangular grey to brown lesions that are parallel to and contained within leaf veins. Infection is favoured by warm, humid weather. Disease symptoms often appear and expand following periods of overcast days, heavy dew or fog. Yield loss occurs due to a loss of photosynthetic area and severity depends on the number of lesions present and how much of the canopy is affected. If lesions are apparent on the ear leaf or above, prior to tassel, significant yield losses are possible. Later developing lesions in the upper canopy likely have less economic impact.
- Northern leaf blight is caused by the fungus *Exserohilum turcicum* L. which overwinter in old corn residue (Munkvold and White, 2016). Lesions are long and narrow, tan-coloured zones that form parallel to leaf edges. Mature lesions are tan or grey and are often described as 'cigar-shaped'. Fungal spores develop on previous crop residue and are blown by the wind or splashed onto leaves. Infection is favoured by moderate temperatures and humid conditions. Six to eighteen hours of free water on the leaf surface is required for fungal infection. So, similar to grey leaf spot, symptoms often appear after days of wet weather or overcast skies. Yield losses of 30% or more are possible when lesions occur in the upper canopy prior to tasseling. However, there is no strong relationship between the area of disease infection and yield loss. The disease can also predispose plants to the stalk rot complex and result in greater lodging.
- Southern leaf blight is a fungal disease caused by *Bipolaris maydis* L. (Baker et al., 2007). The disease thrives in warm and subtropical humid environments and overwinters in previous crop residue. Lesions are typically 5–10 mm wide by 1–2 cm long, in rectangular or oblong tan areas. The appearance of the lesions is very similar to those of grey leaf spot; however, their shape, especially along the edges, is more variable. Similar to most other foliar diseases, the early the occurrence of symptoms the greater the potential economic loss.

6.6 Equipment

Zero-tillage systems depend on specialized seeding equipment that either moves or cuts through residue to place seed consistently. In addition to coulters to cut residue, seed openers are designed to cut residue and open the seed slot. Weight requirement is between 150 and 300 kg per row unit for zero-tillage equipment, which is significantly greater than for conventional seeders. This is an important consideration and one which adds significantly to the cost of zero-tillage seeding equipment. As noted previously, there is a general decrease in the overall operation equipment component of zero-tillage operations, but the seeding equipment is not an area of savings (Pflueger, 2005).

7 Success with zero-tillage maize production

Plan the transition: A key first step is gathering information and learning from other producers and experts with experience in zero-tillage. Current and expected weed pressure and severity in fields should be considered and may influence which fields are most suitable for adoption. Lime application needs and soil compaction issues should be dealt with prior to initiation of the zero-tillage system, if possible.

Start small: Experimenting on a few fields will allow new adopters to learn what works best and define conditions that will be conducive to the success of the system.

Crop rotation is critical: In the recent meta-analysis on the effect of zero-tillage management on maize production, Pittelkow et al. (2015) rated the relative importance of crop rotation as having more than 50% relative importance on final yields. Employing a diverse rotation is the best way to limit negative effects of crop-specific diseases.

Cover crops can be an important addition: Diverse cover cropping systems can bring many benefits and many of those are complimentary with zero-tillage. In particular, cover crops combined with zero-till can limit erosion.

Understand the expected interactions with your soils and climate: Zero-tillage maize production success depends on managing and adapting to field-specific conditions; however, some generalities exist. Zero-tillage is generally most beneficial in drier areas or those that experience acute drought stress during the maize cycle, on medium to coarse-textured soils, and in areas where low soil temperature in the early growing season are of minor concern.

Equipment is important: Equipment engineered for conventional tillage systems is generally not well-adapted for zero-tillage. Zero-tillage seeders need to be significantly heavier, cut or move residue and close the seed slot even when heavy and uneven residue is present. Specialized equipment is required in most cases.

Crop residue management: Appropriate crop rotation can help alleviate the build-up of same-crop residue but regardless of the source of residue, even distribution is very important. In order to maintain uniform conditions for seeding, it is important that combines and harvesters be properly set up to spread residue uniformly over the harvested area.

Hybrid/cultivar selection for field-specific conditions: While most sources have reported no need to specifically test and select hybrids under zero-tillage conditions, there are certain consistent attributes that make some hybrids more suitable for zero-tillage. Uniform and quick emergence, good seedling vigour, and resistance to root and foliar diseases are all advantageous in zero-tillage situations.

8 The future of zero-tillage maize production

Most authorities expect continued expansion of zero-tillage systems, especially in Asia and Africa. Increased availability and affordability of suitable equipment will allow this, as long as access to information and inputs occurs concurrently. Overall, this trend toward increased zero-tillage acres will help reduce the labour requirement for agriculture and could decrease the overall cost of production in many areas. Combined with the soil erosion benefits of zero-tillage, the result should be more sustainable global maize production.

9 Where to look for further information

There are a number of good sources on specific aspects of zero-tillage maize production. Impacts of zero-tillage systems on soils, and guidance on both large and small scale seeding equipment and us can be found in the book *No Tillage Seeding in Conservation Agriculture* by Baker et al. (2007).

Chapter three in *Sustainable Development of Organic Agriculture: Historical Perspectives* edited by Etingoff (2017) summarizes the global spread of conservation agriculture; including extent and discussion of some of the associated advantages and disadvantages.

Current and updated information libraries for zero-tillage production, Conservation Agriculture and direct seeding are available on the web:

http://www.fao.org
http://www.ctic.purdue.edu
http://www.aapresid.org.ar
http://www.act-africa.org
http://www.ecaf.org

10 References

Allmaras, R. R., W. C. Burrows and W. E. Larson. 1964. Early growth of corn as affected by soil temperature. *Soil Sci. Soc. Am. Proc.* 28: 271–5.

Angle, J. S., G. Mc Clung, M. S. Mc Intosh, P. M. Thomas and D. C. Wolf. 1984. Nutrient losses in runoff from conventional and no-till corn watersheds. *J. Environ. Qual.* 13: 431–5. doi:10.2134/jeq1984.00472425001300030021x.

Baker, C. J. and K. E. Saxton (eds). 2007. No-tillage seeding in conservation agriculture. *Cabi.*

Baker, J. M., T. E. Ochsner, R. T. Venterea and T. J. Griffis. 2007. Tillage and soil carbon sequestration – What do we really know? *Agric. Ecosyst. Environ.* 118: 1–5. doi:http://dx.doi.org/10.1016/j.agee.2006.05.014.

Benham, B., D. Vaughan, M. Laird, B. Ross and D. Peek. 2007. Surface water quality impacts of conservation tillage practices on burley tobacco production systems in southwest Virginia. *Water Air Soil Pollut.* 179: 159–66.

Blanco-Canqui, H., and R. Lal. 2008. No-tillage and soil-profile carbon sequestration: An on-farm assessment. *Soil Sci. Soc. Am. J.* 72: 693–701. doi:10.2136/sssaj2007.0233.

Blevins, R. L., D. Cook, S. H. Phillips and R. E. Phillips. 1971. Influence of no-tillage on soil moisture. *Agron. J.* 63: 593–6. doi:10.2134/agronj1971.00021962006300040024x.

Blevins, R. L., and W. W. Frye. 1993. Conservation tillage: An ecological approach to soil .anagement. In Sparks, D. L. (ed.), *Advances in Agronomy*, Academic Press, pp. 33–78.
Bublitz, U. 2014. *Evolucao do plantio driect no Parana*. EMATER-ER, FEBRAPDP, Parana, Brazil.
Butler, J. S., and F. J. Coale. 2005. Phosphorus leaching in manure-amended atlantic coastal plain soils. *J. Environ. Qual.* 34: 370–81. doi:10.2134/jeq2005.0370.
Carpenter, S. R., N. F. Caraco, D. L. Correll, R. W. Howarth, A. N. Sharpley and V. H. Smith. 1998. Nonpoint pollution of surface waters with phosphorus and nitrogen. *Ecol. Appl.* 8: 559–68. doi:10.1890/1051-0761(1998)008[0559:NPOSWW]2.0.CO;2.
Chichester, F. W. 1977. Effects of increased fertilizer rates on nitrogen content of runoff and percolate from monolith lysimeters. *J. Environ. Qual.* 6: 211–17. doi:10.2134/jeq1977.00472425000600020023x.
Clark, E. H. 1987. Soil erosion: offsite environmental effects. In Harlin, J. M. and Bernardi, G. M. (eds), *Soil Loss: Processes, Policies, and Prospects*. New York: Westview, pp. 59–89.
Constantin, J., B. Mary, F. Laurent, G. Aubrion, A. Fontaine, P. Kerveillant, et al. 2010. Effects of catch crops, no till and reduced nitrogen fertilization on nitrogen leaching and balance in three long-term experiments. *Agric. Ecosyst. Environ.* 135: 268–78. doi:http://dx.doi.org/10.1016/j.agee.2009.10.005.
DeFelice, M. S., P. R. Carter and S. B. Mitchell. 2006. Influence of tillage on corn and soybean yield in the United States and Canada. Online. *Crop Manage*. doi:10.1094/CM-2006-0626-01-RS.
Derpsch, R. 2003. Conservation tillage, no-tillage and related technologies. In García-Torres L., Benites, J., Martínez-Vilela, A. and Holgado-Cabrera, A. (eds), *Conservation Agriculture: Environment, Farmers Experiences, Innovations, Socio-economy, Policy*. Springer Netherlands, Dordrecht, pp. 181–90.
Dickey, E. C., and P. J. Jasa. 1983. Row crop planters: Equipment adjustments and performance in conservation tillage. G83-684. University of Nebraska-Lincoln Extension.
Dill, G. M., C. A. CaJacob and S. R. Padgette. 2008. Glyphosate-resistant crops: adoption, use and future considerations. *Pest Manag. Sci.* 64: 326–31. doi:10.1002/ps.1501.
Dinnes, D. L. 2004. Assessments of practices to reduce nitrogen and phosphorus nonpoint source pollution of Iowa's surface waters. USDA-ARS, National Soil Tilth Laboratory.
Eghball, B., and J. E. Gilley. 2001. Phosphorus risk assessment index evaluation using runoff measurements. *J. Soil Water Conserv.* 56: 202–6.
Etingoff, K. (ed). 2017. Sustainable development of organic agriculture: historical perspectives. Apple Academic Press.
Fabrizzi, K. P., F. O. García, J. L. Costa and L. I. Picone. 2005. Soil water dynamics, physical properties and corn and wheat responses to minimum and no-tillage systems in the southern Pampas of Argentina. *Soil and Tillage Res.* 81: 57–69. doi:http://dx.doi.org/10.1016/j.still.2004.05.001.
FAO. 2011. What is conservation agriculture? [Online]. http://www.fao.org/ag/ca/1a.html. (accessed 14 October 2016).
FAO. 2016. AQUASTAT database. Online. http://www.fao.org/nr/water/aquastat/data/query/index.html?lang=en. Verified (accessed 30 September 2016).
Golabi, M. H., D. E. Radcliffe, W. L. Hargrove and E. W. Tollner. 1995. Macropore effects in conventional tillage and no-tillage soils. *J. Soil Water Conserv.* 50: 205–10.
Gupta, S. C., J. B. Swan and E. C. Schneider. 1988. Planting depth and tillage interactions on corn emergence. *Soil Sci. Soc. Am. J.* 52: 1122–7. doi:10.2136/sssaj1988.03615995005200040043x.
Johansen, C., M. E. Haque, R. W. Bell, C. Thierfelder and R. J. Esdaile. 2012. Conservation agriculture for small holder rainfed farming: Opportunities and constraints of new mechanized seeding systems. *Field Crops Res.* 132: 18–32. doi:http://dx.doi.org/10.1016/j.fcr.2011.11.026.
Jones, J. N., J. E. Moody, G. M. Shear, W. W. Moschler and J. H. Lilliard. 1968. The no-tillage system for corn (Zea mays L.). *Agron. J.* 60: 17–20.
Kanwar, R. S., T. S. Colvin and D. L. Karlen. 1997. Ridge, moldboard, chisel, and no-till effects on tile water quality beneath two cropping systems. *J. Prod. Agric.* 10: 227–34. doi:10.2134/jpa1997.0227.

Kaspar, T. C., D. C. Erbach and R. M. Cruse. 1990. Corn response to seed-row residue removal. *Soil Sci. Soc. Am. J.* 54: 1112–17. doi:10.2136/sssaj1990.03615995005400040032x.

Kibet, L. C., A. L. Allen, P. J. A. Kleinman, G. W. Feyereisen, C. Church, L. S. Saporito, et al. 2011. Phosphorus runoff losses from subsurface-applied poultry litter on coastal plain soils. *J. Environ. Qual.* 40: 412–20. doi:10.2134/jeq2010.0161.

Kleinman, P. A., A. Sharpley, L. Saporito, A. Buda and R. Bryant. 2009. Application of manure to no-till soils: phosphorus losses by sub-surface and surface pathways. *Nutr. Cycl. Agroecosys.* 84: 215–27.

Kleinman, P. J. A., A. N. Sharpley, B. G. Moyer and G. F. Elwinger. 2002. Effect of mineral and manure phosphorus sources on runoff phosphorus. *J. Environ. Qual.* 31: 2026–33. doi:10.2134/jeq2002.2026.

Klingman, G. C., and J. M. Spain. 1965. Chemicalization of sod planting. *J. Soil Water Conserv.* 18: 145–52.

Knowler, D., and B. Bradshaw. 2007. Farmers' adoption of conservation agriculture: A review and synthesis of recent research. *Food Policy* 32: 25–48. doi:http://dx.doi.org/10.1016/j.foodpol.2006.01.003.

Lal, R. 1976. No-tillage effects on soil properties under different crops in western Nigeria. *Soil Sci. Soc. Am. J.* 40: 762–8. doi:10.2136/sssaj1976.03615995004000050039x.

Licht, M. A., and M. Al-Kaisi. 2005. Corn response, nitrogen uptake, and water use in strip-tillage compared with no-tillage and chisel plough. *Agron. J.* 97: 705–10.

Liu, W., M. Tollenaar, G. Stewart and W. Deen. 2004. Response of corn grain yield to spatial and temporal variability in emergence. *Crop Sci.* 44: 847–54. doi:10.2135/cropsci2004.8470.

McDowell, L. L., and K. C. McGregor. 1984. Plant nutrient losses in runoff from conservation tillage corn. *Soil Till. Res.* 4: 79–91. doi:http://dx.doi.org/10.1016/0167-1987(84)90018-7.

Moody, J. E., G. M. Shear and J. N. Jones. 1961. Growing corn without tillage. *Soil Sci. Soc. Proc.* 25: 516–17.

Mostaghimi, S., T. A. Dillaha and V. O. Shanholtz. 1988. Influence of tillage systems and residue levels on runoff, sediment, and phosphorus losses. *Trans. of ASAE* 31. doi:10.13031/2013.30677.

Munkvold, G. P., and D. G. White. 2016. *Compendium Of Corn Diseases*, 4th ed., American Phytopathological Society, APS Press.

Norwood, C. A. 2000. Water use and yield of limited-irrigated and dryland corn. *Soil Sci. Soc. Am. J.* 64: 365–70. doi:10.2136/sssaj2000.641365x.

Owens, L. B., and W. M. Edwards. 1993. Tillage studies with a corn-soybean rotation: surface runoff chemistry. *Soil Sci. Soc. Am. J.* 57: 1055–60. doi:10.2136/sssaj1993.03615995005700040029x.

Page, J. B., C. J. Willard and G. W. McCuen. 1947. Progress report on tillage methods in preparing land for corn. *Soil Sci. Soc. Am. J.* 11: 77–80.

Papendick, R. I., M. J. Lindstrom and V. L. Cochran. 1973. Soil mulch effects on seedbed temperature and water during fallow in eastern Washington. *Soil Sci. Soc. Am. J.* 37: 307–14. doi:10.2136/sssaj1973.03615995003700020039x.

Pflueger, B. 2005. *How To Calculate Machinery Ownership and Operating Costs*. EC920w, Cooperative Extension Service, South Dakota State University.

Phillips, S. H., and H. M. Young. 1973. *No-Tillage Farming*, Reiman Associates, Milwaukee, WI, p. 224.

Pimentel, D., C. Harvey, P. Resosudarmo, K. Sinclair, D. Kurz, M. McNair, et al. 1995. Environmental and economic costs of soil erosion and conservation benefits. *Science* 267: 1117–23. doi:10.1126/science.267.5201.1117.

Pimentel, D., and N. Kounang. 1998. Ecology of Soil Erosion in Ecosystems. *Ecosystems* 1: 416–26. doi:10.1007/s100219900035.

Pittelkow, C. M., B. A. Linquist, M. E. Lundy, X. Liang, K. J. van Groenigen, J. Lee, et al. 2015. When does no-till yield more? A global meta-analysis. *Field Crop. Res.* 183: 156–68. doi:http://dx.doi.org/10.1016/j.fcr.2015.07.020.

Randall, G. W., and T. K. Iragavarapu. 1995. Impact of long-term tillage systems for continuous corn on nitrate leaching to tile drainage. *J. Environ. Qual.* 24: 360–6. doi:10.2134/jeq1995.00472425002400020020x.

Rice, C. W., and M. S. Smith. 1984. Short-term immobilization of fertilizer nitrogen at the surface of no-till and plowed soils. *Soil Sci. Soc. Am. J.* 48: 295–7.

Römkens, M. J. M., D. W. Nelson and J. V. Mannering. 1973. Nitrogen and phosphorus composition of surface runoff as affected by tillage method. *J. Environ. Qual.* 2: 292–5. doi:10.2134/jeq1973.00472425000200020028x.

Ross, B. B., P. H. Davis and V. L. Heath. 2001. Water quality improvement resulting from continuous no-tillage practices. Final Report. Colonial Soil and Water Conservation District.

Sharpley, A., and P. Kleinman. 2003. Effect of rainfall simulator and plot scale on overland flow and phosphorus transport. *J. Environ. Qual.* 32: 2172–9. doi:10.2134/jeq2003.2172.

Shipitalo, M. J., and W. M. Edwards. 1998. Runoff and erosion control with conservation tillage and reduced-input practices on cropped watersheds. *Soil Till. Res.* 46: 1–12. doi:http://dx.doi.org/10.1016/S0167-1987(98)80102-5.

Shipitalo, M. J., L. B. Owens, J. V. Bonta and W. M. Edwards. 2013. Effect of no-till and extended rotation on nutrient losses in surface runoff. *Soil Sci. Soc. Am. J.* 77: 1329–37. doi:10.2136/sssaj2013.01.0045.

Soane, B. D., B. C. Ball, J. Arvidsson, G. Basch, F. Moreno and J. Roger-Estrade. 2012. No-till in northern, western and south-western Europe: A review of problems and opportunities for crop production and the environment. *Soil Till. Res.* 118: 66–87. doi:http://dx.doi.org/10.1016/j.still.2011.10.015.

Spargo, J. T., M. M. Alley, R. F. Follett and J. V. Wallace. 2008. Soil carbon sequestration with continuous no-till management of grain cropping systems in the Virginia coastal plain. *Soil Till. Res.* 100: 133–40. doi:http://dx.doi.org/10.1016/j.still.2008.05.010.

Staver, K. W. 2004. Efficient utilization of poultry litter in cash grain rotations. Final Report submitted to: Maryland Grain Producers Utilization Board. MCAE Pub. 2004–03.

Stevenson, J. R., R. Serraj and K. G. Cassman. 2014. Evaluating conservation agriculture for small-scale farmers in Sub-Saharan Africa and South Asia. *Agric. Ecosyst. Environ.* 187: 1–10. doi:http://dx.doi.org/10.1016/j.agee.2014.01.018.

Strohm, J. 1952. This may revolutionize the way you grow corn. *Country Gent.* 122: 26.

Thierfelder, C., J. L. Chisui, M. Gama, S. Cheesman, Z. D. Jere, W. Trent Bunderson, et al. 2013. Maize-based conservation agriculture systems in Malawi: Long-term trends in productivity. *Field Crops Res.* 142: 47–57. doi:http://dx.doi.org/10.1016/j.fcr.2012.11.010.

Thomason, W., R. Grisso, T. K. Balderson, P. Davis, S. Johnson, C. Lawrence, et al. 2006. Deep tillage prior to no-till: research and recommendations. Virginia Tech and Virginia Cooperative Extension Pub. 424-053.

Trigo, E., E. Cap, V. Malach and F. Villarreal. 2009. The case for zero tillage technology in Argentina. *Intl. Food Policy Res. Inst.*

Triplett, G. B., D. M. Van Doren and B. L. Schmidt. 1968. Effect of corn (Zea mays L.) stover mulch on no-tillage corn yield and water infiltration. *Agron. J.* 60: 236–9. doi:10.2134/agronj1968.00021962006000020028x.

Truman, C. C., J. N. Shaw and D. W. Reeves. 2005. Tillage effects on rainfall partitioning and sediment yield from an ultisol in central Alabama. *J. Soil Water Conserv.* 60: 89–98.

Tullberg, J. N., D. F. Yule and D. McGarry. 2007. Controlled traffic farming—From research to adoption in Australia. *Soil Till. Res.* 97: 272–81. doi:http://dx.doi.org/10.1016/j.still.2007.09.007.

Tyler, D. D., and G. W. Thomas. 1977. Lysimeter measurements of nitrate and chloride losses from soil under conventional and no-tillage Corn. *J. Environ. Qual.* 6: 63–6. doi:10.2134/jeq1977.00472425000600010014x.

Unger, P. W., and T. M. McCalla. 1980. Conservation tillage systems. In Brady, N. C. (ed.), *Advances in Agronomy*, Academic Press, pp. 1–58.

USDA. 1985. Conservation tillage: Things to consider. Agricultural Information Bulleting No. 46, Washington, 23pp.

VandenBygaart, A. J., E. Gregorich, G. and D. A. Angers. 2003. Influence of agricultural management on soil organic carbon: a compendium and assessment of Canadian studies. *Can. J. Soil Sci.* 83: 363–80

Verbree, D. A., S. W. Duiker and P. J. A. Kleinman. 2010. Runoff losses of sediment and phosphorus from no-till and cultivated soils receiving dairy manure. *J. Environ. Qual.* 39: 1762–70.
Wade, T., R. Claassen and S. Wallander. 2015. Conservation-practice adoption rates vary widely by crop and region. USDA-ERS Information Bull. 147.
Wagger, M. G., and D. K. Cassel. 1993. Corn yield and water-use efficiency as affected by tillage and irrigation. *Soil Sci. Soc. Am. J.* 57: 229–34. doi:10.2136/sssaj1993.03615995005700010040x.
Wall, P. C. 2007. Tailoring Conservation Agriculture to the Needs of Small Farmers in Developing Countries. *J. Crop Imp.* 19: 137–55. doi:10.1300/J411v19n01_07.
Weil, R. R., and N. C. Brady. 2017. *The Nature And Properties of Soils*, 15th edition. Pearson Education Inc.
West, T. O., and W. M. Post. 2002. Soil organic carbon sequestration rates by tillage and crop rotation. *Soil Sci. Soc. Am. J.* 66: 1930–46. doi:10.2136/sssaj2002.1930.

Chapter 6

Conservation agriculture (CA) for sustainable intensification of maize and other cereal systems: the case of Latin America

Bram Govaerts, International Maize and Wheat Improvement Center (CIMMYT), Mexico; Isabelle François, Consultant, Belgium; and Nele Verhulst, CIMMYT, Mexico

1 Introduction: the need for sustainable soil management

Latin American agriculture of the twenty-first century faces many challenges. Firstly, the soils of Latin America suffer from extensive soil degradation, especially soil erosion, a result of the transfer of agricultural practices and technology from Europe. The conventional European plough-based tillage proved inefficient and even detrimental for tropical soils low in organic matter (Speratti et al., 2015). As a result, neither improved varieties of crops nor other inputs are able to deliver their potential (Sayre and Govaerts, 2011).

Secondly, whilst overall impacts predicted by climate change models vary, extreme rainfall variability is likely to increase in the region. Where excess rainfall occurs, flooding

http://dx.doi.org/10.19103/AS.2016.0001.22

and soil erosion will increase. In low rainfall/drought-prone areas, a lack of precipitation and increased evapotranspiration due to higher temperatures can cause more frequent water stress (Verhulst et al., 2012). In light of the predictions, there is a need to develop resilient crop management systems that are able to cope with both heavy rainfall events and prolonged drought (Verhulst et al., 2011).

A third issue is the increasing concerns about the contribution of agriculture to the global greenhouse gas (GHG) emissions and climate change (Sayre and Govaerts, 2011), including carbon dioxide (CO_2), methane (CH_4) and nitrous oxide (N_2O). Agronomic management practices directly affect the GHG emissions from soil, C sequestration, fuel used for farm operations and the production of inputs.

In recent years, many concerned farmers have begun to adopt and adapt improved crop management practices that lead towards the ultimate vision of sustainable farming systems (Sayre and Govaerts, 2011). These new agricultural practices will not only have to prevent further soil degradation, but also improve the resilience of the system through increased soil organic matter, improved water and nutrient use efficiency and increased flora and fauna biodiversity (Hobbs and Govaerts, 2010). These new management practices and production systems are knowledge-based multiscale interventions that integrate land, water, biodiversity and environmental management to meet the rising food and fibre demands while sustaining ecosystem services and livelihoods. Therefore, the sustainable intensification should be an economically viable and socially acceptable 'production-protection' agriculture that reduces the production risks and protects the multifunctional roles of land to provide for goods, services and livelihoods for the people who depend on land. One of the practices to improve water use, reduce erosion and enhance human labour/animal power efficiencies could be conservation agriculture (CA), both for rainfed as well as irrigated production systems (Govaerts et al., 2008). Speratti et al. (2015) reviewed CA in Latin America. CA has been widely successful in the Southern Cone region of South America. In this region, most of the research into no-till systems started in the 1970s. In Argentina and Uruguay, many fields are under direct seeding, however, not all can be considered 'complete' CA since crop residue is often removed. In Brazil, major actions were taken by the Agricultural Research Institute of Paraná for the dissemination of CA techniques. A common agriculture system used by the indigenous people of Central America, is the slash–mulch system, for example, the legume *Mucuna* is often incorporated in maize crop rotations. The slash–mulch system can be considered a resource-conserving technology and a step towards CA systems, in some cases following all three principles of CA. In Mexico, the International Maize and Wheat Improvement Center (Spanish acronym CIMMYT) started the no-till research in 1973 and since 2001, the Agricultura Sostenible Basada en la Siembra Directa (ASOSID) aimed to assemble an inter-institutional platform project including farmers and their associations, federal and state government agencies located in the region (FIRA, INIFAP); regional, national and international research institutions (CIMMYT and the Centre de Coopération Internationale en Recherche Agronomique pour le Développement, CIRAD); and private companies (Speratti, et al., 2015). The CIMMYT has long-term CA trials that clearly have demonstrated that new, improved wheat and maize varieties can achieve their genetic yield potential only with sound cropping systems management (Govaerts et al., 2006).

As also described in Verhulst et al. (2010), CA combines the following basic principles that have to be combined into a working system and require the integration of needed management elements like fertility management, weed management, post-harvest practices into a context, agro-ecological and value chain-specific production system:

- Reduction in tillage: This can range from zero tillage (ZT) to controlled tillage seeding systems that normally do not disturb more than 20–25% of the soil surface.
- Retention of adequate levels of crop residues and soil surface cover: The objective is to protect the soil from water and wind erosion; to reduce water run-off and evaporation; to improve water productivity and to enhance soil physical, chemical and biological properties associated with long-term sustainable productivity by the retention of sufficient residue on the soil and crop cover. The amount of residues necessary to achieve these ends will vary depending on the biophysical conditions and cropping system.
- Use of crop rotations: The objective is to employ diversified crop rotations to help moderate/mitigate possible weed, disease and pest problems; to utilize the beneficial effects of some crops on soil conditions and on the productivity of subsequent crops; and to provide farmers with economically viable cropping options that minimize risk.

There is a fourth element that needs to be taken into account in the development of a sustainable production alternative. Farmers adopt new practices only when near-term impacts on their well-being can be demonstrated. Thus, sustainable farming systems need to have immediate positive economic impacts, and these must be crystal clear to the farmer. This way, agricultural production can be augmented while reducing threats to the environment (Govaerts, Sayre, and Deckers, 2005).

Long-term experiments can help identify agronomic practices that are more resilient to climatic change and result in high and stable yields (Stirling et al., 2014). CIMMYT's long-term experiments are of great value in evaluating CA as a sustainable production alternative. The vast knowledge and data generated over the past years should be used to solve today's problems. Furthermore, the lessons learned and methodologies developed on the set up of research on agronomic systems per se should be explored in the context of Latin America. The integration of research from local to global and farm to agri-food systems should be considered as a model for other regions.

2 Introduction to CIMMYT's long-term experimental sites

2.1 Rainfed systems in the highlands of central Mexico

In 1990, CIMMYT established a long-term, rainfed cropping trial at El Batán, Mexico (2240 m asl; 19.31°N, 98.50°W), which is representative of the rainfed highlands of central Mexico. Agriculture in these highlands is mainly found in temperate, sub-humid, high valleys (1500–3000 m asl), between 16 and 24°N latitude. Rainfed cropping predominates, with a mean annual rainfall of 350–800 mm (625 mm at El Batán). This occurs during a 4- to 6-month summer wet season followed by a dry and frosty winter (Sayre and Govaerts, 2011). The mean maximum and minimum temperatures are 24 and 6°C, respectively (1991–2009) (Verhulst et al., 2011d). The soil is classified as a Haplic Phaeozem (Clayic) in the World Reference Base system (IUSS Working Group WRB 2007) and as a fine, mixed, thermic Cumulic Haplustoll in the USDA Soil Taxonomy system (Soil Survey Staff, 2003) (Verhulst et al., 2011d). Farm size in the region tends to be small (<5 ha) to medium-scale (5–20 ha). Maize is the major crop; however, wheat, barely, beans, oats and potato are also included. All crops are planted at or just before the onset of the summer rains (Sayre and Govaerts, 2011).

The climate of El Batán also makes it representative of many highland areas in the West Asia and North Africa region, the Southern Cone and Andean Highlands of South America, the central highlands of Ethiopia, the Mediterranean coastal plains of Turkey and the highlands of central Mexico (Van Ginkel et al., 2002). Each area has its specific conditions and problems, but some overall trends are recognizable. The tropical and subtropical highlands, for example, central Mexico, have been densely populated and intensively cropped for centuries resulting in soil erosion and fertility decline. The agricultural system is under stress due to an ever-decreasing area cultivated per household, reduced fodder availability and land degradation. Short, intense rain showers followed by dry periods typify the summer rainy season and the total yearly potential evapotranspiration of 1550 mm markedly exceeds annual rainfall (Verhulst et al., 2011d). Rainfall is unpredictable, next to inadequate; hence, crop production is threatened by chronic soil moisture stress. Precipitation is usually intensive, erratic and short, leading to high run-off and temporal water logging. Cereal grain yields are low (<2 t ha^{-1}). Moreover, fields are often weedy and crops are N deficient, soil structure is poor, and sheet and gully erosion are widespread (Govaerts et al., 2008).

The first rainfed long-term experiment began in 1991. In 1990, the entire experimental area was treated with a broad-spectrum contact herbicide one month after the rains began, then seeded to barley in late July (Fischer et al., 2002a,b). Individual plots are 7.5 m by 22 m and present a micro-topography with a slope of <0.3%. Standard practices include the use of recommended crop cultivars, with maize planted at 75 000 plants ha^{-1} in 75-cm rows and wheat planted in 20-cm rows at 110 kg seed ha^{-1}. Both crops are fertilized using urea at 150 kg N ha^{-1}, with all N applied to wheat before planting and to maize at planting. Weed control is done using appropriate, available herbicides as needed and no disease or insect pest controls are utilized, except for seed treatments applied by commercial seed sources. Planting of both maize and wheat depends on the onset of summer rains but is usually done between 1 and 10 June. The experimental design consists of a randomized complete block with two replications. There are 32 treatments in all. A core set of 16 permanent management practices were based on variation of crop rotation, tillage/planting methods and crop residue management. Crop practices included continuous wheat, continuous maize and rotation of wheat and maize. Residue is kept in the field or removed for fodder. Retained residues are incorporated, if tilled, or left on the soil surface with ZT. A second set of 16 treatments with more complex rotation, residue management and tillage combinations was established/modified in 1996 and later in order to implement already gained knowledge and to test alternative practices more acceptable to the farmer (Verhulst et al., 2011).

A second rainfed long-term experiment began in 1999. Each plot in the experiment measures 6 m × 20 m with 8 raised beds of 75 cm width. Four management practices are studied in 14 treatments with two replicates in a randomized complete block design. The first factor studied is residue management: (1) all above-ground crop residues chopped and kept in the field whether incorporated for conventional beds or retained on the soil surface for permanent raised beds; (2) all crop residues removed by baling for fodder; and (3) crop residues partly removed, wheat residues cut by combine and maize residues cut just below the ear. The second factor studied is tillage: (1) conventional tillage with raised beds formed after each crop and (2) ZT with continued reuse of the existing raised beds (re-shaped if required). The third factor is both phases of the maize–wheat rotation, while tied-ridges and open furrows between the raised beds are the fourth factor.

2.2 Irrigated conditions in arid north-western Mexico

In 1992, with the aim to study common farmer practice, based on extensive tillage with the formation of new beds for each succeeding crop, versus permanent raised beds with different crop residue management, CIMMYT began a long-term sustainability trial at its experiment station near Ciudad Obregón in the Yaqui Valley (north-western Mexico, 38 m asl; 27.33 N, 109.09 W). The Yaqui Valley is agro-climatically representative of areas where 40% of the wheat is produced in the developing world, such as the Indian and Pakistani Punjab and the Nile Valley in Egypt. The trial aims to compare the long-term effects of permanent beds (PBs) with different crop residue management and crop rotation, with the local, commonly used tillage-based practices, for both wheat and maize production. Other practices such as weed control and nutrient management are as locally recommended. Each year, in all treatments, new recommended wheat and maize cultivars have been used and all other management practices applied at recommended levels. Furrow irrigation is used for wheat and maize.

The trial results, which will be discussed at length in the following sections, confirm the benefits of CA for farmers in this region and show how the three main CA practices interact (Govaerts et al., 2005, 2006a,b, 2007a,b).

3 The influence of CA on physical soil quality

3.1 Soil structure and aggregation

Soil structure is a key factor in soil functioning and is an important factor in the evaluation of the sustainability of crop production systems (Verhulst et al., 2010). Soil structure is often expressed as the degree of stability of aggregates (Bronick and Lal, 2005). Soil structural stability is the ability of aggregates to remain intact when exposed to different stresses (Kay et al., 1988) and measures of aggregate stability are useful as a means of assessing soil structural stability (Verhulst et al., 2010).

ZT with residue retention improves dry aggregate size distribution compared to conventional tillage (Fuentes et al., 2012a,b; Govaerts et al., 2007c, 2009a). The effect on water stability of aggregates is even more pronounced, and this was reported for a wide variety of soils and agro-ecological conditions as well in the central highlands (Govaerts et al., 2007c, 2009a; Lichter et al., 2008). In north-western Mexico, plots with permanent raised beds, where residues were at least partly retained, showed the highest amount of macro-aggregates (Verhulst et al., 2011c).

3.2 Hydraulic conductivity and water-holding capacity

Measuring hydraulic conductivity when a residue cover is present in ZT is difficult and might explain the conflicting results reported over the years (Verhulst et al., 2010). However, reduced tillage combined with crop residue retention can improve soil structure and stability thereby facilitating better drainage and water-holding capacity; limiting water logging or drought (Govaerts et al., 2007; Lichter et al., 2008). In the rainfed highlands of Mexico, conductivity was 1.3 times greater in soil under tilled beds than under non-tilled beds ($p < 0.05$) (Patiño-Zúñiga et al., 2009). Water-holding capacity was 1.1 times greater in soils under non-tilled beds than under tilled beds ($p < 0.05$). Tillage will destroy soil aggregates, which will reduce the WHC of the soil. The WHC was also 1.1 times greater in soil from non-tilled beds where the crop residue was retained compared to soil where it was removed or partially removed ($p < 0.05$) (Patiño-Zúñiga et al., 2009).

3.3 Infiltration and run-off

In spite of inconsistent results regarding the effect of tillage and residue management on soil hydraulic conductivity, infiltration is generally higher in ZT with residue retention compared to conventional tillage and ZT with residue removal. This is probably due to the direct and indirect effect of residue cover on water infiltration (Verhulst et al., 2010). Soil macro-aggregate breakdown has been identified as the major factor leading to surface pore clogging by primary particles and micro-aggregates and thus to formation of surface seals or crusts (Le Bissonnais and Arrouays, 1997; Lal and Shukla, 2004). Aggregates are more stable under ZT with residue retention compared to conventional tillage and ZT with residue removal (Govaerts et al., 2007c, 2009a).

In the highlands of Mexico, higher direct infiltration rates were recorded in ZT with residue retention than under conventional tillage, although the infiltration rate in conventional tillage was considerably higher than ZT without residue retention (Govaerts et al., 2009a). Similar results were obtained in bed planting systems in the same area (Govaerts et al., 2007). In the Yaqui Valley, the direct surface infiltration (time-to-pond) was the highest in PBs with all or part of the straw retained (Verhulst et al., 2011d).

3.4 Soil water content and plant available water

Verhulst et al. (2011d) reported that ZT with residue retention generated the highest soil water content, and the effect was more pronounced during the extended, erratic drought periods in rainfed conditions. In the same experiment, Govaerts et al. (2009a) found that ZT with residue retention increased aggregate distribution and stability and direct infiltration, especially compared to ZT without residue retention and to a lesser extent to conventional tillage with or without residue. The increased aggregation and protection of the soil surface by the residue cover allowed the ZT with residue retention practice to maximize profile recharge during rainfall events, which are often afternoon storms with high rainfall intensity (Verhulst et al., 2011d). Hence, the soils under ZT with residue retention generally had higher surface soil water contents compared to tilled soils (Govaerts et al., 2009a). Dendooven et al. (2012) also described that tillage reduced the volumetric water content compared to no-till soil, but only when the crop residue was retained and not when it was removed.

Overall, rainfall was used more efficiently in CA-based practices than for both ZT with residue removal and conventional tillage with or without residue incorporation. ZT with residue retention had higher soil water content than ZT with residue removal and conventional tillage with or without residue, and the effect was more pronounced during periods of drought. Verhulst et al. (2011d) claimed that this higher soil water content was able to buffer the dry episodes during the growing season, resulting in an average yield advantage of approximately 1.5 Mg ha^{-1} for the CA practices over practices involving tillage and ZT with full residue removal. Yields and soil water content were similar in the three CA practices (ZT with full or partial residue retention), showing that removing part of the residue can be a sustainable solution when there is competition for residue. Incorporating residue by tillage was less efficient in terms of soil and water conservation than keeping residue on the soil surface (Verhulst et al., 2011d).

3.5 Soil erosion and soil temperature

As stated before, CA leads to higher aggregate stability as compared to conventionally tilled fields or ZT fields without residue retention. Consequently, this results in lower soil erosion potential in CA (Govaerts et al., 2007).

No-till reduced the soil temperature compared to tilled soil, but only when the crop residue was retained and not when it was removed (Dendooven et al., 2012b).

Similar results were described in north-western Mexico. Soil temperature was higher in conventionally tilled beds (CTBs) where residue was incorporated through tillage than in PBs with residue retention on the surface. The importance of retaining crop residues to soil temperatures can also be seen in the PB systems with different residue management. Soil temperature was significantly higher in the PB where the crop residue was removed than where it was retained (Dendooven et al., 2013).

4 The influence of CA on chemical soil quality

4.1 Soil organic carbon

Soil organic carbon (SOC) has been proposed as a primary indicator of soil quality (Reeves, 1997), especially the SOC concentration of surface soil (Franzluebbers, 2002). The surface soil is the vital horizon that receives much of the seed, fertilizers and pesticides applied to cropland. It is also the layer that is affected by the intense impact of rainfall, and that partitions the flux of gases into and out of the soil. Surface organic matter is essential to erosion control, water infiltration and conservation of nutrients (Verhulst et al., 2010).

Because of the importance of SOC, the influence of the three components of CA on the amount of SOC has been examined extensively.

Dendooven et al. (2012b) found that the SOC was 1.8 times higher in the 0–20 cm layer of CA compared to CT and 1.5 times in the 0–60 cm layer. Approximately 5 times more organic C is left in the field in CA than in CT from 1996 to 2006 (Fuentes et al., 2010). Consequently, the SOC is larger in CA than in CT.

Removal of crop residues in ZT decreased the total C 1.5 times in the 0–60 cm soil layer compared to soil where it was retained. Crop residue retention is important for the soil organic matter content (Govaerts et al., 2007; Dendooven et al., 2012b). The gross SOC turnover was lower in soil with residue retention than without residues (Fuentes et al., 2010).

Although the different components of CA seem to all have their individual influence on SOC, it is also the combination of them that leads to the observed benefits. Fuentes et al. (2010) reported that the SOC content in the 0–10 cm layer was affected by tillage and residue management. The highest SOC content was found in the 0–5 cm layer of the treatment with ZT and residue retention compared other treatments. The soils with ZT and residue retention (both for monoculture and for rotation) showed higher percentages of

SOC and SOC stock than all treatments with conventional tillage (with or without residue retention). Consequently, the combination of ZT with residue retention is what makes aggregates more stable, protects C and thus increases C sequestration and not ZT or residue retention separately (Fuentes et al., 2012).

Similar results were found in the long-term experiment involving irrigated PBs. The soil microbial biomass (SMB) C was significantly higher in PB with all or part of the straw retained than in CTB-straw incorporated and PB-straw burned (Verhulst et al., 2011).

These findings are supported by the literature. Govaerts et al. (2009b) reported in their literature review to determine the influence of the different components comprising CA (reduced tillage, crop residue retention and crop rotation) on SOC. In 7 of 78 cases, the SOC content was lower in zero compared to conventional tillage; in 40 cases it was higher and in 31 of the cases there was no significant difference (Govaerts et al., 2009b).

4.2 Nutrient availability

Nitrogen

The presence of mineral soil N available for plant uptake is dependent on the rate of C mineralization. The literature concerning the impact of reduced tillage with residue retention on N mineralization is inconclusive (Verhulst et al., 2010).

Effects of CA on total N content generally mirrors those observed for total SOC, as the N cycle is inextricably linked to the C cycle. Govaerts et al. (2007c) observed a significantly higher total N under both ZT and permanent raised beds compared to conventional tillage in the highlands of central Mexico. Similarly, increasing the amount of straw retained under permanent raised beds increased total N significantly (Govaerts et al., 2007c). Verhulst et al. (2011c) also concluded that the total N content was significantly higher in PBs with all or part of the straw retained compared to conventionally tilled raised beds where straw was incorporated.

Lichter et al. (2008) reported that permanent raised beds with residue retention resulted in more stable macro-aggregates and increased protection of C and N in the macro-aggregates compared to conventionally tilled raised beds. The total soil N content was found to be 1.2 times larger in no-till beds compared to tilled beds and 1.1 times larger in soil from no-till beds where the crop residue was retained compared to where it was removed or partially removed, which was a significant difference (Patiño-Zúñiga et al., 2009).

Govaerts et al. (2006c) found after 26 cropping seasons in the irrigated maize–wheat system in north-western Mexico that the N mineralization rate was higher in permanent raised beds with residue retention than in conventionally tilled raised beds with all residues incorporated, and also that N mineralization rate increased with increasing rate of inorganic N fertilizer application. However, also the type of residues and the interactions with N management practices determine C and N mineralization (Verachtert et al., 2009). And even after 13 years, N immobilization still occurred in soils with retention of maize residues, in an irrigated maize–wheat rotation system in the north-western Mexico (Govaerts et al., 2006)

Other nutrients

Govaerts et al. (2007c) found that the potassium (K) concentration in both the 0–5 cm and 5–20 cm soil layers increased significantly with increasing residue retention on permanent raised beds. This effect was more pronounced for wheat than for maize. Verhulst et al. (2009) found that most macronutrients, like K, were not affected by the tillage-straw management, but attributed this to the nutrient-rich soil.

Most research in the CIMMYT long-term trials has shown that tillage or crop does not affect extractable calcium (Ca) and magnesium (Mg) levels (Govaerts et al., 2007c; Verhulst et al., 2010), while others have reported different vertical Ca and/or Mg stratification between tillage practices (Duiker and Beegle, 2006; Hulugalle and Entwistle, 1997)

Govaerts et al. (2007c) reported that tillage practice had no significant effect on the concentration of extractable iron (Fe), manganese (Mn) and copper (Cu), but that the concentration of extractable zinc (Zn) was significantly higher in the 0–5 cm layer of permanent raised beds compared to conventionally tilled raised beds with full residue retention in rainfed conditions. Verhulst et al. (2011) reported similar results in irrigated conditions, except for Fe where concentration was lower in CTBs, where straw was incorporated, than in PBs, with straw retained on the field. Residue retention significantly decreased concentrations of extractable Mn in the 0–5 cm layer in permanent raised beds, as reported by Govaerts et al. (2007c) but was not affected in the tests conducted by Verhulst et al. (2011).

4.3 Salinity/sodicity

According to Govaerts et al. (2007c), permanent raised bed planting is a technology that reduces soil sodicity under rainfed conditions. They found the Na concentration to be 2.64 and 1.80 times lower in 0–5 cm and 5–20 cm layer, respectively, in permanent raised beds compared to conventionally tilled raised beds. Furthermore, the Na concentration increased with decreasing amounts of residue retained on the permanent raised beds. None of the management practices in the experiments resulted in Na concentrations that dramatically affected soil dispersion. However, burning the residue resulted in Na concentrations 1.4 times higher than retaining the residue on PBs (Verhulst et al., 2011).

4.4 Acidity

In numerous studies, the pH of the topsoil was found to be lower for ZT than for conventional tillage. Verhulst et al. (2010) have also proposed that the greater SOM accumulation in the topsoil with ZT can lead to acidity from decomposition. Govaerts et al. (2007c) observed a significantly higher pH in the topsoil of the permanent raised beds with full residue retention compared to conventional raised beds with residue retention. Patiño-Zúñiga et al. (2009) observed no difference in pH between soils under conventional or conservation tillage, but results showed that the conductivity was 1.3 times larger under tilled beds than under non-tilled beds. This was supported by Verhulst et al. (2011), who found no difference in pH between CTBs with straw incorporated and PBs with straw retained.

5 The influence of CA on biological soil quality

5.1 Soil microfauna and -flora

Microorganisms are driving many fundamental nutrient-cycling processes, soil structural dynamics, degradation of pollutants and various other services (Bloem et al., 1997). Among the biological properties, soil microorganisms are very sensitive to external perturbations and can act as a sensor for monitoring soil response, and more generally soil quality (Karlen et al., 2006). Changes in tillage, residue and rotation practices are known to induce major shifts in the number and composition of soil fauna and flora, including both pests and beneficial organisms (Andersen, 2003; Bockus and Shroyer, 1998; Sharma et al., 2010).

Soil microbial biomass

The SMB reflects the soil's ability to store and cycle nutrients (C, N, P and S) and organic matter, and has a high turnover rate relative to the total soil organic matter (Carter et al., 1999). Owing to its dynamic character, SMB responds to changes in soil management often before effects are measured in terms of organic C and N (Powlson and Jenkinson, 1981).

In the subtropical highlands of Mexico, residue retention resulted in significantly higher amounts of SMB C and N in the 0–15 cm layer compared to residue removal (Govaerts et al., 2007b). Similarly, in another study carried out in central Mexico, higher SMB with residue retained compared to without residue were recorded (Govaerts et al., 2007a).

The SMB was larger in soil from non-tilled beds where the crop residue was retained compared to where it was removed (Patiño-Zúñiga et al., 2009). The SMB C was significantly higher in PB with all or part of the straw retained than in CTBs with straw incorporated and PB with straw burned in north-western Mexico (Verhulst et al., 2011).

Functional diversity

Functional diversity and redundancy, which refers to a reserve pool of quiescent organisms or a community with vast interspecific overlaps and trait plasticity, are signs of increased soil health, and allow an ecosystem to maintain a stable soil function (Wang and McSorley, 2005). Govaerts et al. (2007b) also reported differences in the community-level physiological profile between soils under maize compared to soils under a wheat crop. This indicates the importance of crop rotation for soil health, but more research is needed to identify the underlying processes and take advantage of them when implementing crop rotations.

Microbial community structure

In a maize/wheat rotation, treatments involving ZT and crop residue retention had the highest levels of microbial diversity and richness compared to conventional tillage with the crop residue removed (Ceja-Navarro et al., 2010a).

Using taxonomic and phylogenetic information it was found that ZT most affected the bacterial communities. The relative abundance of Actinobacteria, Betapreoteobacteria and Gammaproteobacteria was affected by tillage and correlated to the total organic carbon (TOC) and clay content in soil in the central highlands of Mexico (Navarro-Noya et al., 2013). In another long-term experiment in the north-western Mexico, the relative abundance of Actinobacteria also increased in ZT systems; this time with permanent raised beds (Jiménez-Bueno et al., 2016).

Ramirez-Villanueva et al. (2015), working in a long-term field trial in central Mexico, described that CA favoured oligotrophic bacteria, for example, Acidobacteria, Planctomycetes and Verrucomicrobia, while in conventional agriculture bacteria that metabolize easily decomposable organic material were favoured, for example, Actinobacteria.

Crop residue retention resulted in increased populations of Actinomycetes, total bacteria and fluorescent *Pseudomonas*, under both zero and conventional tillage in the highlands of Mexico (Govaerts et al., 2008a). It was found that tillage did not change the most abundant bacterial groups, but it did affect their relative proportion (Núñez et al., 2012).

Residue management had a significant effect on the bacterial community structure when phylogenetic membership and the total enumeration of bacteria were considered. Residue management affected the relative abundance of Bacteroidetes, Betaproteobacteria, Cyanobacteria and Gemmatimonadetes. When no tillage was applied, crop residue management affected the microbial communities more than when conventional tillage was applied (Navarro-Noya et al., 2013). In another the long-term trial of north-western Mexico, application of dried young wheat plants increased the relative abundance of Actinobacteria and Firmicutes in each treatment (Chávez-Romero et al., 2016).

Wheat/maize rotation or crop monoculture did not affect the bacterial community structure. No significant differences in richness, diversity and total abundance of bacteria was found between the treatments. This indicated that even though phylotypes changed, the number and diversity of the bacterial communities were similar (Navarro-Noya et al., 2013).

It can be concluded that practices that include reduced tillage and crop residue retention can be adopted as safer agricultural practices to preserve and improve the diversity of soil bacterial communities (Ceja-Navarro et al., 2010a).

Soil-borne diseases

A reduction in tillage influences different pest species in different ways, depending on their survival strategies and life cycles (Andersen, 1999). Govaerts et al. (2006a, 2007a) found that in the semi-arid and rainfed subtropical highlands of central Mexico, the incidence of *Fusarium* root rot in maize plots with crop rotation with wheat, full residue retention and ZT was moderately increased compared to the conventional practice with tillage. However, there was no direct relation between increased root rot and yield. Govaerts et al. (2006a) concluded that, although root diseases may have affected crop performance, disease affected yield less than other critical plant growth factors such as water availability or micro- and macronutrient status. ZT with rotation and residue retention enhanced water availability, soil structure and nutrient availability more than conventional tillage and, as a result, also gave high yields. ZT and crop retention increased the diversity of microbial life. In the long term, ZT with crop residue retention creates conditions favourable for the development of antagonists and predators, and fosters new ecological stability (Govaerts et al. 2007b, 2008a).

Yield losses due to nematodes can be expected under conventional cropping systems in suboptimal irrigation and semi-arid conditions (Verhulst et al., 2010). In north and central Mexico, the nematode *Pratylenchus thornei* (root lesion nematode) is one of the parasitic nematodes associated with maize (Govaerts et al., 2006c).

In research conducted in the central highlands of Mexico, parasitic nematodes fared better in conventionally tilled situations; under ZT populations decreased. The number of non-plant parasitic nematodes increased under ZT with residue retention, compared to any combination of conventional tillage or residue removal treatments (Govaerts et al., 2006c, 2007a). Residue retention decreased the number of *P. thornei* in both maize and wheat. This result is not surprising, as many of the non-plant parasitic species evaluated feed on bacteria that are likely to increase in number because of the additional soil cover

(Yeates et al., 1999). Concordantly, the incidence of *P. thornei* was markedly less for PBs with full crop residue retention on the soil surface in the long-term permanent raised bed trials being conducted in north-western Mexico (Sayre et al., 2005).

5.2 Soil meso- and macrofauna

Larger soil fauna, in particular soil macrofauna such as earthworms, also play an important role in the soil environment (Turmel et al., 2015). Soil macrofauna populations are generally greater under CA when compared to conventional systems (Govaerts et al., 2009a). Earthworms have been observed to respond positively to crop residue retention and minimum soil disturbance (Turmel et al., 2015). In soils without tillage, biopore networks created by earthworms are not disrupted. This results in a large network of interconnected pores, which can contribute to infiltration and drainage (Verhulst et al., 2011b). Results from the long-term rainfed trial in El Batán indicate that residue retention in tandem with no-tillage and crop rotations (CA) has potential for conserving certain ground-dwelling predators, for example, Araneae (spiders) (Rivers et al., 2016).

However, not all soil macrofauna is beneficial. The risk of increased insect pests with CA, coupled with the high use of pesticides in Mexico and the significant damage caused annually by the fall armyworm in maize, *Spodoptera frugiperda* (J. E. Smith), were reason to study the effects of CA on the arthropod community in the long-term trial located at CIMMYT. The age of the trial at the time of the experiment (22 years) ensured that the different systems had been stabilized over time. The results described greater activity densities of certain predators and predator evenness, a relationship between high soil cover and biological control potential, and no effect on yield of higher numbers of true armyworm in wheat, which may all indicate the potential benefit of CA to the ground-dwelling arthropod community and their beneficial activities (Rivers et al., 2016).

6 Weed management under CA

Weed pressure and crop yield are inversely related; weed management is therefore crucial to achieving the potential yield gains offered by CA systems. Under CA, weed control via tillage is no longer an option, and weed communities and growth dynamics will change compared to under conventional tillage systems. The methods of weed control under CA will likewise need to be adjusted (Nichols et al., 2015).

Where residue is removed from PBs for fodder, the incidence *of Phalaris minor*, the most problematic grass weed, was found to increase when compared to conventional tillage with residue incorporation. Residue burning with PBs dramatically reduces *Phalaris* incidence but the level of incidence of most annual weeds markedly declined with permanent raised beds if an adequate level of crop residue mulch is retained on the soil surface (Sayre et al., 2005). Overall, one of the many economic and environmental benefits of the permanent raised beds is the marked gradual reduction of many weed species, especially annual grasses and broadleaf weeds; reducing the need for selective herbicide applications (Sayre et al., 2005). However, it can be expected that a greater amount of herbicide might be necessary in the first years of transitioning to CA systems, as it may take 4–10 years for yield, soil characteristics and weed populations to reach equilibrium (Nichols et al., 2015). Again, this establishes the tremendous need for long-term experiments such as CIMMYTs, to evaluate new agronomic systems.

In a literature review on the effects of CA on weed dynamics, it was concluded that the principles of CA, particularly crop rotation and surface residue retention, are in themselves methods of weed control. The combined use of all three principles can offer disproportionate advantages, and weed problems are more likely to occur if only one of the CA principles is utilized. For CA, it appears the synergistic effects of utilizing multiple control tactics are even more crucial and their importance cannot be overemphasized (Nichols et al., 2015).

7 The influence of CA on productivity

The effects of soil quality on agricultural productivity are greater in low-input rainfed production systems than in highly productive systems (Scherr, 1999). Govaerts et al. (2006b) determined the soil quality of plots after more than 10 years of different tillage and residue management treatments. There was a direct and significant relationship between the quality status of the soil and the crop yield. ZT with crop residue retention showed the highest crop yields as well as the highest soil quality status (Govaerts et al., 2005, 2006). In contrast, the soil under ZT with crop residue removal showed the poorest soil quality (i.e. low contents of organic C and total N, low aggregate stability, compaction, lack of moisture and acidity) and produced the lowest yields, especially with a maize monoculture (Fuentes et al., 2009; Govaerts et al., 2006b).

The results show that use of the current tillage-based practices with crop residue removal, on these already degraded soils, does not deliver the maximum yield response to, or full return on, the investments into developing new cultivars. Similarly, other inputs, in particular rainfall, are not being efficiently utilized (Govaerts et al., 2005, 2006a,b, 2007a,b; Sayre and Govaerts, 2011).

In the highlands of Mexico, CA resulted in higher yields and net returns per hectare as early as the first and second years after adoption by farmers. The yield difference between CA and the conventional practice was highest when rainfall was lower than average (Romero-Perezgrovas et al., 2014). Additional improvements in grain yield may be attainable if genotypes better adapted to the CA soil environment can be developed (Herrera et al., 2013).

Sayre et al. (2005) reported on a wheat–maize rotation in a long-term sustainability trial under irrigated conditions in north-western Mexico comparing different bed planting systems (conventional tilled beds and permanent raised beds). Residue management varied from full to partial retention, as well as residue burning. Yield differences between management practices only became clear after 5 years (10 crop cycles), with a reduction in the yield of permanent raised beds where all residues had been routinely burned. In contrast to rainfed low rainfall areas, in irrigated agricultural systems, the application of irrigation water appears to 'hide or postpone' the expression of the degradation of many soil properties until they reach a level that no longer can sustain high yields, even with irrigation (Sayre et al., 2005). Lower potential production due to degradation may not show up in intensive, high-input systems as the inputs mask potential degraded processes. The reduced yields reported after five years with PBs where residues were burned were related to a significant decrease in stable macro-aggregation and SMB (Limon-Ortega et al., 2006). In high-input systems, the decreased soil quality status of management practices is reflected in reduced efficiency of inputs (fertilizer, water, biocides and labour) resulting in higher production costs to maintain the same yield levels, rather than in lower yields as such (Scherr, 1999).

8 CA and climate change

8.1 Agriculture and climate change

Agricultural production is one contributor to GHG emissions including carbon dioxide (CO_2), methane (CH_4) and nitrous oxide (N_2O), the three main gases influencing global warming. In fact, agriculture is a major contributor to CH_4 and N_2O emissions, two gases with greater warming potential than CO_2, although emitted at lower concentrations (IPCC, 2014). Agronomic management practices can affect the GHG emissions from soil (Omonode et al., 2007), C sequestration (Govaerts et al., 2009a), fuel used for farm operations and the production of inputs such as inorganic fertilizers (West and Marland, 2002). The net global warming potential (GWP) of an agronomic practice takes into account all these factors. Therefore, the determination of the net GWP should be part of the sustainability evaluation of agronomic practices (Verhulst et al. 2012).

The main GHGs in wheat and maize cropping systems are N_2O and CO_2. In the case of N_2O, N fertilizers are a significant direct source of emissions of this gas in the field and an indirect source through fossil fuel energy consumption associated with manufacturing and transport of fertilizers. In the case of CO_2 there is enormous potential to sequester C through the adoption of CA or no-till systems (Ortiz-Monasterio et al., 2010).

Climate models predict that Mexico will continue to warm during this century. Some emission scenarios predict an increase in annual average temperature by the 2050s ranging from 1.6–2.0°C on the Yucatán peninsula to 3.1–3.5°C in northern Mexico. Annual precipitation is predicted to decrease in most of Mexico, with the largest decrease (10–25%) in the Yucatán peninsula and in north-eastern Mexico (Verhulst et al., 2012). Maize production in countries like Mexico will suffer significantly from these changes in climate (Stirling et al., 2014).

8.2 CA to mitigate climate change

Rainfed conditions in the semi-arid highlands of Mexico

Dendooven et al. (2011, 2012b) evaluated the effect of tillage practice and crop residue management on the net GWP taking into account soil C sequestration, emissions of GHGs from the soil, for example, CO_2, CH_4 and N_2O, fuel used for farm operations (tillage, planting and fertilizer application, harvest) and the production of fertilizer and seeds. Tillage and residue management had little effect on GHGs emitted from the soil. However, the soil organic C content in the 0–60 cm layer was affected strongly by tillage and crop residue management. After nearly 20 years, more C was sequestered in each soil layer of CA than in CT and overall in the 0–60 cm layer of CA contained 36 Mg C ha^{-1} more than CT. This resulted in a negative net GWP for ZT with crop residue retention, whereas with other management practices it was approximately 2000 kg CO_2-C ha^{-1} y^{-1} (Dendooven et al., 2012).

Fuentes et al. (2012b) found that a reduction in soil disturbance combined with residue retention increased C retention in the small and large macro-aggregates of the top soil due to greater aggregate stability and reduced the emissions of CO_2 compared with conventional tillage without residues retention and maize monoculture.

The practice of permanent raised beds with adequate crop residue retention is actively promoted throughout the world for its obvious economic and environmental benefits, but little is known about how GHG emissions might be affected by this agricultural practice (Dendooven et al., 2014). A previous laboratory study showed that dynamics of C and N and GHG emissions were affected. Permanent raised bed planting with crop residue retention decreased emissions of N_2O and CO_2 compared to soil under conventionally tilled raised beds (Patiño-Zúñiga et al., 2009). In field experiments, Dendooven et al. (2013) found that retention of crop residue in PBs increased the emission of CO_2 compared to where it was removed, but tillage did not affect fluxes of CO_2. Emission of CH_4 and N_2O was larger from CBs than from PBs, but crop residue management in PBs had no significant effect on fluxes of CH_4 and N_2O. Concentrations of mineral N were larger in CBs than in PBs, because in a CB most of the organic matter was already mineralized, so less N was immobilized. Removal of crop residue from PBs reduced N immobilization so that the mineral N content increased. In a PB, incorporation and decomposition of the crop residue started and N was immobilized. Maize and wheat crop residues are characterized by a high C:N ratio; so when decomposition started, mineral N was immobilized (Verachtert et al., 2009). However, in a CB most of the organic material was already mineralized, so less N was immobilized. Consequently, as the mineralization of crop residue started later in the PB than in the CB, immobilization of N when crops were fertilized was larger in the PB than in the CB. As such, the inorganic N content in soil was higher in the CB than in the PB. The effect of crop residue on soil N content can also be seen in the crop residue management in the PB. Removing crop residues reduced N immobilization so that the mineral N content increased (Dendooven et al., 2013).

Soil temperature was higher in CBs than in PBs and in PBs with crop residue retained compared with where it was removed. Soil water was better preserved in PBs than in CBs and in PBs where residue was retained than where it was removed.

It was concluded that reduced tillage on PBs and crop residue retention strongly reduced the net GWP of the system compared with the case when beds were remade each year. PBs with residue retention reduced net GWP by 50% compared with conventional beds with residue retention, but the removal of residues from the PBs more than doubled it (Dendooven et al., 2014).

Irrigated conditions in arid north-western Mexico

Studies of the net GWP of agronomic systems in the long-term trials in the Yaqui Valley are ongoing. In high-input systems like the one in the Yaqui Valley, N fertilizers are a significant direct source of emissions of N_2O and NO_x in the field and an indirect source through fossil fuel energy consumption associated with manufacturing and transport of fertilizers (Ortiz-Monasterio et al., 2010).

Adequate fertilizer management can play an important role in reducing N_2O and NO_x emissions in the field and increasing fertilizer efficiency, thereby reducing the necessary fertilizer and associated manufacturing emissions (Verhulst et al., 2012). Monitoring and understanding crop growth and the differences that can occur within one field can be a step towards this goal. Sensor-based N management in wheat and maize is a new technology that uses an optical sensor, which measures the normalized difference vegetative index of the canopy, and can point out differences in crop growth and development within one field (Verhulst et al., 2009b, 2011a).

9 Implementation of CA

9.1 Limiting factors

The environmental and economic benefits of CA are evident in many regions of Latin America; yet its success and adoption are closely tied to the socio-economic issues of each country and region.

The main factors that appear to limit the adoption of CA by small-scale farmers in different regions of Latin America include (Speratti et al., 2015)

- Socio-economic constrains (limited access to financial capital and credit opportunities, inability to take risks, short-term priorities, land tenure).
- Lack of machinery and tools (reliance on manual labour, lack of locally available mechanization solutions).
- Crop residue trade-off (mixed crop-livestock systems, using residue for fodder).
- Agronomic constraints and lack of CA knowledge (poor extension capacity, adaptation of CA principles to different agro-ecological and socio-economic conditions).

In most cases, there are multiple factors limiting CA adoption. The participation of diverse stakeholders working in cooperation is required to ensure the adaptation and adoption of CA.

9.2 Residue trade-off

Mixed (crop-livestock) systems are relatively low-input systems. Much feed comes from crop residues – the non-food residual biomass of crop production (Beuchelt et al., 2015). Since there is high demand for residues as animal feed, farmers in communities practising livestock grazing prefer not to leave residues in the ground for soil conservation. This is one of the main constraints in different regions in Mexico (Van den Broeck et al., 2013; Speratti et al., 2015). Residue may also be removed from the field to be used as biofuel (Turmel et al., 2015). Furthermore, adoption of CA can represent high opportunity costs in terms of foregone sales of residues (Beuchelt et al., 2015).

However, after several years of continued on-farm training and demonstration events conducted through the MasAgro project of CIMMYT, Mexican farmers in the central highlands have begun to see the value of retaining residues on the surface in terms of water infiltration and soil moisture retention. A potential compromise to the residue trade-off is only removing part of the residue for fodder and leaving a part for soil cover (Speratti et al., 2015). It has been shown that ZT treatments with partial residue removal give yields equivalent to treatments with full residue retention (Govaerts et al., 2005). In the highlands of Mexico, between 50% and 70% of the residue can be removed and with retention of the remaining portion an adequate benefit to the soil is provided (Govaerts et al., 2007). In water limited environments, retaining residue can create a positive feedback by increasing yields due to improved soil quality and moisture retention that in turn adds more crop residue to the system that can then be used for feed (Turmel et al., 2015).

Another solution is growing fodder crops or dual-purpose (grain and fodder) crops (Speratti et al., 2015). Forage crops could accumulate more C in soils, compared to grain crops, due to a higher root biomass production stimulated by grazing or mowing (Cairns, et al., 2012). Finally, interdisciplinary research also needs to look into alternative soil cover options through, for example, intercropping, relay cropping or use of cover crops, if CA is to be adopted by not only better-off farmers and thus increase adoption rates (Beuchelt et al., 2015; Turmel et al., 2015).

9.3 The development of the innovation hub concept

In 2013, CA was deployed on 155 million hectares worldwide on both large and small farms (Kassam et al., 2014). However, its success and adoption are closely tied to the socio-economic issues of each country and region. For instance, land tenure can limit the implementation of sustainable soil management practices. Farmers who rent the land will focus mainly on short-term outcome (Pereira de Herrera and Sain, 1999). In Honduras, the expansion of the cattle industry and changes in land markets, distribution and tenure were causing farmers to abandon the sustainable slash-mulch 'abonera' system in the 1990s (Speratti, et al., 2015).

The adaptation and adoption of CA principles have been more successful in the temperate regions of South America, while remaining more challenging for resource-poor farmers in tropical regions (Speratti et al. 2015). Brazil has proven to be a leader in the development of CA practices and technology, working through an effective and innovative network of farmers and their associations, and private and public partnerships. Many regions of Argentina, Uruguay and Panama have adopted CA due to the benefits including soil conservation, reduced production costs and increased soil biodiversity, which enhances environmental equilibrium, improves crop water balance and increases yields (Verhulst et al., 2010). In other regions of Latin America, however, such as Central America and the Andean region, CA adoption has proven more difficult (Speratti et al., 2015). One might argue this could be due to the differences in agroclimatic zone, however, it appears CA has the potential to be adapted and optimized for any region. Therefore, it seems that mainly socio-economic and political issues are limiting the adoption of CA practices in these regions, rather than environmental constraints (Speratti et al., 2015). A review of case studies in Latin America suggests that these socio-economic constraints limiting adoption are access to appropriate machinery, crop residue trade-offs, lack of adaptation of the technology to farmer's agronomic constraints and uncoordinated efforts of stakeholders (Speratti et al., 2015). Of the limitations outlined, most are related to economic constraints, lack of incentives, access to information and traditional practices and beliefs. Moreover, competition for crop residues between animals and soil cover in Central America and the Andean region, as well as in many other regions worldwide, especially rainfed cropping regions, is closely linked to the potential of CA adoption (Speratti et al., 2015). Because of the complex, multi-component nature of CA, it is unlikely that this technology can successfully be scaled out through traditional linear models of research and extension. To support innovation, technology development and further CA adoption, institutional support for research on CA and information dissemination is required (Govaerts et al., 2009).

In Mexico, agricultural development also has to take into account that the country is composed of 32 states with varying agro-ecological areas. This heterogeneity presents a challenge for the development and extension of suitable climate-smart technologies and practices, particularly if these have to be adapted to local conditions. It was clear that complex, multi-component and knowledge-intensive technologies such as CA-based cropping systems and use of improved germplasm could not be successfully scaled out through traditional linear models of research and extension (Hellin et al. 2014; Kassam et al., 2014). What was needed was a change towards a more dynamic agricultural innovation systems approach that reconnects research for development with on-the-ground needs, generates information that helps to align public policies and stimulates further innovations that solve farmers' problems. In light of the geophysical and socio-economic complexity of the country, the level of current and projected agricultural production as well as climatic, environmental and economic challenges, the government acknowledged that an innovation network approach, integrating agricultural production with federal and state authorities, research organizations, and the private sector was greatly needed (Hellin et al., 2014).

As a solution to the poor adoption of CA in different parts of the world, and more specifically in the challenging situation of Mexico, CIMMYT researchers developed the dynamic and interactive concept of innovation *hubs* (Camacho-Villa et al., 2016).

The hub is an effort to foster dynamic innovation systems in Mexico. It focuses on maize- and wheat-based farming systems. The hub is a key element of the 'Sustainable Modernization of Traditional Agriculture' initiative (Modernizacion Sustentable de la Agricultura Tradicional-MasAgro) and other projects implemented in the frame of the sustainable intensification strategy implemented by the CIMMYT, which distinguishes the hub from other efforts in Mexico (Camacho-Villa et al., 2016). Initial definitions conceptualized the hub as a 'potential crop production system in particular world regions serving as a "platform" to develop – with active farmer participation as well as involvement of relevant national agricultural research centers, non-governmental organizations, and private-sector partners – relevant CA-based technologies for immediate use in the selected area' (Sayre, 2006). One of the major lessons learned from experience is that, to achieve the complex system changes leading to widespread adoption of CA, farmer experimentation and adoption must be stimulated by many coordinated activities conducted through collaboration between local partners: farmers, governmental and non-governmental institutions and international research centres (Sayre and Govaerts, 2011). This demands the creation of ties between farmers, markets and organizations that provide services (Díaz-José et al., 2015).

Figure 1 depicts the hub, scientists, technicians, agro-dealers, farmers, financial institutions and more interact in feedback loops and through multi-stakeholder actions to create opportunities for innovation and learning.

One of the projects in the frame of CIMMYT's Sustainable intensification strategy in Latin America is MasAgro. Beginning in 2010, innovation hubs were developed in different agro-ecologies in Mexico. In 2013, in the third year of project operation, MasAgro established 55 CA and agronomic management research trials ('platforms'); 225 on-farm side-by-side innovation demonstration ('modules'); and reached over 200 000 farmers through training, events and technical assistance. Lead farmers and technicians received both intensive and on-farm training to further disseminate technologies and innovations. Farmers are provided with access to input suppliers and markets for their produce.

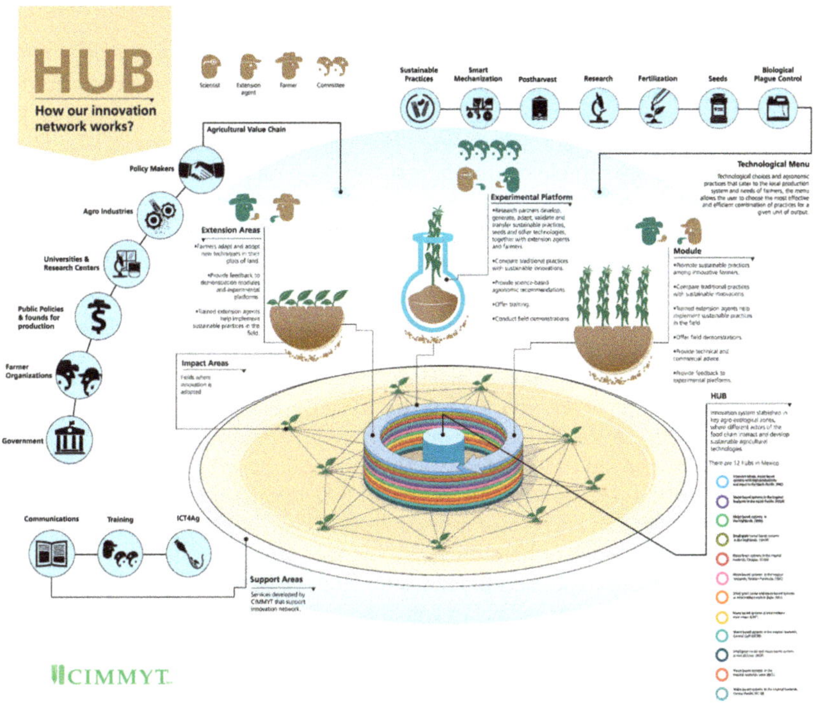

Figure 1 Schematic representation of the hub and its implementation in Mexico.

Through CA, farmers are achieving steady yields, reducing the risk of farming (CIMMYT internal document, 2015). There is an excellent complementarity between the external sources of knowledge (defined as research institutes, universities, and government and extension services), and the internal sources (defined as other farmers, relatives, neighbours and friends) (Camacho-Villa et al., 2016). An example of external and internal sources of knowledge complementing each other is if a farmer hears about a new CA practice from a research institute and talks about it to his/her neighbours. External sources can play different roles such as (i) promoting a dialogue between the developers and users of an innovation through knowledge facilitation; (ii) network brokerage, for example, to organize a platform for a meeting or help obtain public subsidies to promote activities or (iii) managing the innovation process to maintain network stability and coordination and enhance technological change (Díaz-José et al., 2015).

Regular monitoring and evaluation of advances and farmer perceptions, and the adjustments to respond to these help ensure a dynamic and successful development process (Sayre and Govaerts, 2011). In general, to enhance CA adoption rates, a thorough understanding of the site-specific determinants and constraints in the adoption process is needed. Site specificity is required as there are few, if any, universal variables that explain adoption in different contexts. The site-specific determinants and constraints can determine where the efforts of extension on CA should lie (Van den Broeck et al., 2013).

10 Conclusions

Forty years of research show that CA was the better choice at providing continuously higher and more stable yields for both wheat and maize compared to traditional practices in rainfed conditions. The current conventional practices, which are tillage-based and where crop residues are removed, do not result in maximal yields, and are detrimental on already deteriorated soil.

Similarly, other inputs, in particular rainfall, are not being efficiently utilized. In irrigated conditions, yield differences were smaller or absent. Yield differences between management practices only became clear after 5 years (10 crop cycles), with a reduction in the yield of permanent raised beds where all residues had been routinely burned. The degradation of the soil can be masked by the application of irrigation water, until yields drop quite suddenly. Similarly, Sain et al. (1994) found that it took three years after setting up the mucuna rotation in Honduras, for the maize yield of the abonera system to be higher compared to the traditional maize-fallow system without abonera. With regard to the effects of CA on the net GWP of the system, the reduction in soil disturbance combined with residue retention reduced the emissions of CO_2 compared with conventional tillage without residues retention and maize monoculture.

There were highly significant differences in the economic returns between the common farmer practices and the CA-based practices for both wheat and maize. This indicates the clear economic advantage, especially to small- and medium-scale farmers in a developing country setting, if they have the means to adopt appropriate CA technologies. This was also the conclusion of the 2012 study of the United States Agency for International Development (USAID) who examined the benefits of CA in the poorest regions of Ecuador. They found that in one region, reducing tillage without removing crop residue resulted in higher returns than the conventional system. In another region, however, this was not the case, and the principles of CA should be adapted to the condition of the region to make the system productive and viable (Nguema et al., 2013).

Future research requires the understanding of physical, chemical and biological characteristics of the farming system, as well as social, cultural and governance processes. This is required to further develop understanding, as well as for the refinement and catalysis of innovation systems that can generate triple wins (economic, social and environmental) for all actors along the value chain, including society. A sustained network of long-term and on-farm trials in different agro-ecologies is required to achieve this. Modern tools like crowdsourcing, remote sensing, network analysis, big data, dynamic scenario planning, nanotechnology and mechatronics will have to be integrated around the different data streams and system processes around the agricultural innovations systems.

11 Where to look for further information

Further information on the discussed topics can be found on the following websites with knowledge banks on the discussed topics:

http://www.cimmyt.org
http://www.maize.org
http://conservationagriculture.mannlib.cornell.edu

12 Acknowledgements

This chapter is a review of existing literature developed by the CIMMYT Sustainable Intensification team for Latin America. Thanks to the financial support of the CGIAR Research Program MAIZE and the SAGARPA Mexico funded MasAgro (Modernizacion Sustentable de la Agricultura Traditicional) project for making this chapter possible.

13 References

Andersen, A., (1999). Plant protection in spring cereal production with reduced tillage. II. Pests and beneficial insects. *Crop Protection*, 18, pp. 651–7.

Andersen, A., (2003). Long-term experiments with reduced tillage in spring cereals. II. Effects on pests and beneficial insects. *Crop Protection*, 22(1), pp. 147–52.

Beuchelt, T. D., Camacho Villa, C. T., Göhring, L., Hernández Rodríguez, V. M., Hellin, J., Sonder, K. and Erenstein, O., (2015). Social and income trade-offs of conservation agriculture practices on crop residue use in Mexico's central highlands. *Agricultural Systems*, 134, pp. 61–75.

Bloem, J., de Ruiter, P. and Bouwman, L., (1997). Soil food webs and nutrient cycling in agroecosystems. In J. D. Elsas, J. T. Trevors and H. Wellington (eds), *Modern Soil Microbiology*, New York, Marcel Dekker, pp. 245–78.

Bockus, W. W. and Shroyer, J. P., (1998). The impact of reduced tillage on soilborne plant pathogens. *Annual Review of Phytopathology*, 36, pp. 485–500.

Bronick, C. J. and Lal, R., (2005). Soil structure and management: a review. *Geoderma*, 124(1–2), pp. 3–22.

Cairns, J. E., Sonder, K., Zaidi, P. H., Verhulst, N., Mahuku, G., Babu, R., Nair, S. K., Das, B., Govaerts, B., Vinayan, M. T., Rashid, Z., Noor, J. J., Devi, P., Vicente, F. S. and Prasanna, B. M., (2012). *Maize Production in a Changing Climate: Impacts, Adaptation, and Mitigation Strategies*, 1st ed., Elsevier Inc.

Camacho-Villa, T. C., Hellin, J., Govaerts, B., Rendon-Medel, R., Martinez, T. E., Almekinders, C., Guevara-Hernández, F. and Beuchelt, T. C., (2016). Lessons and challenges in the implementation of an agricultural innovation network: The case of the MasAgro hubs. *The Journal of Agricultural Education and Extension*, 22(3) (in press).

Carter, M. R., Gregorich, E. G., Angers, D. A., Beare, M. H., Sparling, G. P., Wardle, D. A. and Voroney, R. P., (1999). Interpretation of microbial biomass measurements for soil quality assessment in humid temperate regions. *Canadian Journal of Soil Science*, 79(4), pp. 507–20.

Ceja-Navarro, J. A., Rivera-Orduna, F. N., Patino-Zuniga, L., Vila-Sanjurjo, A., Crossa, J., Govaerts, B. and Dendooven, L., (2010). Phylogenetic and multivariate analyses to determine the effects of different tillage and residue management practices on soil bacterial communities. *Applied and Environmental Microbiology*, 76(11), pp. 3685–91.

Chávez-Romero, Y., Navarro-Noya, Y. E., Reynoso-Martínez, S. C., Sarria-Guzmán, Y., Govaerts, B., Verhulst, N., Dendooven, L. and Luna-Guido, M., (2016). Dried young wheat plants application favours a limited group of bacteria independent of tillage-crop residue management or N fertilizer application. *Soil & Tillage Research* (in press).

Dendooven, L., Gutiérrez-Oliva, V. F., Patiño-Zúñiga, L., Ramírez-Villanueva, D. A., Verhulst, N., Luna-Guido, M., Marsch, R., Montes-Molina, J., Gutiérrez-Miceli, F. A., Vásquez-Murrieta, S. and Govaerts, B., (2012). Greenhouse gas emissions under conservation agriculture compared to traditional cultivation of maize in the central highlands of Mexico. *Science of the Total Environment*, 431, pp. 237–44.

Dendooven, L., Patiño-Zúñiga, L., Verhulst, N., Boden, K., García-Gaytán, A., Luna-Guido, M. and Govaerts, B., (2013). Greenhouse gas emissions from nontilled, permanent raised, and conventionally tilled beds in the central highlands of Mexico. In M. S. Kang and S. S. Banga (eds), *Combating Climate Change: An Agricultural Perspective*, CRC Press, pp. 283–304.

Dendooven, L., Patiño-Zúñiga, L., Verhulst, N., Boden, K., García-Gaytán, A., Luna-Guido, M. and Govaerts, B., (2014). Greenhouse gas emissions from nontilled, permanent raised, and conventionally tilled beds in the central highlands of Mexico. *Journal of Crop Improvement*, 28(4), pp. 547–74.

Dendooven, L., Patiño-Zúñiga, L., Verhulst, N., Luna-Guido, M., Marsch, R. and Govaerts, B., (2012). Global warming potential of agricultural systems with contrasting tillage and residue management in the central highlands of Mexico. *Agriculture, Ecosystems and Environment*, 152, pp. 50–8.

Díaz-José, J., Rendón-Medel, R., Govaerts, B., Aguilar-Ávila, J. and Muñoz-Rodriguez, M., (2015). Innovation diffusion in conservation agriculture: A network approach. *European Journal of Development Research*, (Cc), pp. 1–16.

Duiker, S. W. and Beegle, D. B., (2006). Soil fertility distributions in long-term no-till, chisel/disk and moldboard plow/disk systems. *Soil and Tillage Research*, 88(1–2), pp. 30–41.

Fischer, R. A., Santiveri, F. and Vidal, I. R., (2002a). Crop rotation, tillage and crop residue management for wheat and maize in the sub-humid tropical highlands I. Wheat and legume performance. *Field Crops Research*, 79(2–3), pp. 107–22.

Fischer, R. A., Santiveri, F. and Vidal, I. R., (2002b). Crop rotation, tillage and crop residue management for wheat and maize in the sub-humid tropical highlands II. Maize and system performance. *Field Crops Research*, 79(2–3), pp. 123–37.

Franzluebbers, A. J., (2002). Soil organic matter stratification ratio as an indicator of soil quality. *Soil and Tillage Research*, 66(2), pp. 95–106.

Fuentes, M., Hidalgo, C., Etchevers, J., de León, F., Guerrero, A., Dendooven, L., Verhulst, N. and Govaerts, B., (2012). Conservation agriculture, increased organic carbon in the top-soil macro-aggregates and reduced soil CO_2 emissions. *Plant and Soil*, 355(1–2), pp. 183–97.

Van Ginkel, M., Trethowan, R. M., Ammar, K., Wang, J. and Lillemo, M., (2002). *Guide to Bread Wheat Breeding at CIMMYT. Wheat Program Special Report No. 5*, Mexico D.F., Mexico.

Govaerts, B., Mezzalama, M., Sayre, K. D., Crossa, J., Nicol, J. M. and Deckers, J., (2006). Long-term consequences of tillage, residue management, and crop rotation on maize/wheat root rot and nematode populations in subtropical highlands. *Applied Soil Ecology*, 32(3), pp. 305–15.

Govaerts, B., Sayre, K. D. and Deckers, J., (2006). A minimum data set for soil quality assessment of wheat and maize cropping in the highlands of Mexico. *Soil and Tillage Research*, 87(2), pp. 163–74.

Govaerts, B., Sayre, K. D. and Deckers, J., (2005). Stable high yields with zero tillage and permanent bed planting? *Field Crops Research*, 94(1), pp. 33–42.

Govaerts, B., Sayre, K. D., Goudeseune, B., De Corte, P., Lichter, K., Dendooven, L. and Deckers, J., (2008). Conservation agriculture as a sustainable option for the central Mexican highlands. *Soil and Tillage Research*, 103(2), pp. 222–30.

Govaerts, B., Sayre, K. D., Lichter, K., Dendooven, L. and Deckers, J., (2007). Influence of permanent raised bed planting and residue management on physical and chemical soil quality in rain fed maize/wheat systems. *Plant and Soil*, 291(1–2), pp. 39–54.

Govaerts, B., Verhulst, N., Castellanos-Navarrete, A., Sayre, K. D., Dixon, J. and Dendooven, L., (2009). Conservation Agriculture and Soil Carbon Sequestration: Between Myth and Farmer Reality. *Critical Reviews in Plant Sciences*, 28(3), pp. 97–122.

Govaerts, B., Verhulst, N., Sayre, K. D., De Corte, P., Goudeseune, B., Lichter, K., Crossa, J., Deckers, J. and Dendooven, L., (2007). Evaluating spatial within plot crop variability for different management practices with an optical sensor? *Plant and Soil*, 299(1–2), pp. 29–42.

Hellin, J., Beuchelt, T., Camacho, C., Govaerts, B., Donnet, L. and Riis-Jacobsen, J., (2014). An innovation systems approach to enhanced farmer adoption of climate-ready germplasm and agronomic practices. Capri Working Paper No. 116, Washington, DC, International Food Policy Research Institute, p. 30.

Herrera, J. M., Verhulst, N., Trethowan, R. M., Stamp, P. and Govaerts, B., (2013). Insights into Genotype × Tillage interaction effects on the grain yield of wheat and maize. *Crop Science*, 53(5), p. 1845.

Hobbs, P. R. and Govaerts, B., (2010). How conservation agriculture can contribute to buffering climate change. In M. P. Reynolds (ed.), *Climate Change and Crop Production*, pp. 177–99.

Hulugalle, N. R. and Entwistle, P., (1997). Soil properties, nutrient uptake and crop growth in an irrigated Vertisol after nine years of minimum tillage. *Soil and Tillage Research*, 42(1–2), pp. 15–32.

IPCC, (2014). *Climate Change 2014: Synthesis Report. Contribution of Working Groups I, II and III to the Fifth Assessment Report of the Intergovernmental Panel on Climate Change [Core Writing Team, R. K. Pachauri and L. A. Meyer (eds)]*, Geneva, Switzerland, 151 pp.

Jiménez-Bueno, N., Valenzuela-Encinas, C., Marsch, R., Ortiz-Gutiérrez, D., Verhulst, N., Govaerts, B., Dendooven, L. and Navarro-Noya, Y., (2016). Bacterial indicator taxa in soils under different long-term agricultural management. *Journal of Applied Microbiology* (in press).

Karlen, D. L., Hurley, E. G., Andrews, S. S., Cambardella, C. A., Meek, D. W., Duffy, M. D. and Mallarino, A. P., (2006). Crop rotation effects on soil quality at three northern corn/soybean belt locations. *Agronomy Journal*, 98, pp. 484–95.

Kassam, A., Friedrich, T., Shaxton, F., Bartz, H., Mello, I., Kienzle, J. and Pretty, J., (2014). The spread of Conservation Agriculture: policy and institutional support for adoption and uptake. *Field Actions Science Reports*, 7.

Kay, B. D., Angers, D. A., Groenevelt, P. and Baldock, J. A., (1988). Quantifying the influence of cropping history on soil structure. *Canadian Journal of Soil Science*, 68(May), pp. 359–68.

Lal, R. and Shukla, M. J., (2004). *Principles of Soil Physics*, New York, Marcel Dekker.

Le Bissonnais, Y. and Arrouays, D., (1997). Aggregate stability and assessment of soil crustability and erodibility: II. Application to humic loamy soils with various organic carbon contents. *European Journal of Soil Science*, 48(1), pp. 39–48.

Limon-Ortega, A., Govaerts, B., Deckers, J. and Sayre, K. D., (2006). Soil aggregate distribution/stability and microbial biomass in a permanent bed wheat-maize planting system after 12 years. *Field Crops Research*, 97, pp. 302–9.

Navarro-Noya, Y. E., Gómez-Acata, S., Montoya-Ciriaco, N., Rojas-Valdez, A., Suárez-Arriaga, M. C., Valenzuela-Encinas, C., Jiménez-Bueno, N., Verhulst, N., Govaerts, B. and Dendooven, L., (2013). Relative impacts of tillage, residue management and crop-rotation on soil bacterial communities in a semi-arid agroecosystem. *Soil Biology and Biochemistry*, 65, pp. 86–95.

Nguema, A., Norton, G. W., Alwang, J., Taylor, D. B., Barrera, V. and Bertelsen, M., (2013). Farm-Level Economic Impacts of Conservation Agriculture in Ecuador. *Experimental Agriculture*, 49(01), pp. 134–47.

Nichols, V., Verhulst, N., Cox, R. and Govaerts, B., (2015). Weed dynamics and conservation agriculture principles: A review. *Field Crops Research*, 183, pp. 56–68.

Núñez, E. V., Bens, F., Valenzuela-encinas, C., Alcántara-Hernandez, R. J., Zavala-díaz De la Serna, F. J., Govaerts, B., Marsch, R., Luna-guido, M., Castro-silva, C., Ruiz-valdiviezo, V. M., Verhulst, N. and Dendooven, L., (2012). Bacterial diversity as affected by tillage practice in a raised bed planting system. *African Journal of Microbiology Research*, 6(43), pp. 7048–58.

Omonode, R. A., Vyn, T. J., Smith, D. R., Hegymegi, P. and Gál, A., (2007). Soil carbon dioxide and methane fluxes from long-term tillage systems in continuous corn and corn–soybean rotations. *Soil and Tillage Research*, 95(1–2), pp. 182–95.

Ortiz-Monasterio, I., Wassmann, R., Govaerts, B., Hosen, Y., Katayyanagi, N. and Verhulst, N., (2010). Greenhouse gas mitigation in the main cereal systems: rice, wheat and maize. In M. P. Reynolds (ed.), *Climate Change and Crop Production*, pp. 151–76.

Pereira de Herrera, A. and Sain, G., (1999). *Adoption of Maize Conservation Tillage in Azuero, Panama*, Mexico D.F., Mexico.

Powlson, D. S. and Jenkinson, D. S., (1981). A comparison of the organic matter, biomass, adenosine triphosphate and mineralizable nitrogen contents of ploughed and direct-drilled soils. *The Journal of Agricultural Science*, 97, pp. 713–21.

Ramirez-Villanueva, D. A., Bello-López, J. M., Navarro-Noya, Y. E., Luna-Guido, M., Verhulst, N., Govaerts, B. and Dendooven, L., (2015). Bacterial community structure in maize residue amended soil with contrasting management practices. *Applied Soil Ecology*, 90, pp. 49–59.

Reeves, D. W., (1997). The role of soil organic matter in maintaining soil quality in continuous cropping systems. *Soil and Tillage Research*, 43(1–2), pp. 131–67.

Rivers, A., Barbercheck, M., Govaerts, B. and Verhulst, N., (2016). Conservation agriculture affects arthropod community composition in a rainfed maize – wheat system in central Mexico. *Applied Soil Ecology*, 100, pp. 81–90.

Romero-Perezgrovas, R., Verhulst, N., De La Rosa, D., Hernández, V., Maertens, M., Deckers, J. and Govaerts, B., (2014). Effects of tillage and crop residue management on maize yields and net returns in the central Mexican highlands under drought conditions. *Pedosphere*, 24(4), pp. 476–86.

Sain, G., Ponce, I. and Borbon, E., (1994). Rentabilidad del sistema de aboneras en el litoral atlantico de Honduras. *Agronomia Mesoamericana*, 5(3), pp. 30–49.

Sayre, K., (2006). *Points for Consideration for the Concept Note Focusing on the Development, Extension and Farmer Adoption of Appropriate Conservation Agriculture (CA) Technologies in CWANA*, Mexico City, Mexico.

Sayre, K. D. and Govaerts, B., (2011). Use of conservation agriculture to improve farming systems in developing countries. In P. Tow, et al. (eds), *Rainfed Farming Systems*, Dordrecht, Springer Netherlands, pp. 861–73.

Sayre, K. D., Limon-Ortega, A. and Govaerts, B., (2005). Experiences with permanent bed planting systems CIMMYT/Mexico. In C. H. Roth, R. A. Fischer, and C. A. Meisner (eds), *Evaluation and Perfomance of Permanent Raised Bed Cropping Systems in Asia, Australia and Mexico*, Griffith, Australia, ACIAR.

Scherr, S. J., (1999). *Soil Degradation A Threat to Developing-Country Food Security by 2020?*, International Food Policy Research Institute, Washington, DC.

Sharma, S. K., Ramesh, A., Sharma, M. P., Joshi, O. P., Govaerts, B., Steenwerth, K. L. and Karlen, D. L., (2010). Microbial community structure and diversity as indicators for evaluating soil quality. In E. Lichtfouse (ed.), *Biodiversity, Biofuels, Agroforestry and Conservation Agriculture*, pp. 317–58.

Soil Survey Staff, (2003). *Keys to Soil Taxonomy*, United States Department of Agriculture, Natural Resources Conservation Service, Washington DC.

Speratti, A., Turmel, M.-S., Calegari, A., Araujo-Junior, C. F., Violic, A., Wall, P. and Govaerts, B., (2015). Conservation agriculture in Latin America. In M. Farooq and K. H. M. Siddique (eds), *Conservation Agriculture*, Springer International Publishing Switzerland, pp. 391–415.

Turmel, M.-S., Speratti, A., Baudron, F., Verhulst, N. and Govaerts, B., (2015). Crop residue management and soil health: A systems analysis. *Agricultural Systems*, 134, pp. 6–16.

Van den Broeck, G., Grovas, R. R. P., Maertens, M., Deckers, J., Verhulst, N. and Govaerts, B., (2013). Adoption of conservation agriculture in the Mexican Bajio. *Outlook on Agriculture*, 42(3), pp. 171–8.

Verachtert, E., Govaerts, B., Lichter, K., Sayre, K. D., Ceballos-Ramirez, J. M., Luna-Guido, M. L., Deckers, J. and Dendooven, L., (2009). Short term changes in dynamics of C and N in soil when crops are cultivated on permanent raised beds. *Plant and Soil*, 320(1–2), pp. 281–93.

Verhulst, N., Govaerts, B., Sayre, K. D., Sonder, K., Romero-Perezgrovas, R., Mezzalama, M. and Dendooven, L., (2012). Conservation agriculture as a means to mitigate and adapt to climate change, a case study from Mexico. In E. Wollenberg et al. (eds), *Climate Change Mitigation and Agriculture*, Oxon: Earthscan, pp. 287–300.

Verhulst, N., Govaerts, B., Verachtert, E., Mezzalama, M., Wall, P. C., Chocobar, A., Deckers, J. and Sayre, K. D., (2010). Improving soil quality for sustainable production systems? In R. Lal and B. A. Stewart (eds), *Advances in Soil Science: Food Security and Soil Quality*, Boca Raton, FL, USA, CRC Press, pp. 137–208.

Verhulst, N., Kienle, F., Sayre, K. D., Deckers, J., Raes, D., Limon-Ortega, A., Tijerina-Chavez, L. and Govaerts, B., (2011). Soil quality as affected by tillage-residue management in a wheat-maize irrigated bed planting system. *Plant and Soil*, 340(1–2), pp. 453–66.

Verhulst, N., Nelissen, V., Jespers, N., Haven, H., Sayre, K. D., Raes, D., Deckers, J. and Govaerts, B., (2011). Soil water content, maize yield and its stability as affected by tillage and crop residue management in rainfed semi-arid highlands. *Plant and Soil*, 344(1), pp. 73–85.

West, T. O. and Marland, G., (2002). A synthesis of carbon sequestration, carbon emissions, and net carbon flux in agriculture: comparing tillage practices in the United States. *Agriculture, Ecosystems & Environment*, 91, pp. 217–32.

Yeates, G. W., Wardle, D. A. and Watson, R. N., (1999). Responses of soil nematode populations, community structure, diversity and temporal variability to agricultural intensification over a seven-year period. *Soil Biology and Biochemistry*, 31(12), pp. 1721–33.

Chapter 7

Precision maize cultivation techniques

Louis Longchamps, Agriculture and Agri-Food Canada, Canada; and Raj Khosla, Colorado State University, USA

1 Introduction

Maize (*Zea mays* L.) has a rich history of cultivation that dates back 7000 years in Mexico (Beadle, 1977). Over the centuries, its cultivation has spread to most parts of the world. Maize cultivation practices vary significantly across agro-climatic regions of the globe, and such practices have changed many times as newer technologies and techniques became available to farmers. The recent advent of information and communication technologies has brought forth new transformations in maize cultivation practices. These technologies are transforming the way maize is cultivated around the world by providing farmers with a suite of novel tools and techniques for improving crop production, enhancing input use efficiency and increasing profitability while achieving environmental sustainability.

The infusion of information and communication technologies into agriculture is certainly new and has led to the development of precision tools and technologies. However, the concept of precision cultivation is in fact not new. In an early report circa 1550, Pedro Cieza de León explains:

> With the dew and moisture they grow by God's will, but the maize could in no way sprout nor the grain ripen if with each seed they did not throw in one or two heads of sardines which they take at sea with their nets. In planting, they put the sardine heads together

http://dx.doi.org/10.19103/AS.2016.0002.07

> with the maize in the same hole which they make to plant the grains, and in this way the maize grows and yields abundantly. (Rowe, 1969)

This technique, observed in Peru around 1550 by the Spanish conquistador and chronicler Pedro Cieza de León, enabled each maize plant to access the right amount of fertilizer and moisture at the right place and at the right time, hence encompassing three of the five basic principles of precision cultivation practices, which translated into an abundant yield as described by de León. A similar technique developed and demonstrated by International Crop Research Institute for Semi-Arid Tropics (ICRISAT) delivered a micro-dose of fertilizer to each maize plant still proves beneficial today (Hove et al., 2008). While such precision maize cultivation (PMC) techniques would be beneficial in hand-cultivated fields, it is not practical in large-scale agricultural systems. In a small-scale context, farmers use their eyes as sensors and their hands as precision tools, but in a large-scale context, advanced techniques and technologies are necessary to achieve the same ends. Nevertheless, precision cultivation practices, both for large- or small-scale farming environments, build on the concept of spatial and temporal variability observed in farms. It is logical to think that as the field size increases, so does the inherent spatial variability of soil properties (Beckett and Webster, 1971; Wilding and Drees, 1983). Yet, such spatial and temporal variability is reported worldwide across large- (e.g. 120 ha field in Illinois) and small-scale (e.g. <1 ha field in China) farming environments (Kravchenko and Bullock, 2000; Cao et al., 2012).

Precision technologies such as global positioning systems (GPS), geographic information systems (GIS), remote sensing (RS), geo-statistics and many others have enabled and empowered practitioners and farmers to not only observe spatial and temporal variability, but to precisely characterize variability and create maps that are actionable. The term 'precision cultivation' refers to the management of crop by embracing the concept described by the five 'R's'. Application of the 'Right input', at the 'Right time', in the 'Right amount', at the 'Right place' and in the 'Right manner'.

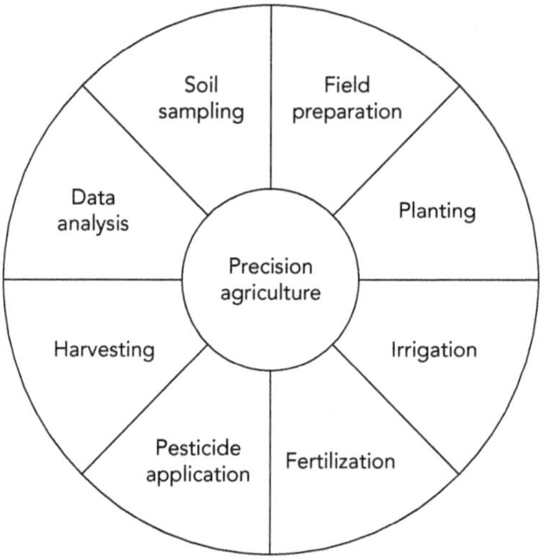

Figure 1 Schematic of different stages of maize cultivation: each segment refer to a different precision activity.

PMC is still in its stage of infancy as only a few techniques are currently adopted by the majority of maize producers. Yet, tools such as precision planters, sprayers and irrigation systems that are necessary for implementing PMC are now commercially available. Today, one among many limitations to the adoption of PMC techniques is the ability to reliably measure and manage variability in soil and crop properties at every location in a field. This chapter provides an overview of the current state of PMC techniques at different stages of maize cultivation (Fig. 1). A section on pre-planting will provide the current knowledge on the spatial variability of soil properties and how it influences PMC techniques. The following sections include planting and the current state of variable-rate seeding, nutrition and various approaches for site-specific fertilizer management, irrigation and the current knowledge of precision irrigation, different aspects of precision pest control and weed management, the opportunities enabled by digital yield maps and big data management, and finally economic aspects and future perspectives of PMC.

2 Pre-planting

2.1 Soil sampling

Soil sampling and analysis is an age-old technique that provides current status of soil properties (Thompson et al., 2005). Soil testing may enable identification of yield-limiting factors in soil and may allow farmers to make information-based management decisions (Miao et al., 2006). In particular, decisions related to nutrient management rely greatly on soil testing (Davis and Westfall, 2009). Traditionally, however, soil testing is performed to determine the average nutrient status of the field, following which a uniform fertilizer application is recommended for the entire field. Precision management of nutrients mandates a departure from traditional sampling practice that is focused on central tendency. Instead, precision management relies heavily on measuring the dispersion in soil properties across a field to account for variability associated with locations in the field (Zhu et al., 2004). Hence failing to use appropriate survey methodology for quantifying spatial patterns may result in inaccurate or even misleading information (Franzen and Peck, 1995). Review of literature reflects upon the evolution of various soil sampling practices suitable for PMC (Kerry et al., 2010).

Grid sampling is a survey method that intends to gather information from an area, avoiding over-dispersed or over-concentrated sampling locations. A study conducted in 2015 over 261 agriculture retailers spread across 33 maize-growing states in the United States observed that 75% of the respondents were offering grid soil sampling services (Erickson and Widmar, 2015). Grid soil sampling is accomplished by dividing a survey area (in this case a maize field) into grids and acquiring one geocoded sample within each grid (Fig. 2). The location of the sample within each grid cell should be assigned randomly or be unaligned to attain the highest quality of the semi-variogram model (Fig. 3). A semi-variogram is a model developed by geo-statisticians that provides information about the distance (*range*) beyond which the soil property is no longer auto-correlated, the variance of the soil property values in the field (*sill*), and the micro-scale variability or measurement error (*nugget*). Semi-variograms enable the interpolation of sampled data to create a continuous surface of parameters of interest. Including randomness in the sampling locations provides a suite of distances (or lags) between sample pairs used to model the semi-variogram. Another important consideration

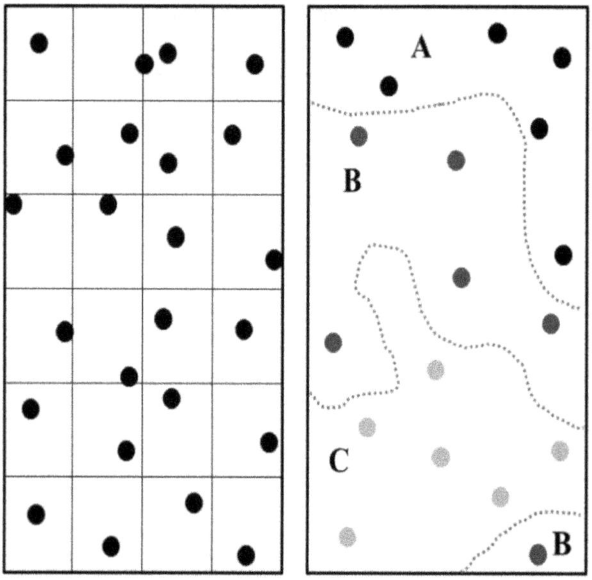

Figure 2 Schematic of two sampling techniques: random within grid (left) and stratified random sampling (right).

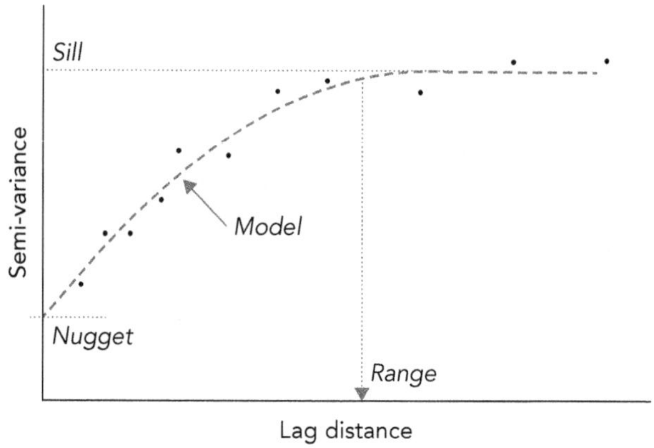

Figure 3 Schematic of a semi-variogram fitted with a spherical model and showing the nugget, the sill and the range.

in designing a sampling plan is to acquire samples at distances that capture spatial dependency existing in the field (Khosla et al., 2010). If samples are acquired at distances larger than the spatial range, such a design would fail to describe the spatial structure of soil properties and semi-variogram would show a pure nugget effect (i.e. a straight

Table 1 Summary of a literature review on the spatial range of selected soil properties along with the corresponding sampling density required to accurately model the spatial pattern

Soil property	Min* (m)	Mean (m)	Max (m)	Sampling density** (sample/ha)	Number of occurrences
Al	262	262	262	0.9	1
Bulk density	106	118	129	4.5	2
Ca	150	206	294	1.5	3
CEC	35	45	55	30.9	2
Clay	18	139	220	3.2	8
Cu	786	786	786	0.1	1
EC	50	98	163	6.6	3
Fe	326	326	326	0.6	1
Mg	257	429	736	0.3	3
Mn	1032	1032	1032	0.1	1
Na	30	36	42	48.2	2
Nitrogen (N)	65	77	89	10.5	2
N Uptake	24	66	90	14.2	3
pH	35	163	320	2.3	12
Phosphorus (P)	30	108	294	5.3	7
Potassium (K)	35	165	574	2.3	7
Sand	6	138	421	3.3	9
Silt	73	106	131	5.6	4
Soil depth	79	156	232	2.6	2
Soil moisture	79	119	200	4.4	5
SOM	45	169	360	2.2	11
SWC at 10 cm	143	143	143	3.1	1
SWC at 20 cm	201	201	201	1.5	1
SWC at 50 cm	139	139	139	3.2	1
SWC at 80 cm	53	53	53	22.2	1
SWC at 120 cm	53	53	53	22.2	1
SWC at 150 cm	37	37	37	45.7	1
Texture	70	70	70	12.8	1
Yield	17	124	305	4.0	9
Zn	137	137	137	3.3	1

* Caution should be taken when comparing spatial range computed with different models because the model (e.g. Gaussian or spherical) has a great influence on the spatial range value (R. Reich, pers. comm.).
** Sampling density is calculated by using the mean spatial range to create a grid and assuming that sampling grid should be 0.4 times the spatial range (Kerry et al., 2010) and assuming a systematic sampling pattern.
Min= Minimum, Max=Maximum.

line indicating no relationship between semi-variance and sampling distance). A 'rule of thumb' for determining soil sample interval suggests collecting samples at distances that are about 0.4 times the spatial range (Kerry et al., 2010). The spatial range of soil properties is both property-specific and site-specific. According to the 2015 survey study, the most common grid soil sampling service (among 72% of agriculture retailers) is to collect one sample per ha (Erickson and Widmar, 2015). Soil survey companies and farmers have adopted this sampling density because grid soil sampling at higher resolution is cost prohibitive in a commercial farming context (Khosla et al., 2010). Table 1 summarizes the spatial range and sampling density of various soil properties reported in literature, which indicates that a 1 ha grid size is too coarse to accurately model the spatial pattern of most soil properties.

2.1.1 Stratified soil sampling

A soil sampling method that is often preferred over random or grid soil sampling for a better resource-use efficiency (e.g. time, labour and cost) is stratified soil sampling (Robert, 2002). This soil sampling technique, also referred to as 'smart sampling', employs pre-existing or known sources of variability (e.g. knowledge about history of the field, map of soil electrical conductivity, field topography, bare soil imagery of the field, yield map, etc.) to guide the assignment of soil sampling locations in the field (Fig. 2). Ancillary information is used to identify and delineate homogeneous areas within the field (Pennock et al., 2007). In theory, decreasing the variability of the surveyed area would logically reduce the number of observations required to accurately and reliably characterize the area (R. Reich, pers. comm.). For the same area, it is thus possible to produce a map with the same accuracy and reliability as random within grid but with fewer samples (Shaner et al., 2008).

In summary, soil sampling, testing and analysis continues to be the best way to acquire information about soil for plant nutrition decisions in particular. When surveying a field to create a map, it is important to account for anticipated spatial range of the soil property being mapped to successfully characterize the field in un-sampled areas. Knowledge of geo-statistics is important to create reliable maps and take full advantage of acquired soil samples (Khosla et al., 2010). Knowledge of the field can further help delineate strata and reduce the amount of samples required.

2.2 Management zones

A widely accepted definition of 'management zone' is an area (contiguous or not) of a crop field that possesses homogenous soil and landscape properties influencing the targeted management practice (Doerge, 1999; Schepers et al., 2004; Moral et al., 2010; Khosla et al. 2010). A review of literature by Khosla et al. (2010), and updated for the purpose of this chapter, reported that management zones have been used for various crop management practices such as fertilization (e.g. N, P, K, Ca, Mg and Fe), irrigation, manure amendment, weed control, pest control and tillage (Table 2). The most frequently used data layers for the purpose of delineating management zones are soil electrical conductivity, grain yield maps, field micro-topography, soil texture and soil organic matter (SOM) content (Table 8.1; Khosla et al., 2010). Depending on the purpose of management zones, a variety of information can be used for delineation. The use of management zones capturing within-field soil variability to divide the field into new management units is a proven

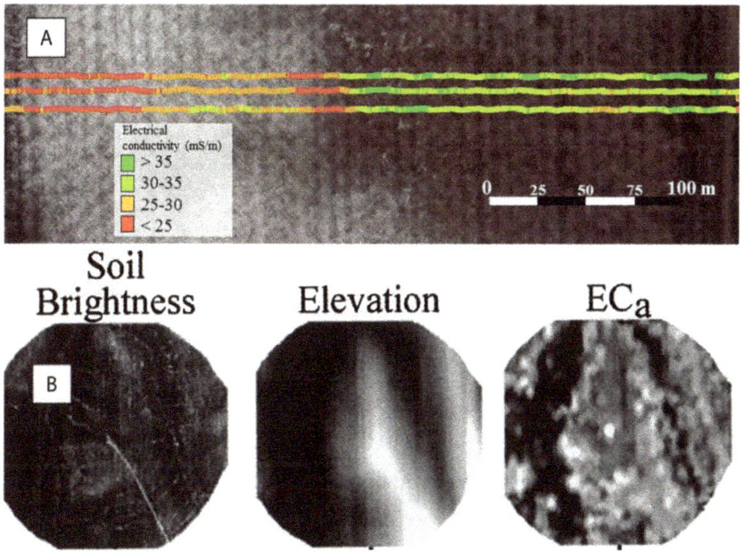

Figure 4 Example of a field in Ault, Colorado, where bare soil satellite imagery (red, green and blue: RGB) and electrical conductivity reflect similar spatial patterns in the field (A). Example of three different types of data from the same field showing similar spatial pattern (From Schepers et al., 2004) (B).

approach to implement PMC techniques for enhancement of productivity, profitability and environmental stewardship (Khosla et al., 2002; Koch et al., 2004, Roudier et al., 2010).

In order to be useful for crop management, zones must display significant agronomic differences for the targeted yield-limiting factor (Schepers et al., 2004; Lawes and Robertson, 2011). If, in a particular field, zones are difficult to delineate (e.g. no clear limit between the zones), are highly fragmented and/or show little agronomic differences, delineating management zones for such fields may not be the best approach for PMC. Indeed, management zones are meant to highlight agronomic differences existing from one area to another in a field and thus to address spatial macro-variability. Research has shown that the use of site-specific management zones that are simple to develop, functional and economically feasible increases the potential of adoptability by farmers (Li et al., 2007). The absence of consensus on the best data layers to be used for the purpose of delineating zones may potentially be explained by the differences that exist from one site to another and to the similarity in spatial patterns that often exists among the varied types of data used (Fig. 4). Indeed, it has often been observed that when clear management zones exist in a field, these can be delineated the same way using different sources of information (e.g. several examples exist where bare soil imagery shows similar zones as electrical conductivity; see Fig. 4; Derby et al., 2007).

2.2.1 Apparent electrical conductivity

Apparent electrical conductivity (EC_a) is a soil property that can be characterized at high sampling density in a relatively short period and at relatively low cost. For these

Table 2 A list of properties used in various techniques of delineating site-specific management zones

Category		Properties used to delineate management zones	Number of occurrences
I. Soil properties			[118]*
	Chemical	Soil organic matter and soil carbon	17
		Nitrogen	13
		Phosphorous	11
		Potassium (6), magnesium (2) and calcium (3)**	11
		Soil pH	6
		Cation exchange capacity	4
		Grid sampling	2
		Gypsum requirement	1
		Targeted sampling	1
	Physical	Soil texture	23
		Soil type	7
		Soil colour	6
		Soil moisture content	5
		Water content/holding capacity	5
		Bulk density	2
		Aggregate stability in water test	1
		Hard pan	1
		Penetration resistance	1
		Soil depth	1
II. Landscape attributes			[30]
		Topography	24
		Aspect	2
		Curvature	2
		Other	2
III. Crop properties			[39]
		Yield map (spatial)	25
		Yield map (temporal)	9
		Shoot density	3
		Ground-based leaf area index (LAI)	1
		Protein content (wheat)	1

Table 2 continued

Category	Properties used to delineate management zones	Number of occurrences
IV. Sensing		[65]
	Soil electrical conductivity sensor	31
	Satellite imagery and other platforms	16
	Normalized differential vegetative index (NDVI)	13
	Digital photography (crop canopy)	4
	Ultrasonic sensing	1
V. Management practice		[4]
	No tillage, chisel plough, with and without traffic	1
	Tillage depth	1
	Tillage force	1
	Weeding, mulching and traffic aisles	1
VI. Weed and pest management		[5]
	Weed population	3
	Soya bean cyst nematode densities	1
	Silverleaf whitefly population	1
VII. Subjective approach		[7]
	Self-made zones for soil management	7
VIII. Modelling		[3]
	Crop/simulation/GIS	3
Total		271

*Total of each category is shown in brackets.
** Number of occurrences for each property is in parenthesis.

reasons, EC_a mapping is currently offered by 14% of the 261 agriculture retailers surveyed and such offering is forecast to double in the next few years (Erickson and Widmar, 2015). The two dominant factors influencing electrical conductivity are the soil solution volumetric fraction and its conductivity, the soil solution being the only conducting phase in the soil (Friedman, 2005). The soil properties affecting the soil solution's conductivity include but are not limited to salinity, clay content and type, cation exchange capacity, organic matter and soil temperature (Corwin and Lesch, 2005). Soil salinity is considered a dynamic soil property and thus EC_a maps acquired in saline soils (i.e. electrical conductivity of the saturate extract (EC_e) of 4 dS/m or higher) are considered questionable for the purpose of delineation of zones as compared to EC_a data acquired for non-saline soils (i.e. $EC_e < 4$ dS/m; Johnson et al., 2005). On the other hand, in non-saline soils, soil texture is the main factor influencing soil solution conductivity and thus is likely more stable in time and hence more adapted for management zones delineation (Corwin and Lesch, 2005). Management zones delineated using EC_a have shown significant differences in maize yield (Johnson et al., 2003; Kitchen et al., 2003), SOM, texture (Cambouris et al., 2006; Shaner et al., 2008), phosphorus and pH (Cambouris et al., 2006). While soil EC_a mapping appears as a powerful technique for characterizing maize fields, caution should be exercised on the timing of data acquisition with EC_a. Mapped EC_a data should be used as a guide (Brevik et al., 2006) and validated with soil survey, historical productivity and local agronomic knowledge (Grisso et al., 2005).

2.2.2 Yield maps

GPS-enabled digital yield maps illustrate grain yield harvested from every square meter of a field (Khosla and Flynn, 2008). Yield map is a sum total reflection of all 'yield-influencing' biotic and abiotic factors and thus represents an integrative layer of high-resolution spatial information about the field (Bunselmeyer and Lauer, 2015). When coupled with historical yield maps of the same field, it provides a reflection of temporal variability in the field and often serves as a logical data layer to delineate productivity potential management zones (Hopkins, 2010). While it may appear logical to use yield maps to assess spatial and temporal variability in crop productivity, research has indicated that delineating zones based solely on yield maps is not necessarily reliable (Hornung et al., 2006). For example, depending on the total precipitation received during a crop-growing season, spatial patterns in yield map can change significantly every year (Schepers et al., 2004). A zone highlighting lower elevation areas within a field may show the lowest yield in a particularly wet year because of waterlogging in that area. Conversely, the same area of the field may correspond to the highest yield in a relatively dry year because of availability of soil moisture in that area. While single-year yield maps may not be a consistently reliable data layer for delineating soil-specific management zones, multiple years of data from the same field can alleviate such limitation and assist with delineation of zones. In fact, it can be argued that the temporal variability of yield is often more important than the spatial variability (Whelan and McBratney, 2000; Bunselmeyer and Lauer, 2015). Bunselmeyer and Lauer (2015) observed that 4 to 24 years of yield maps are required to detect consistent sub-field yield patterns, depending on tillage and crop rotations. It is thus suggested to use yield maps with caution for the purpose of delineation of management zones and characterizing patterns over several years (e.g. more than four years).

2.2.3 Topography

The use of topography for delineating management zones is intuitive because of its influence on the spatial distribution (i.e. erosion and deposition process) of several yield-limiting factors such as surface (A) horizon thickness, nitrogen, phosphorus, SOM, clay and water content (Verity and Anderson, 1990; Kravchenko and Bullock, 2000; Thelemann et al., 2010). Maize yield under rain-fed environments is often negatively correlated with variability in field elevation (Kravchenko and Bullock, 2000). Geographic information system software enables terrain analysis of digital elevation models and respective computation of several topographic attributes such as slope, aspect, flow accumulation, distance to flow accumulation, wetness index or sediment transport index among others (Marques da Silva and Silva, 2008). Topography has been successfully used to delineate zones for N management in maize (Khosla et al., 2002; Koch et al., 2004; Jaynes et al., 2011). It is important to consider that the vertical accuracy of the position obtained with global navigation satellite systems (GNSS) is roughly half of the horizontal accuracy (Hofmann-Wellenhof et al., 2007). Therefore, when surveying a field for micro-topography, it is more appropriate to use high-resolution tools such as real-time kinematic positioning systems or LIDAR as opposed to non-differential GNSS, especially if the range of elevation within a field is small.

2.2.4 Remote sensing

Remote sensing is a compelling tool for PMC because it is non-destructive, it can be acquired on large areas and it provides fundamental biophysical information beyond human vision among other advantages (Jensen, 2009). Bare soil reflectance imagery has been reported in literature as a data layer for delineating management zones (Khosla et al., 2002). Bare soil image is correlated with different soil properties such as surface texture, SOM, salinity and moisture, among others (Weidong et al., 2002; Metternicht and Zinck, 2003; Mzuku et al., 2005). Certain considerations should be accounted for when using bare soil imagery to characterize soil properties for agricultural use, notably imagery acquired from soil surface may not reflect soil properties in the root zone and certain surface conditions affect apparent soil reflectance (e.g. bedding, tillage or presence of crop residues) (Barnes et al., 2003). For this reason, a model developed for one site may not be representative of another site. It is thus recommended to use remote sensing data in conjunction with other sources of information when delineating management zones to achieve the best results. Another consideration is the use of 'no-till' cultivation practice that is increasing (35.5% of the cropland in the United States in 2009 was managed under 'no till'), which may render impractical the use of bare soil imagery for management zones delineation (Horowitz, 2011).

2.3 Characterizing spatial soil variability at the pre-planting stage

There is a comprehensive body of literature that exists on soil spatial variability and on how to characterize it in management zones using electrical conductivity, yield maps, topography, remote sensing and other sources of information. This section has discussed the current state of knowledge on spatial variability of soil in maize fields and different measures that can be undertaken to characterize this variability at the pre-planting stage of the crop-growing season. It is important to understand that assessing and characterizing spatial soil variability and translating it into actionable information remains one of the main challenges of PMC today.

3 Planting

3.1 Precision planting

Planting is one among the most important operations regulating the success of any row crop cultivation including maize. Over the years, farmers, crop practitioners and scientists have come to realize that the precision with which the seed is planted can have a great impact on the resulting yield at the end of the season. Maize cultivation has greatly evolved since seeds were hand planted in the field. Planting is now accomplished with great precision, handling seeds one by one, and placing them at a specific depth and spacing. From seed depth (Molatudi and Mariga, 2009) to stand uniformity (Andrade and

Abbate, 2005) and from planting density (Sangoi, 2001) to planting configuration (Robles et al., 2012), all have varying degrees of influence on final yields.

The phenological plasticity of maize is limited as compared to other crops such as soybean (*Glycine max* (L.) Merr.), and this increases the importance of precision planting (Vega et al., 2001). For instance, plant skip on the maize row can result in as much as 27% yield loss (Doerge et al., 2015). The plant skip may be due to germination problems or planter's malfunction. Therefore, the seed placement is important to ensure the highest germination and emergence rate. Maize planter equipment manufacturing companies understand the importance of this operation and hence offer a variety of tools that enable farmers to plant with a precision never attained before in the history of maize planting (Li et al., 2016) at such a large scale. Yet, certain aspects of seed placement that impact yield have not yet been addressed by seeder development engineers. One example is the angle in which the seed is placed in the soil. A study conducted in Oklahoma concluded that the orientation of the maize seed could result in up to 463 kg/ha yield increase as compared to random orientation placement (Torres 2012). Increased yield was explained by improved light interception and better radiation-use efficiency. A team of engineers from the same institution have demonstrated that despite challenges such as variable kernel shape within and across maize hybrids, an orientation mechanism could be successfully added to a planter row unit to achieve significantly higher rate of leaf azimuth perpendicular to crop row as compared to random orientation (Koller et al., 2013). It is likely that such features that increase planting precision will be added to commercial planters in the near future.

Variable-rate seeding consists of varying the maize seed rate across a field, based on soil physical and chemical properties such that it matches with the capacity of soil to supply optimal nutrition and moisture to the crop. It is based on observations that soil properties vary in space. Several factors may influence optimal seed rate such as yield goal, soil fertility, soil depth, soil type, climate zone, hybrid, hybrid maturity time, seed germination rate, and irrigation, among others (Roozeboom et al., 2007). Because of the complexity of factors and interactions among factors that influence relationship between maize seed rate and grain yield, it is unlikely that a single PMC technique will enable the development of a prescription for seed maps for every field. It is rather a thorough understanding of this phenomenon paired with on-site trials that will enable farmers and crop practitioners to achieve profitable variable-rate seeding.

3.2 Case study

In a study conducted in Colorado on variable-rate maize seeding, it was observed that the optimal seed rate increases with corresponding yield potential (Fig. 5). These results seem to indicate that where conditions were conducive to higher yield, the optimal plant population was higher and where conditions were less conducive to yield, the optimal plant population was lower. These observations are consistent with results obtained by Doerge et al. (2015). Across the three years of this study (2011, 2012 and 2013), there was change in the scale of variability across low, medium and high yield classes ranging from as little as 1.25t/ha in 2011 to as high as 4.4t/ha in 2013 (Fig. 5). However, the trend in yield classes at each geo-location in the field were consistent. Interestingly further investigation of corresponding selected soil properties of the high and low yield classes did not show significant agronomic differences, which further illustrates the complexity of delineating management zones for the purpose of variable-rate seeding.

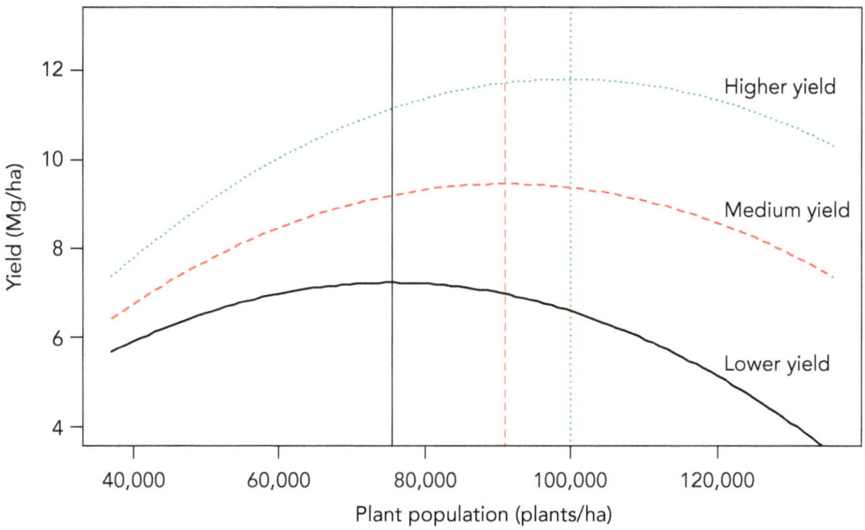

Figure 5 Relationship between yield and plant population for three yield categories. Optimal plant population is highlighted by the vertical lines for each yield category.

4 Nutrition

If there is one aspect of PMC that has received the highest amount of attention among scientists and practitioners, it is nutrition. This is also where the largest opportunity for PMC lies, especially nitrogen (N) (Robert, 2002). N is the most important nutrient supplied to maize to reach full yield potential (Miao et al., 2007) and, on the other hand, excess N leaching in the biosphere is an environmental concern (Rockström et al., 2009). PMC offers the possibility to optimize N rate at every location of a field to reach its highest yield with lowest impact on the environment (Raun et al., 2002). Nitrogen management for maize cultivation should aim at replenishing the inorganic N pool to meet maize N needs without exceeding the crop requirement, thus leading to a balance between the N input and the maize N uptake (Shanahan et al., 2008; Ma and Biswas, 2015). Since the late 1990s, several PMC techniques have been studied to improve N management.

4.1 Nitrogen management

4.1.1 Management zones

The concept of management zones has been described previously. In 2015, 49% of the 261 respondents of a survey of agricultural dealers mentioned that they were offering soil sampling by zone (Erickson and Widmar, 2015). These zones represented either soil mapping units, yield, electrical conductivity or other factors. Soil sampling by management zones is projected to increase in coming years. There is no consensus on the techniques to delineate management zones for precision N management in maize and numerous layers of information can contribute to the delineation of zones (Khosla et al., 2010). A group of scientists have developed an open access software program called Management Zones Analyst (MZA), which provides a quick and automated procedure to assist with delineating management zones (Fridgen et al., 2004). The MZA software input consists of any quantitative data available for a specific field (e.g. topographic attributes, electrical conductivity, yield, imagery, etc.). The MZA software integrates all sources of information, performs a cluster validity function to produce the fuzziness performance index and performs normalized classification entropy to assist in choosing the most appropriate number of zones in a field (Fridgen et al., 2004).

A case study, conducted in Colorado on maize, classified fields into zones of high, medium and low productivity potential based on bare soil imagery, topography and farmer's historical knowledge about the field (Khosla et al., 2002; Koch et al., 2004; Hornung et al., 2006). Zones delineated with this method exhibited significant differences in soil properties (Mzuku et al., 2005). Once delineated, these zones were managed based on their productivity potential by synchronizing variable application of N and manure (Moshia et al., 2014, 2015). Results have shown that application of nutrients using site-specific management zones in maize cultivation enhanced nutrient-use efficiency, nutrient uptake, grain yield, net $ return and reduction in environmentally sensitive nutrient loads into soil on a field-scale basis (Khosla et al., 2002; Koch et al., 2004; Hornung et al., 2006). Variable-rate manure application using management zones showed that it is possible to improve soil fertility and consequently yield of low productivity potential management zones (Moshia et al., 2015).

The use of management zones is among the oldest PMC techniques because it is intuitive to maize producers, who often can tell a high from a low-yielding area in their fields. Yet, the use of management zones by maize producers is still marginal because of the difficulty to divide fields into management units that have significant agronomic differences (Ma and Biswas, 2015). For this reason, soil and crop scientists still need to research technologies and techniques that will standardize, simplify and increase the reliability of management zones delineation providing actionable information for economically and environmentally optimum nutrient management.

4.1.2 Proximal soil sensing

Proximal soil sensing is a technique that uses field-based sensors to obtain a signal when the sensor is in contact with or at a short distance (within 2 m) from the soil (Viscarra Rossel et al., 2011). While proximal soil-sensing data may contain larger variance or bias than conventional laboratory data, the use of these sensors is cheaper, faster and simpler, which enables the acquisition of large amount of data over extended areas. Proximal soil sensors thus allow the creation of soil maps in a timely manner (Viscarra Rossel et al., 2010). Multiple proximal soil-sensing techniques have been investigated to enable high-resolution mapping of soil properties (Viscarra Rossel et al., 2011; IUSS, 2016). While some sensors are stationary, others can be used on mobile platforms commonly referred to as 'on-the-go', which provide the best efficiency for soil mapping. The proximal soil sensors use electrical and electromagnetic, optical and radiometric, mechanical, acoustic, pneumatic and electrochemical measurement concepts (Adamchuk et al., 2004). A suite of 'on-the-go' proximal soil sensors are commercially available, notably soil reflectance sensors (OpticMapper, Veris Technologies, Salina, KS; Lund and Maxton, 2011), electromagnetic induction sensors (EM38-MK2, Geonics Limited, Mississauga, Canada), electrical resistivity with contact electrodes (Veris 3100, Veris Technologies, Salina, KS) and electrochemical sensors (Soil pH Manager, Veris Technologies, Salina, KS; Schirrmann et al., 2011), among others. These on-the-go soil sensors offer new possibilities for more thorough soil survey executed in a shorter amount of time.

A case study conducted in eight fields across Midwestern states of the United States evaluated a proximal multi-sensor (i.e. optical, electrochemical, electrical and GPS) platform (Kweon, 2012). This system was equipped with an optical sensor for soil reflectance measurement (>200 readings ha^{-1}), ion selective electrodes for pH measurement (10 to 20 readings ha^{-1}) and coulter electrodes for EC_a measurement (>200 readings ha^{-1}). Kweon (2012) demonstrated that the root mean squared error of prediction of this system was 2.78 g kg^{-1} for SOM, 1.20 $cmol_c$ kg^{-k} for CEC, and 0.26 for pH, which was significantly below the range of these soil properties in these fields. This high-resolution field characterization exhibited spatial variability at a range that was shorter than the range of commercial soil sampling density (e.g. 1 sample ha^{-1}), which further supports the need for on-the-go soil-sensing systems.

4.1.3 Crop canopy sensing

It has been argued that the use of soil-based information alone may not be sufficient to significantly improve N-use efficiency in maize cultivation because it does not account for crop variability (Shanahan et al., 2008). The idea behind crop canopy sensing is to capture the in-season 'health status' of the plant and apply crop inputs to remediate spots in the fields where crop health status is below par. Crop sensors are used to detect changes in crop canopy before they are visible to the naked eye (Schepers et al., 2006) and at a much faster pace and larger scale than a human could achieve. Spectral properties of the crop canopy highlight the leaf content of numerous pigments such as chlorophyll (green), carotenoids (orange) and anthocyanin (purple), which reflect the physiological status of the plant (Gitelson et al., 2006). To some extent, the optical properties of the crop canopy are the result of all factors influencing plant growth and physiological status, and thus, can be used in a decision support system (DSS) as a good integrator of information on the crop (Samborski et al., 2009). For that matter, crop canopy sensing is a powerful tool for PMC (Fig. 6). In an early field study, Markwell et al. (1995) reported a strong relationship ($r^2 = 0.96$) between maize leaf chlorophyll content and leaf optical properties. Multiple factors can result in decreased leaf chlorophyll concentration (i.e. chlorosis), notably dehydration, flooding, freezing, ozone, herbicides, competition, disease, insects and nutrient deficiency, among others (Carter and Knapp, 2001). Hence, the leaf optical properties can be similar for a deficiency in nitrogen, iron, sulphur, magnesium and manganese (Chappelle et al., 1992; Peñuelas et al., 1994; Blackmer et al., 1996). However, in general, chlorosis is related to plant nutrition, notably nitrogen, magnesium and iron, and nitrogen is by far the most limiting nutrient in maize production (Wann, 1930; Davis and Westfall, 2009). Therefore, N availability is probably the most prominent factor influencing within field leaf chlorophyll content (Blackmer and Schepers, 1995). Soil nitrogen levels change at short range (Table 1) in a crop field and crop canopy sensors allow characterization of micro-variability in crop N status (Fig. 7.). However, caution must be exercised in discerning crop canopy signature that is associated with N variability versus other soil or crop properties.

Multiple sensor packages are now commercially available to detect in-season changes in crop canopy health status. The large majority of commercial sensors use reflectance, which is the ratio of the intensity of reflected radiation to that of incident radiation. Two types of sensors exist for on-the-go N fertilization in maize: the sprayer boom-mounted sensors and the tractor roof-mounted sensors. Examples of sprayer boom-mounted sensors include the GreenSeeker (Trimble Navigation Ltd., Sunnyvale, CA), the CropCircle (Holland

Figure 6 Variability map of a circular field in eastern Colorado planted in maize where spatial variability in crop canopy colour is apparent.

Scientific, Lincoln, NE) and the OptRx (AgLeader Technology, Ames, IA). Examples of tractor roof-mounted sensors include the Yara N-sensor (Yara UK Limited, Grimsby, UK) and the CropSpec (Topcon Precision Agriculture, Livermore, CA). In addition, there are a few handheld sensors that are commercially available such as GreenSeeker Handheld by Trimble Inc. All of these sensors exploit the large difference in reflectance existing between the red and the near-infrared wavebands of the maize reflectance spectrum. By using the normalized difference vegetation index (NDVI) and other vegetation indices involving wavebands within the red to near-infrared wavebands, these sensors can detect a stressed canopy (Cao et al., 2015).

In order to provide actionable information, sensor readings are processed with an algorithm and converted into N recommendations. However, several factors such as maize hybrid, seasonal variation, water status, irradiation or growth stage can influence the actual N requirement of a plant despite a similar sensor output value (Samborski et al., 2009). Several solutions can improve the decision making by accounting for those sources of error. The most common strategy is to use a 'N-rich strip' or a reference area within the field where non-limiting amount of pre-plant N fertilizer has been applied to reach maximum health status considering local conditions. This reference area provides the highest sensor values that can be expected from the sensor in a specific day for a specific field. By comparing sensor values from other locations to sensor values from the N-rich reference area, it is possible to provide a more accurate N recommendation (Teal et al., 2006; Dellinger et al., 2008; Solari et al., 2008; Bausch and Brodahl, 2011). A variation of this technique consists in establishing ramp calibration strips in the field or stepped application rate (e.g. from 0 to 280 kg N/ha in increments of 20 kg) of pre-plant N across the field (Raun et al., 2008). By sensing crop canopy that received different levels of pre-plant N, it is possible to determine the lowest N rate required to achieve maximum canopy production and optimal grain yield. This technique has the advantage of providing detailed information of maize N needs in a particular field and in a particular season. Both the reference strip and the ramp calibration strip techniques require an additional operation of pre-plant N application. To avoid this additional operation, another technique called 'virtual reference strip' has been developed. This concept was developed by Holland (2009) and consists of establishing transects through the field prior to application of fertilizer. Crop canopy information is acquired over these reference transects for field-specific calibration of N prescription algorithm. While this requires an additional operation prior to fertilizer application, it is considered less time and resources consuming than establishing a reference area within each field. These three normalization procedures enable the N recommendation algorithm to account for various sources of error and thus should be considered when implementing precision maize fertilizer application based on crop canopy sensing.

It is important to use the right tool to implement precision maize fertilizer application based on crop canopy sensing. Certain sensors such as the GreenSeeker were originally developed for wheat (*Triticum aestivum* L.) production (Rutto and Arnall, 2004) and thus may not be fully adapted to maize production at early growth stages. Indeed, the GreenSeeker has two attributes that benefit its use for wheat production to the expense of its benefit for use in maize at early growth stages. The first attribute is that its field of view is wide (i.e. 60 cm). In wheat, this is an advantage because it provides data over a larger crop canopy area. However, in maize, it results in a large portion of the field of view covering soil rather than crop canopy early in the season, resulting in more noise in the sensor output. The second attribute is that it uses the peak of chlorophyll absorption spectrum in the red band to calculate the NDVI. This makes the sensor more sensitive to small variations of leaf chlorophyll content and more powerful to detect small changes. However, this also causes the NDVI to saturate when irradiation in the red bands are completely absorbed (red reflectance = 0) by the canopy. For these reasons, the GreenSeeker sensor is recommended for use in maize between the growth stages V8 (8-leaf) and V12 (12-leaf). Prior to V8, there is too much noise and after V12, the NDVI saturates. In practice, farmers start applying in-season N fertilizer from the V3–V4 (3-, 4-leaf) growth stage of maize, which makes the GreenSeeker sensor non-adapted for this practice. It must be noted, however, that though the handheld version of the GreenSeeker sensor, also referred to as 'pocket sensor', lacking precision in comparison with the standard GreenSeeker sensor, is a robust tool to implement N fertilizer rate algorithms based on crop sensing (Crain et al., 2012). Such handheld sensors are gaining popularity in small-scale resource-poor farming environments in the developing world (Crain et al., 2012).

4.1.4 Fusion of information

Fast computer processing capabilities are now readily accessible. Hence, on-the-go fusion of information is becoming a realistic and promising way of conducting variable-rate nutrient management. The fusion of information is a common way of delineating management zones, which often integrate more than one data layer (Khosla et al., 2010). While these data layers often account for soil properties, it would be logical to also account for crop properties (Shanahan et al., 2008; Khosla et al., 2010). In a study integrating soil-based management zones with crop-based active canopy sensing to direct N application in maize, Roberts et al. (2012) found that sensor-based algorithm may benefit from adjustment for each management zones. In other words, they found that a crop sensor-based algorithm may be suitable for medium productivity levels, but not for high and low productivity levels.

A case study conducted in Colorado in 2015 aimed at comparing three different N management practices in maize: (1) a farmer's approach which consists of a uniform application, (2) a variable-rate N application approach based on management zones and (3) a variable-rate N application approach based on crop sensing adapted for each management zone (MZRS). Management zones were delineated using bare-soil imagery, topography and historical yield. Remote sensing (NDVI) data was acquired at the V8 (8-leaf) growth stage of maize using a GreenSeeker sensor platform. For N application, farmer's approach received a flat rate of N; the management zone approach received a higher rate of N on high productivity potential management zones and a lower rate of N on low productivity management zones. For the MZRS approach, the N rates were linked to NDVI values acquired within each management zones and were ramped up (within high productivity zone) or down (within low productivity zone). Results showed that the MZRS precision N management approach increased NUE and profitability and maintained yield while reducing N loadings in the biosphere (Fig. 7). It is thus logical to hypothesize that significant improvements in precision N management can be achieved by adjusting N rate algorithms for each zone independently.

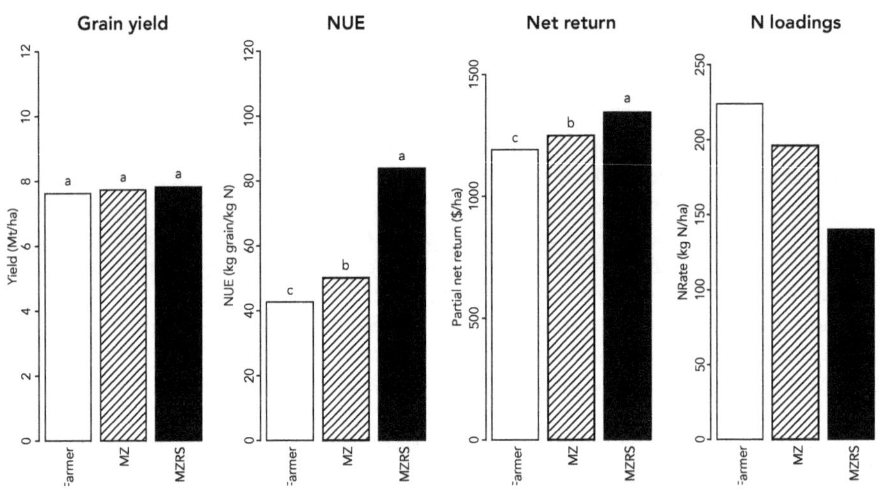

Figure 7 Bar graphs of four different parameters (grain yield, nitrogen use efficiency or NUE; net return and nitrogen loadings) that were evaluated in this study against three unique approaches of precision N management: farmer's uniform N management (Farmer), variable-rate N management based on management zones (MZ) and variable-rate N management based on remote sensing within management zones (MZRS). Different letters indicate significant (α = 0.05) differences among treatments.

4.1.5 Decision support systems

Another way to merge information from multiple sources is to develop complex models integrating weather data, soil characteristics and crop growth models under the form of a decision support system (DSS). With the increasing number of meta-data analysis applied to agriculture and the precision in soil, crop and weather monitoring, farmers now have access to reliable DSSs for nutrient management in maize. AmaizeN is a DSS developed in New Zealand that helps farmers planning N fertilizer application by integrating crop production and environmental impact components and providing estimates of crop yield, N-fertilizer application rate and economic returns (Li et al. 2009). Another tool called SCAN – standing for Soil, Crops and Atmosphere for Nitrogen – enables site-specific N rate recommendations. The SCAN system builds on multiple years of field trials in Ontario and Quebec (Canada) and on meta-analysis of the soil and weather parameters' influence on maize N response (Tremblay et al. 2012). While the SCAN system integrates complex datasets (e.g. satellite imagery and weather data) and uses advanced analysis approaches (e.g. fuzzy interference), it provides a simple interface for farmers to use with minimal inputs (Hmielowski, 2016). The Nutrient Expert (NE) is a DSS, developed by the International Plant Nutrition Institute for smallholder farmers of Asia and Africa, that integrates information on site characteristics, climate, price of farm inputs and crop produce to provide N recommendations (Pampolino et al., 2012). A case study conducted over 82 farmer's fields, two seasons and two tillage systems in southern India compared productivity and economic return of N management using three different approaches: NE, farmer's practices and state recommendations (Satyanarayana et al., 2013). Figure 8 illustrates the potential improvement in yield production when using a DSS over conventional approaches.

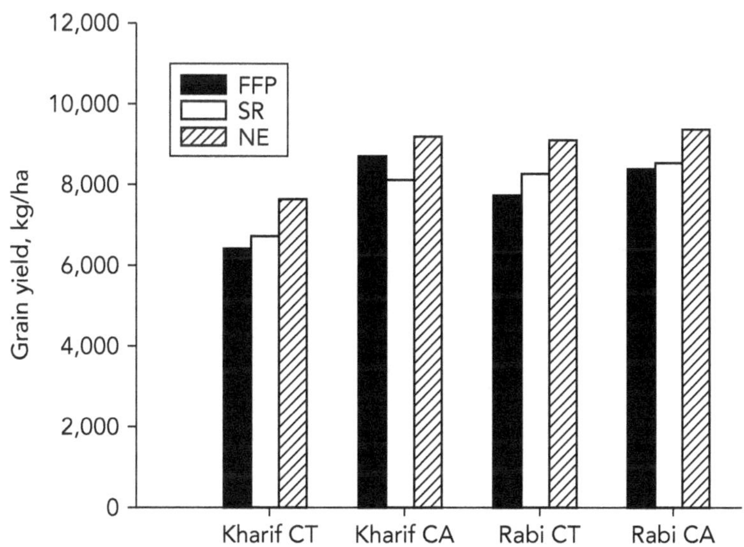

Figure 8 Grain yield comparison between farmer's practice (FFP), state recommendations (SR) and Nutrient Expert (NE) over two seasons (Kharif/Fall and Rabi/Spring) and over two tillage systems (conventional (CT) and conservation (CA)). Adapted from Satyanarayana et al. 2013.

4.1.6 Other nutrients (P, K, micro-elements, lime)

While variable-rate N management is an important PMC technique, spatial management of other nutrients has also been reported to be beneficial. Similar to N, studies have found that within-field spatial variability of phosphorus, potassium and micronutrients (e.g. Mn, B) can affect maize yield (Robert et al., 1999; Kravchenko and Bullock, 2000; Sawchik and Mallarino, 2007; Tekin et al., 2011). It is thus logical to think that PMC techniques accounting for P, K and micronutrients could be beneficial. A case study conducted in a maize field in Belgium in 2007 investigated the possibility of using an on-the-go visible and near-infrared soil sensor mounted in the front of the planter to predict soil-extractable P content and modulate starter P rate on-the-go (Maleki et al., 2008). It was found that the applied P rate was 5% lower using this variable-rate technique as compared to a uniform rate and that the yield was 4% higher using the variable-rate technique. This indicates that PMC techniques aiming to manage other nutrients than N may also increase productivity while reducing nutrient loadings in the biosphere.

4.2 Temporal synchrony

Another important aspect of precision nutrient management in maize cultivation is the synchrony between the crop nutritional needs and the timing of nutrient application (Fig. 9). It has been argued that among the most important factors to close the yield gap between potential yield and actual yield is to apply nutrients with better synchrony between N fertilizer application and maize plant uptake during both the vegetative and reproductive stages (Ciampitti and Vyn, 2014). Although it seems logical to think that supplying N just prior to rapid N uptake by the plant increases productivity due to little

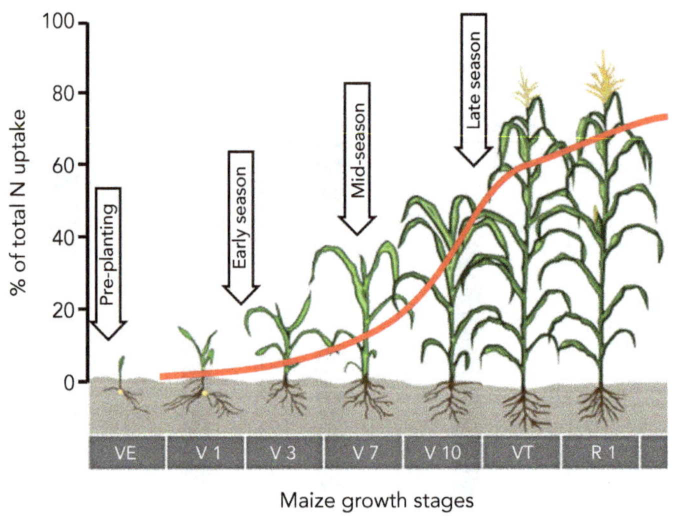

Figure 9 Illustration of synchrony between maize N uptake (red line) and N fertilizer application (arrows) from pre-planting to the reproductive stage. Adapted from University of Illinois Extension, 1999 and University of Missouri, Plant Science Extension 2006.

N loss through leaching and denitrification, few scientific evidences support this idea (Zebarth et al., 2009). In fact, the relationship between N application timing and yield seems to be greatly influenced by soil properties and weather conditions (Gehl et al., 2005; Viswakumar et al., 2008). For instance, a sandy soil exposed to frequent precipitations will benefit more from split N fertilizer application than a finer texture soil in a dryer climate. Splitting the N budget in half remains a coarse approach considering the smooth and gradual changes in maize N uptake along the crop-growing season (Abendroth et al., 2011). Few studies have compared the effect of 'spoon feeding' maize (i.e. frequent applications of small amounts of N) to a single or split N application (Lamm et al., 2001; Lamm and Trooien, 2003).

5 Precision maize irrigation

Maize is a water-exhaustive crop, consuming 500 to 800 mm of water depending on climate (FAO, 2016). In order to deliver sufficient amounts of water to the crop, farmers often apply higher irrigation rates than crop water needs, thus insuring that water is not a limiting factor for maximum yield. Certain irrigation systems such as flooding or furrow irrigation are inherently inefficient (40–80% efficiency) for delivering water to crop, further reducing the water productivity (Howell, 2003). With high energy cost and increasing water scarcity in irrigated agricultural areas, farmers are looking for alternative ways to increase water-use efficiency. Certain strategies have been used to increase the efficiency of flooding and furrow irrigation systems, notably precision field levelling, surge irrigation and reuse of irrigation water runoff (Perlman, 2016). More advanced irrigation systems such as drip irrigation or spray irrigation give a higher level of precision for water delivery to the plants (70–95% efficiency) and thus increase water productivity (Howell, 2003).

One aspect of precision irrigation that has received significant attention over the years and has enabled improvement in irrigation efficiency is irrigation scheduling. DSSs for irrigation scheduling help farmers take two important decisions: *How much water to apply?* and *When to apply water*? (Rhoads and Yonts, 1991). Irrigation scheduling for maize often use a water balance approach accounting for the water present in the soil and the crop water requirement. This system aims at initiating irrigation only when water stored in the soil and precipitations cannot meet the crop water requirements (Andales et al., 2011).

Sensor-based irrigation scheduling is an approach that has mostly been used in horticultural crops and is beginning to interest maize producers (Krupnik et al., 2017). This precision cultivation technique is based on soil moisture sensors estimating soil water content at a high frequency (e.g. every 15 minutes) and transferring the data over the Internet to either initiate irrigation or inform the farmer about the water status of the soil. Using these systems, farmers can maintain soil water content in an optimal zone to avoid droughts and/or over-saturation of the soil profile. Sensor-based irrigation scheduling is currently the most advanced precision irrigation technique that is commercially available.

In recent years, new opportunities have become available for increasing water productivity. A component of precision irrigation that has received little attention in scientific literature is the spatial variability of crop water needs and available soil water. When characterizing a field with strong topographic features, it is easy to envision that lower areas of the field may have more water available to the crop than relatively higher areas of the field. Such a spatially variable environment may justify implementation of variable-rate irrigation system. However, what may not be as straightforward is the spatial variability in soil water content that exists in levelled fields. A case study conducted in Colorado assessed the spatial variability of soil water content in two precision levelled maize fields (Longchamps et al., 2015). The study consisted of measuring soil water content with neutron probes at a high density (about 9 locations per hectare), at five depths ranging from 15 to 135 cm and across the entire crop-growing season of maize. The major conclusions of this study were (i) spatial variability of soil water content existed in both fields with certain areas of the field having as high as 300% or more soil water content than other areas, (ii) the scale of spatial variability was much larger in deep soil profile than at the soil surface and (iii) spatial patterns can be stable in time across the crop-growing season. These conclusions suggest that spatial management of maize irrigation water could potentially be beneficial, even in precision levelled fields. Studies conducted in New Zealand have compared variable-rate irrigation to uniform irrigation in maize and have observed water savings from 5 to 21% without impairing grain yield (Hedley et al., 2010a,b).

As scientific research on site-specific irrigation accumulates, and the impacts of precision irrigation are found to be significantly improving water-use efficiency, it is possible that farmers may want to install drip irrigation systems on the basis of soil variability rather than zones that are independent of soil properties. Such a system would allow a high level of control for irrigation volume, timing and placement, and would lead to much higher water-use efficiency.

6 Pesticides

PMC techniques that are commercially available for spot spraying of pesticides are rare. The challenge of spot-spraying pesticides is limited by the ability to reliably map distribution of pest populations in order to achieve a level of 'safety' that would be similar to blanket application of pesticides.

6.1 Weeds

Weeds tend to grow in spatially aggregated patterns across fields, thus leaving weed-free areas that do not require herbicide treatment (Wiles et al., 1992; Cardina et al., 1997; Clay et al., 2006; Longchamps et al., 2016). To achieve site-specific weed management, two broad approaches have been reported in literature: (i) applying herbicide according to a prescription map and (ii) spot-spraying weeds using weed detection sensors. The first approach requires an additional step that consists of mapping weed species and their spatial distribution in the field, followed by herbicide application across the field. The second approach is implemented in a single step and requires the use of sensors that could detect weeds while in motion. This technique mandates use of a direct injection-enabled herbicide sprayer to keep the pure chemical (herbicide) separated from water tank. In both approaches the overall goal is to ensure weed control and to maximize crop yield, while reducing the cost of herbicides and any negative herbicide impact on the environment.

A case study conducted in Quebec, Canada, mapped spatial distribution of annual weeds for 19 site-years on maize fields (Longchamps et al., 2016). It was found that only 6 site-years out of 19 presented a spatially aggregated pattern of weed distribution that was conducive to spot spraying of herbicides. Longchamps et al. (2016) concluded that spot-spraying herbicide may not be profitable in every field and a field pre-evaluation may be required to implement spot spraying only in fields where weeds are spatially aggregated.

Even fields showing spatial aggregation contained a certain level of randomness possibly due to wind-transported seeds and the like, and that a base rate (e.g. 50% of the dose) of herbicide may still be required outside weed patches to provide adequate protection and control from weeds across the entire field.

Commercial spot spraying of weeds in maize fields is not being currently implemented possibly due to reasons mentioned above, and also because of the engineering challenge of distinguishing crop from weeds reliably and efficiently (Longchamps et al., 2010). However, with increasing computer power, resolution of optical sensors and with the advent of unmanned aerial vehicles (UAVs) in agriculture, it is anticipated that this PMC technique will soon become more efficient and reliable.

6.2 Arthropods

One aspect of PMC that is underused is precision pest management. Similar to other soil and crop properties, the spatial distribution of pests often shows a certain level of aggregation (Gent et al., 2004; Davidson et al., 2007; Schwartz et al., 2010). For instance, Ellsbury et al. (1998) characterized spatial distribution of western (*Biabrotica virgifera virgifera* LeConte) and northern (*Biabrotica barberi* Smith & Lawrence) corn rootworm adult emergence and they found that the range of spatial dependence could be 180 to 550 m for western corn rootworm and 172 to 281 m for northern corn rootworm. They suggested that information on spatial distribution of rootworm, coupled with maps of soil properties, could be used to derive site-specific soil insecticide application where pest damages are more likely and where insecticides may be more efficient. It has been demonstrated that treating only the infested or more susceptible portions of a field can reduce pesticide use by 30% or more (Merrill et al., 2011). Scouting fields to produce infestation maps may be cost, labour and time prohibitive. Remote sensing has the potential to accurately map pest infestation because of the distinctive phenotypic characteristics of pest impact on crop canopy such as leaf curling, wilting, chlorosis or necrosis of photosynthetic plant parts; stunted growth; or in some cases reduction in leaf area due to severe defoliation (Prabhakar et al., 2012). Another technique is to delineate management zones related to pest distribution (Davidson et al., 2007). However, in both instances the decision to treat the pest infestation is after the damage has occurred to the crop. Future advances in pest detection would include asymptomatic detection of pest via remote and proximal sensing. This would enable decisions to treat portions of the field that are most susceptible to pest infestation prior to damage to the crop (Martenelli et al., 2014; Lopez-Lopez et al., 2016).

A case study conducted in four fields in Colorado assessed the possibility to control the three insect pests western corn rootworm, *Diabrotica virgifera virgifera* LeConte; European corn borer, *Ostrinia nubilalis* (Hübner); and western bean cutworm, *Richia albicosta* (Smith) using productivity potential management zones (Davidson et al., 2007). They found that larvae and adults of the three insect pests were more common in high-productivity management zones. They concluded that management zones seemed to correlate well with abundance of certain insect pests and might prove useful for managing insects in a site-specific manner.

6.3 Soil applied pesticide degradation

Another aspect of pesticide management that could benefit from PMC techniques is the spatial pattern of soil-applied pesticide bioactivity. Soil properties such as soil organic carbon, clay content, water content and pH may influence the biodegradation and plant absorption of herbicides (Williams 'II' et al., 2001). Consequently, even though soil-applied pesticides are applied uniformly, their efficiency may be spatially variable across the field. A case study conducted in three dryland fields of northeastern Colorado assessed the rate of atrazine and metolachlor dissipation across management zones delineated with shallow (30 cm) electrical conductivity (Bridges et al., 2008; Shaner et al., 2011). It was found that in two out of three fields, atrazine and metolachlor dissipation rate was different across zones. It was found that the atrazine half-life differed by about one day from one zone to another. Although this may not be actionable information, if soil properties were mapped with a higher level of precision (e.g. higher resolution and different soil properties), it is possible that more significant differences could be highlighted and lead to profitable changes in pesticide management. One possible PMC technique is to apply higher doses where pesticide bioactivity may be lower or apply a second dose using spot spraying in these areas.

7 Harvest

Yield mapping for maize production is often the first tool employed by farmers interested in precision agriculture. A yield map created by yield monitoring systems is an easy way to observe spatial variability that exists in a particular field and data collected over years also provide assessment of temporal changes in yield. Yield maps provide a wealth of information on how the crop responds to management practices and more importantly the difference in response from one location of the field to another and over years. It is a direct reflection of management decisions and serves as an invaluable feedback mechanism on economic return. In the 1990s, the first grain yield maps were developed with the hypothesis that the difference in grain yield across the field would be consistent over the years (Blackmore et al., 2003). However, farmers and scientists rapidly observed that it was not the case and realized that a single year of data may not be a good indicator of the long-term trends in yields observed in a field (Panneton et al., 2001). Indeed, multiple factors, other than stable factors such as soil productivity potential, may influence the yield, notably, weed patches, pest infestations, planting operation, emergence rate and climatic conditions among others. It is thus generally accepted that if yield maps are used to characterize the spatial pattern of productivity potential in a field, multiple years of data are required (Panneton and Brouillard 2002; Diker et al., 2004). Nevertheless, a yield map remains a powerful tool to characterize, understand and assess the spatial dynamics existing in maize fields.

A yield map is generated by a yield monitoring system that is composed of a suite of sensors: a GPS, a yield sensor, a grain moisture sensor, a ground speed sensor and a header sensor (Fig. 10). When the combine harvester header is down, the header sensor triggers the yield recording system including the GPS, speed sensor, grain flow meter and grain moisture sensor. Grain is harvested by the header and when it travels through the grain elevator, it is monitored by a grain flow sensor. Different types of grain flow sensors exist, and they all need to be calibrated to accurately report actual grain yield. While in the grain elevator, a sub-sample of the grain is diverted to the grain moisture sensor. Because the grain is monitored in the clean grain elevator, there is a delay (lag) between the time when the grain is harvested from its location in the field, and the time the grain volume

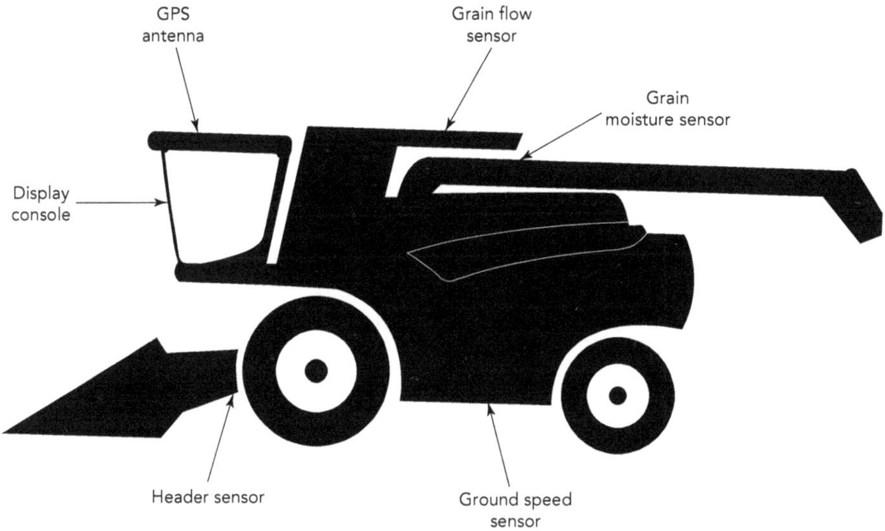

Figure 10 Schematic of the suite of sensors employed in the creation of a maize grain yield map.

is monitored and recorded in the system. This lag and other sources of error must be accounted for prior to analysis of yield map.

To account for possible errors introduced during yield mapping, it is important to clean the yield map. Otherwise, information contained in a yield map may be misleading if used in a decision-making process (Lyle et al., 2013). Khosla and Flynn (2008) developed and described a detailed process to remove the sources or error in yield maps including removing unneeded data, removing low GPS resolution data, correcting for grain moisture, removing excessively low- and high-yield values, adjusting for combine speed variation, adjusting for lag in grain flow and correcting for GPS antenna offset and the like. Lyle et al. (2013) conducted a literature review on yield map post-processing and identified four main sources or error, notably the harvesting dynamics of the combine harvester, the continuous measurement of moisture and yield, the accuracy of positioning information and those induced by the harvester operator. Because of the complexity associated with cleaning a grain yield map, scientists and engineers are developing algorithms to automate this process (Sudduth and Drummond, 2007). Although certain yield mapping software companies have started to embed a yield map post-processing component in their systems, no solution currently exists to perfectly clean a yield map without involving manual editing (Shearer et al., 2004). Calibration of yield monitor and manual editing of yield maps may be a deterrent for using yield maps and a motivation to exploit other potential sources of information for yield mapping such as remote sensing.

Yield estimation by remote sensing is a powerful tool, especially where yield monitoring systems are not available. Satellite imagery now offers the possibility to acquire multiple layers of information in a single season on an entire farm and with sub-meter pixel resolution. Multiple studies have reported high correlation between satellite imagery data and maize grain yield (Shanahan et al., 2001; Chang et al., 2003; Sakamoto et al., 2013). While satellite imagery may contain an error of accuracy for maize grain yield, they can provide enough precision to reliably study the spatial patterns in grain yield across a field.

A calibration process may further improve the accuracy of such maps, but no cleaning and post-processing of the map is required to ensure that the data is valid at every location of the field. This may represent a significant advantage over yield maps generated by yield monitoring systems if one is not ready to meticulously post-process raw yield maps.

8 Data analysis

8.1 Big data

Employing information and communication technologies on farm produces large amounts of data. In one crop cycle, a farmer can obtain more than two million data points from private and public sources for a single field alone (Table 3). While this is a wealth of data, it is a challenge to extract actionable information from such 'big data'. The National Science Foundation defines big data as *large, diverse, complex, longitudinal, and/or distributed data sets generated from instruments, sensors, Internet transactions, email, video, click streams, and/or all other digital sources* (NSF-NIH Interagency Initiative, 2012). It has been argued by many in the media and the agricultural industry that big data and big data analytics hold the key to sustainable intensification of agriculture in the future (Shearer, 2014). Exploiting big data involves multiple steps such as acquisition, information extraction and cleaning, data integration, modelling and analysis, and interpretation and deployment (Jagadish et al., 2014). Because of the complexity of such an integrated approach, the potential benefits from big data rely on analytics. The science of big data analytics applied to agriculture is still at its infancy stage and the broad-scale adoption of big data science in agriculture is hampered by numerous societal questions about data governance, privacy and ownership (Russom 2011; Bertot and Choi, 2013). Nevertheless, big data collection and analysis has potential to be a powerful tool for PMC and help producers make a giant leap towards evidence-based farm management.

> *When you're trying to reach a goal, data not only tells you if you're succeeding, but it also suggests which activities you should do more of in order to improve your results.*
>
> Bill Gates

One potential example of big data benefits for maize production is to help producers take a decision on when to plant, which maize variety, and with how much starter fertilizer, based on data accumulated over the years. For instance, based on past performance of maize hybrids under archived weather conditions and planting date, if a weather pattern (temperature and precipitations) similar to a previous year is recognized, the management practices associated with areas where maize performed the best could be replicated at a larger scale on the farm. In other words, instead of relying solely on best management practices developed in small-scale experiments often conducted far from the farm, farmers could now integrate information collected at a large scale on their own farm in the decision system.

Another component of big data applied to agriculture that has tremendous potential for improving maize cultivation practices is the use of open data systems or big linked data for benchmarking. Benchmarking consists, as it suggests, in comparing on-farm productivity to productivity of farms of the same agricultural context. It can be employed for decision making by farmers, policy makers and companies involved in agri-business

Table 3 Example of dataset collected over an entire cropping cycle for a quarter section circular field (64.7 ha), excluding machinery data, market data and other data such as crop scouting

Period	Data	No. of columns/ bands	Frequency of data acquisition during the crop season	Number of data points/pixel in the field	Total
Pre-planting	Soil samples	30	Once	65	1950
Pre-planting	Bare soil imagery (5m pixel)	5	Once	25,900	129,500
Pre-planting	Electrical conductivity (18 m pass spacing)	6	Once	8,000	48,000
Pre-planting	Elevation	4	Once	8,000	32,000
Planting	As applied map of seeds	15	Once	7,000	105,000
Planting	As applied map of fertilizer	10	Once	7,000	70,000
Fertilizing	As applied side dress N	10	Once	7,000	70,000
Irrigation	Irrigation depth	1	12 times	5,400	64,800
Pest control	As applied pesticides	10	3 times	7,000	210,000
Harvest	Yield map	12	Once	97,500	1,170,000
Entire	Soil moisture from probe	3 (depths)	(14,400) Every 15 min. from May to Sept.	2	86,400
Entire	Weather data	14	4320 (Hourly)	1	60,480
				Total	**2,048,130**

and agro-inputs, banks, insurance companies and service providers (FAO-AIMS, 2016). For instance, open data systems could inform a farmer of logical reasons why yield in certain fields was lower than regional average by comparing differences in management practices. For example, similar management practices were used, but lower yielding field used a different maize hybrid. Big linked data benchmarking could thus inform about best crop management practices, best hybrids, best inputs or best timing of operations to implement for current weather pattern in a certain region based on wealth of data generated by all farms in this region. All in all, big data would integrate G x E x M (genotype by environment by management) interactions across small- and large-scale farming which logically has innumerable benefits to maize production and agriculture in general to be more productive, profitable, efficient and sustainable.

8.2 Economics

The adoption of PMC techniques can be profitable for multiple reasons. This may include but is not limited to higher yields, savings of farm inputs, better quality of the harvested product and reduction in loss of inputs into the biosphere (Heege, 2013). Case studies

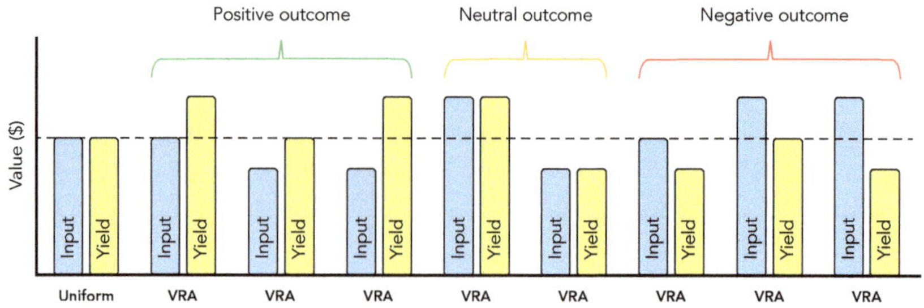

Figure 11 Different possible economic outcomes from the implementation of variable-rate application (VRA) of crop inputs as compared to uniform application. The VRA scenarios are in comparison with the conventional practice as benchmark, where uniform rate of inputs is applied to the crop and yield is realized.

comparing precision agriculture to conventional agriculture found the adoption of PMC techniques to be profitable (Plant et al., 2000; Watson et al., 2003; Koch et al., 2004). However, benefits depend on multiple factors such as crop, annual acres on which soil and crop characterization equipment is used, input cost, crop value and level of variability existing in the field managed with precision among others (Godwin et al., 2003; Griffin and Lowenberg-DeBoer, 2005; Heege, 2013). The economic outcomes due to implementation of variable-rate application of inputs (Fig. 11) can be positive, neutral or negative depending on the situation (Whelan and McBratney, 2000). Lowenberg-DeBoer and Swinton (1997) designed a chart for the economic analysis of on-farm implementation of site-specific crop management. The main decision factor relies on a positive gross margin that is larger than the annual capital cost of field characterization (information) and input application equipment (variable-rate technology). Lowenberg-DeBoer and Swinton (1997) also suggested accounting for net revenue risk and environmental benefits. While the gross margin and the cost of input application equipment are relatively easy to measure, the other factors of the equation (i.e. field characterization, net revenue risk and environmental benefits) are more elusive. The difficulty to get a full picture of the benefits associated with the implementation of PMC techniques may explain the low adoption of this approach by maize producers.

One way to estimate profitability associated with adoption of precision agricultural techniques is to measure profitability at the farm level rather than at the field level. Schimmelpfennig and Ebel (2016) used data from the 2010 USDA's Agricultural Resource Management Survey (1507 respondents), which includes information about the use of information technology on the farm to assess the cost savings from precision agriculture for maize producers. They tested five technologies (i.e. yield monitoring, yield map, auto-steering, GPS and variable-rate technology) used alone or in combinations (six scenarios). They found that on average, variable production cost per acre for maize often decreased when adopting precision agriculture technologies in eight out of eleven cases. Interestingly, Schimmelpfennig and Ebel (2016) suggested that it may be more profitable to adopt precision agricultural technologies in a package (e.g. yield monitoring, yield mapping, auto-steering and variable-rate technology) rather than sequentially as it is often done. Although adoption of PMC techniques may be slower than anticipated, multiple examples of its profitability at the field and at the farm scale exist (Bramley, 2009; Koch et al., 2004).

The 'holy grail' of multiple precision agricultural companies is to devise a system that would map economic return at every square meter of the field (Matthew Powell, AgWorld Inc., Australia, pers. comm.). While this may appear as a simple process in theory, it is difficult to implement because of the error (location error, value accuracy) contained in every data layer and because of the variability in the cost and value of different inputs, outputs and operations. Massey et al. (2008) conducted a case study to investigate the possibility of using long-term (10-year) yield maps for the development of a profit map and identification of profitability zones within a field. They found that profitability maps can be useful to identify locations of the field where net profit was negative for different reasons such as topsoil erosion. They were also able to measure the impact of tree line or drainage ways on profitability. Overall, this case study suggested that profit mapping can be valuable for producers and that profit mapping standards and protocols should be developed by the precision agriculture community (Massey et al., 2008).

9 Future trends

PMC and precision agriculture in general is a sector that is oriented towards the future. Since its early days in the late 1980s, precision agriculture has inspired a futuristic conception of farming. With the advent of more powerful and cheaper technologies, it is now possible to envisage farming in an entirely new way. In this section there is a series of techniques and technologies that are thought to have potential for inclusion in the future of PMC.

9.1 Unmanned aerial vehicles

The need for Unmanned Aerial vehicles (UAVs) in agriculture comes from the cost associated with frequent and detailed crop scouting for pests, nutrient deficiencies and other anomalies (e.g. plugged irrigation nozzle). It is expected that in the near future, UAVs will be employed in multiple operations on a farm such as crop scouting, pest distribution mapping, crop loss assessment and bare soil imagery among others (Ehsani and Maja, 2013). The Federal Aviation Agency's Small UAV Rule is effective since 29 August 2016, which enables commercial implementation of UAV operations on the farm (FAA, 2016). It is anticipated that with increased commercial use, this technology will mature, which will lead to improved design, lower system price, enhanced imagery procedures, looser regulations and more practical information dissemination to farmers (Zhang and Kovacs, 2012).

9.2 Sensor fusion

Sensor fusion consists of integrating information from multiple sensors with the objective of achieving a specific task that cannot be done successfully with a single sensor. For instance, Klose et al. (2008) proposed the use of sensor fusion to distinguish maize from weeds, combining spectral and geometric attributes of the plants. Sensor fusion is also expected to conduct soil surveys that are more thorough by combining multiple sensor types such as optical, electrical and electrochemical among others in a single survey operation (Adamchuk et al., 2011). As the price of sensing technologies decreases, it is logical to observe an increase in sensor fusion systems for maize production.

9.3 Wireless sensor networks

With increasing pressure for enhancing crop water productivity, it is expected to observe an increase in water saving technologies use on the farm. Multiple studies have investigated the potential of wireless sensor network (WSN) to better manage irrigation on farms (Kim et al., 2008; Vellidis et al., 2008). These systems have been found to assist in determining the timing and amount of irrigation water to apply. Other potential use of WSN is scouting for insects. Automatic insect detection traps using machine vision can assist farmers in monitoring pest invasion in maize fields (Loughlin, 2013). It is anticipated that miniaturization of the WSN nodes will further increase the dissemination and the adoption of this technology for multiple applications in maize production (Ruiz-Garcia et al., 2009).

9.4 Internet of things

Internet of things (IoT) can be defined as a network of Internet-enabled components that are linked to a web service enabling interaction between the components (Zhao et al., 2010). For instance, a farm manager can have access to the data from soil moisture probes equipped with cellular antenna. Based on these data, decision could be made to initiate an irrigation event from the farm office. The IoT adds convenience to farm management and enables farmers to accomplish multiple tasks (e.g. initiating irrigation while in office placing orders) simultaneously (Liu et al., 2015). With increasing government regulations in the food industry, it is also anticipated that the IoT will assist with food traceability (Badia-Melis et al., 2015).

9.5 Nanotechnology

PMC involves an increased level of precision timing and placement of crop inputs. For instance, it is long documented that banding starter N fertilizer at 5 cm to the side of the row and 5 cm below the soil surface increases yield compared to broadcast fertilizer (Ritchie et al., 1996). Nanotechnology is a new field of research applied to agriculture that has the potential to take crop input timing and placement precision at a whole new level. Research is being conducted to create capsules that can release N only when it recognizes signals from the plant that fertilizer is needed via a 'smart delivery' system (Solanki et al., 2015). Nanotechnology also has the potential to offer a new generation of pesticides that will be transported to specific sites in the plant, which will enable effective pathogen control using very low doses (Ahmed et al., 2013). The unprecedented level of precision offered by this new technology is expected to contribute to PMC techniques in the future.

9.6 Spatial structure of soil microbiome

As medical science discovers the effects of microbiome on human health, agricultural science also explores the effect of soil microbial diversity on plant growth. It has been documented that soil microbiome may affect productivity by influencing plant hormones, imposition or suppression of diseases, nutrient availability and tolerance to abiotic stresses (Bakker et al., 2015). Soil biome is spatially variable in crop fields both horizontally and vertically (Torsvik et al., 2002) and thus may potentially explain spatial variability existing in crop growth. While the current way to measure organic matter in soils relies on destructive loss-on-ignition protocol, we could perhaps envision that a DNA analysis using PCR be routinely conducted in the future to characterize soil microbiome.

9.7 Integrated systems

As it was outlined in previous sections (i.e. Sections 4.1.4 and 8.2), it is sometimes beneficial to integrate multiple sources of information or multiple techniques to achieve higher profitability. It is anticipated that in the future, PMC techniques and technologies will be integrated in a seamless manner to create a synergy that will allow farmers to increase their efficiency, productivity and profitability (Kitchen, 2008). For instance, on-the-go fertilizer application may include on-the-go crop sensing as well as soil-sensing components integrated with soil maps preloaded in the on-board computer that would allow optimization of N rates based on agronomic, economic and environmental factors.

9.8 On-farm field trials

Newer farm machinery acquired by maize producers in the high-income countries is often equipped with sensors that enable creation of as-applied maps or yield maps, which can potentially be employed in on-farm field trial experiments (Whelan et al., 2012; Schimmelpfennig and Ebel, 2016). Farmers often give higher credit to field trials conducted on their farm or closeby, rightfully arguing that their soils, sun exposure or microclimate, among other factors, are unique to their situation (Joel Schneekloth, Colorado State University Extension, pers. comm.). While it is impractical to conduct on-farm field trials on every farm because of the resources (e.g. time and money) required to do so, it may be advantageous to conduct trials at a larger scale using farm machinery capable of producing maps. For instance, in a field where a farmer has observed large variability, different rates of N or seed could be applied in long experimental strips, and when as-applied map is merged with yield map, it would be feasible to identify optimal rate based on local conditions. Over the years, such on-farm field trials conducted at a large scale with results shared among farmers and coupled with ancillary and weather data could result in precise 'best management practices guidelines' that are specific to each agro-ecosystem of the country. It is anticipated that with larger numbers of farm machinery equipped with sensors, more on-farm trials will be conducted and software will be developed to automatically process, merge and analyse the data collected.

9.9 Robotics

The negative impact of compaction induced by large farm machinery on agricultural fields has reduced the maize productivity and farm profitability (Tim Chamen et al., 2015). Moreover, advances in technology now enable deployment of autonomous machines capable of achieving tasks that used to require time and skills. For instance, research is being conducted to develop farm robots that can control weeds without involving human labour (Slaughter et al., 2008). Developments in robotics applied to agriculture include the design of fleets of robots operating in swarms (Emmi et al., 2014). It is anticipated that robots will be used in maize production to alleviate negative impacts of compaction and increase productivity.

10 Conclusion

As the world's second largest crop in terms of production measured in metric tons, maize cultivation is intimately bound to human's fate on planet earth. PMC techniques can help farmers maintain or increase their productivity while reducing their impact on the biosphere. PMC is an approach that is integrated by nature; involves collection of data, analysis and conversion of data to actionable information; decision making; implementation of farm operations and feedback on the decisions taken. This philosophy of maize production is not new; however, it is now implemented at a level that is often beyond the capacity of a single person. For this reason, it is expected that PMC techniques will be adopted by farmers who are open to new technologies and who are eager to collaborate with agronomists, people from the industry and scientists to achieve a new standard of maize cultivation for the greater good of our planet.

11 References

Abendroth, L. J., R. W. Elmore, M. J. Boyer and S. K. Marlay. Corn Growth and Development. Iowa State University, University Extension, 2011. http://publications.iowa.gov/18932/1/Corn%20growth%20and%20development001.pdf.

Adamchuk, V. I, J. W. Hummel, M. T. Morgan and S. K Upadhyaya. 'On-the-Go Soil Sensors for Precision Agriculture'. *Computers and Electronics in Agriculture* 44, no. 1 (July 2004): 71–91. doi:10.1016/j.compag.2004.03.002.

Adamchuk, Viacheslav, I., Raphael, A. Viscarra Rossel, Kenneth, A. Sudduth and Peter Schulze Lammers. 'Sensor Fusion for Precision Agriculture', 2011. http://www.intechopen.com/books/sensor-fusion-foundation-and-applications/sensor-fusion-for-precision-agriculture.

Ahmed, F., N. Arshi, S. Kumar, S. S. Gill, R. Gill, N. Tuteja and B. Heun Koo. 'Nanobiotechnology: Scope and Potential for Crop Improvement'. In *Crop Improvement Under Adverse Conditions*, edited by Narendra Tuteja and Sarvajeet Singh Gill, 245–69. New York: Springer, 2013. http://link.springer.com/chapter/10.1007/978-1-4614-4633-0_11.

Andales, A. A., J. L. Chávez and T. A. Bauder. Irrigation Scheduling: The Water-Balance Approach. Colorado State University Extension, 2011. http://extension.colostate.edu/docs/pubs/crops/04707.pdf.

Andrade, F. H., and E. Pablo, Abbate. 'Response of Maize and Soybean to Variability in Stand Uniformity'. *Agronomy Journal* 97, no. 4 (2005): 1263. doi:10.2134/agronj2005.0006.

Badia-Melis, R., P. Mishra and L. Ruiz-García. 'Food Traceability: New Trends and Recent Advances. A Review'. *Food Control* 57 (2015): 393–401.

Bakker, M. G., J. M. Chaparro, D. K. Manter and J. M. Vivanco. 'Impacts of Bulk Soil Microbial Community Structure on Rhizosphere Microbiomes of Zea Mays'. *Plant and Soil* 392, no. 1–2 (21 March 2015): 115–26. doi:10.1007/s11104-015-2446-0.

Barnes, E. M., K. A. Sudduth, J. W. Hummel, S. M. Lesch, D. L. Corwin, G. Yang, C. S. T. Doughtry and W. C. Bausch. 'Remote-and Ground-Based Sensor Techniques to Map Soil Properties'. *Photogrammetric Engineering and Remote Sensing* 69, no. 6 (2003): 619–30.

Barnhisel, R. I., M. J. Bitzer, J. H. Grove and S. A. Shearer. 'Agronomic Benefits of Varying Corn Seed Populations: A Central Kentucky Study'. *Precision Agriculture*, no. precisionagricu3 (1996): 957–65.

Bausch, W. C., and M. K. Brodahl. 'Strategies to Evaluate Goodness of Reference Strips for in-Season, Field Scale, Irrigated Corn Nitrogen Sufficiency'. *Precision Agriculture* 13, no. 1 (19 June 2011): 104–22. doi:10.1007/s11119-011-9230-9.

Beadle, G. W. 'Origin of Zea Mays'. In *Origins of Agriculture*, Reed CE., 615–35. The Hague, Netherland: Mouton Publishers, 1977. http://agris.fao.org/agris-search/search.do?recordID=US201302413227.

Beckett, P. H. T., and R. Webster. 'Soil Variability: A Review'. *Soils and Fertilizers* 34, no. 1 (1971): 1–15.
Bertot, J. C., and H. Choi. 'Big Data and E-Government: Issues, Policies, and Recommendations'. In *Proceedings of the 14th Annual International Conference on Digital Government Research*, 1–10. ACM, 2013. http://dl.acm.org/citation.cfm?id=2479730.
Blackmer, T. M., and J. S. Schepers. 'Use of a Chlorophyll Meter to Monitor Nitrogen Status and Schedule Fertigation for Corn'. *Journal of Production Agriculture* 8, no. 1 (1995): 11–12.
Blackmer, T. M., J. S. Schepers, G. E. Varvel and E. A. Walter-Shea. 'Nitrogen Deficiency Detection Using Reflected Shortwave Radiation from Irrigated Corn Canopies'. *Agronomy Journal* 88, no. 1 (1996): 1. doi:10.2134/agronj1996.00021962008800010001x.
Blackmore, S., R. J. Godwin and S. Fountas. 'The Analysis of Spatial and Temporal Trends in Yield Map Data over Six Years'. *Biosystems Engineering, Precision Agriculture – Managing Soil and Crop Variability for Cereals* 84, no. 4 (April 2003): 455–66. doi:10.1016/S1537-5110(03)00038-2.
Bramley, R. G. V. 'Lessons from Nearly 20 Years of Precision Agriculture Research, Development, and Adoption as a Guide to Its Appropriate Application'. *Crop and Pasture Science* 60, no. 3 (2009): 197–217.
Brevik, E. C., T. E. Fenton and A. Lazari. 'Soil Electrical Conductivity as a Function of Soil Water Content and Implications for Soil Mapping'. *Precision Agriculture* 7, no. 6 (12 October 2006): 393–404. doi:10.1007/s11119-006-9021-x.
Bridges, M., W. B. Henry, D. L. Shaner, R. Khosla, P. Westra and R. Reich. 'Spatial Variability of Atrazine and Metolachlor Dissipation on Dryland No-Tillage Crop Fields in Colorado'. *Journal of Environment Quality* 37, no. 6 (2008): 2212. doi:10.2134/jeq2007.0568.
Bunselmeyer, H. A., and J. G. Lauer. 'Using Corn and Soybean Yield History to Predict Subfield Yield Response'. *Agronomy Journal* 107, no. 2 (2015): 558. doi:10.2134/agronj14.0261.
Cambouris, A. N., M. C. Nolin, B. J. Zebarth and M. R. Laverdière. 'Soil Management Zones Delineated by Electrical Conductivity to Characterize Spatial and Temporal Variations in Potato Yield and in Soil Properties'. *American Journal of Potato Research* 83, no. 5 (September 2006): 381–95. doi:10.1007/BF02872015.
Cao, Q., Y. Miao, J. Shen, W. Yu, F. Yuan, S. Cheng, S. Huang, H. Wang, W. Yang and F. Liu. 'Improving in-Season Estimation of Rice Yield Potential and Responsiveness to Topdressing Nitrogen Application with Crop Circle Active Crop Canopy Sensor'. *Precision Agriculture* 17, no. 2 (12 August 2015): 136–54. doi:10.1007/s11119-015-9412-y.
Cao, Q., C. Zhenling, X. Chen, R. Khosla, T. H. Dao and Y. Miao. 2012. 'Quantifying Spatial Variability of Indigenous Nitrogen Supply for Precision Nitrogen Management in Small Scale Farming'. *International Journal of Precision Agriculture* 13 (2012): 45–61.
Cardina, J., G. A. Johnson and D. H. Sparrow. 'The Nature and Consequence of Weed Spatial Distribution'. *Weed Science* 45, no. 3 (1997): 364–73.
Carter, G. A., and A. K. Knapp. 'Leaf Optical Properties in Higher Plants: Linking Spectral Characteristics to Stress and Chlorophyll Concentration'. *American Journal of Botany* 88, no. 4 (1 April 2001): 677–84.
Chang, J., D. E. Clay, K. Dalsted, S. Clay and M. O'Neill. 'Corn (Zea Mays, L.) Yield Prediction Using Multispectral and Multidate Reflectance'. *Agronomy Journal* (2003). http://agris.fao.org/agris-search/search.do?recordID=US201300940460.
Chappelle, E. W., M. S. Kim and J. E. McMurtrey III. 'Ratio Analysis of Reflectance Spectra (RARS): An Algorithm for the Remote Estimation of the Concentrations of Chlorophyll A, Chlorophyll B, and Carotenoids in Soybean Leaves'. *Remote Sensing of Environment* 39, no. 3 (March 1992): 239–47. doi:10.1016/0034-4257(92)90089–3.
Ciampitti, I. A., and T. J. Vyn. 'Understanding Global and Historical Nutrient Use Efficiencies for Closing Maize Yield Gaps'. *Agronomy Journal* 106, no. 6 (2014): 2107. doi:10.2134/agronj14.0025.
Clay, D. E., S. A. Clay and K. Reitsma. Best Management Practices for Corn Production in South Dakota. South Dakota State University, College of Agriculture and Biological Sciences, AgBio Communications, 2009.
Clay, S. A., B. Kreutner, D. E. Clay, C. Reese, J. Kleinjan and F. Forcella. 'Spatial Distribution, Temporal Stability, and Yield Loss Estimates for Annual Grasses and Common Ragweed (Ambrosia

Artimisiifolia) in a Corn/soybean Production Field over Nine Years'. *Weed Science* 54, no. 2 (1 March 2006): 380–90.

Corwin, D. L., and S. M. Lesch. 'Apparent Soil Electrical Conductivity Measurements in Agriculture'. *Computers and Electronics in Agriculture, Applications of Apparent Soil Electrical Conductivity in Precision Agriculture* 46, no. 1–3 (March 2005): 11–43. doi:10.1016/j.compag.2004.10.005.

Crain, J., I. Ortiz-Monasterio and B. Raun. 'Evaluation of a Reduced Cost Active NDVI Sensor for Crop Nutrient Management'. *Journal of Sensors* (2010). http://dx.doi.org/10.1155/2012/582028.

Davidson, S., F. Peairs and R. Khosla. 'Insect Pest Densities Across Site-Specific Management Zones of Irrigated Corn in Northeastern Colorado'. *Journal of Economic Entomology*, 100 (2007): 781–9.

Davidson, S. A., F. B. Peairs and R. Khosla. 'Insect Pest Densities Across Site-Specific Management Zones of Irrigated Corn in Northeastern Colorado'. *Journal of Economic Entomology* 100, no. 3 (1 June 2007): 781–9. doi:10.1093/jee/100.3.781.

Davis, J. G., and D. G. Westfall. Fertilizing Corn. Colorado State University Extension, 2009. http://www.ext.colostate.edu/pubs/crops/00538.html.

Dellinger, A. E., J. P. Schmidt and D. B. Beegle. 'Developing Nitrogen Fertilizer Recommendations for Corn Using an Active Sensor'. *Agronomy Journal* 100, no. 6 (2008): 1546. doi:10.2134/agronj2007.0386.

Derby, N. E., F. X. M. Casey and D. W. Franzen. 'Comparison of Nitrogen Management Zone Delineation Methods for Corn Grain Yield'. *Agronomy Journal* 99, no. 2 (2007): 405. doi:10.2134/agronj2006.0027.

Diker, K., D. F. Heermann and M. K. Brodahl. 'Frequency Analysis of Yield for Delineating Yield Response Zones'. *Precision Agriculture* 5, no. 5 (2004): 435–44.

Doerge, T. 'Defining Management Zones for Precision Farming'. *Crop Insights* 8, no. 21 (1999): 1–5.

Doerge, T., K. O'Bryan and D. Gardner. 'Putting Variable-Rate Seeding to Work on Your Farm'. Field Facts, Pioneer Hi-Bred, Johnston, IA., 6, no. 12 (2006).

Doerge, T., M. Jeschke and P. Carter. 'Planting Outcome Effects on Corn Yield'. *Crop Insights* 25, no. 1 (2015): 1–7.

Duvick, D. N. 'The Contribution of Breeding to Yield Advances in Maize (Zea Mays, L.)'. edited by BT – *Advances in Agronomy* 86: 83–145. Academic Press, 2005. http://www.sciencedirect.com/science/article/pii/S006521130586002X.

Ehsani, R., and J. M. Maja. 'The Rise of Small UAVs in Precision Agriculture'. *Resource Magazine* 20, no. 4 (2013): 18–19.

Ellsbury, M. M., W. D. Woodson, S. A. Clay, D. Malo, J. Schumacher, D. E. Clay and C. G. Carlson. 'Geostatistical Characterization of the Spatial Distribution of Adult Corn Rootworm (Coleoptera: Chrysomelidae) Emergence'. *Environmental Entomology* 27, no. 4 (1 August 1998): 910–17. doi:10.1093/ee/27.4.910.

Emmi, L., M. Gonzalez-de-Soto and P. Gonzalez-de-Santos. 'Configuring a Fleet of Ground Robots for Agricultural Tasks'. In ROBOT2013: First Iberian Robotics Conference, edited by Manuel, A. Armada, Alberto Sanfeliu, and Manuel Ferre, 505–17. Advances in Intelligent Systems and Computing 252. Springer International Publishing, 2014. http://link.springer.com/chapter/10.1007/978-3-319-03413-3_37.

Erickson, B., and D. A. Widmar. 'Precision Agricultural Services Dealership Survey Results'. Sponsored By CropLife Magazine And The Center For Food And Agricultural Business. Purdue University, 2015. http://agribusiness.purdue.edu/files/file/2015-crop-life-purdue-precision-dealer-survey.pdf.

FAA. 'Unmanned Aircraft Systems', 2016. https://www.faa.gov/uas/.

FAO. 'FAO – Water Development and Management Unit – Crop Water Information: Maize', 2016. http://www.fao.org/nr/water/cropinfo_maize.html.

FAO-AIMS. 'Big Data: Unlocking the Future for Agriculture!' ORG. FAO IMS Activity, 2016. http://aims.fao.org/activity/blog/big-data-unlocking-future-agriculture.

Federico M., R. Scalenghe, S. W. Davino and A. M. Dandekar. 'Advanced Methods of Plant Disease Detection'. *A Review in Agronomy for Sustainable Development* 35, no. 1 (2014):1–25.

Fleming, K. L., D. G. Westfall, D. W. Wiens, L. E. Rothe, J. E. Cipra and D. F. Heermann. 'Evaluating Farmer Developed Management Zone Maps for Precision Farming'. *Precision Agriculture acsesspublicati*, no. precisionagric4a (1999): 335–43. doi:10.2134/1999.precisionagproc4.c29.

Franzen, D. W., and T. R. Peck. 'Field Soil Sampling Density for Variable Rate Fertilization'. *Journal of Production Agriculture* 8, no. 4 (1995): 568–74.

Fridgen, J. J., N. R. Kitchen, K. A. Sudduth, S. T. Drummond, W. J. Wiebold and C. W. Fraisse. 'Management Zone Analyst (MZA)'. *Agronomy Journal* 96, no. 1 (2004): 100–8.

Friedman, S. P. 'Soil Properties Influencing Apparent Electrical Conductivity: A Review'. *Computers and Electronics in Agriculture, Applications of Apparent Soil Electrical Conductivity in Precision Agriculture* 46, no. 1–3 (March 2005): 45–70. doi:10.1016/j.compag.2004.11.001.

Gehl, R. J., J. P. Schmidt, L. D. Maddux and W. B. Gordon. 'Corn Yield Response to Nitrogen Rate and Timing in Sandy Irrigated Soils'. *Agronomy Journal* 97, no. 4 (2005): 1230. doi:10.2134/agronj2004.0303.

Gent, D. H., H. F. Schwartz and R. Khosla. 2004. Distribution and incidence of iris yellow spot virus in Colorado and its relation to onion plant populations and yield. *Journal of Plant Diseases* 88 (5): 446–52.

Gitelson, A. A., G. P. Keydan and M. N. Merzlyak. 'Three-Band Model for Noninvasive Estimation of Chlorophyll, Carotenoids, and Anthocyanin Contents in Higher Plant Leaves'. *Geophysical Research Letters* 33, no. 11 (1 June 2006): L11402. doi:10.1029/2006GL026457.

Godwin, R. J., T. E. Richards, G. A. Wood, J. P. Welsh and S. M. Knight. 'An Economic Analysis of the Potential for Precision Farming in UK Cereal Production'. *Biosystems Engineering* 84, no. 4 (2003): 533–45.

Griffin, T. W., and J. Lowenberg-DeBoer. 'Worldwide Adoption and Profitability of Precision Agriculture Implications for Brazil'. *Revista de Politica Agricola* 14, no. 4 (2005): 20–37.

Grisso, R. D., D. L. Holshouser, M. M. Alley and W. E. Thomason. 'Precision Farming Tools. Soil Electrical Conductivity', 1 September 2005. https://vtechworks.lib.vt.edu/handle/10919/51377.

Havlin, J. L., J. D. Beaton, S. L. Tisdale and W. L. Nelson. *Soil Fertility and Fertilizers: An Introduction to Nutrient Management*. no. Ed.6 (1999): x + 499p.

Hedley, C. B., S. Bradbury, J. Ekanayake, I. J. Yule and S. Carrick. 'Spatial Irrigation Scheduling for Variable Rate Irrigation'. In *Proceedings of the NZ Grasslands Association* 72 (2010): 97–102.

Hedley, C. B., S. Bradbury, J. Ekanayake, I. J. Yule, S. Carrick et al. 'Spatial Irrigation Scheduling for Variable Rate Irrigation'. In *Proceedings of the New Zealand Grassland Association* 72 (2010): 97–102. New Zealand Grassland Association. http://www.growsmart.com/stuff/contentmgr/files/0/06e853467ac748d7dcbc564b0a16d9cd/pdf/spatial_irrigation_scheduling_for_variable_rate_irrigation.pdf.

Heege, H. J. *Precision in Crop Farming Site Specific Concepts and Sensing Methods: Applications and Results. Springer eBook Collection*. Springer, New York, 2013. https://login.ezproxy.csupueblo.edu/login?url=http://dx.doi.org/10.1007/978-94-007-6760-7.

Hofmann-Wellenhof, B., H. Lichtenegger and E. Wasle. *GNSS – Global Navigation Satellite Systems: GPS, GLONASS, Galileo, and More*. Springer Science & Business Media, 2007.

Holland, K. Sensor-Based Chemical Management for Agricultural Landscapes. US20090007485 A1, filed 3 July 2008, and issued 8 January 2009. http://www.google.com/patents/US20090007485.

Hopkins, J. 'USDA Economic Research Service – Tailored Reports: Crop Production Practices', 2010. http://www.ers.usda.gov/data-products/arms-farm-financial-and-crop-production-practices/tailored-reports-crop-production-practices.aspx.

Hornung, A., R. Khosla, R. Reich, D. Inman and D. G. Westfall. 'Comparison of Site-Specific Management Zones'. *Agronomy Journal* 98, no. 2 (2006): 407. doi:10.2134/agronj2005.0240.

Horowitz, J. *No-till Farming Is a Growing Practice*. DIANE Publishing, 2011.

Hove, L., N. Mashingaidze, S. J. Twomlow, P. Nyathi, M. Moyo, W. Mupangwa and E. Masvaya. 'Micro Doses, Mega Benefits Promoting Fertilizer Use in Semi-Arid Zimbabwe'. Monograph, 2008. http://oar.icrisat.org/423/.

Howell, T. A. 'Irrigation Efficiency'. *Encyclopedia of Water Science*. Marcel Dekker, New York, 2003, 467–72.

Hmielowski, T. 'Translating Research Findings into Practice.' *Crops, Soils, Agronomy News* 61, no. 9 (2016): 4–6.

Inman, D., R. Khosla and T. Mayfield. 'On-the-Go Active Remote Sensing for Efficient Crop Nitrogen Management'. *Sensor Review* 25, no. 3 (1 September 2005): 209–14. doi:10.1108/02602280510606499.

IUSS. 'IUSS Working Group on Proximal Soil Sensing Techniques'. ProximalSoilSensing.org, 2016. http://www.proximalsoilsensing.org/techniques/.

Jagadish, H. V., J. Gehrke, A. Labrinidis, Y. Papakonstantinou, J. M. Patel, R. Ramakrishnan and C. Shahabi. 'Big Data and Its Technical Challenges'. *Communications ACM* 57, no. 7 (July 2014): 86–94. doi:10.1145/2611567.

Jaynes, D. B., T. C. Kaspar and T. S. Colvin. 'Economically Optimal Nitrogen Rates of Corn: Management Zones Delineated from Soil and Terrain Attributes'. *Agronomy Journal* 103, no. 4 (2011): 1026–35.

Jensen, J. R. *Remote Sensing of the Environment: An Earth Resource Perspective 2/e*. Pearson Education India, 2009.

Johnson, C. K., R. A. Eigenberg, J. W. Doran, B. J. Wienhold, B. Eghball and B. L. Woodbury. 'Status of Soil Electrical Conductivity Studies by Central State Researchers'. *Transactions of the ASAE* 48, no. 3 (2005): 979–89.

Johnson, C. K., D. A. Mortensen, B. J. Wienhold, J. F. Shanahan and J. W. Doran. 'Site-Specific Management Zones Based on Soil Electrical Conductivity in a Semiarid Cropping System'. *Agronomy Journal* 95, no. 2 (2003): 303–15.

Kerry, R., M. A. Oliver and Z. L. Frogbrook. 'Sampling in Precision Agriculture'. In *Geostatistical Applications for Precision Agriculture*, 35–63. Springer, New York, 2010. http://link.springer.com/chapter/10.1007/978-90-481-9133-8_2.

Khosla, R., K. Fleming, J. A. Delgado, T. M. Shaver and D. G. Westfall. 'Use of Site-Specific Management Zones to Improve Nitrogen Management for Precision Agriculture'. *Journal of Soil and Water Conservation* 57, no. 6 (1 November 2002): 513–18.

Khosla, R., and B. Flynn. 'Understanding and Cleaning Yield Monitor Data'. *Soil Science Step-by-Step Field Analysis*, no. soilsciencestep (2008): 113–30.

Khosla, R., D. G. Westfall, R. M. Reich, J. S. Mahal and W. J. Gangloff. 'Spatial Variation and Site-Specific Management Zones'. In *Geostatistical Applications for Precision Agriculture*, 195–219. Springer, New York, 2010. http://link.springer.com/chapter/10.1007/978-90-481-9133-8_8.

Kim, Y., R. G. Evans and W. M. Iversen. 'Remote Sensing and Control of an Irrigation System Using a Distributed Wireless Sensor Network'. *IEEE Transactions on Instrumentation and Measurement* 57, no. 7 (2008): 1379–87.

Kitchen, N. R., S. T. Drummond, E. D. Lund, K. A. Sudduth and G. W. Buchleiter. 'Soil Electrical Conductivity and Topography Related to Yield for Three Contrasting Soil–crop Systems'. *Agronomy Journal* 95, no. 3 (2003): 483–95.

Kitchen, N. R. 'Emerging Technologies for Real-Time and Integrated Agriculture Decisions'. *Computers and Electronics in Agriculture, Emerging Technologies For Real-time and Integrated Agriculture Decisions* 61, no. 1 (April 2008): 1–3. doi:10.1016/j.compag.2007.06.007.

Klose, R., A. Ruckelshausen, M. Thiel and J. Marquering. 'Weedy–a Sensor Fusion Based Autonomous Field Robot for Selective Weed Control'. In 66th International Conference Agricultural Engineering/AgEng, Stuttgart-Hohenheim, VDI-Verlag, Conference Proceedings, 167–72, 2008. https://www.hs-osnabrueck.de/fileadmin/HSOS/Homepages/COALA/Veroeffentlichungen/2008-VDI-Weedy.pdf.

Koch, B., R. Khosla, W. M. Frasier, D. G. Westfall and D. Inman. 'Economic Feasibility of Variable-Rate Nitrogen Application Utilizing Site-Specific Management Zones'. *Agronomy Journal* 96, no. 6 (2004): 1572–80.

Koller, A. A., R. K. Taylor, P. R. Weckler, M. D. Buser and W. R. Raun. 'Design, Performance Prediction, and Validation of a Seed Orienting Corn Planter'. In 2013 Kansas City, Missouri, 21 July to 24

July 2013, 1. American Society of Agricultural and Biological Engineers, 2013. http://elibrary.asabe.org/abstract.asp?aid=43491.

Kravchenko, A. N., and D. G. Bullock. 'Correlation of Corn and Soybean Grain Yield with Topography and Soil Properties'. *Agronomy Journal* 92, no. 1 (2000): 75. doi:10.2134/agronj2000.92175x.

Krupnik, T. J., U. Schulthess, Z. A. Ahmed and A. J. McDonald. Sustainable crop intensification through surface water irrigation in Bangladesh? A geospatial assessment of landscape-scale production potential. *Land Use Policy* 60 (2017): 206–22.

Kweon, G. 'Toward the Ultimate Soil Survey: Sensing Multiple Soil and Landscape Properties in One Pass'. *Agronomy Journal* 104, no. 6 (2012): 1547. doi:10.2134/agronj2012.0156.

Lamm, F. R., T. P. Trooien, H. L. Manges and H. D. Sunderman. 'Nitrogen Fertilization FOR Subsurface Drip–Irrigated Corn'. *Transactions of the ASAE* 44, no. 3 (2001): 533.

Lamm, F. R., and T. P. Trooien. 'Subsurface Drip Irrigation for Corn Production: A Review of 10 Years of Research in Kansas'. *Irrigation Science* 22, no. 3–4 (9 September 2003): 195–200. doi:10.1007/s00271-003-0085-3.

Lawes, R. A., and M. J. Robertson. 'Whole Farm Implications on the Application of Variable Rate Technology to Every Cropped Field'. *Field Crops Research* 124, no. 2 (2011): 142–8.

Li, F. Y., P. R. Johnstone, A. Pearson, A. Fletcher, P. D. Jamieson, H. E. Brown and R. F. Zyskowski. AmaizeN: A decision support system for optimizing nitrogen management of maize. *NJAS-Wageningen Journal of Life Sciences* 57, no. 1 (2009): 93–100.

Li, Y., Y. Bingxin, C. Tao, Y. Yiming, H. Xiantao, L. Quanwei, L. Zhijie, Y. Xiaowei and Z. Dongxing. 'Global overview of research progress and development of precision maize planters.' *International Journal of Agricultural and Biological Engineering* 9, no. 1 (2016): 9.

Li, Y., Z. Shi, F. Li and H.-Y. Li. 'Delineation of Site-Specific Management Zones Using Fuzzy Clustering Analysis in a Coastal Saline Land'. *Computers and Electronics in Agriculture* 56, no. 2 (April 2007): 174–86. doi:10.1016/j.compag.2007.01.013.

Li, Y., Y. Bingxin, C. Tao, Y. Yiming, H. Xiantao, L. Quanwei, L. Zhijie, Y. Xiaowei and Z. Dongxing. 'Global overview of research progress and development of precision maize planters.' *International Journal of Agricultural and Biological Engineering* 9, no. 1 (2016): 9.

Liu, Y., H. Wang, J. Wang, K. Qian, N. Kong, K. Wang, L. Zheng, Y. Shi and D. W. Engels. 'Enterprise-Oriented IoT Name Service for Agricultural Product Supply Chain Management'. *International Journal of Distributed Sensor Networks* 7 (January 2015): doi:10.1155/2015/308165.

Longchamps, L., R. Khosla, R. Reich and D. W. Gui. 'Spatial and Temporal Variability of Soil Water Content in Leveled Fields'. *Soil Science Society of America Journal* 79, no. 5 (2015): 1446. doi:10.2136/sssaj2015.03.0098.

Longchamps, L., B. Panneton, R. Reich, M.-J. Simard and G. D. Leroux. 'Spatial Pattern of Weeds Based on Multi-Species Infestation Maps Created by Imagery'. *Weed Science* 64, no. 3 (2016).

Longchamps, L., B. Panneton, G. Samson, G. Leroux and R. Thériault. 'Discrimination of Corn, Grasses and Dicot Weeds by Their UV-Induced Fluorescence Spectral Signature'. *Precision Agriculture* 11, no. 2 (Avril 2010): 181–97. doi:10.1007/s11119-009-9126-0.

Loughlin, D. 'Developments in the World of Insect Detection'. *International Pest Control* 55, no. 2 (April 2013): 88–90.

Lowenberg-DeBoer, J. Economics of Variable Rate Planting for Corn. Purdue University, Department of Agricultural Economics, 1998. http://ageconsearch.umn.edu/bitstream/28685/1/sp98-02.pdf.

Lowenberg-DeBoer, J., and S. M. Swinton. 'Economics of Site-Specific Management in Agronomic Crops'. *The State of Site-Specific Management for Agriculture acsesspublicati*, no. thestateofsites (January 1997): 369–96. doi:10.2134/1997.stateofsitespecific.c16.

Lund, E. D., and C. R. Maxton. 'Proximal Sensing of Soil Organic Matter Using the Veris® OpticMapperTM'. In 2nd Global Workshop on Proximal Soil Sensing, Montreal, Quebec, Canada, 15–19 2011. http://adamchukpa.mcgill.ca/gwpss/Papers/GWPSS_2011_Lund.pdf.

Lyle, G., B. A. Bryan and B. Ostendorf. 'Post-Processing Methods to Eliminate Erroneous Grain Yield Measurements: Review and Directions for Future Development'. *Precision Agriculture* 15, no. 4 (20 November 2013): 377–402. doi:10.1007/s11119-013-9336-3.

Lyon, D. 'How Do Plant Populations Affect Yield?' Press Releases from Panhandle Research and Extension Center, 1 January 2009. http://digitalcommons.unl.edu/panpressrel/19.

Ma, B.-L., and D. K. Biswas. 'Precision Nitrogen Management for Sustainable Corn Production'. In *Sustainable Agriculture Reviews*, edited by Eric Lichtfouse and Aakash Goyal, 33–62. Sustainable Agriculture Reviews 16. Springer International Publishing, 2015. http://link.springer.com/chapter/10.1007/978-3-319-16988-0_2.

Maleki, M. R., A. M. Mouazen, B. D. Ketelaere, H. Ramon and J. D. Baerdemaeker. 'On-the-Go Variable-Rate Phosphorus Fertilisation Based on a Visible and near-Infrared Soil Sensor'. *Biosystems Engineering* 99, no. 1 (January 2008): 35–46. doi:10.1016/j.biosystemseng.2007.09.007.

Manuel L.-L., R. Calderón, V. González-Dugo, P. Zarco-Tejada and E. Fereres. Early Detection and Quantification of Almond Red Leaf Blotch Using High-Resolution Hyperspectral and Thermal Imagery. *Remote Sensing* 8, no. 4 (2016): 276; doi:10.3390/rs8040276.

Markwell, J., J. C. Osterman and J. L. Mitchell. 'Calibration of the Minolta SPAD-502 Leaf Chlorophyll Meter'. *Photosynthesis Research* 46, no. 3 (January 1995): 467–72. doi:10.1007/BF00032301.

Marques da Silva, J. R., and L. L. Silva. 'Evaluation of the Relationship between Maize Yield Spatial and Temporal Variability and Different Topographic Attributes'. *Biosystems Engineering* 101, no. 2 (October 2008): 183–90. doi:10.1016/j.biosystemseng.2008.07.003.

Massey, R. E., D. B. Myers, N. R. Kitchen and K. A. Sudduth. 'Profitability Maps as an Input for Site-Specific Management Decision Making'. *Agronomy Journal* 100, no. 1 (2008): 52. doi:10.2134/agrojnl2007.0057.

Merrill, S. C., S. M. Walter, F. B. Peairs and J. A. Hoeting. 'Spatial Variability of Western Bean Cutworm (Lepidoptera: Noctuidae) Pheromone Trap Captures in Sprinkler Irrigated Corn in Eastern Colorado'. *Environmental Entomology* 40, no. 3 (1 June 2011): 654–60. doi:10.1603/EN10076.

Metternicht, G. I., and J. A Zinck. 'Remote Sensing of Soil Salinity: Potentials and Constraints'. *Remote Sensing of Environment* 85, no. 1 (25 April 2003): 1–20. doi:10.1016/S0034-4257(02)00188-8.

Miao, Y., D. J. Mulla, J. A. Hernandez, M. Wiebers and P. C. Robert. 'Potential Impact of Precision Nitrogen Management on Corn Yield, Protein Content, and Test Weight'. *Soil Science Society of America Journal* 71, no. 5 (2007): 1490. doi:10.2136/sssaj2005.0396.

Molatudi, R. L., and I. K. Mariga. 'The Effect of Maize Seed Size and Depth of Planting on Seedling Emergence and Seedling Vigour'. *Journal of Applied Sciences Research* 5, no. 12 (2009): 2234–37.

Moral, F. J., J. M. Terrón and J. R. Silva. 'Delineation of Management Zones Using Mobile Measurements of Soil Apparent Electrical Conductivity and Multivariate Geostatistical Techniques'. *Soil and Tillage Research* 106, no. 2 (2010): 335–43.

Moshia, M. E., R. Khosla, J. G. Davis, D. G. Westfall and K. Doesken. 'Precision Manure Management on Site-Specific Management Zones: Topsoil Quality and Environmental Impact'. *Communications in Soil Science and Plant Analysis* 46, no. 2 (19 January 2015): 235–58. doi:10.1080/00103624.2014.967862.

Moshia, M. E., R. Khosla, L. Longchamps, R. Reich, J. G. Davis and D. G. Westfall. 'Precision Manure Management across Site-Specific Management Zones: Grain Yield and Economic Analysis'. *Agronomy Journal* 106, no. 6 (2014): 2146. doi:10.2134/agronj13.0400.

Mzuku, M., R. Khosla and R. Reich. 'Bare Soil Reflectance to Characterize Variability in Soil Properties'. *Communications in Soil Science and Plant Analysis* 46, no. 13 (20 July 2015): 1668–76. doi:10.1080/00103624.2015.1043463.

Mzuku, M., R. Khosla, R. Reich, D. Inman, F. Smith and L. MacDonald. 'Spatial Variability of Measured Soil Properties across Site-Specific Management Zones'. *Soil Science Society of America Journal* 69, no. 5 (2005): 1572–9.

NSF-NIH Interagency Initiative. 'Core Techniques and Technologies for Advancing Big Data Science and Engineering (BIGDATA)', 2012. Nsf12499. http://www.nsf.gov/pubs/2012/nsf12499/nsf12499.pdf.

Pampolino, M. F., Witt, C., Pasuquin, J. M., Johnston, A. and Fisher, M. J. 'Development Approach and Evaluation of the Nutrient Expert Software for Nutrient Management in Cereal Crops'. *Computers and Electronics in Agriculture* 88 (2012): 103–10.

Panneton, B., and M. Brouillard. 'Use of Fuzzy Mapping to Extract Management Zones from Yield Maps'. *AIC*, 2002.

Panneton, B., M. Brouillard and T. Piekutowski. 'Integration of Yield Data from Several Years into a Single Map'. In Proc III ECPA-European Conference on Precision Agriculture, Montpellier, France, 18–21 June 2001.

Pennock, D., T. Yates and J. Braidek. 'Soil Sampling Designs'. *Soil Sampling and Methods of Analysis* (2007): 1–14.

Peñuelas, J., J. A. Gamon, A. L. Fredeen, J. Merino and C. B. Field. 'Reflectance Indices Associated with Physiological Changes in Nitrogen- and Water-Limited Sunflower Leaves'. *Remote Sensing of Environment* 48, no. 2 (1 May 1994): 135–46. doi:10.1016/0034-4257(94)90136-8.

Perlman, H. 'Irrigation: Irrigation Techniques, USGS Water-Science School', 2016. http://water.usgs.gov/edu/irmethods.html.

Plant, R. E., G. Stuart Pettygrove and W. R. Reinert. 'Precision Agriculture Can Increase Profits and Limit Environmental Impacts'. *California Agriculture* 54, no. 4 (July 2000): 66–71. doi:10.3733/ca.v054n04p66.

Prabhakar, M., Y. G. Prasad and Mahesh, N. Rao. 'Remote Sensing of Biotic Stress in Crop Plants and Its Applications for Pest Management'. In *Crop Stress and Its Management: Perspectives and Strategies*, edited by B. Venkateswarlu, Arun, K. Shanker, Chitra Shanker, and M. Maheswari, 517–45. Springer, Dordrecht, The Netherlands, 2012. http://link.springer.com/chapter/10.1007/978-94-007-2220-0_16.

Raun, W. R., J. B. Solie, G. V. Johnson, M. L. Stone, R. W. Mullen, K. W. Freeman, W. E. Thomason and E. V. Lukina. 'Improving Nitrogen Use Efficiency in Cereal Grain Production with Optical Sensing and Variable Rate Application'. *Agronomy Journal* 94, no. 4 (2002): 815–20.

Raun, W. R., J. B. Solie, R. K. Taylor, D. B. Arnall, C. J. Mack and D. E. Edmonds. 'Ramp Calibration Strip Technology for Determining Midseason Nitrogen Rates in Corn and Wheat'. *Agronomy Journal* 100, no. 4 (2008): 1088. doi:10.2134/agronj2007.0288N.

Rhoads, F. M., and C. D. Yonts. 'Irrigation Scheduling for Corn–why and How'. *National Corn Handbook*. 20 (1991).

Ritchie, K. B., E. G. Hoeft, E. D. Nafziger, L. C. Gonzini, J. I. Warren and W. L. Banwart. 'Starter Fertilizers for No-till Corn'. *Better Crops Plant Food* 80 (1996): 6–9.

Robert, P. C. 'Precision Agriculture: A Challenge for Crop Nutrition Management'. *Plant and Soil* 247, no. 1 (2002): 143–9.

Robert, P. C., R. H. Rust, W. E. Larson, A. M. Coelho, J. W. Doran and J. S. Schepers. 'Irrigated Corn Yield as Related to Spatial Variability of Selected Soil Properties'. In ACSESS Publications. American Society of Agronomy, Crop Science Society of America, Soil Science Society of America, 1999. https://dl.sciencesocieties.org/publications/books/abstracts/acsesspublicati/precisionagric4a/441.

Roberts, D. F., R. B. Ferguson, N. R. Kitchen, V. I. Adamchuk and J. F. Shanahan. 'Relationships between Soil-Based Management Zones and Canopy Sensing for Corn Nitrogen Management'. *Agronomy Journal* 104, no. 1 (2012): 119. doi:10.2134/agronj2011.0044.

Robles, M., I. A. Ciampitti and T. J. Vyn. 'Responses of Maize Hybrids to Twin-Row Spatial Arrangement at Multiple Plant Densities'. *Agronomy Journal* 104, no. 6 (2012): 1747. doi:10.2134/agronj2012.0231.

Rockström, J., W. Steffen, K. Noone, Å. Persson, F. S. Chapin, E. F. Lambin, T. M. Lenton, et al. 'A Safe Operating Space for Humanity'. *Nature* 461, no. 7263 (24 September 2009): 472–5. doi:10.1038/461472a.

Roozeboom, K. L., S. Duncan, D. L. Fjell, R. L. Vanderlip, D. L. Devlin, K. A. Janssen, B. L. S. Olson, C. R. Thompson, D. F. Leikam and D. B. Mengel. 'Corn Production Handbook'. C-560. Kansas State Univ., Manhattan. Corn Production Handbook. C-560. Kansas State Univ., Manhattan, 2007.

Rowe, J. H. 'The Sunken Gardens of the Peruvian Coast'. *American Antiquity* 34, no. 3 (1969): 320–5. doi:10.2307/278415.

Ruiz-Garcia, L., L. Lunadei, P. Barreiro and J. I. Robla. 'A Review of Wireless Sensor Technologies and Applications in Agriculture and Food Industry: State of the Art and Current Trends'. *Sensors* (Basel, Switzerland) 9, no. 6 (16 June 2009): 4728–50. doi:10.3390/s90604728.

Russom, P. 'Big Data Analytics'. TDWI Best Practices Report, Fourth Quarter, 2011, 1–35.

Rutto, E., and D. B. Arnall. The History of the GreenSeekerTM Sensor. Sillwater, Oklahoma: Oklahoma State University Cooperative Extension Service, 2004. http://www.nue.okstate.edu/GreenSeeker/PSS-2260web.pdf.

Sakamoto, T., A. A. Gitelson and T. J. Arkebauer. 'MODIS-Based Corn Grain Yield Estimation Model Incorporating Crop Phenology Information'. *Remote Sensing of Environment* 131 (2013): 215–31.

Samborski, S. M., N. Tremblay and E. Fallon. 'Strategies to Make Use of Plant Sensors-Based Diagnostic Information for Nitrogen Recommendations'. *Agronomy Journal* 101, no. 4 (2009): 800. doi:10.2134/agronj2008.0162Rx.

Sangoi, L. 'Understanding Plant Density Effects on Maize Growth And Development: An Important Issue To Maximize Grain Yield'. *Ciência Rural* 31, no. 1 (February 2001): 159–68. doi:10.1590/S0103-84782001000100027.

Satyanarayana, T., K. Majumdar, M. Pampolino, A. M. Johnston, M. L. Jat, P. Kuchanur, D. Sreelatha, J. C. Sekhar, Y. Kumar, R. Maheswaran and R. Karthikeyan. 'Nutrient ExpertTM: A Tool to Optimize Nutrient use and Improve Productivity of Maize'. *Better Crops-South Asia* 97, no. 1 (2013): 21–4.

Sawchik, J., and A. P. Mallarino. 'Evaluation of Zone Soil Sampling Approaches for Phosphorus and Potassium Based on Corn and Soybean Response to Fertilization'. *Agronomy Journal* 99, no. 6 (2007): 1564. doi:10.2134/agronj2006.0151.

Schepers, A., R., J. F. Shanahan, M. A. Liebig, J. S. Schepers, S. H. Johnson and A. Luchiari. 'Appropriateness of Management Zones for Characterizing Spatial Variability of Soil Properties and Irrigated Corn Yields across Years'. *Agronomy Journal* 96, no. 1 (2004): 195–203.

Schepers, J. S., D. Francis and G. Varvel. 'Chlorophyll Meters'. USDAARS, Gov. Print. Office, Washington, DC, 2006.

Schimmelpfennig, D., and R. Ebel. 'Sequential Adoption and Cost Savings from Precision Agriculture'. *Journal of Agricultural and Resource Economics* 41, no. 1 (2016): 97–115.

Schirrmann, M., R. Gebbers, E. Kramer and J. Seidel. 'Soil pH Mapping with an on-the-Go Sensor'. *Sensors* 11, no. 1 (2011): 573–98.

Shanahan, J. F., N. R. Kitchen, W. R. Raun and J. S. Schepers. 'Responsive in-Season Nitrogen Management for Cereals'. *Computers and Electronics in Agriculture, Emerging Technologies For Real-time and Integrated Agriculture Decisions*, 61, no. 1 (April 2008): 51–62. doi:10.1016/j.compag.2007.06.006.

Shanahan, J. F., J. S. Schepers, D. D. Francis, G. E. Varvel, W. W. Wilhelm, J. M. Tringe, M. R. Schlemmer and D. J. Major. 'Use of Remote-Sensing Imagery to Estimate Corn Grain Yield'. *Agronomy Journal* 93, no. 3 (2001): 583. doi:10.2134/agronj2001.933583x.

Shaner, D. L., R. Khosla, M. K. Brodahl, G. W. Buchleiter and H. J. Farahani. 'How Well Does Zone Sampling Based on Soil Electrical Conductivity Maps Represent Soil Variability?' *Agronomy Journal* 100, no. 5 (2008): 1472. doi:10.2134/agronj2008.0060.

Shaner, D., M. Stromberger, R. Khosla, A. Helm, B. Bosley and N. Hansen. 'Spatial Distribution of Enhanced Atrazine Degradation across Northeastern Colorado Cropping Systems'. *Journal of Environmental Quality* 40 (2011): 46–56.

Shearer, S. A., M. Veal, T. S. Stombaugh, S. F. Higgins, T. G. Mueller, C. R. Dillon, J. P. Fulton, D. J. Mulla and et al. 'Post Processing Correction to Improve the Accuracy of Yield Monitor Data in Grain Crops'. In Proceedings of the 7th International Conference on Precision Agriculture and Other Precision Resources Management, Hyatt Regency, Minneapolis, MN, USA, 25–28 July 2004, 173–86. Precision Agriculture Center, University of Minnesota, Department of Soil, Water and Climate, 2004. http://www.cabdirect.org/abstracts/20063164334.html.

Shearer, S. A. 'Big Data: The Future of Precision Agriculture'. Presented at the The InfoAg Conference, Union Station, St. Louis, MO, 29 June 2014. http://past.infoag.org/abstract_papers/papers/paper_233.pdf.

Schwartz, H., D. H. Gent, S. M. Fichtner, R. Khosla, L. A. Mahaffey, M. A. Camper and W. S. Cranshaw. 'Spatial and Temporal Distribution of *Iris yellow spot virus* and Thrips in Colorado Onion Fields'. *Journal of Plant Health Progress*. PHP-2010-0820-01-RS (2010).

Slaughter, D. C., D. K. Giles and D. Downey. 'Autonomous Robotic Weed Control Systems: A Review'. *Computers and Electronics in Agriculture* 61, no. 1 (April 2008): 63–78. doi:10.1016/j.compag.2007.05.008.

Solanki, P., A. Bhargava, H. Chhipa, N. Jain and J. Panwar. 'Nano-Fertilizers and Their Smart Delivery System'. In *Nanotechnologies in Food and Agriculture*, edited by Mahendra Rai, Caue Ribeiro, Luiz Mattoso, and Nelson Duran, 81–101. Springer International Publishing, 2015. http://link.springer.com/chapter/10.1007/978-3-319-14024-7_4.

Solari, F., J. Shanahan, R. Ferguson, J. Schepers and A. Gitelson. 'Active Sensor Reflectance Measurements of Corn Nitrogen Status and Yield Potential'. *Agronomy Journal* 100, no. 3 (2008): 571. doi:10.2134/agronj2007.0244.

Sudduth, K. A., and S. T. Drummond. 'Yield Editor'. *Agronomy Journal* 99, no. 6 (2007): 1471. doi:10.2134/agronj2006.0326.

Teal, R. K., B. Tubana, K. Girma, K. W. Freeman, D. B. Arnall, O. Walsh and W. R. Raun. 'In-Season Prediction of Corn Grain Yield Potential Using Normalized Difference Vegetation Index'. *Agronomy Journal* 98, no. 6 (2006): 1488.

Tekin, A. B., H. Gunal, K. Sindir and Y. Balci. 'Spatial Structure of Available Micronutrient Contents and Their Relationships with Other Soil Characteristics and Corn Yield'. *Fresenius Environmental Bulletin* 20, no. 3a (2011): 783–92.

Thelemann, R., G. Johnson, C. Sheaffer, S. Banerjee, H. Cai and D. Wyse. 'The Effect of Landscape Position on Biomass Crop Yield'. *Agronomy Journal* 102, no. 2 (2010): 513. doi:10.2134/agronj2009.0058.

Tim, C. W. C., A. P. Moxey, W. Towers, B. Balana and P. D. Hallett. 'Mitigating Arable Soil Compaction: A Review and Analysis of Available Cost and Benefit Data'. *Soil and Tillage Research, Soil Structure and its Functions in Ecosystems: Phase Matter & Scale Matter*, 146, Part A (March 2015): 10–25. doi:10.1016/j.still.2014.09.011.

Tollenaar, M., A. A. Dibo, A. Aguilara, S. F. Weise and C. J. Swanton. 'Effect of Crop Density on Weed Interference in Maize'. *Agronomy Journal* 86, no. 4 (1994): 591. doi:10.2134/agronj1994.00021962008600040003x.

Torres, G. M. 'Precision Planting of Maize (Zea Mays L.)'. Oklahoma State University, 2012. https://shareok.org/handle/11244/6544.

Torsvik, V., L. Øvreås and T. F. Thingstad. 'Prokaryotic Diversity--Magnitude, Dynamics, and Controlling Factors'. *Science* 296, no. 5570 (10 May 2002): 1064–6. doi:10.1126/science.1071698.

Tremblay, N., Y. M. Bouroubi, C. Bélec, R. W. Mullen, N. R. Kitchen, W. E. Thomason, S. Ebelhar, D. B. Mengelf, W. R. Raung, D. D. Francish, E. D. Voriesi and I. Ortiz-Monasterio. 'Corn Response to Nitrogen is Influenced by Soil Texture and Weather'. *Agronomy Journal* 104, no. 6 (2012): 1658–71.

Van Roekel, R. J., and J. A. Coulter. 'Agronomic Responses of Corn to Planting Date and Plant Density'. *Agronomy Journal* 103, no. 5 (2011): 1414. doi:10.2134/agronj2011.0071.

Vega, C. R. C., F. H. Andrade, V. O. Sadras, S. A. Uhart and O. R. Valentinuz. 'Seed Number as a Function of Growth. A Comparative Study in Soybean, Sunflower, and Maize'. *Crop Science* 41, no. 3 (2001): 748. doi:10.2135/cropsci2001.413748x.

Vellidis, G., M. Tucker, C. Perry, C. Kvien and C. Bednarz. 'A Real-Time Wireless Smart Sensor Array for Scheduling Irrigation'. *Computers and Electronics in Agriculture* 61, no. 1 (April 2008): 44–50. doi:10.1016/j.compag.2007.05.009.

Verity, G. E., and D. W. Anderson. 'Soil Erosion Effects on Soil Quality and Yield'. *Canadian Journal of Soil Science* 70, no. 3 (1990): 471–84.

Viscarra R. R. A., V. I. Adamchuk, K. A. Sudduth, N. J. McKenzie and C. Lobsey. 'Chapter Five – Proximal Soil Sensing: An Effective Approach for Soil Measurements in Space and Time'. In *Advances in Agronomy*, edited by Donald, L. Sparks, Volume 113: 243–91. Academic Press, 2011. http://www.sciencedirect.com/science/article/pii/B9780123864734000051.

Viscarra Rossel, R. A., A. B. McBratney, and B. Minasny. Proximal Soil Sensing. Springer, 2010. http://books.google.com/books?hl=en&lr=&id=xpcaRHfieewC&oi=fnd&pg=PR5&dq=proximal+soil+sensing&ots=Z9iFCgU2xa&sig=2Nk8PQH7vxMOrzl0pe6UFn6Ouiw.

Viswakumar, A., R. W. Mullen, A. Sundermeier and C. E. Dygert. 'Tillage and Nitrogen Application Methodology Impacts on Corn Grain Yield'. *Journal of Plant Nutrition* 31, no. 11 (15 October 2008): 1963–74. doi:10.1080/01904160802403102.

Wann, F. B. 'Circular No. 85-Chlorosis Yellowing of Plants: Cause and Control', 1930. http://digitalcommons.usu.edu/uaes_circulars/77/.

Watson, S. E. Segarra, S. Machado, E. Bynum, T. Archer, K. Bronson and et al. 'Precision Farming in Irrigated Corn Production: An Economic Perspective'. In *Selected Paper Prepared for Presentation at the Southern Agricultural Economics Association Annual Meeting*, Mobile, Alabama, 2003. http://ageconsearch.umn.edu/bitstream/35053/1/sp02wa01.pdf.

Weidong, L., F. Baret, G. Xingfa, T. Qingxi, Z. Lanfen and Z. Bing. 'Relating Soil Surface Moisture to Reflectance'. *Remote Sensing of Environment* 81, nos. 2–3 (August 2002): 238–46. doi:10.1016/S0034-4257(01)00347-9.

Whelan, B. M., and A. B. McBratney. 'The "Null Hypothesis" of Precision Agriculture Management'. *Precision Agriculture* 2, no. 3 (November 2000): 265–79. doi:10.1023/A:1011838806489.

Whelan, B. M., J. A. Taylor and A. B. McBratney. 'A "small Strip" Approach to Empirically Determining Management Class Yield Response Functions and Calculating the Potential Financial "net Wastage" Associated with Whole-Field Uniform-Rate Fertiliser Application'. *Field Crops Research* 139 (November 2012): 47–56. doi:10.1016/j.fcr.2012.10.012.

Wilding, L. P., and L. R. Drees. 'Spatial Variability and Pedology'. Capítulo 4: Wilding, L. P., Smeck, N. E. and Hall, G. F. (Eds), *Pedogenesis and Soil Taxonomy I. Concepts and Interactions*. Developments in Soil Science A 11 (1983).

Wiles, L. J., G. W. Oliver, A. C. York, H. J. Gold and G. G. Wilkerson. 'Spatial Distribution of Broadleaf Weeds in North Carolina Soybean (Glycine Max) Fields'. *Weed Science* 40, no. 4 (1992): 554–7.

Williams II, M. M., D. A. Mortensen, A. R. Martin and D. B. Marx. 'Within-Field Soil Heterogeneity Effects on Herbicide-Mediated Crop Injury and Weed Biomass'. *Weed Science* 49, no. 6 (1 November 2001): 798–805. doi:10.1614/0043-1745(2001)049[0798:WFSHEO]2.0.CO;2.

Wright, R. J., T. A. Devries, L. J. Young, K. J. Jarvi and R. C. Seymour. 'Geostatistical Analysis of the Small-Scale Distribution of European Corn Borer (Lepidoptera: Crambidae) Larvae and Damage in Whorl Stage Corn'. *Environmental Entomology* 31, no. 1 (1 February 2002): 160–67. doi:10.1603/0046-225X-31.1.160.

Zebarth, B. J., C. F. Drury, N. Tremblay and A. N. Cambouris. 'Opportunities for Improved Fertilizer Nitrogen Management in Production of Arable Crops in Eastern Canada: A Review'. *Canadian Journal of Soil Science* 89, no. 2 (1 May 2009): 113–32. doi:10.4141/CJSS07102.

Zhang, C., and J. M. Kovacs. 'The Application of Small Unmanned Aerial Systems for Precision Agriculture: A Review'. *Precision Agriculture* 13, no. 6 (31 July 2012): 693–712. doi:10.1007/s11119-012-9274-5.

Zhao, Ji-chun, Jun-feng Zhang, Yu Feng and Jian-xin Guo. 'The Study and Application of the IOT Technology in Agriculture'. In 2010 3rd IEEE International Conference on Computer Science and Information Technology (ICCSIT), 2: 462–5, 2010. doi:10.1109/ICCSIT.2010.5565120.

Chapter 8

Improving nutrient management for sustainable intensification of maize

Kaushik Majumdar, International Plant Nutrition Institute – South Asia, India; Shamie Zingore, International Plant Nutrition Institute – Sub-Saharan Africa, Kenya; Fernando García and Adrian Correndo, International Plant Nutrition Institute – Latin America – Southern Cone, Argentina; Jagadish Timsina, University of Melbourne, Australia; and Adrian M. Johnston, International Plant Nutrition Institute, Canada

1 Introduction

Maize (*Zea mays* L.), a crop of worldwide economic importance, provides approximately 30% of the food calories to more than 4.5 billion people in 94 developing countries (Jat et al., 2013). Production of maize surpassed rice in 1996 and wheat in 1997, and is now close to 850 million tonnes (Mt); and maize production is increasing at twice the annual rate of rice and thrice that of wheat (Fischer et al., 2014). Factors driving this increase are the increasing demands for maize for food, feed and bio-fuel, and its resilience to biotic and abiotic stresses affecting most of the intensive production systems of the world. The demand for maize is expected to double worldwide, from about 558 Mt in 1995 to over 1100 Mt in 2050 (Dixon et al., 2008). With new analysis (Gerland et al., 2014) showing that, contrary to previous literature, world population can be expected to increase from the current 7.2 billion to 9.6 billion in 2050 and 10.9 billion in 2100, the demand for

http://dx.doi.org/10.19103/AS.2016.0002.08

maize may increase further. The increase in maize demand will be acute in Asia, where an 87% rise in demand from 162 Mt in 1995 to 303 Mt in 2020 is projected (Timsina et al., 2011). The human and environmental costs of expanding agricultural lands are such that, in reality, most of the necessary production gains going forward must be achieved on existing farmland through intensification of production (Foley et al., 2011).

The International Maize and Wheat Improvement Center (CIMMYT) identified eight *Mega maize environments* across the world, with yield potential ranging from 6 Mg ha^{-1} in the dry lowland tropical environment (Coastal Eastern Africa, Central America, India, North-East Brazil) to 14 Mg ha^{-1} in the wet temperate environment (US Corn Belt, Western Europe and Argentina) (Fischer et al., 2014). The top five producers of maize in the world are the United States, China, Brazil, Mexico and Argentina, with an average yield of 10.7, 6.0, 5.2, 3.3 and 6.6 Mg ha^{-1}, respectively (www.faostat3.fao.org), and with varying extent of yield gaps across the countries (Table 1). Neumann et al. (2010) also showed variable yield gaps across major maize production areas around the world, suggesting the possibility of further intensification. Cassman (1999) suggested that meeting future food demand while minimizing expansion of cultivated area will depend primarily on continued intensification, and average farm yields of cereals must reach 70–80% of the yield potential ceiling within 30 years. The author, referencing other studies (Tollenaar et al., 1994), further suggested that much of the observed genetic gain in yield during the past 30 years could be attributed to greater stress resistance rather than an increase in yield potential, hinting at a larger influence of crop management on future yield gains and efforts to reduce yield gaps. Mueller et al. (2012) found that the global yield variability of several crops, including maize, are heavily influenced by fertilizer application, irrigation and climate. They further suggested that closing yield gaps to 100% of attainable yield could increase worldwide maize production by 64%. Closing maize yield gaps to 50% of the attainable yield (≈2.3 Mg ha^{-1}) in Sub-Saharan Africa (SSA) primarily requires addressing nutrient deficiencies, but addressing yield gaps of 75% of the attainable yield (≈3.5 Mg ha^{-1}) would require an increase in irrigation and nutrient use in most areas. Furthermore, in case of

Table 1 Summary of global maize farm yields (FY), potential yields (PY) and yield gaps of maize in 2010, and respective annual rates of change over the past 20 years

	Estimated yield (Mg ha^{-1}) and yield gap (%)			Rate of change (% per annum)		
Region	FY	PY	Yield gap	FY	PY	Yield gap
Iowa, USA	11.4	15	36	1.8	0.8	−1.0
China	5.3	10.4	96	0.7	0.8	0.1
Eastern Africa	1.4	7.1	400	0.8	1.5	0.7
Western Africa	1.7	5.3	200	1.9	1.5	−0.4
Brazil	4.0	12	200	2.6	0.8	−1.8
Argentina	7.4	15	100	2.4	1.1	−1.3
Italy	9.5	14	47	0.5	Na	Na
Ukraine	4.4	12	170	2.8	Na	Na

Source: Modified from Fischer et al., 2014.

major cereals like maize, wheat and rice, 73% of underachieving areas could close their yield gaps by solely focusing on nutrient inputs (with 18%, 16% and 35% increase in N, P_2O_5 and K_2O application over baseline global consumption, respectively). Jointly increasing irrigated area and nutrient application could close yield gaps in all underachieving areas, with 30%, 27% and 54% increase in N, P_2O_5 and K_2O application, respectively, and with a 25% increase in irrigated hectares.

Improved nutrient management not only provides significant opportunities to increase maize production across growing environments but also helps minimize the environmental impacts of intensification through reduction in nutrient imbalances and inefficiencies.

2 Nutrient management challenges in maize

The extent of maize yield gaps and their association with nutrient constraints is not monolithic, but varies with regions (Mueller et al., 2012), as well as within regions (Chen et al., 2014). The differential nutrient constraints in major maize-growing areas arise due to crop management strategies specific to growing environments. For example, maize is grown on different soil types, in different ecologies (rain-fed or irrigated) and seasons (winter, spring, summer), as mono-crop or inter-crop, or in multiple cropping sequences where more than one crop is grown in one year, under no- or conventional-tillage practices, in smallholder systems or in large-scale farming and so on. Consequently, nutrient management challenges have different extent and complexities due to the large variability in growing environments, and would require site-specific solutions to address them. The challenges associated with maize production in some of the major maize ecologies are described in the following sections.

2.1 South Asia

Maize area, production and productivity in South Asia grew significantly during the past decade (2003–11). Suitability of maize cultivation in the three cropping seasons of South Asia, higher productivity, profitability and lesser water requirement as compared to winter rice and wheat; its resilience to abiotic stresses such as terminal heat stress; and fewer pest and disease problems were the major drivers of such change in South Asia. Maize is predominantly grown by smallholder farmers in the region under a diverse combination of soils, seasons, cropping systems, tillage and ecologies, which makes nutrient management decisions difficult. For example, maize experience phosphorus (P) deficiency in rice–maize (R-M) systems of South Asia due to anaerobic to aerobic transition of the growing environment during the cropping cycle (Timsina et al., 2010) and K deficiency due to high extraction by grain and stover (Timsina et al., 2013). In flooded rice, fixed P in the soil solubilizes immediately after flooding, leading to a flush of available P supply to rice (Kirk et al., 1990). Subsequent drying, however, reduces its availability to maize for which strong crop responses to P fertilizer are expected. In systems of low P fertility, the repeated anaerobic–aerobic transition in the R-M system increases P extraction, further lowering fertility. Phosphorus availability in maize–wheat cropping system, however, does not follow a similar trend as both crops are grown under non-flooded condition. Similarly, Majumdar et al. (2012) illustrated the variable nitrogen (N) availability to spring and winter maize grown under no-till (ZT) and conventional tillage (CT) in India. The authors recorded lower yield in

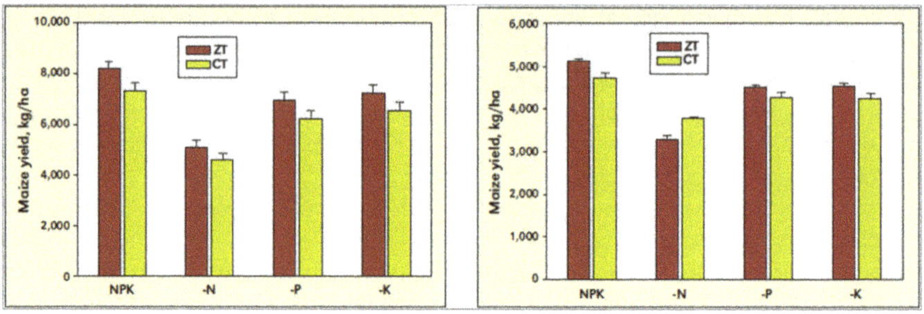

Figure 1 Average yields (kg ha^{-1}) of winter (left) and spring (right) maize in omission plot trials under zero-tillage (ZT) and conventional-tillage (CT) systems. The bars represent the standard errors of the means.

Source: Majumdar et al., 2012.

N omission plots under ZT spring maize as compared to conventionally tilled maize, but not in winter maize (Fig. 1). This is probably due to either greater immobilization of available N, losses of N through leaching and denitrification, lower mineralization of soil organic N or some combination of these factors (Moschler and Martens, 1975) that reduced the N availability to ZT spring maize, particularly in the initial growth phase of the crop. However, due to longer duration of winter maize, the mineralization of the immobilized N under ZT systems might have supplied more N in the later growth stages, thereby increasing the yield over conventionally grown maize.

The above examples highlight the complexity of nutrient decision support that is often beyond the knowledge of smallholder farmers and field extension agents of South Asia. Consequently, high yield gaps in farmers' fields due to imbalanced and inadequate fertilizer application leading to nutrient mining and low nutrient use efficiencies are some of the major challenges to the sustainability of maize production in the region. The lack of farm- and farmer-specific fertilizer application strategies and decision support tools to implement improved practices at large scale exacerbate the challenges.

2.2 Sub-Saharan Africa

In SSA, maize production has expanded from about 9 million ha (M ha) in 1961 to 27 M ha in 2013, while total production has increased from 8 to 47 Mt over the same period. The increased production of 500% was mainly supported by area expansion, with less than 30% of the increase attributable to productivity gains, as average maize yields have remained less than 2 Mg ha^{-1}. Regional production of maize and other food crops has lagged behind population growth, resulting in declining per capita production and increasing food deficits and malnutrition over the past six decades (Smale and Jayne, 2003). Maize is mostly grown on smallholder farms of less than 2 ha in SSA with very low use of fertilizer or organic resources (Giller et al., 2011). Despite maize accounting for 40% of the fertilizer used, average nutrient application in maize production is very low at less than 30 kg ha^{-1} (Fig. 2) (FAO, 2015). The nutrient use trends also show a growing inclination to reinforce unbalanced nutrient use, with N application increasing notably between 2009 and 2013,

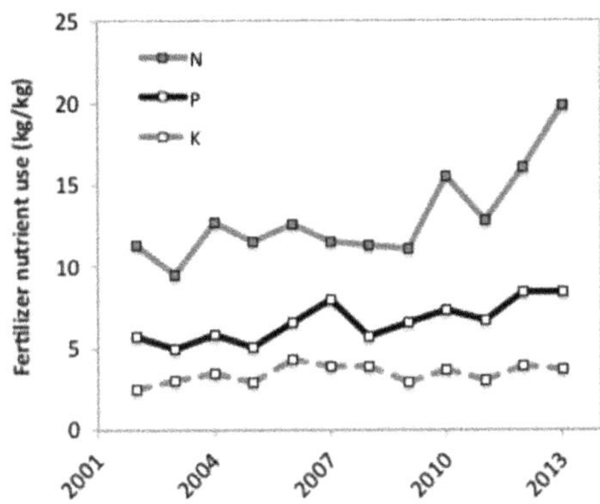

Figure 2 Fertilizer nutrient use (kg ha^{-1}) in Sub-Saharan Africa from 2002 to 2013 based on FAO Statistics.

Source: FAO, 2015.

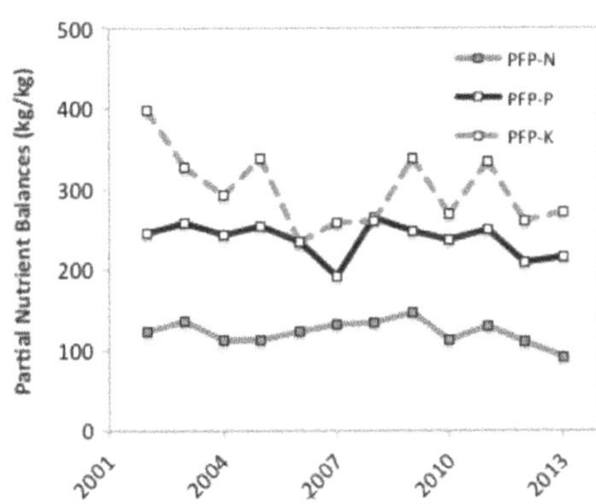

Figure 3 Partial factor productivity (kg kg^{-1}) of nutrient use in Sub-Saharan Africa from 2002 to 2013 based on FAO Statistics.

Source: FAO, 2015.

while K application remained low and static during the same period. Cultivation of maize with low and unbalanced nutrient application is the underlying agronomic cause of soil fertility decline and remains the most important factor contributing to low productivity and large yield gaps (Snapp et al., 1998). Nutrient balances for SSA are largely negative, and losses of the most limiting nutrient, N, are estimated at 20–70 kg ha^{-1} annually (Stoorvogel and Smaling, 1998). These large negative balances are caused by over-exploitation of soil nutrient stocks as farmers use low levels of nutrients, coupled with the removal of nutrients in harvested produce and N losses through leaching and erosion (Craswell and Vlek, 2013). Partial factor productivity (PFP; kg maize grain produced per kg nutrient application) for maize in SSA is high, although the trends are decreasing gradually due to increasing trend of fertilizer use (Fig. 3). The high PFP (>100 kg kg^{-1}N and >200 kg kg^{-1} P and K) is associated with low rates of fertilizer use and high rates of nutrient mining.

2.3 Latin America – Southern Cone

Total maize grain production in the Southern Cone of Latin America (Argentina, Bolivia, Chile, Paraguay and Uruguay) increased from 5.6 Mt in 1961 to 39.5 Mt in 2013. This increase in production has resulted from increases in planted area from 3.4 M ha in 1961 to 6.6 M ha in 2013, and increase in grain yield from 1.6 to 6.0 Mg ha^{-1}. During the same period, the annual increase in planted area, grain yield and grain production have been of 1.3%, 2.5% and 3.8%, respectively.

Argentina is the leading maize producer in the region, with 74% of the total area and 81% of the grain production, followed by Paraguay (16% and 10% of area and production, respectively), Bolivia, Chile and Uruguay. Chile leads the region in achieving highest grain yield with 11 Mg ha^{-1} in the most recent years, mostly because of high radiation and large thermal amplitude of high latitudes and irrigation. Yields have been variable in the past five years in Argentina, Uruguay, Paraguay and Bolivia, averaging 6.5, 4.4, 3.5 and 2.6 Mg ha^{-1}, respectively. Maize is one of the main crops in the region receiving fertilizers in large areas. In the recent years, 75–95% of the maize area has been fertilized in Argentina with average rates of 70–80 kg N and 10–16 kg P ha^{-1} (Fertilizar AC, www.fertlizar.org). Increases in fertilizer consumption since 1993 have resulted in decreases in the PFP for N and in the partial nutrient balance (PNB) for P, between 1993 and 2012 (Fig. 4). These performance indicators suggest that N and P rates are still behind desired sustainability standards for the maize cropping systems. Estimations of PNB and PFP in the 2010/11 season indicate that N and P application rates should be decreased in Chile and Uruguay, respectively, while N rates should be increased in Uruguay (Table 2).

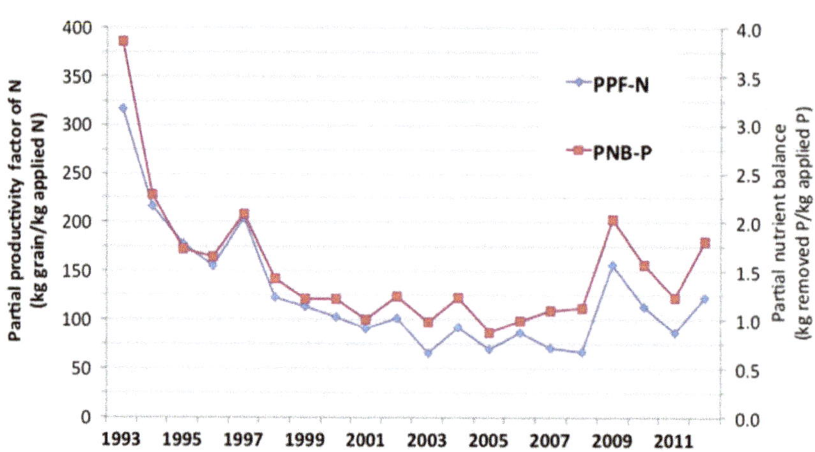

Figure 4 PPF (same as PFP) for N and PNB for P of maize in Argentina between 1993 and 2012.

Table 2 Estimations of partial nutrient balance (PNB) and partial factor productivity (PFP) of N and P for maize in the 2010/11 season in Argentina, Chile and Uruguay. Elaborated from data of Heffer (2013) and Mancassola and Casanova (2015)

Country	Nutrient	Argentina	Chile	Uruguay
Average application (kg ha^{-1})	N	80	286	29
	P	16	32	15
Grain yield (kg ha^{-1})		7804	11081	3538
Nutrient removal (kg ha^{-1})	N	100	139	46
	P	20	33	9
Partial nutrient balance (kg ha^{-1})	N	1.26	0.49	1.58
	P	1.27	1.03	0.60
Partial factor productivity (kg kg^{-1})	N	98	39	121
	P	489	348	235

The Pampas region of Argentina, the temperate central-eastern plains, produces approximately 80% of the total maize production of the country (Hall et al., 1992). The main challenges of maize nutrient management for the region are adjustment of N diagnosis and recommendations in early and late-season maize; calibrating soil P tests and recommendations according to farmer's decisions; improving sulphur (S) diagnosis and determination of optimum rates; micronutrient deficiencies; variable nutrient management in different maize regions with high intra-field variations (topography and others) and alternatives for right source, time and placement to increase N, P and S nutrient efficiency and effectiveness of N, P and S.

2.4 China

Maize plays a significant role in expanding overall cereal grain production in China. Ranking as the most widely planted crop, it had planted area of about 29.5% of the food crops of the country, or 32.5 M ha, in 2010 (Xu et al., 2014). A recent study (Fischer et al., 2014) has shown that the maize yield gap in China is about 100% (Table 1). The study identified the major nutrient-related constraints as imbalanced and inadequate fertilizer use, delayed application and low use of organic manures. Notwithstanding the above, a number of studies have now documented over application of N and P fertilizer, driven by the high yield goals with increasing maize price, leading to nutrient accumulation in the soil and the consequent environmental problems (Xu et al., 2014; Zhao et al., 2006). In a survey of farmer practices across the three temperate maize regions, Chen et al. (2014) reported average N application rates of 257 kg ha^{-1}, well above the average of 160 kg ha^{-1} used in the high-yielding regions like Iowa. With such high N rates, it is not surprising that recovery of N by crops is less than 30% of that applied (Ju et al., 2009). Sun et al. (2012) showed considerable regional and cropping system variation in N losses to the environment from three major river valleys that may be masked in national-scale studies (Table 3). A recent study in North Eastern and North Central China suggested that

Table 3 Estimated N losses from fertilizer N in three main river valleys in China in 1995

	Nitrogen loss in river valley (Mt)			
River valley	Denitrification	N transported into water bodies	NH_3 volatilization	Total loss
Yangtze river valley	1.77–2.61	2.87	1.32	5.96–6.80
Yellow river valley	0.24–0.53	0.65	0.17	1.06–1.35
Pearl river valley	0.39–0.49	0.62	0.29	1.30–140

Source: Modified from Sun et al., 2012.

about 72.4% (*n* = 848), 60.5% (*n* = 717) and 72.6% (*n* = 989) of all the observations for agronomic efficiency of N, P and K (AEN, AEP and AEK) values for maize were lower than 15 kg kg^{-1} (Xu et al., 2014). Thus closing yield gaps and optimizing nutrient management for higher efficiency are the twin nutrient-related challenges faced by China for sustainably intensifying smallholder maize systems in the country.

3 Improved nutrient management

Maize extracts large amounts of mineral nutrients from the soil due to large grain and stover yields. Proper nutrient management of an exhaustive crop like maize should aim to supply fertilizers adequate for the demand of the crop and apply it in ways that minimize loss and maximize the efficiency of use (Timsina et al., 2010). The amount of fertilizer required depends on many factors, including the indigenous supply of each nutrient, which can be of appreciable quantity. Thus, emphasis must be placed on supplying the nutrients through fertilizers in a timely manner to fill the gap between nutrient requirements for target yields and nutrient availability from indigenous sources such as soil organic matter (SOM), farmyard manure (FYM), composts and crop residues on a site-specific basis. Fertilizer is the dominant source of nutrients and is required to increase the yield of crops, but should be applied in such a quantity that it becomes profitable and will have a least adverse effect on the environment. Improving our understanding of nutrient uptake timing and rates, partitioning and remobilization of nutrients between maize plant parts, provides opportunities to optimize fertilizer sources, rates, application timings and methods, following the '4R Nutrient Stewardship Principles' (IPNI, 2012). Two major aspects are important while developing fertilizer recommendations for high-yielding maize production systems: (1) assessment of the amount of a given mineral nutrient that needs to be acquired by the plant during the growing season, referred to as 'total nutrient uptake' and (2) the amount of nutrient transported out of the field with grain and straw/stover, referred to as 'removed with harvested product' (Bender et al., 2013). Providing the nutrients as and when required by the crop and replenishing the nutrients exported out of the field with harvested products ensure sustainability of production systems. Further improvement of fertility practices requires matching in-season nutrient uptake with availability. Bender et al. (2013) illustrated the nutrient uptake pattern of modern maize varieties that is critical for optimizing source, rate, time and method of fertilizer application

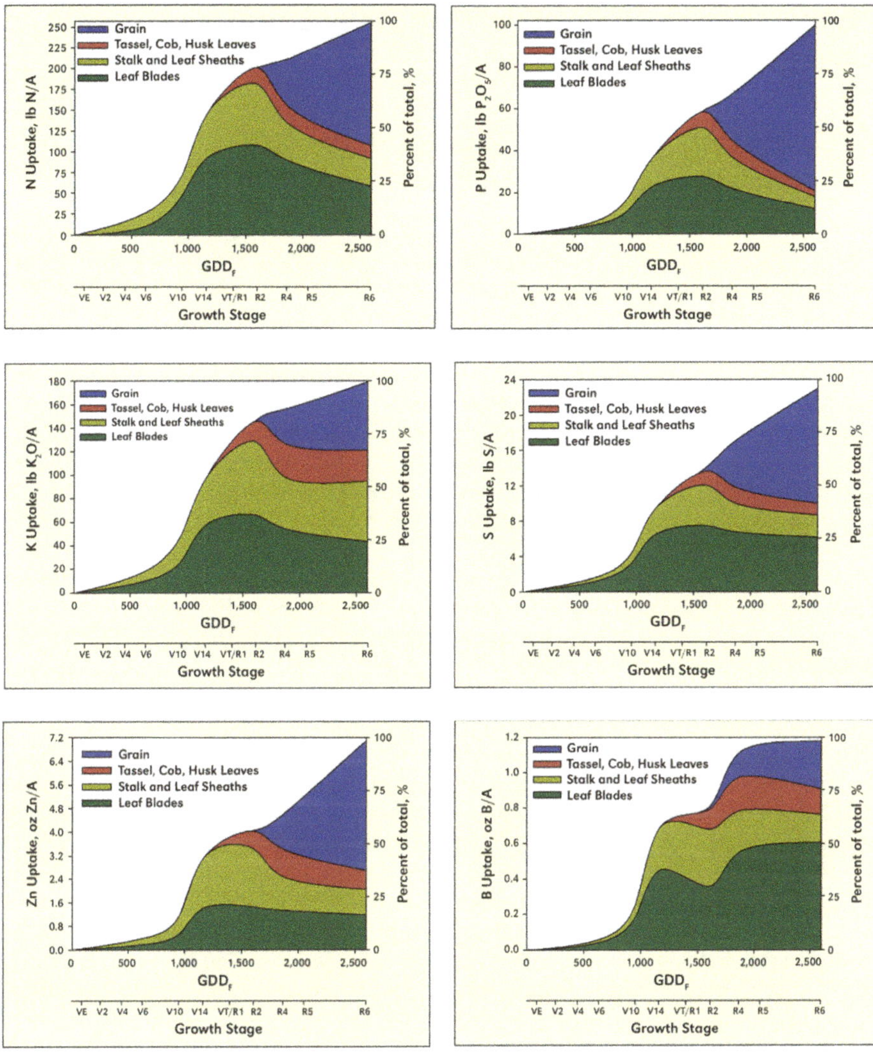

Figure 5 Total maize N, P, K, S, Zn and B uptake and partitioning across four plant stover fractions: leaf, stalk, reproductive and grain tissues. Each value is a mean of six hybrids across two site-years at Urbana, Illinois (2010), and DeKalb, Illinois (2010). GGDF = growing degree days (Fahrenheit).

Source: Bender et al., 2013.

to match demand and supply (Fig. 5). The nutrient management strategies based on the 4R Nutrient Stewardship Principles (IPNI, 2012) are at the core of site-specific solutions to the complex maize nutrition requirements and provides the opportunity to achieve economic, social and environmental benefits out of agronomic management in diverse environments.

4 Case studies in improving nutrient management: South Asia

4.1 Optimizing nutrient management in maize using a fertilizer decision support tool

Imbalanced fertilizer application is identified as one of the major reasons for low maize productivity in South Asia. Despite the proven economic, social and environmental benefits of balanced fertilization, its adoption at the farm level is low. The lack of appropriate tools and implementation mechanisms has been a major hindrance that restricted the wide-scale adoption of balanced fertilization. The International Plant Nutrition Institute (IPNI) and its partner organizations in South Asia have jointly developed a user-friendly dynamic nutrient management tool, Nutrient Expert® (NE), that can generate farm-specific fertilizer recommendation for maize. The tool is based on the site-specific nutrient management (SSNM) principles (Pampolino et al., 2012) and utilizes information of the growing environment to provide science-based balanced fertilizer recommendations for maize that are tailored to a particular location, cropping system and farmer resource availability.

The NE tool development in India was followed by large-scale on-farm validation across different growing environments of maize. The NE-based recommendations were compared to the existing fertilizer recommendation practices such as farmers' fertilization practices (FFP) and local (or state) recommendations (SR). The comparative data of different treatments from the validation trials for maize are given in Table 4 (Majumdar et al., 2014).

The NE-based fertilizer recommendation for maize improved yield and nutrient use efficiency (NUE) as compared to FFP and SR (Table 4) across multiple locations in India. The modal N, P_2O_5 and K_2O application rates in FFP were 138, 23 and 0 kg ha^{-1}, respectively, which highlighted the generally imbalanced practices adopted by farmers even in maize that removes a large amount of K from a field. In addition, the timing of N application recommended by NE provided improved NUE for the maize crop. Maize yield in the FFP was reasonably high (7800 kg ha^{-1}) but the low (12.4 kg grain kg^{-1} nutrient) PFP suggested inefficient management of nutrients by farmers. The NE recommendations for individual farm fields were developed using information on attainable yield, cropping

Table 4 Modal values of fertilizer application rates, yield and partial factor productivity (PFP) in different treatments in maize validation trials in India

Parameter	FFP (*n* = 482)	SR (*n* = 296)	NE (*n* = 510)	NE-FFP	NE-SR
N, kg ha^{-1}	138 (27–550)*	100 (80–280)	130 (90–257)	−8	30
P_2O_5, kg ha^{-1}	23 (0–280)	60 (22–75)	37 (17–92)	14	−23
K_2O, kg ha^{-1}	0 (0–352)	50 (0–75)	56 (18–143)	56	6
Grain yield, kg ha^{-1}	7800 (1024–11 766)	4200 (1051–10 785)	8400 (2337–12 460)	600	4200
PFP, kg grain kg^{-1} nutrient	12.4 (7–78)	21 (6–46)	27.5 (10–62)	13.1	6.5

*Values in parentheses are ranges.

system nutrient balance based on nutrient input and off-take from the field, previous crop history, local constraints and so on that allowed optimization of nutrient sources, application rates and timing of application. The current study showed that location-specific fertilizer recommendation from the NE tool provided a significant opportunity for optimizing nutrient management in maize, which may improve farm profitability and environment stewardship of applied nutrients.

4.2 Integrating bio-physical and socio-economic determinants for farm and farmer-specific fertilizer recommendations

Fertilizer, or any other input, used by farmers is strongly influenced by farmer resource endowment and other socio-economic determinants. Farm typologies reflect the potential access of different farmer groups to resources such as fertilizer for managing their crops. Categorizing the types of farmers, based on socio-economic parameters, also allows realistic evaluation of the constraints and opportunities that exist within farmer groups for appropriate agricultural interventions (Banerjee et al., 2014). It may be argued that the 'one recommendation fits all' strategy for fertilizer, common in South Asia, is often not accepted by farmers mainly because it does not match the bio-physical requirement of the farm and the resource endowment of the farmer. A recent study in Telangana State of India assessed the interrelation between socio-economic determinants of maize farmers, maize yield and fertilizer use (Madhavi et al., 2015). The study showed that farm size, farmers' age, maize-growing experience and off-farm income strongly influence fertilizer application and maize yield in the studied districts. Fewer years of experience in maize farming was found to be a major cause of inappropriate P application and consequent loss in productivity and profitability.

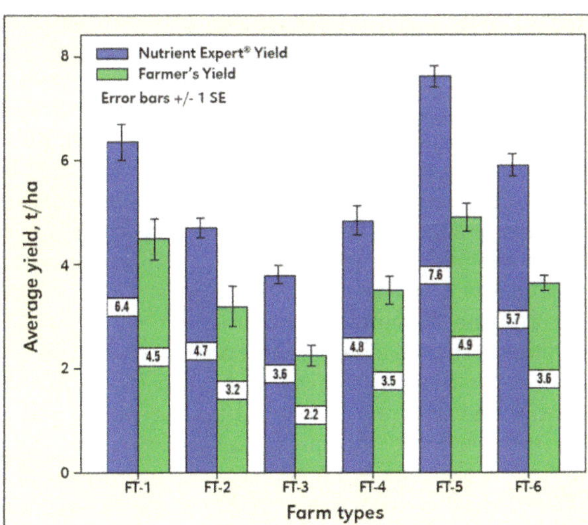

Figure 6 Comparison of maize yields across different farm types under NE and FFP.
Source: Banerjee et al., 2015.

A similar study in smallholder maize systems in Eastern India (Banerjee et al., 2015) utilized socio-economic and crop management data from a rapid rural survey in a multivariate statistical analysis to classify 127 farmers into six farm types. The identified farm types were then used for generating and implementing farm-specific fertilizer recommendation using the NE tool. The comparative on-farm trials in 127 locations, assessing the NE-based recommendation and the FFP, showed that the NE-based farm-specific fertilizer recommendation increased maize yield by 42, 47, 70, 38, 55 and 63% over FFP in farm types 1–6 (Fig. 6). The results highlighted the potential benefit of integrating bio-physical and socio-economic factors while developing fertilizer recommendations for smallholder maize farmers.

As mentioned earlier, lack of easy-to-use fertilizer decision support tools is one of the hindrances for the adoption of balanced fertilization in South Asia. Farmers' access to soil testing is also very limited. The NE tool fills the critical gap of research-for-development by providing support to extension agents to develop and promote science-based, site-specific fertilizer recommendation even in the absence of a soil test.

5 Case studies in improving nutrient management: Sub-Saharan Africa (SSA)

Key approaches for adapting 4R practices to support sustainable maize intensification in the smallholder farming systems in SSA include (i) targeted management of fertilizer and organic resources under highly variable soil fertility conditions; (ii) fertilizer rates that optimize crop productivity and farmers' income; (iii) timing of N application under highly variable inter- and intra-seasonal rainfall; and (iv) spot application of fertilizer for low fertilizer rates and for P management in P-fixing soils.

5.1 Targeted management of fertilizer and organic resources under highly variable soil fertility conditions

The use of mineral fertilizers in most countries in SSA has been mainly promoted through blanket N and P recommendations that are based on agro-ecological zones. Improving the blanket recommendations to account for inter- or intra-farm soil fertility variability is necessary to maximize the benefits of projected increases in fertilizer use. A distinctive feature that characterizes smallholder farming systems in much of SSA is the wide diversity of farming households and marked heterogeneity for soil properties at short ranges due to management and inherent soil factors (Zingore et al., 2007; Tittonell et al., 2005). Attempts to develop site-specific fertilizer recommendations in SSA have mainly focused on multi-location fertilizer response trials in sites representative of soil units that potentially give variable responses for fertilizer sources and rates. Multi-location trials conducted in a range of studies across SSA revealed that baseline yields, and yields for different fertilizer treatments, increased with increasing soil fertility status (Zingore et al., 2007). In Malawi, application of N alone gave the largest yield increase for all treatments and for a range of soil fertility conditions, confirming that N was the most limiting nutrient (Fig. 7). Addition of P also led to a significant increase in yields on the high fertility fields, but in medium fertility fields, in addition to N, application of K, Ca, Zn and B was also required to significantly increase crop yields above the N treatment. On the depleted soils, baseline yields were very low, and were increased to less than 1 Mg ha^{-1} by applying N and to less

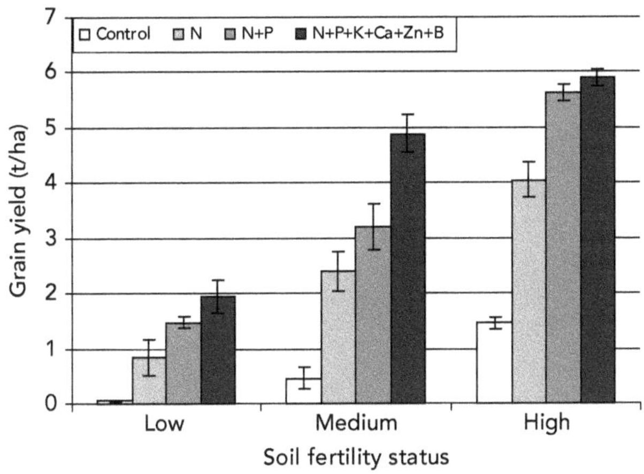

Figure 7 Maize yield response to nutrient application under variable soil fertility conditions in Southern Malawi.

Source: Zingore, 2011.

than 2 Mg ha^{-1} by applying N and P. The effects of K and secondary and micronutrients were very small. Under such conditions, balanced fertilizer application and investment in organic resources is required to increase productivity by concurrently addressing deficiencies of multiple nutrients and soil degradation.

Despite the high levels of soil nutrient mining in SSA, very limited research has been conducted to establish the extent of deficiencies of K, and secondary and micronutrients, and their requirements for maize production areas. Potassium, Ca, Mg, S, Zn, Mn, B and Mo have been observed to limit crop growth, especially in infertile sandy soils that have limited reserves and are continuously cropped (Vanlauwe et al., 2015). Soil mapping programmes in Ethiopia and other countries have established the wide occurrence of S, Zn and B deficiency, while significant maize responses to S, Zn, B, Mg and Mo have been observed across the continent (Wendt and Rijpma, 1997; Weil and Mughogho, 2000). Recognizing the need for balanced nutrient management, research and development programmes have been initiated in a number of countries (e.g. Malawi, Kenya, Ethiopia, Rwanda and Tanzania) to support the development of fertilizer blends containing K, secondary and micronutrients and avail them at large scale as alternatives to N and P fertilizers.

In areas with relatively high livestock densities, the use of cattle manure plays a key role in maize production and maintenance of soil fertility. In North-East Zimbabwe, fields that frequently receive manure are more productive and more fertile than fields with no history of manure application (Zingore et al., 2008). Studies in Zimbabwe and elsewhere in SSA have reported on the beneficial effects of manure on crop yields, both in the short term and in the long term (Mtambanengwe and Mapfumo, 2005). Results from studies in Zimbabwe showed that application of manure alone increased maize productivity above control in fields with low and medium fertility, but not in fields with high soil fertility (Fig. 8). Annual addition of manure at 10 Mg ha^{-1} has been recommended to double maize yields

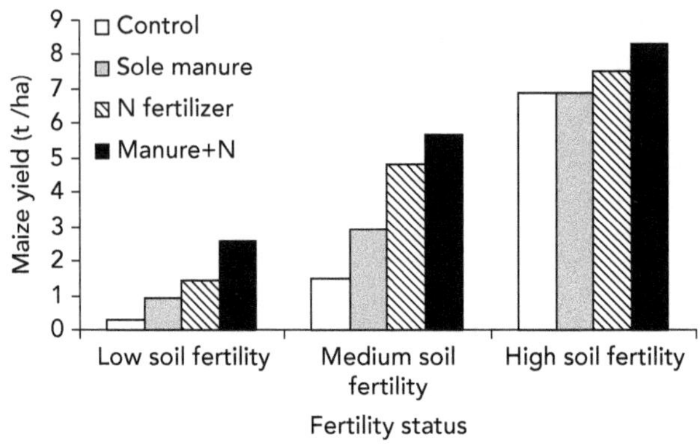

Figure 8 Effects of manure and fertilizer application in maize productions as influenced by soil fertility status in Eastern Zimbabwe.

from an average of 1 Mg ha^{-1} without nutrient inputs to about 2 Mg ha^{-1} (Grant, 1981). The average amounts of cattle manure available (<1 t per household) are far below the amounts that are required to double maize yields. There is therefore need to apply manure in combination with fertilizers, not only because of potential synergic benefits but also because of the limited availability of both types of nutrient resources (Vanlauwe and Giller, 2006). When combined with mineral N fertilizers, manure led to increased productivity above fertilizer treatments alone, and the absolute increase in maize yields decreases with increasing soil fertility status (Fig. 8).

The role of manure in maize production intensification is particularly important on degraded fields where crops respond poorly to fertilizer application due to other limiting factors not addressed by fertilizers (Zingore et al., 2008). Under such conditions, application of large amounts of manure for several seasons may be necessary to restore SOM and associated soil physical fertility, and balance soil nutrients to facilitate good fertilizer response (Bayu et al., 2005). A major factor limiting the effectiveness of manure in smallholder farming systems is the low N contents (<1.2%), high C:N ratios and high lignin concentrations that typically exceed 15%. This results in N immobilization that is compounded when manure is used in combination with low-quality crop residues. The immediate effects derived from manure are related to its effects on the supply of multiple nutrients, reduction of Al saturation in acidic soils as well as improving the soil structure, rather than the provision of N (Bayu et al., 2005). Hence, manure is commonly applied as a basal fertilizer, with N being applied as a top dressing. Long-term use of manure allows build-up of SOM, and when SOM is raised to certain levels, net N mineralization can occur (Giller et al., 1997). In some cases, though, immediate N availability was observed and this has been attributed to high concentrations of ammonium in the manure.

5.2 Nitrogen rates for optimization of crop productivity and economic returns

Nutrient management recommendations must target appropriate rates for nutrients limiting crop productivity. Soil analysis results are used as a standard method to determine nutrient application rates, but a very small proportion of smallholder farmers can afford and have access to soil analysis services. Alternatively, proxy indicators for soil fertility identifiable to farmers based on soil type, management history and current yields have been used to provide guidelines for site-specific nutrient application rates. Several studies have shown that complex differences in soil fertility and crop response to fertilizer associated with soil types and farmer management practices can be summarized into four categories that

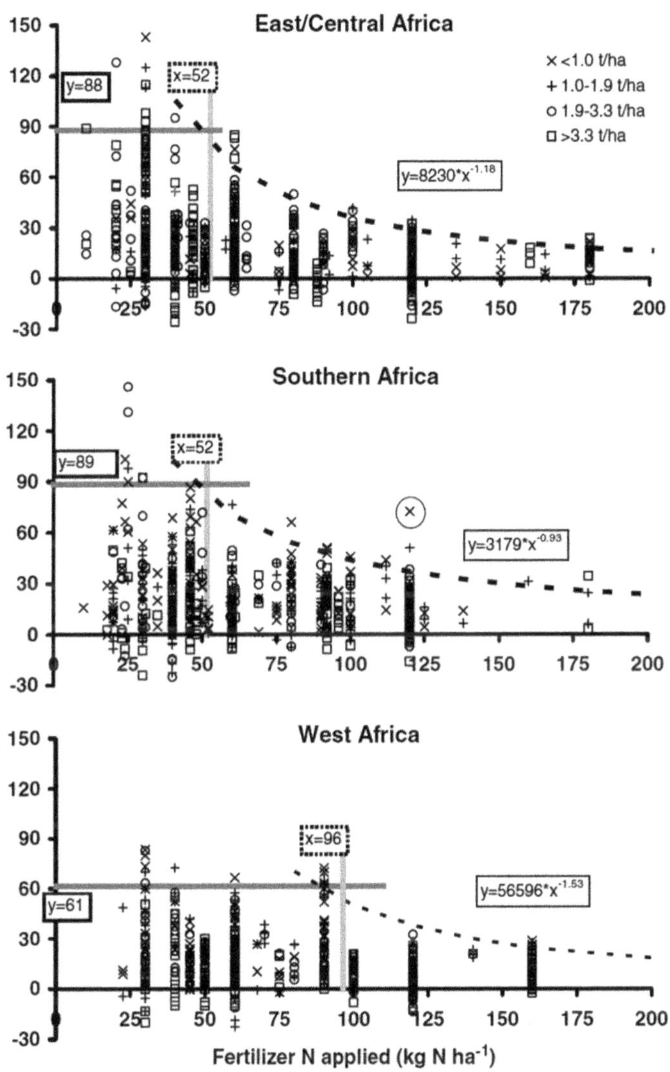

Figure 9 Agronomic efficiency of fertilizer N (N-AE) at different N application rates in a range of maize production environments in SSA.

Source: Adapted from Vanlauwe et al. (2011).

can be used to guide nutrient sources and rates as follows: (i) low responsive fertile soils that are nutrient saturated and require low rates of fertilizer for 'maintenance fertilization'; (ii) high responsive fertile soils that require moderate to high rates of N and P; (iii) high responsive infertile soils that require moderate to high rates of N and P, and addition of K and/or secondary and micronutrients; and (iv) low responsive degraded infertile soils that respond poorly to fertilizer application.

Results from Zimbabwe show that returns diminish rapidly on most soils at nutrient application rates greater than 60 kg N ha^{-1} and 10 kg P ha^{-1} (Zingore et al., 2007). Likewise, a countrywide study in Malawi revealed that agronomic and fertilizer use efficiency was highest at rates between 30–60 kg ha^{-1} N, and recommended a downward revision of fertilizer recommendations from 92 kg ha^{-1} N to 35 kg ha^{-1} N when farmers produce maize for sale (Benson, 1999). Recommendations of application rates for maximizing yields are unattractive to cash-constrained farmers. Meta-analysis of N response data from trials from a range of geographic and agro-ecological conditions across SSA by Vanlauwe et al. (2011) showed highly variable agronomic N use (AE_N) efficiency values for application rates of 0–180 kg ha^{-1}. The upper boundary of AE_N was only maintained up to 52 kg ha^{-1} N in East and Southern Africa and up to 96 kg ha^{-1} N in West Africa (Fig. 9). The relatively higher N rates that supported the upper boundary was likely due to higher rainfall in maize production sites in West Africa. The relatively low fertilizer rates to achieve the most efficient and economic application rates suggest the occurrence of other yield-limiting factors that diminish responses even when yields are far below attainable yields. The low N application threshold levels to maintain the maximum upper boundary of AE_N also indicate that economic returns to fertilizer application diminish rapidly at low rates in smallholder farming systems, indicating the need to focus on N application at moderate rates for smallholder farmers whose main goal is to maximize the rate of returns rather than maximize income. It is therefore plausible, in general, to promote a phased process for maize production intensification by first increasing N use from the current 25 kg ha^{-1} N to about 60 kg ha^{-1} (with balanced application of other nutrients) with a view to further increase fertilizer application and yield targets when other yield-limiting factors are addressed, especially in sites that have high potential for maize intensification. It should, however, be noted that the rates of fertilizer use are very variable, and the use of average values often masks areas where fertilizer rates are already relatively very high.

5.3 Timing of N application under highly variable inter- and intra-seasonal rainfall

Maize production by smallholder farmers in SSA is mostly done under rain-fed conditions with a very small area under irrigation. An increasing occurrence of drought presents a major risk for intensification. Full rates of P and K and basal N are usually applied at planting, while additional N is further applied in one or two top dressings. Therefore, flexible methods of applying N depending on the prevailing rainfall are required to optimize fertilizer use efficiency. To manage in variable rainfall environments, Piha (1993) devised and successfully tested a flexible system of fertilization, in which theoretically optimum rates of P, K and S are applied based on target yields at planting, while N is applied as a series of split applications, which are adjusted during the season according

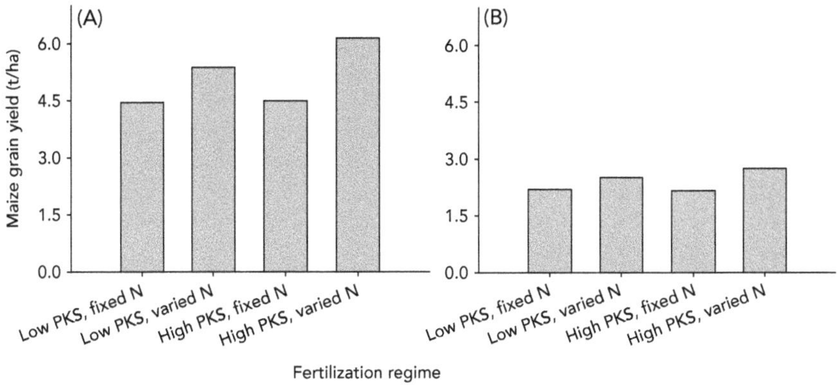

Figure 10 Effects of fixed and variable N application strategies on maize productivity for a high agro-potential site (A) and a low agro-potential site (B) in central Zimbabwe.

Source: Kurwakumire et al., 2014.

to the degree of water stress observed. This system optimizes resource use efficiency during good rainfall seasons, while ensuring minimum losses under drought due to reduced fertilizer inputs. This fertilization strategy is anchored by simple principles of how the different nutrients behave in soils. Because N is highly mobile and easily hand applied as top dressings, a crop N requirement under variable rainfall conditions is best met by adjusting top dressings according to seasonal rainfall pattern. Immobile P and K nutrients, that are not taken up by the crop in the current season in the event of reduced yields due to drought, remain in the soil and can benefit the subsequent crop.

The flexible system of fertilization, in which optimum rates of P, K and S fertilizers are applied based on yield potential in an average rainfall season, while N is applied as a series of split applications adjusted to the evolving rainfall pattern in any one season, resulted in more efficient maize production (Fig. 10). Trials over five years on farmers' fields gave 25–42% higher yield, and 21–41% higher profit than did existing fertilizer recommendations, that increase farmers' susceptibility to risk. The fertilization strategy optimizes N use efficiency during good rainfall seasons, while ensuring minimum losses under low moisture or drought.

5.4 Spot application of fertilizer for low fertilizer rates and P in P-fixing soils

The fertilizer micro-dosing technology, based on spot application of low amounts of fertilizer in the planting hills, has been widely evaluated and promoted as an affordable and low risk technology to improve crop productivity and fertilizer use efficiency, particularly in semi-arid maize production systems (Twomlow et al., 2010). Spot application of small amounts of fertilizer to maize ensures placement in the main rooting zone and enhances its uptake by the crop, with further enhancements obtained when combined with physical soil management practices aimed at water conservation. In N deficient soils, maize yields were increased by 30–50% through application of low N rates of 17 kg ha^{-1} (Twomlow et al., 2010). However, the yield intensification strategy based on micro-dosing fertilizer application is associated with mining of nutrients, especially N, and is unsustainable for maintaining high-yield levels in the long term. The focus of placement without addressing the right source and rate requirements can be used at best as a stopgap measure to stimulate productivity in the short term while building the capacity of farmers to transit to more sustainable nutrient management systems that encompass complete 4R practices.

Similarly, under the low P input systems that are characteristic of smallholder farming in SSA, the right placement of applied P can drastically increase phosphorus use efficiency (PUE) leading to increased yields. Spot placement of P fertilizers in planting holes results in higher PUE compared to broadcasting and incorporation, particularly in P-fixing soils that are widespread in East Africa. Studies in the region have reported that spot placement resulted in higher maize yields than broadcasting and incorporation at P rates <50 kg ha^{-1} (van der Eijk et al., 2006). This implies that for the resource scarce smallholder farmers in SSA, spot application of small amounts of P offers the best option for improving PUE.

6 Case studies in improving nutrient management: Latin America – Southern Cone

6.1 Adjusting N diagnosis and recommendations for early and late-season maize

Nitrogen is the nutrient that most often limits the production of maize in Argentina. Research in Argentina has generated calibrations based on N availability for facilitating the formulation, interpretations and recommendations of fertilizers. Figure 11 shows that the available N index {nitrate-N content in the soil at 0–60 cm (Ns) before planting plus fertilizer N (Nf)} could explain much of the variation of the maize yield in the Pampas region. To improve the calibrations, regional models have been adapted explaining local variability that could arise from soil, climate and management factors. In general, available N (Ns+Nf) of 120–220 kg ha^{-1} would result in maize yield of 10–12 Mg ha^{-1} (Alvarez et al., 2003; García et al., 2010). Recent research has shown that the use of soil N mineralization potential index would improve the prediction of N needs for maize (Sainz Rozas et al., 2008).

Other tools for diagnosing N needs and/or status of maize could include use of software with model simulations calibrated at different cropping regions (Satorre et al., 2006); soil nitrate-N test at V5–V6 stage (Espósito et al., 2006; Pagani et al., 2008) and methods that assess the status of N in actively growing crops with local sensors such as the Minolta SPAD 502 (Ferraris and Couretot, 2011), or remote sensors that estimate the normalized development vegetation index (NDVI) such as the Greenseeker® (Melchiori, 2012). The use of sensors allows in-season intervention improving N use efficiency.

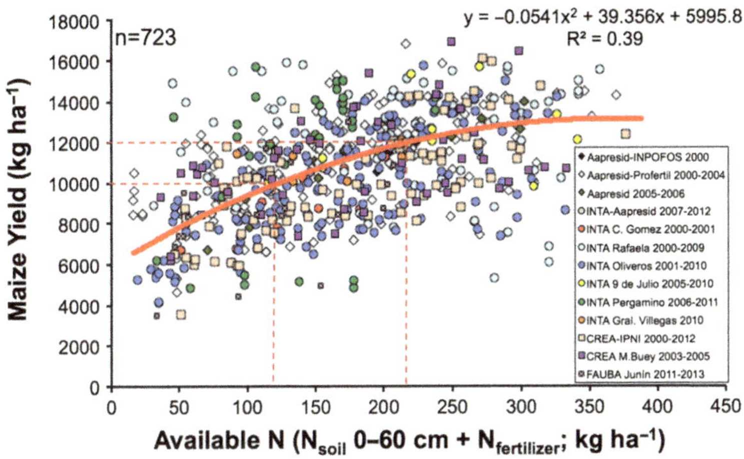

Figure 11 Maize yield as a function of N availability, defined as pre-plant nitrate-N 0–60 cm + fertilizer N, in the Pampas region of Argentina. Data of 192 experiments (723 points) distributed in Buenos Aires, Santa Fe, Cordoba, Entre Rios and La Pampa provinces during the period 2000–12. The dotted lines indicate the availability of N required to achieve yields of 10 and 12 Mg ha^{-1}.

6.2 Calibrating soil P tests and recommendations according to farmers' decisions

After N, P is the most frequently deficient nutrient for maize in Argentina. Diagnosis of soil P fertility is based on pre-plant soil testing determining the level of Bray 1 P in the top soil (0–20 cm). Figure 12 shows calibration for the Pampas, where the critical range is estimated between 15 and 18 mg kg^{-1}, below which 77% of the cases responded to P addition, and above which 66% did not respond to P.

From soil Bray P levels, the criteria for P fertilization can be defined as 'sufficiency', prioritizing the immediate crop, or 'construction and maintenance', prioritizing the build-up of soil resources. Phosphorus fertilization of grain crops in Argentina was historically done following the philosophy of sufficiency, resulting in decreased soil Bray 1 P levels and negative P balance. With the build-up criterion, research has shown a positive balance of about 10 kg ha^{-1} P, which could increase the soil P level by 4 mg kg^{-1} in soils of the northern Pampas (Ciampitti et al., 2011). Farmer's decision on P fertilization, however, depends on several other factors such as land tenure, capital availability, technology adoption and so on.

6.3 Improving S diagnosis and determination of optimum rates

Sulphur was detected as a deficient nutrient in most of the maize-growing regions by the mid-1990s. Deficiencies have been attributed to a decrease of SOM because of continuous cropping for more than 100 years, adoption of no-tillage, high frequency of

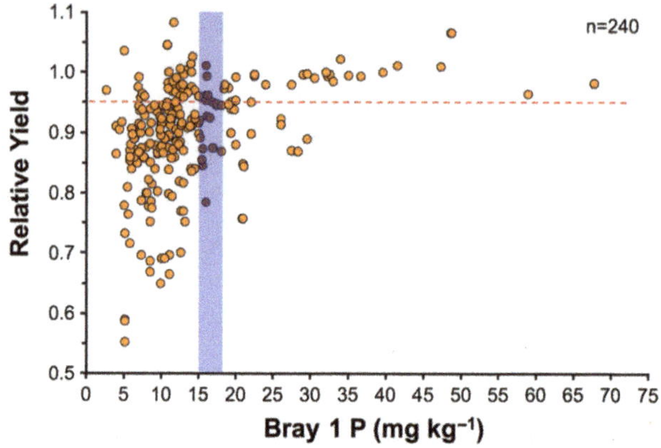

Figure 12 Relationship between the relative yield of maize and the soil Bray 1 P concentration (0–20 cm) in the Pampas region (n = 240) of Argentina (Buenos Aires, Santa Fe, Cordoba and Entre Rios provinces) during the period 1997–2012. The vertical strip indicates the critical range of P Bray (15–18 mg kg^{-1}) according to the method of Cate and Nelson to reach 95% of relative yield.

soybean in the rotation and increasing crop yields. Soils have been extensively tested for S across the region but only sporadic correlations were found between sulphate-S at planting and yield response. Thresholds of 10 mg kg^{-1} of sulphate-S (0–20 cm) at pre-plant were reported (García et al., 2010), below which the response is highly probable. The recommended S dose varies between 5 and 20 kg ha^{-1}, depending on the level of expected yields and farming history. Generally, the S response is saturated above 10–15 kg ha^{-1}. Additionally, it is important to mention two characteristics of the S response: (i) it is common to observe a positive relationship with N response (Cordone et al., 2001; Fontanetto et al., 2010; García et al., 2010; Pagani et al., 2012) and (ii) S fertilizer has shown residual effects on yields of succeeding crops in the rotation (Fontanetto et al., 2003; García et al., 2010).

6.4 Exploring micronutrient deficiencies for the different cropping systems

The intensification of agriculture in the past 20 years has resulted in decreased levels of micronutrients in soils (Sainz Rozas et al., 2013). Thus, several cases of Zn deficiency and fertilizer responses in maize have been registered in recent years. Yield responses to micronutrients range from 5 to 10%, in fields under a long farming history, low SOM, sandy soils, and/or soil Zn levels (0–20 cm, DTPA extractable Zn) below 1 mg kg^{-1} (Michiels and Ruffo, 2012). Responses to micronutrients are commonly observed either with the application of solid or liquid chemical or physical mixtures, or with seed and foliar treatments. Other nutrients such as magnesium and boron have been reported as deficient in some areas; however, further field experimentation would be needed for confirmation.

6.5 Defining variable nutrient management for different regions with high intra-field variations

Precision agriculture practices have defined management zones based on site-specific soil and crop properties, which contributed to a more efficient use of N and P. For the sub-humid Pampas, Gregoret et al. (2011) developed a model for diagnosis of N availability and recommendation of N fertilizer rates according to management zones for dry land maize grown on Haplustolls. The model considered variability between zones by including soil available water content at planting and available N (soil nitrate-N at 0–200 cm). In the same region, Espósito (2013) proposed an econometric model that allowed optimizing variable rate applications of N in maize, where the main parameter is a composite topographic index affected by the predicted climate for the year (El Niño Southern Oscillation – ENSO).

The use of remote sensors for variable N rate applications is another alternative to manage intra-field variations. Many authors reported the ability of remote sensing models to predict N deficiencies in crops (Holland and Schepers, 2010; Melchiori, 2010). In Argentina, remote sensing experiences in maize have allowed the validation of procedures, and the development of yield prediction models based on the canopy reflectance. The variable rate models based on remote sensing could be one of the key technologies to increase the N use efficiency. However, doubts remain about the generalization of recommendation models, and the complexity of techniques as well as the costs of the specific equipment and devices that might limit the rate of the adoption of this technology (Melchiori, 2012).

6.6 Proposing alternatives for right source, time and placement to increase N, P and S nutrient efficiency and effectiveness

Urea is the main source of N in Argentina with 62% of the N fertilizer market. However, urea ammonium nitrate, with 16% of the market share, is frequently used in maize, especially in V4–V6 applications. The use of urease and/or nitrification inhibitors is still limited; however, field research has shown interesting results in different regions and conditions (Sainz Rozas et al., 1999; Ferraris and Couretot, 2011; Fontanetto et al., 2011). Research has shown that N applications at V5–V6 stage are more efficient than pre-plant or planting applications (Sainz Rozas et al., 1999, 2001; Pagani et al., 2008). Farmers applying N at this stage mostly dibble UAN, and in some cases ammonium thiosulphate, which also provides S. Split applications up to V10–V12 stage have also shown that synchronicity of N supply with demand would greatly increase N use efficiency (Melchiori, 2012) by reducing N losses by leaching and/or denitrification (Sainz Rozas et al., 2001).

For P, broadcast applications at about 45–60 days before planting have shown similar efficiencies to banding at planting time when the rate of application of phosphate fertilizer is over 20 kg ha^{-1} of P and/or the level of Bray P is above 8 mg kg^{-1} (Barbagelata, 2011). Broadcast P applications under no-tillage should be carefully analysed for fields with steep slopes and near streams or other surface waters to avoid potential P runoff losses.

Sulphate-S sources such as ammonium sulphate, potassium sulphate, potassium-magnesium sulphate and calcium sulphate have shown similar efficiencies, as either broadcast or banding applications (Torres Duggan et al., 2012). Elemental S is of low solubility and requires undergoing oxidation processes before plants could use it.

7 Case studies in improving nutrient management: use of the Nutrient Expert® tool in China

Soil testing has been promoted and used as a means of fertilizer recommendations in China and has helped to increase maize grain yield and NUE. Soil test-based fertilizer recommendation in smallholder systems poses several challenges and uncertainties, including taking representative soil samples, identifying a suitable analytical method and establishing a good correlation and calibration between soil test value and crop response. Besides being labour-intensive and time-consuming, the challenge of selecting a science-based fertilizer recommendation philosophy to use with the soil test data often affects the outcome. An alternative method of fertilizer recommendation is using the NE for maize. In China, NE was developed based on analysis of relationships among yield response, indigenous nutrient supply, agronomic efficiency and relative yield from datasets collected from 2001 to 2010. The tool was developed to help crop advisors from the public and private sector as well as scientists and extension specialists in large-scale implementation of site-specific fertilizer recommendations. On-farm field validation was conducted in 408 farmers' fields located in Hebei, Henan, Shandong, Shanxi, Jilin, Liaoning and Heilongjiang provinces in 2010–12 to test the efficacy of the NE in different growing environments of China (Xu et al., 2014). Results indicated that NE recommendations matched the yield and profit of farmers' practice (FP) in six of the seven provinces and exceeded the yield and profit in the seventh province. The NE tool reduced N application by 67.8 kg ha^{-1} (30.2%) and P_2O_5 by 7.0 kg ha^{-1} (11.2%), and increased K_2O use by 18.9 kg ha^{-1} (38.9%) compared with FP across all sites. The average gross profit over fertilizer cost in the NE recommendation was 3049 US$ ha^{-1} compared with 2928 US$ ha^{-1} in FP for all sites, and gross profit improved from 22 to 276 US$ ha^{-1} with an average of 121 US$ ha^{-1} (4.1%). Fertilizer recommendation based on NE significantly improved N use efficiency compared to FP for each province ($P < 0.05$). The average AEN, recovery efficiency of N (REN) and PFPN for NE were 12.2, 0.302 and 65.7 kg kg^{-1} as compared to corresponding values of 8.3, 0.200 and 48.5 kg kg^{-1} for FP. The results showed that fertilizer recommendations based on the NE tool could maintain grain yield and profitability and improve NUE through 4R nutrient stewardship. Considering the ease of use, and its applicability in diverse maize-growing areas in China, the Nutrient Expert® tool is a promising approach for optimizing nutrient use in maize when soil testing is not timely or not available.

8 Research priorities

8.1 South Asia

In addition to maize grown in annual sequences with other crops, several crops are also grown as inter-crop with maize in South Asia. For example, potato and winter irrigated maize, legumes and rain-fed maize, vegetables and maize in peri-urban areas. As maize cultivation becomes more widespread and intensive in the region, and particularly in Bangladesh and South India, an emerging issue of great importance is how to sustain the productivity of maize and maize-based cropping systems through integrated soil fertility management (ISFM) strategies. Continuous production of high-yielding maize will lead to rapid depletion of mineral nutrients from soil because of greater nutrient uptake and removal by maize than other cereals such as rice or wheat. So there is a need to understand more about the extent and rate of nutrient depletion and soil physical degradation in the intensified maize systems in South Asia. To push the achieved grain yields even higher up the yield potential curve will require larger amounts of nutrients, their better management and overall soil stewardship. Further research to fill the knowledge gaps in sustainable intensification of maize may need fine-tuning using 4R strategies in different maize-growing environments and under diverse management practices.

Right source: Research on identifying appropriate fertilizer sources for maize has been limited in South Asia. For long-duration crops like maize, where nitrogen is applied in multiple splits, use of slow release/enhanced efficiency fertilizer provides the opportunity to supply nutrient over an extended period of time while saving labour cost. However, the availability and use of enhanced efficiency fertilizer is limited in the region and needs to be prioritized to increase N use efficiency and profitability of maize farming.

Right rate: As for the right source, very limited research has been done to determine the appropriate rates for maize grown in sequential cropping or in intercropping. The criticality of such information can be envisioned if we consider maize and legumes growing together. Excess N application in such intercropping system would restrict pod formation in legumes while lower N rate will adversely affect maize growth and yield. Similarly, estimation of right P rate in high P-fixing soils needs further study considering that large portion of the applied P gets fixed because of the broadcast application while fixed P is slowly released from the soil over an extended period of time.

Right time: The application time of fertilizer in maize under sequential cropping, or in inter- or relay cropping is another researchable issue with large knowledge gaps. The maize hybrid that matures in 100 days in the rainy season takes 125 days to mature as a spring crop, while it takes around 160 days in winter. Estimation of the right time of N application becomes a difficult choice as the high nutrient requiring physiological stages of maize appear at different days after sowing in different seasons. Although maize requires nutrients throughout the growing season, the most effective application time that ensures profitability and high NUE in each season needs further research.

Right place: Mechanization is limited in smallholder systems of South Asia. Consequently, fertilizers are applied as broadcast though sub-surface drilling provides a great opportunity to increase NUE. However, that would require the development of machinery suitable for land holdings that are smaller than 0.5 acres.

As maize is grown in sequence with other crops in South Asia, research on optimizing 4R in different cropping systems should be prioritized:

- 4Rs for cropping systems: for diverse cropping systems of the region and when crops are grown in sequence under anaerobic and aerobic growing environment
- 4Rs for conservation agriculture systems: conventional versus zero-till crop establishment and under differential residue retention
- 4Rs for climate change adaptation/mitigation: for adaptation/mitigation of climate change impacts such as drought, excess precipitation, heat stress and progressive N limitation

Developing effective communication/dissemination strategies for smallholder systems is another critical area of research not only for South Asia but also in smallholder maize systems across continents. Transfer of knowledge to smallholder farmers for accruing the benefits of research has been a major challenge that has not been addressed adequately. This will require coordinated efforts from bio-physical and social scientists to devise the right way of communicating with smallholder farmers who are often constrained by resources to venture beyond subsistence farming.

8.2 Sub-Saharan Africa

There are emerging positive trends in fertilizer use and crop productivity in several countries in SSA (Sanchez, 2010) that have been mainly driven by improved policies and macroeconomic growth. There are opportunities for the generation of research outputs that support the evaluation and implementation of 4R practices to consolidate the gains in maize productivity that have been achieved in recent years. Key areas for future research include the following:

- Spatial analysis approaches are useful in complementing technology development processes, whether for site-specific fertilizer recommendations based on multi-site fertilizer trials or for identifying scaling-up recommendation domains for targeting of crop and water management practices for maize production intensification. Recent developments in soil fertility analysis using infrared spectroscopy allow for rapid, accurate and cost-effective large-scale soil analysis and surveillance. Research to apply infrared spectroscopy in soil and plant sample analysis offers opportunities to accelerate data collections for accurate diagnosis of soil fertility constraints and recommendation of technological options.
- To achieve the much needed impact, efforts to restore soil fertility and improve productivity in SSA will require concerted investments in availing inputs at affordable prices and building capacity among agricultural service providers and farmers to implement ISFM technologies suited to site-specific, socio-economic and bio-physical conditions.
- Major research gaps still exist in the development of effective and practical technologies to diagnose and map extensive areas of non-responsive degraded soils that give poor response to fertilizer application. There is currently limited understanding of the key soil factors underlying the non-responsiveness in degraded soils, and there is a need for research activities that address the chemical, physical and biological processes limiting crop response to management. Such information will be valuable in determining the most effective technologies for restoring productivity in degraded soils.

8.3 Latin America – Southern Cone

Improved nutrient management for maize in the Southern Cone region of Latin America would lead to improved fertilizer efficiency and effectiveness contributing to sustainable crop production systems. This would come from adopting many of the process- and input-based technologies already proven through research and field evaluation. A quick list would include soil sampling, use of local recommendation systems, site-specific management according to topographic variation, in-season N applications supported by pre-plant soil sampling plus local and remote sensing, among others.

Table 5 Grain yield, water productivity, N use efficiency and N_2O-N emissions at two field experiments comparing ecological intensification (EI) and farmers' practices (FP) at the Northern (Paraná) and Southern Pampas (Balcarce) of Argentina

	Paraná		Balcarce	
Indicator	EI	FP	EI	FP
Grain yield (kg ha^{-1})	8777[a]	6124[b]	9743	8052
Water productivity (kg mm^{-1})	6.1[a]	4.8[b]	9.9[a]	8.3[b]
Physiological N use efficiency (kg grain kg^{-1} applied N)	28.9[a]	23.5[b]	45.8[a]	41.3[b]
N_2O-N emissions (kg N_2O-N ha^{-1})	–	–	0260	0255

[a,b] Significant differences between treatments within the same location (Paraná or Balcarce) with 5% of probability.

Source: Adapted from Caviglia et al. (2012); Picone et al. (2013); and O. Caviglia, R. Rizzalli and L. Picone (pers. comm.).

Further research should explore responses to nutrients other than N, P and S. Soil acidification would be a constraint in the coming years as soil pH and base saturation have been declining in several areas because of continuous cropping, increasing grain yields and use of N fertilizers. Better knowledge and information on the relationship between in-season status of the crops and alternatives to correct their nutritional disorders would also be a significant contribution.

A major issue in the search for sustainability of agro-ecosystems of the region involves the evaluation of the whole farming system regarding soil and management practices and their interactions: rotations, use of cover crops, varieties, planting date, plant population, weeds, diseases and pests control, nutrient management, tillage systems, water management and so on. Improved radiation and water- and nutrient use efficiencies could be achieved through a more intensive use of land by including winter season crops and/or through better agronomic practices for each specific crop (Caviglia and Andrade, 2010; Caviglia et al., 2012). Preliminary results from cropping systems trials at the northern and southern Pampas have shown improvements on performance indicators such as grain yields, water productivity, physiological N use efficiency and N_2O-N emissions under ecologically intensified systems compared to farmer's current practices (Caviglia et al., 2012; Picone et al., 2013) (Table 5). Right soil and crop management interacts with 4R nutrient management in achieving better sustainability indicators for the main resources and inputs involved in the cropping system.

9 Conclusions

Maize cultivation around the world is increasing rapidly, driven by the demand from food, feed and industrial sectors, as well as for its resilience to abiotic and biotic stresses associated with climate change. Increase in global maize production has largely been achieved through area expansion, with limited contribution from increase in crop productivity. There are clear evidences that maize is replacing traditional crops in several geographies as market opportunities open up for farmers. Sustainably increasing productivity of maize systems in traditional and new areas are important to spare expansion of agriculture to natural ecosystems as we strive to meet future demands. This chapter highlights evidence of large management yield gaps in the major maize producing areas in the world. Appropriate fertilizer use is crucial for closing such yield gaps, as plant nutrient demand of maize is high. Maize is grown both in smallholder systems as well as in large-scale farming operations. Both groups of farmers are looking for critical guidance to rationalize their nutrient management to improve productivity and profitability in an environmentally sustainable manner. High spatial diversity among farms, lack of access to soil testing, prevalence of blanket recommendations and low resource endowment of farmers create significant challenge for nutrient management in smallholder maize systems. Addressing such challenges would require empowering farmers and their advisors with decision tools that help develop farm-specific fertilizer recommendations, integrating bio-physical characteristics of farms and socio-economic determinants of farmers. In the mechanized large-scale farming operations in the developed countries, striving for high maize yields in an environment where strict regulations associated with agricultural nutrient use are in place may require fine-tuning of the 4R strategies for increasing nutrient use efficiencies and reducing loss of nutrients to the environment. Significant knowledge gap exists in the area of transfer of nutrient management technologies to the farming community. This is particularly true for the smallholder farmers in Asia and Africa. Overcoming such challenges may require significant investment for not only devising the best way to communicate with farmers but making precision nutrient management technologies scale-neutral for fitting into different scale of farming operations.

10 Where to look for further information

Nutrient management for intensification of maize is a widely researched topic across the globe. Excellent papers and monographs are published regularly in international peer-reviewed journals. Besides those resources, readers may explore the following institutions for further information on maize nutrient management research trends or for collaboration opportunities:

- International Maize and Wheat Improvement Center, www.cimmyt.org
- CGIAR Research Program on Maize, www.maize.org
- International Plant Nutrition Institute (IPNI), www.ipni.net
- International Institute of Tropical Agriculture, www.iita.org
- Food and Agriculture Organization of the United Nations, www.fao.org
- International Fertilizer Industry Association (IFA), www.fertilizer.org
- Website of Asian Maize Conference held every 4 years

11 References

Alvarez, R., Steinbach, H., Alvarez, C. Y. and Grigera, S. 2003. Recomendaciones para la fertilización nitrogenada de trigo y maíz en la pampa ondulada. Informaciones Agronómicas 18. INPOFOS Cono Sur. Acassuso, Buenos Aires, Argentina, pp. 14–19.

Banerjee, H., Goswami, R., Chakraborty, S., Dutta, S., Majumdar, K., Satyanarayana, T., Jat, M. L. and Zingore, S. 2014. Understanding biophysical and socio-economic determinants of maize (*Zea mays* L.) yield variability in eastern India. *NJAS – Wageningen Journal of Life Sciences*. http://dx.doi.org/10.1016/j.njas.2014.08.001.

Banerjee, H., Goswami, R., Dutta, S., Chakraborty, S. and Majumdar, K. 2015. Farm typology-based phosphorus management for maize in West Bengal. *Better Crops-South Asia*, 9: 8–11.

Barbagelata, P. 2011. Fertilización fosfatada para trigo y maíz en siembra directa: Diagnóstico de fertilidad y estrategias de fertilización. Actas Simposio Fertilidad 2011. IPNI-Fertilizar AC, pp. 90–7.

Bayu, W., Rethman, N. F. G. and Hammes, P. S. 2005. The role of manure in sustainable soil fertility management in sub-Saharan Africa. *Journal of Sustainable Agriculture*, 25: 113–36.

Bender, R. R., Haegele, J. W., Ruffo, M. L. and Below, F. E. 2013. Modern Corn Hybrids' Nutrient Uptake Patterns. *Better Crops*, 97(1): 7-10

Benson, T. 1999. Validating and strengthening the area-specific fertilizer recommendations for hybrid maize grown by Malawian smallholders. A research report of the results of the nationwide1997/98 Maize Fertilizer Recommendations Demonstration, Ministry of Agriculture, Malawi.

Cassman, K. G. 1999. Ecological intensification of cereal production systems: Yield potential, soil quality, and precision agriculture. *Proc. Natl. Acad. Sci. U.S.A.*, 96: 5952–9.

Caviglia, O. and Andrade, F. 2010. Sustainable intensification of agriculture in the Argentinean Pampas: Capture and efficiency of environmental resources. *The Americas Journal of Plant Science and Biotechnology*, 3(1): 1–8.

Caviglia, O. P., Rizzalli, R. H., Van Opstal, N. V., Barbieri, P., Melchiori, R. J., Cerrudo, A., Gregorutti, V. C., Monzon, J. P., Barbagelata, P. A., Martinez, J. J., García, F. and Andrade, F. H. 2012. Productividad y eficiencia en el uso de agua y nitrógeno en sistemas intensificados. Proceedings XIX Latin American Soil Science Congress-XXIII Argentinean Soil Science Congress. Mar del Plata, Buenos Aires, Argentina. SLCS-AACS.

Chen, X., Cui, Z., Fan, M., Vitousek, P., Zhao, M., Ma, W., Wang, Z., Zhang, W., Yan, X., Yang, J., Deng, X., Gao, Q., Zhang, Q., Guo, S., Ren, J., Li, S., Ye, Y., Wang, Z., Huang, J., Tang, Q., Sun, Y., Peng, X., Zhang, J., He, M., Zhu, Y., Xue, J., Wang, G., Wu, L., An, N., Wu, L., Ma, L., Zhang, W. and Zhang, F. 2014. Producing more grain with lower environmental costs. *Nature*. doi: 10.1038/nature13609.

Ciampitti, I. A., García, F. O., Picone, L. I. and Rubio, G. 2011. Phosphorus budget and soil extractable dynamics in field crop rotations in Mollisols. *Soil Science Society of America Journal*, 75: 131–42.

Cordone, G. F., Martínez, F., Capurro, J. Y. and Abrate, R. 2001. Fertilización de maíz con nitrógeno y azufre en el centro-sur de la provincia de Santa Fe. VII Congreso Nacional de Maíz. AIANBA. Pergamino, Buenos Aires, Argentina.

Craswell, E. T. and Vlek, P. L. G. 2013. Mining of nutrients in African soils due to agricultural intensification. In R. Lal and B. A. Stewart (eds), *Principles of Sustainable Soil Management in Agroecosystems. Advances in Soil Science*, CRC press: Boca Raton, LA, pp. 401–22.

Dixon, J., Braun, H. J. and Crouch, J. H. 2008. Overview: transitioning wheat research to serve the future needs of the developing world. CIMMYT-IFPRI publication. CIMMYT, Mexico D.F.

Espósito, G. 2013. Análisis de la variabilidad espacio-temporal de la respuesta al nitrógeno en maíz mediante un modelo econométrico mixto espacial (MEME). Doctoral Thesis, Graduate School, Facultad de Ciencias Agropecuarias, Universidad Nacional de Cordoba, p. 108.

Espósito, G., Castillo, C. and Balboa, R. 2006. Calibración y validación de un método de diagnóstico de fertilización nitrogenada en maíz para el sur de Córdoba (Argentina). *RIA*, 35(3): 45–63.

FAO, 2015. FAOSTAT. Food and Agriculture Organization of the United Nations Website. http://faostat.fao.org.

Ferraris, G. N. Y. and Couretot, L. 2011. Fertilización nitrogenada de maíz bajo tres escenarios productivos. Proyecto Regional Agrícola, CERBAN. Maíz en Siembra Directa, Aapresid, pp. 33–8.

Fischer, R. A., Byerlee, D. and Edmeades, G. O. 2014. Crop yields and global food security: will yield increase continue to feed the world? ACIAR Monograph No. 158. Australian Centre for International Agricultural Research: Canberra, xxii + 634 pp.

Foley, J. A., Ramankutty, N., Brauman, K. A., Cassidy, E. S., Gerber, J. S., Johnston, M., Mueller, N. D., O'Connell, C., Ray, D. K., West, P. C., Balzer, C., Bennett, E. M., Carpenter, S. R., Hill, J., Monfreda, C., Polasky, S., Rockström, J., Sheehan, J., Siebert, S., Tilman, D. and Zaks, D. P. M. 2011. Solutions for a cultivated planet. *Nature,* 478: 337–42.

Fontanetto, H., Gambaudo, S., Keller, O., Albretch, J., Meroi, G., Weder, E., Gianinetto, G. Y. and Sillón, M. 2011. Fertilización nitrogenada en maíz. Maíz en Siembra Directa. Aapresid, pp. 39–42.

Fontanetto, H., Keller, O., Albretch, J., Boschetto, H., Negro, C., Belotti, L. Y. and Giailevra, D. 2010. Fertilización combinada de N y S. Maíz en Siembra Directa, Aapresid, pp. 85–90.

Fontanetto, H., Vivas, H. S., Albretch, R. Y. and Hotián, J. L. 2003. La fertilización con N, P y S y su residualidad en una secuencia agrícola de la región central de Santa Fe: Efecto sobre el rendimiento de granos. Simposio Fertilidad '2003: El Fósforo en la Agricultura Argentina, Rosario, Santa Fe, INPOFOS Cono Sur, pp. 91–3.

García, F. O., Boxler, M., Minteguiaga, J., Pozzi, R., Firpo, L., Ciampitti, I. A., Correndo, A. A., Bauschen, F., Berardo, A. Y. and Reussi Calvo, N. 2010. La Red de Nutrición de la Región CREA Sur de Santa Fe: Resultados y conclusiones de los primeros diez años 2000–2009. AACREA, p. 64.

Gerland, P., Raftery, A. E., Ševčíková, H., Li, N., Gu, D., Spoorenberg, T., Alkema, L., Fosdick, B. K., Chunn, J., Lalic, N., Bay, G., Buettner, T., Heilig, G. K. and Wilmoth, J. 2014. World population stabilization unlikely this century. *Science*, 346(6206): 234–7.

Giller, K. E., Cadish, G., Ehaliotis, C., Adams, E., Sakala, W. D. and Mafongoya, P. L. 1997. Building soil nitrogen capital in Africa. In R. J. Buresh, P. A. Sanchez and F. Calhoun (eds), Replenishing Soil Fertility in Africa. Soil Science Society of America Special Publication No. 51. Soil Science Society of America, Madison, Wisconsin, USA, pp. 151–92.

Giller, K. E., Tittonell, P., Rufino, M. C., Wijk, M. T. V., Zingore, S., Mapfumo, P., Adjei-Nsiah, S., Herrero, M., Chikowo, R., Corbeels, M., Rowe, E. C., Baijukya, F., Mwijage, A., Smith, J., Yeboah, E., Burg, W. J. V. D., Sanogo, O. M., Misiko, M., Ridder, N. D., Karanja, S., Kaizzi, C., K'ungu, J., Mwale, M., Nwaga, D., Pacini, C. and Vanlauwe, B. 2011. Communicating complexity: Integrated assessment of trade-offs concerning soil fertility management within African farming systems to support innovation and development. *Agricultural Systems*, 104: 191–203.

Grant, P. M. 1981. The fertilization of sandy soils in peasant agriculture. *Zimbabwean Agricultural Journal,* 78: 169–75.

Gregoret, C., Díaz Zorita, M., Dardanelli, J. and Bongiovanni, R. 2011. Regional model for nitrogen fertilization of site-specific rainfed corn in haplustolls of the central Pampas, Argentina. *Precision Agric.,* 12: 831–49.

Hall, A. J., Rebella, C. M., Ghersa, C. M. and Culot, J. Ph. 1992. Field crop systems of the Pampas. In C. J. Pearson (ed.), *Field Crop Ecosystems*. Ecosystems of the World, vol. 18. Elsevier, Amsterdam, pp. 413–50.

Heffer, P. 2013. *Assessment of Fertilizer Use by Crop at the Global Level.* AgCom/13/39. IFA. Paris, France.

Holland, K. H. and Schepers, J. S. 2010. Derivation of a variable rate nitrogen application model for in-season fertilization of corn. *Agronomy Journal,* 102: 1415–24.

IPNI. 2012. 4R Plant Nutrition: A Manual for Improving the Management of Plant Nutrition. In T. W. Bruulsema, P. E. Fixen and G. D. Sulewski (eds), International Plant Nutrition Institute, Norcross, GA, USA.

Jat, M. L., Satyanarayana, T., Majumdar, K., Parihar, C. M., Jat, S. L., Tetarwal, J. P., Jat, R. K. and Saharawat, Y. S. 2013. Fertilizer best management practices for maize systems. *Indian Journal of Fertilizers,* 9(4): 80–94.

Ju, X.T., Xing, G.X., Chen, X.P., Zhang, S.L., Zhang, L.J., Liu, X.J., Cui, Z.L., Yin, B., Christie,P., Zhu, Z.L. and Zhang, F.S. 2009. Reducing environmental risk by improving N man-agement in intensive Chinese agricultural systems. *Proceedings of the National Academy of Sciences U.S.A.*106, 3041–6.

Kirk, G. J. D., Yu, T. R. and Choudhury, F. A. 1990. *Phosphorus Chemistry in Relation to Water Regime.* Phosphorus Requirements for Sustainable Agriculture in Asia and Oceania. IRRI, Los Banos, pp. 211–23.

Kurwakumire, N., Chikowo, R., Mtambanengwe, F., Mapfumo, P., Snapp, S., Johnston, A. and Zingore, S. 2014. Maize productivity and nutrient and water use efficiencies across soil fertility domains on smallholder farms in Zimbabwe. *Field Crops Research,* 164: 136–47.

Madhavi, A., Balaguruvaiah, D., Shankariah, M., Manjulatha, G., Kiran Reddy, G., Prabhakar Reddy, Pavanchandra Reddy, Srinivas, A., Suresh, K., Satyanarayana, T., Dutta, S. and Majumdar, K. 2015. Comparative study on yield variability and P fertilizer use trends in the established and emerging maize growing districts of Telangana. *Better Crops-South Asia*, 9: 23–6.

Majumdar, K., Satyanarayana, T., Dutta, S., Pampolino, M., Jat, M. L., Shahi, V., Iftikar, W., Govil, V. and Singh, V. K. 2014. On-farm performance of Nutrient Expert® for maize: fertiliser recommendation, yield, and nutrient use efficiency. *Better Crops-South Asia*, 8: 21–3.

Majumdar, K., Jat, M. L. and Shahi, V. B. 2012. Effect of spatial and temporal variability in cropping seasons and tillage practices on maize yield responses in eastern India. *Better Crops-South Asia,* 6: 4–6.

Mancassola, V. and Casanova, O. 2015. Balance de nutrientes de los principales productos agropecuarios de Uruguay para los años 1990, 2000 y 2010. Informaciones Agrónomicas de Hispanoamérica 17: 2–13. IPNI. Acassuso, Buenos Aires, Argentina.

Melchiori, R. J. M. 2010. Enfoques alternativos para el diagnóstico de fertilidad de suelos: una mirada desde el cultivo para el manejo de nitrógeno. *Proceedings of the XXII Argentine Soil Science Congress.* 31 May to 4 June 2010. Rosario, Argentina. AACS.

Melchiori, R. J. M. 2012. Fertilización variable con N: herramientas, criterios y recomendaciones. 19th ISTRO Conference – IV SUCS Meeting. Striving for Sustainable High Productivity. 24–28 September 2012. Montevideo, Uruguay.

Michiels, C. L. Y. and Ruffo, M. L. 2012. El zinc limita el rendimiento del maíz en la región pampeana argentina. Actas CD XIX Congreso Latinoamericano – XXIII Congreso Argentino de la Ciencia del Suelo. 16–20 de Abril de 2012. Mar del Plata, Buenos Aires. Argentina.

Moschler, W. W. and Martens, D. C. 1975. Nitrogen, phosphorus, and potassium requirements in no-tillage and conventionally tilled corn. *Soil Science Society of America Journal,* 39: 886–91.

Mtambanengwe, F. and Mapfumo, P. 2005. Organic matter management as an underlying cause for soil fertility gradients on smallholder farms in Zimbabwe. *Nutrient Cycling in Agroecosystems,* 73: 227–43.

Mueller, N. D., Gerber, J. S., Johnston, M., Ray, D. K., Ramankutty, N. and Foley, J. A. 2012. Closing yield gaps through nutrient and water management. *Nature.* doi: 10.1038/nature11420.

Neumann, K., Verburg, P. H., Stehfest, E., Müller, C. 2010. The yield gap of global grain production: A spatial analysis. *Agricultural Systems.* doi:10.1016/j.agsy.2010.02.004.

Pagani, A., Echeverría, H. E., Andrade, F. H. and Sainz Rozas, H. R. 2012. Effects of nitrogen and sulfur application on grain yield, nutrient accumulation, and harvest indexes in maize. *Journal of Plant Nutrition*, 35: 1080–97.

Pagani, A., Echeverría, H. E., Sainz Rozas, H. and Barbieri, P. A. 2008. Dosis óptima económica de nitrógeno en maíz bajo siembra directa en el sudeste bonaerense. *Ciencia del Suelo,* 26(2): 183–93.

Pampolino, M. F., Witt, C., Pasuquin, J. M., Johnston, A. and Fisher, M. J. (2012). Development approach and evaluation of the Nutrient Expert software for nutrient management in cereal crops. *J. Computers and Electronics in Agri.,* 88: 103–10.

Picone, L., Videla, C., Bayer, C., Rizzalli, R. and García, F. 2013. Emisiones de óxido nitroso y dióxido de carbono en maíz bajo sistemas de producción con distinta intensificación. *Proceedings Symposium Fertility 2013.* IPNI LASC-Fertilizar AC, pp. 257–65.

Piha, M. 1993. Optimising fertilizer use and practical rainfall capture in semi-arid environment with variable rainfall. *Experimental Agriculture,* 29: 405–15.

Sainz Rozas, H. R., Echeverria, H. E. and Picone, L. I. 2001. Denitrification in maize under no–tillage: effect of nitrogen rate and application time. *Agron. J.,* 65: 1314–23.

Sainz Rozas, H. R., Echeverria, H. E., Studdert, G. A. and Andrade, F. H. 1999. No-tillage maize nitrogen uptake and yield: Effect of urease inhibitor and application time. *Agron. J.,* 91: 950–5.

Sainz Rozas, H., Calviño, P. A., Echeverría, H. E., Barbieri, P. A. and Redolatti, M. 2008. Contribution of anaerobically mineralized nitrogen to the reliability of planting or pre-sidedress soil nitrogen test in maize. *Agron. J.*, 100: 1020–5.

Sainz Rozas, H., Eyherabide, M., Echeverría, H. E., Angelini, H., Larrea, G. E., Ferraris, G. N. and Barraco, M. 2013. Cuál es el estado de la fertilidad de los suelos argentinos?. En F. García y A. Correndo (eds), Simposio Fertilidad 2013: Nutrición de Cultivos para la Intensificación Productiva Sustentable. 22–3 de Mayo de 2013. Rosario. IPNI Cono Sur-Fertilizar AC, pp. 62–72.

Sanchez, P. 2010. Tripling yields in tropical Africa. *Nature Geoscience*, 3: 299–300.

Satorre, E. H., Micheloud, J. R., Belligoi, J. and Cavasassi, J. L. 2006. MAICERO: Nuevos Criterios para el Diagnóstico y Manejo de la Fertilización del Cultivo de Maíz en Argentina. Convenio AACREA y PROFERTIL S.A., software de aplicación agronómica.

Smale, M. and Jayne, T. 2003. Maize in eastern and southern Africa: seeds of success in retrospect. EPTD Discussion Paper No. 97. Washington: Environment and Production Technology Division, International Food Policy Research Institute.

Snapp, S. S., Mafongoya, P. L. and Waddington, S. 1998. Organic matter technologies for integrated nutrient management in smallholder cropping systems of southern Africa. *Agriculture, Ecosystems and Environment*, 71: 185–200.

Stoorvogel, J. J. and Smaling, E. M. A. 1998. Research on soil fertility decline in tropical environments: integration of spatial scales. *Nutrient Cycling in Agroecosystems*, 50: 151–8.

Sun, B., Zhang, L., Yang, L., Zhang, F., Norse, D. and Zhu, Z. 2012. Agricultural non-point source pollution in china: causes and mitigation measures. *AMBIO*, 41: 370–9.

Timsina, J., Buresh, R. J., Dobermann, A. and Dixon, J. 2011. *Rice-Maize Systems in Asia: Current Situation and Potential*. Los Banōs (Philippines), International Rice Research Institute and International Maize and Wheat Improvement Center, p. 232.

Timsina, J., Jat, M. L. and Majumdar, K. 2010. Rice-maize systems of South Asia: current status, future prospects and research priorities for nutrient management. *Plant Soil*, 335: 65–82.

Timsina, J., Singh, V. K. and Majumdar, K. 2013. Potassium management in Rice-maize systems in South Asia. *Journal of Soil Science and Plant Nutrition*. doi: 10.1002/jpln.201200253.

Tittonell, P., Vanlauwe, B., Leffelaar, P. A., Shepherd, K. D. and Giller, K. E. 2005. Exploring diversity in soil fertility management of smallholder farms in western Kenya: II. Within-farm variability in resource allocation, nutrient flows and soil fertility status. *Agriculture Ecosystems and Environment*, 110: 166–84.

Tollenaar, M., McCullough, D. E. and Dwyer, L. M. 1994. In G. A. Slafer (ed.), *Genetic Improvement of Field Crops*, Dekker, New York, pp. 183–236.

Torres Duggan, M., Melgar, R., Rodríguez, M. B., Lavado, R. S. and Ciampitti, I. A. 2012. Sulfur fertilization technology in the Argentine Pampas region: a review. *Rev. Agronomía and Ambiente*, 32(1–2): 61–73. FA-UBA, Buenos Aires, Argentina.

Twomlow, S., Rohrbach, D., Dimes, J., Mupangwa, W., Ncube, B., Hove, L., Moyo, M., Mashingaidze, N. and Maphosa, P. 2010. Micro-dosing as a pathway to Africa's Green Revolution: evidence from broad-scale on-farm trials. *Nutrient Cycling in Agro-ecosystems*, 88: 3–15.

Vanlauwe, B., Descheemaeker, K., Giller, K. E., Huising, J., Merckx, R., Nziguheba, G., Wendt, J. and Zingore, S. 2015. Integrated soil fertility management in sub-Saharan Africa: unravelling local adaptation. *Soil*, 1: 491–508.

Vanlauwe, B. and Giller, K. E. 2006. Popular myths around soil fertility management in sub-Saharan Africa. *Agriculture, Ecosystems and Environment*, 116: 34–46.

Vanlauwe, B., Kihara, J., Chivenge, P., Pypers, P., Coe, R. and Six, J. 2011. Agronomic use efficiency of N fertilizer in maize-based systems in sub-Saharan Africa within the context of integrated soil fertility management. *Plant Soil*, 339: 35–50.

Van der Eijk, D., Janssen, B. H. and Oenema, O. 2006. Initial and residual effects of fertilizer phosphorus on soil phosphorus and maize yields on phosphorus fixing soils. A case study in south–west Kenya. *Agriculture, Ecosystems and Environment*, 116: 104–20.

Weil, R. R. and Mughogho, S. K. 2000. Sulfur nutrition in maize in four regions of Malawi, *Soil Science Society of America Journal*, 92: 649–56.

Wendt, J. W. and Rijpma, J. 1997. Sulphur, zinc, and boron deficiencies in the Dedza Hills and Thiwi-Lifidzi regions in Malawi. *Tropical Agriculture*, 74: 81–9.

Xu, X., He, P., Pampolino, M. F., Johnston, A. M., Qiua, S., Zhao, S., Chuan, L. and Zhou, W. 2014. Fertilizer recommendation for maize in China based on yield response and agronomic efficiency. *Field Crops Research*, 157: 27–34.

Zhao, R. F., Chen, X. P., Zhang, F. S., Zhang, H., Schroder, J. and Romheld, V. 2006. Fertilization and nitrogen balance in a wheat–maize rotation system in North China. *Agronomy Journal*, 98: 938–45.

Zingore, S. 2011. Maize productivity and response to fertilizer use as affected by soil fertility variability, manure application, and cropping system. *Better Crops with Plant Food*, 95: 4–6.

Zingore, S., Delve, R. J., Nyamangara, J. and Giller, K. E. 2008. Multiple benefits of manure: The key to maintenance of soil fertility and restoration of depleted sandy soils on African smallholder farms. *Nutrient Cycling in Agroecosystems*, 80: 267–82.

Zingore, S., Murwira, H. K., Delve, R. J. and Giller, K. E. 2007. Influence of nutrient management strategies on variability of soil fertility, crop yields and nutrient balances on smallholder farms in Zimbabwe. *Agriculture Ecosystems and Environment*, 119: 112–26.

Chapter 9

Crop rotation: a sustainable system for maize production

Bao-Luo Ma, Ottawa Research and Development Centre, Agriculture and Agri-Food Canada; and Zhigang Wang, Inner Mongolia Agricultural University, China

1 Introduction: escalating global food demand and environmental challenges

The challenge of securing the global food supply and environmental sustainability is increasingly recognized by policy makers and scientific communities (Ma and Wu, 2016). Considering both global population growth and increased consumption of animal protein, global food demand is likely to double by 2050 against a backdrop of a changing climate and an unsustainable use of land, water, energy and labour (Tilman et al., 2011). As the major caloric source of human beings, production of cereal crops such as maize (or corn; *Zea mays* L.) has grown fast to fulfil food demand during the past century (Wu and Ma, 2015), with modern agricultural practices such as intensive mono-cropping, genetic improvements and the heavy use of agrochemicals for nutrient supply and pest/disease control.

It was believed that the invention of chemical fertilizers and pesticides could forever replace crop rotation without loss of yield quantity and quality, but that opinion has changed (Bullock, 1992). Many studies have demonstrated that simplification of mono-cropping systems has been developed at the expense of agricultural sustainability

http://dx.doi.org/10.19103/AS.2016.0002.10

Published by Burleigh Dodds Science Limited..

(Lichtfouse et al., 2009; Wu and Ma, 2015). For example, the injudicious use of chemical fertilizers, such as nitrogen (N) fertilizer, is one of the fundamental causes of environmental pollution, such as eutrophication, soil acidification, accumulation of reactive N and greenhouse gas (GHG) emissions (Davidson et al., 2014; Guo et al., 2010; IPCC, 2014). More and more agricultural scientists, policy makers and agri-food producers worldwide have now recognized the importance of developing self-sustainable, energy-efficient and highly diversified agricultural systems.

2 The simplification of cropping systems and associated problems for sustainability

Crop rotation or growing a sequence of different plant species on the same land in different growing seasons has been practised for thousands of years since ancient Chinese, Greek and Roman times (Karlen et al., 1994). The British Agricultural Revolution (1750–1880) was, in a large part, associated with the development of the Norfolk four-crop rotation, which was observed to greatly increase crop yields and livestock productivities due to improving soil fertility and reducing fallow (Overton, 1996). With the influence of settlers from Europe, and the discovery of biological N fixation by legumes, 3- to 7-year crop rotation was prevalent in North America (Karlen et al., 1994) and other agricultural regions of the world since the early part of the twentieth century, especially in the dry climatic regions. The yield benefit to maize from crop rotation is well documented. Long-term experimental results from Missouri (Johnson, 1927) showed that, after 30 years of rotation (without fertilizers but 6.8 t ha^{-1} manure application annually in both rotation and continuous cropping systems), maize yield increased by 60.4% in the 6-year rotation (maize, oat, wheat, clover and timothy) and 30.8% in the 3-year rotation (maize, wheat and clover) over that of the continuous mono-cropping. Results from Ohio (Johnson, 1927) with fertilizers use instead of manure showed an increased grain yield of rotational maize, 29.9% above that of continuous mono-cropping. In Delaware, maize yields increased by 157% in a rotation of maize (including a cover crop of rye and vetch, soybean, wheat, clover and timothy), as compared to continuous maize mono-cropping, when no fertilizer or manure was applied (Johnson, 1927). During that period, the main consideration on crop rotation was the need for some sort of legume to provide N for successive crops or green manure crops to maintain fertility of the soils.

After World War II, innovations of modern agricultural practices have led to a simplification of the components of maize-based agricultural systems globally, which is characterized by the reduction of extended crop rotation and increasing extensive mono-cropping maize (MM) (Karlen et al., 1994; Wu and Ma, 2015). Deen et al. (2016) summarized seven main drivers leading to this simplification of crop rotations currently practised. First, the discovery of the Haber–Bosch process in 1909 enabled artificial N fixation and production of ammonia. This large-scale industrial fixation of N has displaced the main value of legumes from the rotation and enabled crop production without livestock, leading to an 800% increase in agricultural consumption of N during the twentieth century (Tilman et al., 2001). Second, the role of crop rotations for weed suppression or control of pests and diseases has largely been eliminated by the introduction of synthetic herbicides and pesticides. For instance, herbicides such as 2,4-D and atrazine could selectively kill weeds without injuring the crop plants. This greatly reduced the need for complex rotations. Third, genetic improvement has led to crop plant

Published by Burleigh Dodds Science Limited.

varieties with better tolerance to biotic (diseases, pests, weeds, etc.) and abiotic stresses (drought, cold, heat, crowding, etc.) (Tollenaar and Lee, 2004). This has further reduced the crop rotation complexity. The introduction of genetically modified crops with herbicide tolerance or resistance to insects has reduced the reliance on crop rotation for herbicides/pesticides-based controls for weeds, pests and diseases. Fourth, with the increased use of synthetic fertilizers that replace the nutrients from livestock manure, livestock production is increasingly separated from crop production systems at both the farm and regional scale (Beaulieu et al., 2001; Boschma et al., 2015). This has resulted in the removal of forages from crop rotations, as well as increased production of grain maize, soybean and maize silage (Rankin, 2015). Fifth, advancements in mechanization further promoted the adoption of short-term rotations. A reduced number of crops in rotations enable farmers to own fewer pieces of equipment and reduce capital investment in equipment, and specialize in the production of fewer crops and take advantage of economies for large-scale operations (Johnston, 2013; McGranahan, 2014). Sixth, other factors such as government policies, favourable economics driven partly by growing demand for a narrow spectrum of crops, market-driven commodity programmes that emphasized short-term profit, increased prevalence of rented land, public and private research and development efforts devoted to genetic improvement of maize and soybean, and increased food and industrial uses for both maize and soybean oils and various by-products, all drove the diversity of crop rotation to a simplified mode (Karlen, 2004; Porter et al., 2003).

Simplification of maize-based cropping system in favour of MM or short rotations is generally profitable in the short time. However, the change may lead to some negative consequences in the long run. For example, in the corn belt of northeast China, the largest maize-growing region in Asia, long-term intensified maize mono-cropping, without any organic amendment but only inorganic fertilizers, has caused a large reduction in soil organic matter (SOM) and degraded soil aggregate structure (Wang et al., 2010). Owing to long-term maize mono-cropping with intensive conventional tillage, a plow pan is formed in the shallow layer (often <15 cm), resulting in diverse physical and chemical limitations for root growth (Wang et al., 2014, 2015). In addition, long-term, continuously excessive application of N fertilizer, especially urea, has caused significant soil acidification (Guo et al., 2010). Deterioration in soil quality caused directly or indirectly by mono-cropping has not only induced maize yield stagnation, but also limited resource use efficiency. The negative effects of mono-cropping on the environment have been widely noted, including but not limited to, the contamination of land and freshwater, increasing GHG emissions and loss of biodiversity (Yu and Qiao, 2016; Liang et al., 2016), even though some of these effects are site-specific.

From a long-term perspective, the negative effect of mono-cropping is expected with the duration of simple rotation. First, the longer the period the land is under a simple rotation or mono-cropping, the greater probability of yield reductions because infestations of weeds, insects or diseases are expected, as external inputs such as pesticides and improved genetics can only reduce but cannot eliminate the impact of these pests. The loss of resistance of crop varieties to specific diseases, and emerging of bugs, pathogens and weeds to effective pesticides and herbicides has been often reported. Second, simple crop rotations or mono-cropping are associated with reduction of SOM and overall soil quality (Congreves et al., 2015). For example, reduction in SOM will lead to the deterioration of soil aggregate stability (Congreves et al., 2015; Munkholm et al., 2013) and increase soil erosion (Stonehouse et al., 1998). Reduced SOM also results in reduced soil biological activities, which in return influence soil fertility (Tiemann et al., 2015). Third, yield of crops under simple rotations or mono-cropping appeared to be more vulnerable to abnormal moisture conditions, particularly drought (Gentry et al., 2013). In hot and dry years, diversification of maize–soybean rotations with additional crop species and conservation tillage has increased maize yield by 7% and soybean yield by 22%, respectively (Gaudin et al., 2015). Yield reduction and variability of simple rotations or mono-cropping associated with drought may be exacerbated by climate change. Global climate change is inevitable with warmer, drier summers and more variable precipitation patterns in the mid latitude regions (IPPC, 2014). In contrast, the increase in yield potential of maize and soybean would increase crop water requirements (Richards, 2000). Soils associated with more complex crop rotations tend to have greater plant water availability. Fourth, simple rotation or mono-cropping impacts negatively on soil quality and soil erosion. Crops like maize and soybean are much less effective in reducing soil erosion, consequently reducing water infiltration and nutrient retention, in contrast to perennial species (Asbjornsen et al., 2014). Integration of deep-rooted perennial plants into mono-cropping systems dominated by maize and soybean fields resulted in a 95% reduction in sediment export, a 90% reduction in total phosphorus (P) export and a 85% reduction in total N export (Helmers et al., 2012). Fifth, mono-cropping or simple maize–soybean rotations requires greater N inputs than more complex rotations that include forages, which means increasing risk of N loss to the environment. Gentry et al. (2013) identified three factors that are predictors of yield penalty under continuous maize production system: (1) the unfertilized continuous maize yield, which reflects net soil N mineralization; (2) the years in continuous maize – a measure of the effects of accumulated maize residue in continuous maize systems; and (3) the difference between continuous maize and soybean maize delta yields (maximum yield – unfertilized yield) – a measure of the interaction between yearly weather patterns and crop rotation. They pointed out that N availability was one of the primary causative agents of the continuous maize yield penalty, which was induced by maize crop residue accumulation. The larger carbon (C)/N ratio of maize crop residues and the greater quantity of biomass remaining after maize harvest compared with soybean production explains the observation of reduced net soil N mineralization in continuous maize systems (Kaboneka et al., 1997; Gentry et al., 2001). Additionally, because of crop residue-induced N immobilization and the unpredictable nature of organic N mineralization in continuous maize, growers often apply an additional 45 kg N ha^{-1} or more to assure sufficient crop N availability (Blackmer et al., 1997; Ding et al., 1998). Sixth, simple crop rotations are associated with greater tillage requirements (McLaughlin et al., 2002; Pittelkow et al., 2014), reduced fuel use efficiency (Liang et al., 2013) and lower yield potential, increased N fertilizer requirements and lower SOM, which result in greater GHG emissions and C footprints (Meyer-Aurich et al., 2006; Ma et al., 2012; Liang et al., 2016). Seventh, simple rotations or mono-cropping are associated with larger weed seedbanks, increased weed pressure and rapidity of herbicide resistance. Herbicide resistance is now widespread for many commonly used herbicides (Shaner, 2014) and represents an emerging challenge to simple crop rotations or mono-cropping (Mortensen et al., 2012).

3 Yield enhancement in maize–legume rotation systems

During the past four decades, many studies have clearly illustrated the yield advantage of maize in a legume rotation system over continuous maize mono-cropping (Howard et al.,

Published by Burleigh Dodds Science Limited.

1998; Katsvairo and Cox, 2000; Ma et al., 2003; Varvel and Wilhelm, 2003; Gentry et al., 2013). Currently, the demand for nutritional and safer grain is increasing, and adoption of crop rotation practices with organic amendment and conservation tillage to produce healthy grains is gaining popularity. Many farmers, researchers and policy makers worldwide now recognize the importance of developing innovative cropping systems that can guarantee higher crop yields while minimizing further deterioration of the environment (Robertson et al., 2014; Ma and Wu, 2015; Ashworth et al., 2016; Deen et al., 2016). As the most popular crop in North America, maize, when grown in rotation with legume crops such as alfalfa (*Medicago sativa* L.) or soybean (*Glycine max* L.), often produces greater yields and requires less N fertilizer than continuous mono-cropping (Edwards et al., 1988; Ma et al., 2003; Advient-Borbe et al., 2007). In a recent long-term crop rotation study, Ashworth et al. (2016) showed that across the 12 years of a 4-year cycle, maize–soybean–maize–soybean rotation was the highest yielding (10.1 Mg ha^{-1}) during Phase III, which was greater than continuous maize during all phases. The increased yield has been attributed to the beneficial effects of crop rotation, including a better balance of plant nutritional factors (Chan et al., 2013; Ma et al., 2016), improved soil physical properties, interrupted insect populations, increased root activity, a shift in soil mycorrhizal populations, reduced disease severity (Reid et al., 2001) and enhanced seasonal N mineralization (Howard et al., 1998; Smith and McSorley, 2000; Ma et al., 2003; Wu et al., 2008). More importantly, legume–cereal rotation systems not only maximize cereal and legume productivity, but also optimize agricultural sustainability across a far more complex landscape of production, which has recently caused unprecedented changes in our environment (Davis et al., 2012; IPCC, 2014). Thus, this practice has been regarded as a promising strategy for reducing GHG emissions, and lowering the negative environmental impact for a sustainable agricultural development (Davis et al., 2012; Ma et al., 2012).

In eastern Ontario, a long-term rotation experiment was carried out at the Central Experimental Farm of Agriculture and Agri-Food Canada, Ottawa, Ontario. The soil in the initial year contained an average of 340, 270 and 390 g kg^{-1} of clay, silt and sand, respectively, and had a water pH of 6.5 (Ma et al., 2003). Maize in biennial rotation with soybean (SM), and maize in biennial rotation with forage legume (FM; alfalfa 1992–2010 or red clover 2011–15) were compared to continuous MM. All the rotation phases (i.e. maize–soybean and soybean–maize, maize forage legume and forage legume–maize) were included so that MF, MS and MM presented each year. In maize, preplant N fertilizer was applied as urea at rates of 0, 100 kg N ha^{-1} and fall-applied manure (50 t ha^{-1} fresh weight) either uncomposted stockpiled (SML) or incompletely composted (RML), respectively in all cropping systems. In the MM system, high rates of N, 200 kg N ha^{-1}, and 100 t ha^{-1} of stockpiled (SMH) or partially composted (RMH) manure treatments (Ma et al., 1999a) were also tested. No N fertilizer was applied on alfalfa (or red clover since 2011) or soybean plots, but the entire field, including alfalfa and soybean plots, was fertilized with ample P and potassium (K) fertilizers according to the soil test recommendations prior to maize planting each year (Ma et al., 2012; 2016). Several trends can be clearly drawn based on the actual yields of the well-fertilized treatment (200 kg N ha^{-1}) in MM, and the relative yields of other treatments in MS and MF (Fig. 1) as well as in MM (Fig. 2) systems: (1) grain yields of maize were highly influenced by weather conditions; predominately rainfall amounts and distribution during the maize growing seasons. Over the past 24 years, maize yields under the well-fertilized treatment varied from as low as 3.1 Mg ha^{-1} to as high as 10.7 Mg ha^{-1} in MM system (Fig. 1; actual yields of maize under the well-fertilized treatment –200 kg N ha^{-1}). (2) With the application of a suboptimum rate of N fertilizer

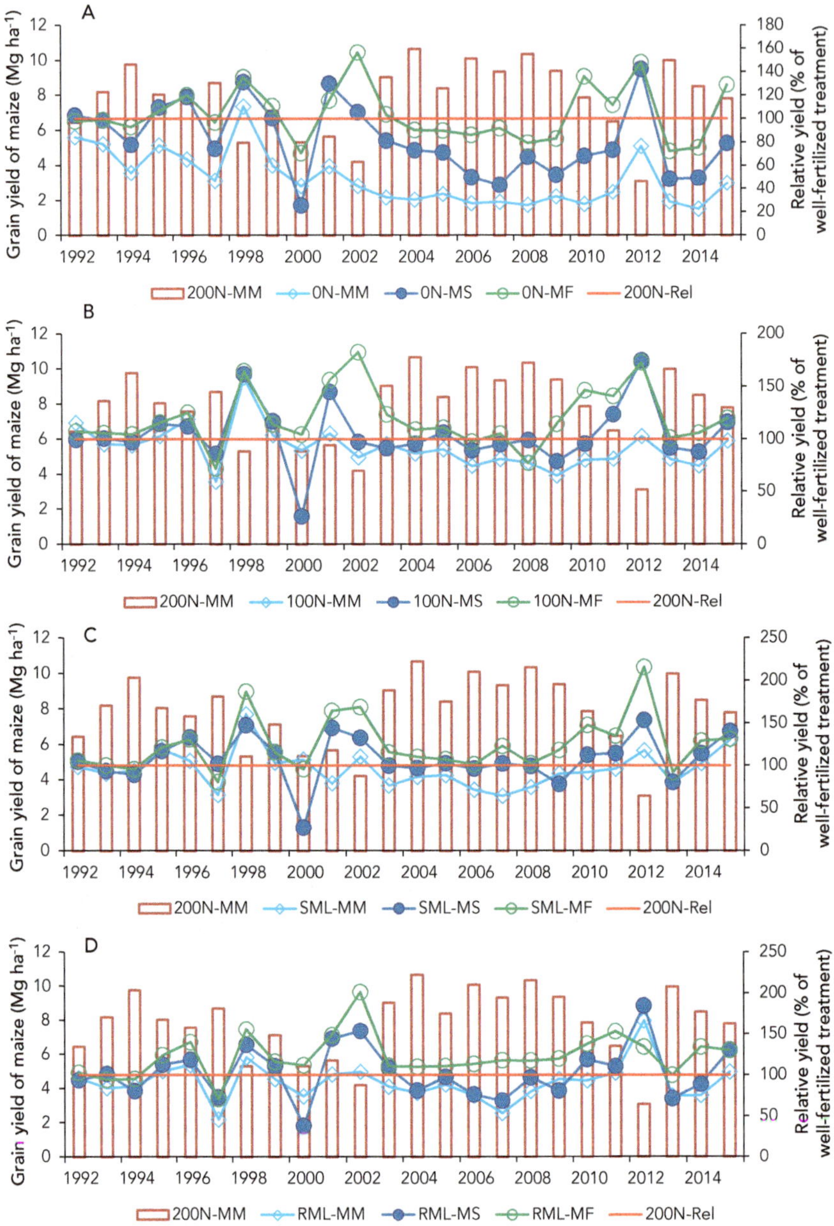

Figure 1 The grain yield of well-fertilized maize and the relative grain yields of rotational maize as a percentage of the well-fertilized (200 kg N ha^{-1}) continuous maize monoculture from 1992 to 2015. 0N, 100N, without N or with preplant N fertilizer at 100 kg N ha^{-1}; SML, RML, fall-applied stockpiled or partially composted (rotted) manure at 50 t ha^{-1} (fresh weight); MM, continuous maize monoculture; MS, maize following soybean in a biennial rotation without fertilizer N application; MF, maize following forage legume (alfalfa or red clover) in a biennial rotation without fertilizer N application.

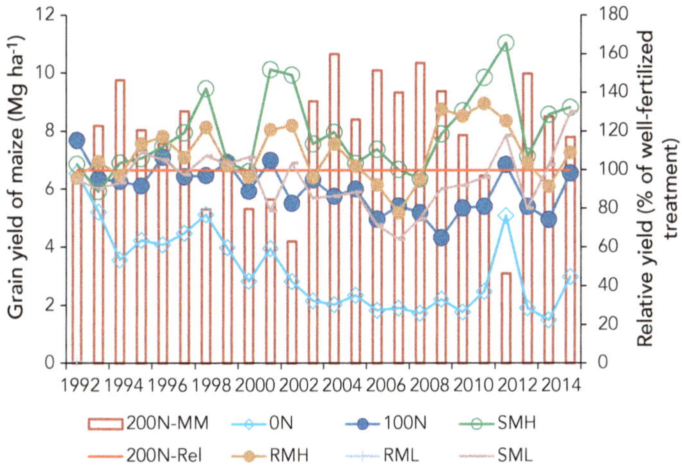

Figure 2 The grain yield of well-fertilized maize and the relative grain yields of continuous MM under different soil amendment, as a percentage of the well-fertilized (200 kg N ha^{-1}) continuous maize monoculture from 1992 to 2015. 0N, 100N, 200N, without N or with preplant N fertilizer at 100 or 200 kg N ha^{-1}, respectively; SML, RML, fall-applied stockpiled or partially composted (rotted) manure at 50 t ha^{-1} (fresh weight); SMH, RMH, fall-applied stockpiled or rotted manure at 100 t ha^{-1} (fresh weight).

(100 kg N ha^{-1}), the relative yields were on average 104% in MS and 120% in MF (Fig. 1B), compared to those of MM under the well-fertilized treatment –200 kg N ha^{-1}. (3) Without N application, maize yields declined dramatically in MM, but 80–100% yields of maize were achievable in MF or MS (Fig. 1A). (4) The highest and stable yields of maize were obtained in MF and MS rotation systems amended with stockpiled (Fig. 1C) or partially composted manure (Fig. 1D). (5) Repeated amendments with high rate of stockpiled or partially composted manure treatments produced higher yields, especially in years with poor weather conditions for maize growth. For example, insufficient crop heat units and frequent rainstorms in year 2000, and growing season of 2012 with severe drought stress, dramatically reduced maize yields of the well-fertilized treatment, while decent grain yields of maize were achieved in all manure-applied plots in these environments (Fig. 2). Overall, this indicates that both continuous improvement in maize yield and sustaining the environment are feasible with a well-planned crop rotation and better nutrient cycling efficiency in the agroecosystems.

Canola (*Brassica napus* L.) was developed by Canadian plant breeders as a healthy human oil and nutritious animal feed, using traditional plant breeding techniques, and is now a major cash crop in western Canada. Canola production is far less common in eastern Canada. However, recent operationalization of a canola and soybean processing and refining plant in Bécancour, Quebec, has brought producers closer to market, which has significantly brightened the prospects of canola production in eastern Canada (Ma et al., 2015). Research efforts have been made to develop knowledge and technology required for expanding canola production in eastern Canada. One of the research activities was to investigate how the new crop would fit into the existing cropping systems. A 4-year phase rotation (canola–wheat–maize–soybean) was initiated in 2011 and is ongoing. Some preliminary data showed that grain yields of rotational maize were significantly affected by the preceding crops, with order of soybean > canola > wheat > maize. During the first 4 years, maize yields increased by 13–41% when following soybean, and by 8–39% ($p < 0.0001$) when following wheat, compared to MM. The yield ratio (rotational canola yield as a fraction of the average mono-cropping canola yield) was also affected by the previous 2-year sequence of crops, with canola yield increased by 2–65% when following the maize–soybean sequence, compared to mono-cropping.

Using a meta-analysis approach, Ma and Wu (2016) recently assembled 148 pairs of data from long-term crop rotation experiments and quantified the maize–legume rotation effect, in comparison with continuous maize mono-cropping system (MM). They reported that compared with MM, maize yield in the maize-grain legume (soybean, field pea, etc.) system increased by 28.7% with a confidence interval (CI) of 22.1–35.6%; in the maize–green manure rotation, maize yield increased by an average of 39.9% with CI of 28.2–54.0%.

4 The impact of crop rotation on soil quality

It is not easy to estimate the contribution of maize–legume rotation to improve soil quality parameters. Kay (1990) reported that crops with deep roots, which could penetrate soil layers and increase the number of larger soil pores in rotation, could enable the shallow root crop to obtain water and nutrients from deeper soil layer; and plant species with greater rhizodeposition in rotation could enhance soil C persistence and improve soil structure. In contrast, Glab et al. (2013) indicated that crop rotation did not play a significant role in soil structure modification even the crops differed in terms of their impact on soil porosity. Bullock (1992) showed that abandonment of multiyear rotations in favour of a simple biennial rotation has generally resulted in a degradation of soil structure. For example, a maize–soybean rotation leads to greater yields of both maize and soybean, but this rotation often degrades soil aggregation faster than the well-fertilized continuous maize, due to less crop residue returns from soybean (Power, 1990). The primary cause for this response appears to be related to soil organic carbon content (SOC). West and Post (2002) illustrated that a change from continuous maize to maize–soybean rotation did not result in changes of SOC, however, enhancing rotation complexity could sequester an average of 200 ± 120 kg C ha^{-1} yr^{-1}. They speculated that following the initiation of an enhancement in rotation complexity, SOC could reach a new equilibrium in approximately 40–60 years.

Maize in rotation with legume crops is generally beneficial to aggregate soil stability and formation of favourable soil structure (Robinson et al., 1994; Kay, 1990). Compared to the soil properties under continuous maize, soil aggregate stability in most years was the highest under continuous alfalfa or where a legume (either alfalfa or red clover) was grown in rotation with maize (Raimbault and Vyn, 1991). Hussain et al. (1988) also concluded that crop rotation can increase soil aggregation over time, based on geometric mean diameter (GMD) values of soil particle size. They reported average GMD values of 150, 211 and 311 μm, respectively, for continuous maize mono-cropping, maize following soybean and an oat/sweet clover mixture following maize.

The ability of non-leguminous winter cover crops, spring barley (*Hordeum vulgare* L.), fall rye (*Secale cereale* L.) and annual ryegrass (*Lolium multiflorum* Lam.), to affect SOC, total and dilute acid extractable polysaccharides and aggregate stability, expressed as mean weight diameter (MWD), was assessed in a study involving both a field trial and a laboratory incubation experiment (Liu et al., 2005). The researchers found that all the cover

Published by Burleigh Dodds Science Limited.

crop amendments increased MWD and percentages of water stable 2–6 mm aggregates at all incubation periods. Soil aggregate stability highly correlated with dilute acid extractable polysaccharides both in the field and incubation experiments. This study suggests that the dilute acid extractable polysaccharide fraction represents active binding agents under short-term cover crops. It has been shown that soil aggregate stability can be increased under 8-month non-leguminous cover crops in the intensively cultivated soils (Liu et al., 2005).

Maize–legume rotation also influences soil erodibility. Hussain et al. (1988) reported that the rate of splash detachment from continuous maize field was higher than from soil under crop rotation (48 vs 31 mg cm^{-3}). Over a 9-year period, 793 Mg ha^{-1} of soil was lost from the continuous maize mono-cropping, compared to loss of 202 Mg ha^{-1} in a maize–oat–sweet clover rotation (Johnston et al., 1942). On the contrary, Bullock (1992) indicates that 22-year maize–soybean rotations can result in greater soil erosion than continuous maize. For example, over an 18-year period, soil loss from a 22-year maize–soybean rotation was 45% higher than that from continuous maize (van Doren et al., 1984). This often occurs because the amount of surface residue following soybean is very low (Power, 1990). The contradictory reports of crop rotation on soil quality parameters are likely associated with the weather, geographic and soil conditions under the individual investigations.

Soil quality is a complex trait, being influenced by soil physical, chemical and biological properties. Therefore, it is difficult in quantifying and integrating the numerous attributes in soil quality assessment. Using the Cornell Soil Health Assessment (CSHA) method, Congreves et al. (2015) appraised the overall soil quality and showed the highest CSHA score for maize in rotation with alfalfa, and the lowest score in MM.

Manure is a by-product in agriculture production and animal husbandry. Appropriate use of farm manure is not only important for improving crop productivity, but also has a large impact on the environment. For example, in the Pakistan-administered Kashmir region, it was found that combination of fertilizer N and poultry manure increased dry matter and grain yields of maize, and improved SOM, total N, available P and available K, compared with the control and fertilizer N alone treatment (Abbasi et al., 2010). The application of composted poultry manure has also enhanced maize water use efficiency and thus displayed the greater marginal benefits, compared with those of inorganic fertilizer treatments (Farhad et al., 2013).

5 The impact of crop rotation on soil fertility and nutrient use efficiency

The benefit of crop rotations, especially maize–legume rotation on improving nutrient use efficiencies (mainly N and P) and reducing losses of N to surface and groundwater has been illustrated by a number of studies (Ding et al., 1998; van Kessel and Hartley, 2000). Several earlier studies showed that soybean and alfalfa, which do not require supplemental N inputs, can effectively use or 'scavenge' residual N remaining in the soil from previous crops such as maize (Johnson et al., 1975; Mathers et al., 1975; Muir et al., 1976). Research has shown that nitrate can be utilized by the crop from any depth where plant roots penetrate. Olson et al. (1970) reported that soil nitrate concentration was reduced by 34–82% at a depth of 1.2–1.5 m in a maize–alfalfa rotation, compared to continuous maize field. They illustrated that the decrease in solution nitrate was directly proportional to the number of years in alfalfa production. Soybean can also scavenge residual soil N, but was not as effective as alfalfa because of the shallower rooting depth of soybean compared to alfalfa (Havlin et al., 1990; Karlen et al., 1991).

Besides N nutrition, crop rotation often promotes changes in various nutrient sources and, subsequently, affects availability of these nutrients to the plant. Several studies have demonstrated better synchronization between N/P mineralization and crop uptake in rotation cropping systems than in mono-cropping (Ma et al., 1999a,b, 2003, 2016; Varvel and Wilhelm, 2003; Wu et al., 2008). Total N uptake and N use efficiency (NUE) of maize, especially during the grain-filling stage, were significantly higher under maize–legume rotations than in continuous maize cropping system (Ma et al., 2003), resulting in superior yield performance (Franke et al., 2008; Gentry et al., 2013). The greater accumulation of N or more N credit resulting from the preceding alfalfa crop in the rotation is the main contribution to a superior maize yield in a maize–alfalfa rotation than in continuous maize or even a maize–soybean rotation. In a long-term sunflower–wheat–pea rotation, Rodríguez-Lizana et al. (2010) illustrated distinct rates of crop residue decompositions: at the end of a 3-year decomposition cycle, 13.5, 6.7 and 8.5 kg N ha^{-1}; 2.9, 0 and 0.7 kg P ha^{-1}; and 9.9, 22.5 and 78 kg K ha^{-1}, respectively released from pea, wheat and sunflower crop residues. This indicates that while the direct contribution of N release from crop residue to succeeding crops may not be significant, the large differences in K release pattern will play a significant role in K nutrition and other nutrients in balance.

Root exudates of some legume varieties can help release unavailable P into the soil for recovery by the legumes themselves or the rotated crops (Liu et al., 2005; Sinclair and Vadez, 2012). Copeland and Crookston (1992) observed that maize in a 2-year rotation with soybean accumulated significantly more P than did maize in continuous mono-cropping. This indicates that maize yield increases associated with crop rotation may have been due to improved general plant nutrition. In the same study, they reported that K and total micronutrient content increased for maize in rotation with soybean as compared to continuous mono-cropping. The general improvement in plant nutrition appears to be associated with an improvement in maize root function, and causal agents such as mycorrhizae (Liu et al., 2002). The same research group (Copeland et al., 1993) also demonstrated an increased water use by first-year rotated maize or increased water use efficiency of rotated soybean, as compared to continuous mono-cropping. Their data appear to indicate that rotation increased root surface and/or root activity which in turn improved water relations and increased grain yield.

The benefit of maize–legume rotation is also shown in smallholder farming systems of sub-Saharan Africa and mountain regions of Nepal. In Zimbabwe, for example, maize yield was up to 61% greater after jack-bean [*Canavalia ensiformis* (L.) DC] than that of continuous maize, and the number of weeds was deceased by up to 94% in maize–black sunn hemp (*Crotalaria juncea* L.) rotation (Mhlanga et al., 2015). However, the use of pigeon pea [*Cajanus cajan* (L.) Millsp.] as a cover crop in a maize–cover crop rotation system displayed higher densities of *Bidens pilosa* compared with maize mono-cropping (control treatment). This was likely due to the large N requirement of rotational maize to produce more seeds (Mhlanga et al., 2016). Integrated weed management suitable for rotation types is needed to achieve both the yield benefit and conservation agriculture in the specific rotation systems.

An important consideration in improving both yield and nutrient use efficiency for sustainable crop production is nutrient balance. Taking N and P nutrients as an example, the increased yields of maize under legume–maize rotation systems can be partially explained by the fact that P uptake in the maize–soybean and maize–forage rotation systems was

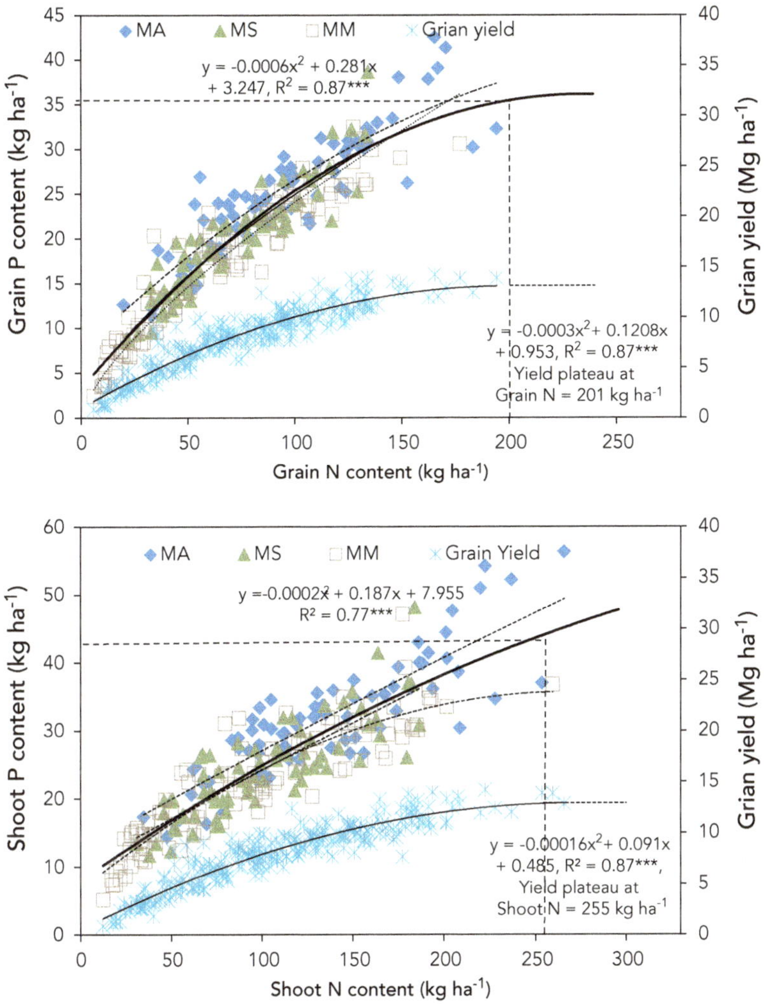

Figure 3 Relationships between grain and shoot N and P contents, and between grain yield and N contents in grain and shoot, derived from a 3-year study of the same rotation experiment (adapted from Ma et al., 2016). MA, maize–alfalfa rotation; MS, maize-soybean rotation; MM, maize monoculture. The dotted curve lines are non-linear regression for MA (n = 72), MS (n = 72) and MM (n = 108), respectively; solid curve lines are non-linear regression for all data (n = 252); dashed lines indicate the critical N and P contents in grain and shoot that attained maximum grain yield by a plateau trend between yield and N content; significant at $^{***}p < 0.001$.

better synchronized with N uptake resulting in a suitable N:P ratio for greater yields (Ma et al., 2016). In eastern Ontario, Canada, grain yields of maize reached a plateau of 13.1 Mg ha^{-1} when shoot and grain N accumulations were at 255 and 201 kg N ha^{-1}, respectively. Correspondingly, at the plateau yield level, shoot accumulated 43 kg P ha^{-1} and grain contained 36 kg P ha^{-1} (Fig. 3; Ma et al., 2016), or a N:P ratio of 6. In a study on maize nutrition in response to crop rotation and N fertilizer rates, Riedell et al. (2009) revealed significant interactions of N rate and rotation on N, P, K, Ca, Mg and Zn concentrations in plant shoots. Using long-term rotation experiments as the platform, it is of fundamental importance to apply current innovative technologies such as metabolomics for gaining insight into plant nutrient uptake and nutrient balance and exploit the full benefit of site-specific rotation systems.

Because there is a high N availability in maize cropping as a result of the rotation effect, Ma et al. (2003) suggested that fertilizer application for maize following soybean should be reduced by 60–70 kg N ha^{-1} to attain a better economic performance, under rainfed eastern Canadian conditions. N fertilizer replacement value (FRV) is defined as the amount of inorganic N fertilizer required achieving the same yield in continuous maize as that attained by non-N-fertilized maize that followed a preceding legume crop (Hesterman et al., 1987; Ma et al., 2003). Fox and Piekielek (1988) estimated the FRV values of 146–187 kg N ha^{-1} for three forage legumes (alfalfa, birdsfoot trefoil and red clover) in Pennsylvania, United States. Similarly, Blevins et al. (1990) illustrated FRV values of 75 and 65 kg N ha^{-1} for hairy vetch and big flower vetch, respectively. In Lancaster, Vanotti and Bundy (1995) reported an FRV value of 153 kg N ha^{-1} for alfalfa from a long-term experiment. In general, the average FRV for grain legume is approximately 62 kg N ha^{-1}, whereas a significantly higher FRV (149 kg N ha^{-1}) is estimated for green manure or forage legume (Ma et al., 2016). Taking full advantage of N credit from the preceding legume to maize crop will help crop producers to consider the amount of N needed to avoid excessive fertilizer for optimum yield. This reduction in N application to maize following a legume crop will prevent the inefficient use of N, which results in N losses through leaching, denitrification or volatilization. This also helps mitigate the pollution of reactive N to surface and groundwater, and contribute to clean water and clean air for the benefit to the general public.

6 Additional benefits of maize–legume rotation

6.1 Benefits of crop rotation on the control of diseases, insects and weeds

Crop rotation is an efficient and fundamental tool for integrated pest management that farmers are well aware of and are frequently in practice (Mora and Moreno, 1984; Munkwold, 2003). Francis and Clegg (1990) stated that 'the greater the differences between crops in a rotation sequence, the better cultural control of pests can be expected'. One of the reasons for reduction in crop yields in mono-cropping systems is the increased infestation of some soil pathogens (Crookston et al., 1990). For example, in North America, next to European corn borer (*Ostrinin nubilalis*), corn rootworm (*Diabrotica* spp.) is the most destructive and widespread insect pest of maize, causing extensive crop damage (Ma et al., 2009). Both the northern and western rootworm species are univoltine, and for the most part, go through one diapause (one generation) per year. The adult females enter the grove to lay eggs in the soil near the base of maize plant in later summer, and these eggs survive the winter. The eggs hatch the following spring, and larvae feed on maize roots from early June to August. The pest is more problematic in fields with continuous maize mono-cropping and under no-till cultivation.

Published by Burleigh Dodds Science Limited.

The use of crop rotation to suppress diseases and pests is considered to be one of the best and cost-effective practices (Munkwold, 2003). However, before rotating into another crop, several factors that limit the effectiveness of crop rotations in suppressing diseases and insects should be taken into consideration. The botanical classification of crop species should be figured out before deciding on the rotation crops. Farmers should be aware of how long the pathogen can survive in the crop residue/soil and which additional plant species (mainly weed species and cover crops) can be a host for the pathogens. Attention should also be paid to weeds that can host diseases harmful to the legumes or maize, and herbicides can be rotated and used effectively for weed control during the rotation period. Suppression of weeds also helps preventing the reintroduction of the pathogen to a clean field when the rotated crop is planted again (Krupinsky et al., 2002). Apparently, there is challenge to harness the N benefit and avoid accumulation of specific weed seed bank in a maize–legume cover crop rotation system (Thierfelder et al., 2012, 2013).

6.2 Reducing environmental costs

Global warming resulting from GHG emissions of agricultural origin is currently considered an important environmental issue (IPCC, 2014; Ma et al., 2010). Assessing the sustainability of a production system should consider both the economic crop productivity and the environmental impact. Consumers and the public want to know the C cost and wish to cut C footprints in their choice of food products. Farmers are urged to adopt effective management practices to improve productivity and at the same time to lower the C footprints of agricultural production on the farm (Gan et al., 2011). Large-scale farm operations appear willing to adopt beneficial management practices that deliver ecosystem services, and the general public appears willing to pay producers for the delivery of specific services, such as cleaner lakes in the Midwestern US (Robertson et al., 2014). A fair credit system and effective government regulations and policies are needed to promote the adoption of beneficial management practices.

A new concept of specific emissions or crop yield-based emissions was created to link the quantities of N_2O emissions with crop yields (Ma et al., 2010). Considering agriculture production as a system, Gan et al. (2011) defined C footprint as the total GHGs per unit grain basis (kg CO_2eq kg^{-1} grain) to quantify the impact of GHGs on environmental sustainability. Diversification of cropping systems, particularly with the inclusion of grain legumes in cereal-based crop rotations enhances plant-available N in soil, leading to increased crop productivity because resources can be used by individual crops more efficiently (Robertson et al., 2000; Zentner et al., 2004; Adviento-Borbe et al., 2007). It also may reduce pesticide applications, generate higher inputs of crop residual C, leading to an increase in soil C storage (Lal, 2004; Gan et al., 2011), hence reducing CO_2 emissions and lowering C footprint effectively (Gan et al., 2014). A detailed procedure for calculation of C footprints and related literature on this topic can be found in a recent review by Liang et al. (2016).

6.3 The benefits of crop rotation to biodiversity and ecological services

Crop rotations have more biological diversity as compared with mono-croppings. This diversity can be divided into temporal and spatial components. Temporal diversity results from a sequence of crops being grown in a given field (i.e. crop change with time), and has been an important tool in breaking pest cycles, reducing soil erosion and increasing yields. The second component, spatial diversity, results from greater numbers of crops being grown at a given time on the landscape. Higher biodiversity richness induced by crop rotation may be associated with reduced infestations of diseases, insects and weeds (Gurr et al., 2003; Stoner, 2007), significant reductions in nutrient leaching losses and improved soil fertility (Russell, 2002) and increased crop productivity (Ma et al., 2003, 2012).

In a recent study, Thom et al. (2016) reported that rotations between alternating crops promote biodiversity by providing a habitat for a variety of beneficial insects. On-farm biodiversity with introducing specialty oilseeds into current crop rotations could also provide abundant floral resources for pollinating insects (Thom et al., 2016). Crop rotations that improve the biodiversity of cropping systems can make crop ecosystems more stable and thereby sustain the long-term productivity. Thus, a more diversified strategy of integrating crop and small-scale livestock production by the farm operator into a moderate-sized rotational cropping system will meet production needs and utilize opportunities beneficially.

Farming is a systems enterprise with multiple moving parts and sometimes complex interactions (Robertson et al., 2014). Including legume crops in a maize-dominated rotation system can not only achieve direct N credit benefits, but also indirect rotational benefits due to suppression on diseases, insects and weeds. Altogether, maize–legume rotation displays ecological services benefits due to the reduced use of chemical fertilizers and agrochemicals (pesticides and herbicides). Therefore, this system could be readily managed to contribute to clean water, biocontrol and other biodiversity benefits, climate stabilization and long-term soil fertility, thereby helping meet society's need for agriculture that is economically and environmentally sustainable (Robertson et al., 2014). Such ecological services benefit requires a systems-level socioecological understanding of the agroecosystems and should be assessed and appraised by the whole society.

In the Loess Plateau of Northern China, unprecedented rising air temperature during the past three decades has shortened maize and wheat growth durations (Mo et al., 2016), and crop productivity is facing great challenge of drought, heat and other abiotic stresses. Several recently developed dryland farming techniques, aimed at increasing harvesting rain water and adjusting thermal balance, have substantially contributed to the optimization of maize phenological patterns and improvement of grain yields (Mo et al., 2013). Integrating dryland crop production with leguminous forage and natural grassland, as well as animal husbandry into a multifunction system has led to sustaining long-term development on the semiarid area (Jiang et al., 2006).

The sustainable agricultural development in the area of ecologically fragile regions such as the Karst regions of Southwest China means the exploitation of agro-ecosystems on sustainable basis with an aim to maintain healthy and sustainable balance between the supply and demand of needs of inhabitants. This requires the integrated process of agricultural production, ecological construction and agricultural sustainable development. Sustainable agricultural development can also promote the sustainable development of the ecological environment (Anjum et al., 2014).

6.4 The economic benefit of crop rotation

A major challenge for maize production systems is to maximize economic returns while minimizing the environmental impact through an appropriate crop rotation system (Ma et al., 2016). Crop rotation will have an impact on the returns to farmers, and the response will be different on different soils and in different regions. Rao and Mathuva (2000) at the

International Centre for Research in Agroforestry compared pigeon pea-based systems with maize mono-cropping and suggested that the legume–maize rotation is 32–49% more profitable than continuous maize cropping. In our long-term rotation site, a stage 2 drought occurred in 2012 (Ma et al., 2015). Grain yields of maize were dramatically reduced in all treatments, but the highest reduction (60%) in yield was the continuous maize mono-cropping with well-fertilized N treatment, where the yield of rotational maize was up to 72% of the long-term average (Fig. 1). This indicates that rotational maize would be more tolerant to environmental vulnerability and generate greater economic benefit than MM, especially in situations with catastrophic climate event. The economic challenge of crop rotation we faced would likely focus on maintaining profitability, while producing crops with practices that prevent soil degradation and loss of N and other nutrients to surface and groundwater resources. The relative importance of cropping systems would likely change as more data on long-term and environmental benefits of crop rotations become available, especially if the information caused major changes in public policies regarding land use and crop production practices.

7 Summary and future trends

This chapter identified the unsustainable problems induced by the simplification of crop rotation, and highlighted the role of a legume–maize crop rotation as a system with high-yielding potential and as a more cost-effective and environmental-friendly strategy for sustainable maize production, compared with continuous maize mono-cropping, accompanied by application of large amounts of inorganic fertilizer.

In modern crop production systems, N is the main costly input and often causes environmental problems when it is used inappropriately. N application in maize crop rotation systems should also be re-examined. Grain legumes, such as soybean, contribute less N than forage or green manure legumes to the subsequent maize because most of the N fixed biologically by grain legumes is translocated to the grain, which is removed from the field. Hence, the N requirement of maize is seldom met, and additional N based on appropriate soil amendments or other organic sources is required to exploit the potential of the maize crop in favourable seasons. Conversely, forage legumes such as alfalfa or red clover had a larger FRV than grain legumes, resulting in greater grain yields of maize in the subsequent season. Other benefits of maize–legume rotation, including the improvement of soil quality, reduced infestation of diseases, pests and weeds, promotion of soil biodiversity and ecological services are also illustrated in this review.

However, the physiological/biological mechanisms responsible for those benefits need to be further understood. Over the past three decades, scientific breakthroughs in molecular biology and dramatic advances in analytical sciences have provided researchers with new tools to assess the molecular basis for the benefits of rotational maize production. Future progress in this area will lead to the development and large-scale adoption of crop production practices that are economically viable, environmentally sustainable and socially acceptable.

8 Where to look for further information

Readers are referred to the recent monograph titled *Crop Rotations: Farming Practices, Monitoring and Environmental Benefits*, edited by Bao-Luo Ma, and *Sustainable*

Agriculture Reviews – Cereals, edited by Eric Lichtfouse and Aakash Goyal for further information. A review article by Robertson et al. has also provided invaluable thoughts on the benefits of crop rotation from the view point of ecosystem services.

9 Acknowledgements

This work was a joint contribution between Agriculture and Agri-Food Canada (AAFC) and Inner Mongolia Agricultural University. The project was financially supported, in part, by the Canola Council of Canada and AAFC Growing Forward II Canola/Flax Cluster Project J-000353. We thank Dr Chang Liang, a senior research scientist of Environment Canada, who critically reviewed the final version of this chapter and provided us with invaluable suggestions. AAFC-ORDC contribution no. 16-066.

10 References

Abbasi, M. K., Khaliq, A., Shafiq, M., Kazmi, M. and Ali, I. (2010), 'Comparative effectiveness of urea N, poultry manure and their combination in changing soil properties and maize productivity under rainfed conditions in northeast Pakistan', *Pak. J. Agric. Sci.*, 46, 211–30.

Adviento–Borbe, M. A. A., Haddix, M. L., Binder, D. L., Walters, D. T. and Dobermann, A. (2007), 'Soil greenhouse gas fluxes and global warming potential in four high–yielding maize systems', *Glob. Change Biol.*, 13, 1972–88.

Anjum, S. A., Ullah, E., Xue, L.-L., Wang, L.-C., Tanveer, M. and Nadeem, M. (2014), 'Potential measures for sustainable agricultural development in Karst regions of South west China', *J. Food Agric. Environ.*, 12, 464–8.

Asbjornsen, H., Hernández-Santana, V., Liebman, M., Bayala, J. and Chen, J. (2014), 'Targeting perennial vegetation in agricultural landscapes for enhancing ecosystem services', *Renew. Agric. Food Syst.*, 29, 101–25.

Ashworth, A. J., Allen, F. L., Saxton, A. M. and Tyler, D. D. (2016), 'Long-term corn yield impacted by cropping rotations and bio-covers under no-tillage', *Agron. J.*, 108, 1495–1502.

Beaulieu, M. S., Bédard, F. and Lanciault, P. (2001), 'Distribution and Concentration of Canadian Livestock, Agriculture and Rural Working Paper Series Working Paper', No. 47, Statistics Canada, Ottawa, ON.

Blackmer, A. M., Voss, R. D. and Mallarino, A. P. (1997), 'Nitrogen fertilizer recommendations for corn in Iowa', PM 1714, Iowa State University Extension, Ames, IA.

Blevins, R. L., Herbek, J. H. and Frye, W. W. (1990), 'Legume cover crops as a nitrogen source for no–till corn and grain sorghum', *Agron. J.*, 82, 769–72.

Boschma, M., Joaris, A. and Vidal, C. (2015), 'Concentration of livestock production', http://ec.europa.eu/agriculture/envir/report/en/live_en/report.htm.

Bullock, D. G. (1992), 'Crop rotation', *Crit. Rev. Plant Sci.*, 11, 309–26.

Chan, Y. K., McCormick, W. A. and Ma, B. L. (2013), 'Fertilizer effects on soil archaeal abundance at two Ontario experimental farms during three consecutive cropping seasons', *Appl. Soil Ecol.*, 68, 26–35.

Congreves, K. A., Hayes, A., Verhallen, E. A. and Van Eerd, L. L. (2015), 'Long-term impact of tillage and crop rotation on soil health at four temperate agroecosystems', *Soil Till. Res.*, 152, 17–28.

Copeland, P. J., Allmaras, R. R., Crookston, R. K. and Nelson, W. W. (1993), 'Corn-soybean rotation effects on soil water depletion', *Agron. J.*, 85, 203–10.

Published by Burleigh Dodds Science Limited.

Copeland, P. J. and Crookston, R. K. (1992), 'Crop sequence affects nutrient composition of corn and soybean grown under high fertility', *Agron. J.*, 84, 503–9.

Crookston, R. K., Kurle, J. E., Copeland, P. J., Ford, J. H. and Lueschen, W. E. (1990), 'Rotational cropping sequence affects yield of corn and soybean', *Agron. J.*, 83, 108–13.

Davidson, E. A., Galloway, J. N., Millar, N. and Leach, A. M. (2014), 'N-related greenhouse gases in North America: Innovations for sustainable future', *Curr. Opin. Environ Sustain.*, 9, 1–8.

Davis, A. S., Hill, J. D., Chase, C. A., Johanns, A. M. and Liebman, M. (2012), 'Increasing cropping system diversity balances productivity, profitability and environmental health', *PLoS ONE*, 7(10), 1–8.

Deen, W., Martin, R. C., Hooker, D. and Gaudin, A. (2016), 'Crop rotation trends: past, present and future benefits and drivers', in B. L. Ma (ed.), *Crop Rotations: Farming Practices, Monitoring and Environmental Benefits*. Nova Science Publisher, Inc., New York, NY, pp. 35–50.

Ding, W., Hume, D. J., Vyn, T. J. and Beauchamp, E. G. (1998), 'N credit for soybean to a following maize crop in central Ontario', *Can. J. Plant Sci.*, 78, 29–33.

Edwards, J. H., Thurlow, D. L. and Eason, J. T. (1988), 'Influence of tillage and crop rotation on yields of corn, soybean, and wheat', *Agron. J.*, 80, 76–80.

Farhad, W., Cheema, M. A., Saleem, M. F., Radovich, T., Abbas, F., Hammad, H. M. and Wahid, M. A. (2013), 'Response of maize hybrids to composted and non-composted poultry manure under different irrigation regimes', *Int. J. Agric. Biol.*, 15, 181–90.

Fox, R. H. and Piekielek, W. P. (1988), 'Fertilizer N equivalence of alfalfa, birdsfoot trefoil, and red clover for succeeding corn crops', *J. Prod. Agric.*, 1, 313–17.

Francis, C. A. and Clegg, M. D. (1990), 'Crop rotations in sustainable production systems', in C. A. Edwards, R. Lal, P. Madden, R. H. Miller and G. House (eds), *Sustainable Agricultural Systems. Soil and Water Conservation Society.* Ankeny, Iowa, pp. 107–22.

Franke, A. C., Laberge, G., Oyewole, B. D. and Schulz, S. (2008), 'A comparison between legume technologies and fallow, and their effects on maize and soil traits, in two distinct environments of the West African Savannah', *Nutr. Cycl. Agroecosyst.*, 82, 117–35.

Gan, Y. T., Liang, B. C., Chai, Q., Lemke, R. L., Campbell, C. A. and Zentner, R. P. (2014), 'Improving farming practices reduces the carbon footprint of spring wheat production', *Nature Commun.*, 11(5), 1–13.

Gan, Y. T., Liang, B. C., Hamel, C., Cutforth, H. and Wang, H. (2011), 'Strategies for reducing the carbon footprint of field crops for semiarid areas: A review', *Agron. Sustain. Develop.*, 31, 643–56.

Gaudin, A. C. M., Tolhurst, T. N., Ker, A. P., Janovicek, K., Tortora, C., Martin, R. C. and Deen, W. (2015), 'Increasing crop diversity mitigates weather variations and improves yield stability', *PLoS ONE*, 10(2), e0113261. doi:10.1371/journal.pone.0113261.

Gentry, L. F., Below, F. E., David, M. B. and Bergerou, J. A. (2001), 'Source of the soybean N credit in maize production', *Plant Soil*, 236, 175–84.

Gentry, L. F., Ruffo, M. L. and Below, F. E. (2013), 'Identifying factors controlling the continuous corn yield penalty', *Agron. J.*, 105, 295–303.

Glab, T., Ścigalska, B. and Labuz, B. (2013), 'Effect of crop rotations with triticale (× *triticosecale* Wittm.) on soil pore characteristics', *Geoderma*, 202–3, 1–7.

Guo, J. H., Liu, X. J., Zhang, Y., Shen, J. L., Han, W. X., Zhang, W. F., Christie, P., Goulding, K. W. T., Vitousek, P. M. and Zhang, F. S. (2010), 'Significant acidification in major Chinese croplands', *Science*, 327, 1008–10.

Gurr, G. M., Wratten, S. D. and Luna, J. M. (2003), 'Multi-function agricultural biodiversity: pest management and other benefits', *Basic Appl. Ecol.*, 4, 107–16.

Havlin, J. L., Kissel, D. E., Maddus, L. E., Claassen, M. M. and Long, J. H. (1990), 'Crop rotation and tillage effects on soil organic carbon and nitrogen', *Soil Sci. Soc. Am. J.*, 54, 448–52.

Helmers, M. J., Zhou, X., Asbjornsen, H., Kolka, R. and Tomer, M. D. (2012), 'Sediment removal by prairie filter strips in row-cropped ephemeral watersheds', *J. Environ. Qual.*, 41, 1531–9.

Hesterman, O. B., Russelle, M. P., Sheaffer, C. C. and Heichel, G. H. (1987), 'Nitrogen utilization from fertilizer and legume residues in legume–corn rotations', *Agron. J.*, 79, 726–31.
Howard, D. D., Chambers, A. Y. and Lessman, G. M. (1998), 'Rotation and fertilization effects on corn and soybean yields and soybean cyst nematode populations in a no–tillage system', *Agron. J.*, 90, 518–22.
Hussain, S. K., Mielke, L. N. and Skopp, J. (1988), 'Detachment of soil as affected by fertility management and crop rotations', *Soil Sci. Soc. Am. J.*, 52, 1463–8.
IPCC-Intergovernmental Panel on Climate Change (2014), 'Fifth assessment report: Climate change 2014: Mitigation of climate change', Working Group III and the 39th Session of the IPCC, Berlin, Germany.
Jiang, H. M., Li, F. M. and Jiang, J. P. (2006), 'Soil carbon pool and effects of soil fertility in seeded alfalfa fields on the semi-arid Loess Plateau in China', *Soil Biol. Biochem.*, 38, 2350–8.
Johnson, J. W., Welch, L. R. and Kurtz, L. T. (1975), 'Environmental implications of N fixation by soybeans', *J. Environ. Qual.*, 4, 303–6.
Johnson, T. C. (1927), 'Crop rotation in relation to soil productivity', *J. Am. Soc. Agron.*, 19, 518–27.
Johnston, C. A. (2013), 'Agricultural expansion: land use shell game in the U.S. Northern Plains', *Landscape Ecol.*, 29, 81–95.
Johnston, J. R., Browning, G. M. and Russelle, M. B. (1942), 'The effect of cropping practices on aggregation, organic matter content, and loss of soil and water in the Marshall silt loam', *Agron. J.*, 7, 105–13.
Kaboneka, S., Sabbe, W. E. and Mauromoustakos, A. (1997), 'Carbon decomposition kinetics and nitrogen mineralization from corn, soybean, and wheat residues', *Commun. Soil Sci. Plant Anal.*, 28, 1359–73.
Karlen, D. L. (2004), 'Cropping systems: Rain-fed maize-soybean rotations of North America', in R. M. Goodman (ed.), *Encyclopedia of Plant and Crop Science*. Marcel Dekker, New York, NY, pp. 358–62.
Karlen, D. L., Berry, E. C., Colvin, T. S. and Kanwar, R. S. (1991), 'Twelve-year tillage and crop rotation effects on yields and soil chemical properties in northeast Iowa', *Commun. Soil Sci. Plant Anal.*, 22, 1985–2003.
Karlen, D. L., Varvel, G. E., Bullock, D. G. and Cruse, R. M. (1994), 'Crop rotations for the 21st Century', *Adv. Agron.*, 53, 1–45.
Katsvairo, T. W. and Cox, W. J. (2000), 'Tillage rotation management interaction in corn', *Agron. J.*, 92, 493–500.
Kay, B. D. (1990), 'Rates of change of soil structure under different cropping systems', *Adv. Soil Sci.*, 12, 1–52.
Krupinsky, J. M., Bailey, K. L., McMullen, M. P., Gossen, B. D. and Turkington, T. K. (2002), 'Managing plant disease risk in diversified cropping systems', *Agron. J.*, 94, 198–209.
Lal, R. (2004), 'Carbon emission from farm operations', *Environ. Int.*, 30, 981–90.
Liang, A., McLaughlin, N. B., Ma, B. L., Gregorich, E. G., Morrison, M. J., Burtt, S. D. and Patterson, B. S. (2013), 'Changes in moldboard plough draught and tractor fuel consumption on continuous corn after 18 years of organic and inorganic N amendments', *Int. J. Hydrogen Energy*, 52, 89–95.
Liang, B. C., Herath, A. W. and Ma, B. L. (2016), 'Carbon footprints in crop rotation systems', in B. L. Ma (ed.), *Crop Rotations: Farming Practices, Monitoring and Environmental Benefits*. Nova Science Publisher, Inc., New York, NY, pp. 177–207.
Lichtfouse, E., Navarrete, M., Debaeke, P., Souchère, V., Alberola, C. and Menassieu, J. (2009), 'Agronomy for sustainable agriculture: A review', *Sustain. Agric. Dev.*, 29, 1–6.
Liu, A., Hamel, C., Costa, C., Elmi, A., Ma, B. L. and Smith, D. L. (2002), 'Concentrations of K, Ca and Mg in maize colonized by arbuscular mycorrhizal fungi under field conditions', *Can. J. Soil Sci.*, 82, 271–8.
Liu, A., Ma, B. L. and Bomke, A. A. (2005), 'Effects of cover crops on soil aggregate stability, total organic C and polysaccharides', *Soil Sci. Soc. Am. J.*, 69, 2041–8.

Ma, B. L., Biswas, D. K., Herath, A. W., Whalen, J. K., Ruan, S. Q., Caldwell, C., Earl, H., Vanasse, A., Scott, P. and Smith, D. L. (2015), 'Growth, yield and yield components of canola as affected by nitrogen, sulphur and boron fertilization', *J. Plant Nutr. Soil Sci.*, 178, 658–70.

Ma, B. L., Dwyer, L. M. and Gregorich, E. G. (1999a), 'Soil nitrogen amendment effects on nitrogen uptake and grain yield of maize', *Agron. J.*, 91, 650–6.

Ma, B. L., Dwyer, L. M. and Gregorich, E. G. (1999b), 'Soil nitrogen amendment effects on seasonal nitrogen mineralization and nitrogen cycling in maize production', *Agron. J.*, 91, 1003–9.

Ma, B. L., Liang, B. C., Biswas, D. K., Morrison, M. J. and McLaughlin, N. B. (2012), 'The carbon footprint of maize production as affected by nitrogen fertilizer and maize-legume rotations', *Nutr. Cycl. Agroecosyst.*, 94, 15–31.

Ma, B. L., Meloche, F. and Wei, L. (2009), 'Agronomic assessment of Bt trait and seed or soil-applied insecticides on the control of corn rootworm and yield', *Field Crops Res.*, 111, 189–6.

Ma, B. L., Subedi, K. D. and Liu, A. (2006), 'Variations in grain nitrogen removal associated with management practices in corn production', *Nutr. Cycl. Agroecosyst.*, 76, 67–80.

Ma, B. L., Wu, T. Y., Tremblay, N., Deen, W., McLaughlin, N. B., Morrison, M. J., Gregorich, E. G. and Stewart, G. (2010), 'Nitrous oxide fluxes from corn fields: On-farm assessment of the amount and timing of nitrogen fertilizer', *Global Change Biol.*, 16, 156–70.

Ma, B. L. and Wu, W. (2016), 'Crop productivity and environment impact in a maize-legume rotation system: A review', in B. L. Ma (ed.), *Crop Rotations: Farming Practices, Monitoring and Environmental Benefits*. Nova Science Publisher, Inc., New York, NY, pp. 1–33.

Ma, B. L., Ying, J., Dwyer, L. M., Gregorich, E. G. and Morrison, M. J. (2003), 'Crop rotation and soil nitrogen amendment effects on maize production in Eastern Canada', *Can. J. Soil Sci.*, 83, 483–95.

Ma, B. L., Zheng, Z. M., Morrison, M. J. and Gregorich, E. G. (2016), 'Nitrogen and phosphorus nutrition and stoichiometry in the response of maize to various N rates under different rotation systems', *Nutr. Cycl. Agroecosyst.*, 104, 93–105.

Mathers, A. C., Stewart, B. A. and Blair, B. (1975), 'Nitrate removal from soil profiles by alfalfa', *J. Environ. Qual.*, 4, 403–5.

McGranahan, D. A. (2014), 'Ecologies of scale: multifunctionality connects conservation and agriculture across fields, farms, and landscapes', *Land*, 3, 739–69.

McLaughlin, N. B., Gregorich, E. G., Dwyer, L. M., Ma, B. L. and Topp, G. C. (2002), 'Effect of organic and inorganic soil N amendments on mouldboard plow draft', *Soil Till. Res.*, 64, 211–19.

Meyer-Aurich, A., Weersink, A., Janovicek, K. and Deen, W. (2006), 'Cost efficient rotation and tillage options to sequester carbon and mitigate GHG emissions from agriculture in Eastern Canada', *Agric. Ecosyst. Env.*, 117, 119–27.

Mhlanga, B., Cheesman, S., Chanhan, B. S. and Thierfelder, C. (2016), 'Weed emergence as affected by maize (*Zea mays* L.)-cover crop rotations in contrasting arable soils of Zimbabwe under conservation agriculture', *Crop Prot.*, 81, 47–56.

Mhlanga, B., Cheesman, S., Maasdorp, B., Mupangwa, W. and Thierfelder, C. (2015), 'Contribution of cover crops to the productivity of maize-based conservation agriculture systems in Zimbabwe', *Crop Sci.*, 55, 1791–1805.

Mo, F., Sun, S., Liu, X., Wang, J., Zhang, X., Ma, B. L. and Xiong, Y. C. (2016), 'Phenological responses of spring wheat and maize to changes in crop management and rising temperatures from 1992 to 2013 across the Loess Plateau', *Field Crops Res*, 196, 337–47.

Mo, F., Zhou, H., Wang, J., Zhao, H., Zhang, H., Wu, S., Chen, Y., Yang, T., Deng, H., Asfa, B., Wang, R., Simon, N. N., Li, F. and Xiong, Y. (2013), 'Development and application of micro-field rain-harvesting technologies', *Trans. Chin. Soc. Agric. Eng.*, 29, 1–17 (in Chinese).

Mora, L. E. and Moreno, R. A. (1984), 'Cropping pattern and soil management influence on plant diseases: I. Diplodia macrospora leaf spot of maize', *Turrialba*, 34, 35–40.

Mortensen, D. A., Egan, J. F., Maxwell, B. D., Ryan, M. R. and Smith, R. G. (2012), 'Navigating a critical juncture for sustainable weed management', *BioScience*, 62, 75–84.

Muir, J., Boyce, J. S., Seim, E. C., Mosher, P. N., Deibert, E. J. and Olson, R. A. (1976), 'Influence of crop management practices on nutrient movement below the root zone in Nebraska soils', *J. Environ. Qual.*, 5, 255–9.

Munkholm, L. J., Heck, R. J. and Deen, B. (2013), 'Long-term rotation and tillage effects on soil structure and crop yield', *Soil Till. Res.*, 127, 85–91.

Munkwold, G. P. (2003), 'Cultural and genetic approaches to managing mycotoxins in maize', *Ann. Rev. Phytopathol.*, 41, 99–116.

Olson, R. J., Hensler, R. F., Attoe, O. J., Witzel, S. A. and Peterson, L. A. (1970), 'Fertilizer nitrogen and crop rotation in relation to movement of nitrate nitrogen through soil profiles', *Soil Sci. Soc. Am. Proc.*, 34, 448–52.

Overton, M. (1996), 'Agricultural Revolution in England: The transformation of the agrarian economy 1500-1850', Cambridge University Press. ISBN 978-0-521-56859-3, p. 206.

Pittelkow, C. M., Liang, X., Linquist, B. A., Jan van Groenigen, K., Lee, J., Lundy, M. E., van Gestel, N., Six, J., Venterea, R. T. and van Kessel, C. (2014), 'Productivity limits and potentials of the principles of conservation agriculture', *Nature*, 517, 365–8.

Porter, P. M., Huggins, D. R., Perillo, C. A., Quiring, S. R. and Crookston, R. K. (2003), 'Organic and other management strategies with two- and four-year crop rotations in Minnesota', *Agron. J.*, 95, 233–44.

Power, J. F. (1990), 'Legumes and crop rotations', in C. A. Francis, C. B. Flora and L. D. King (eds), *Sustainable Agriculture in Temperate Zones*. Wiley, New York, NY, pp. 178–204.

Raimbault, B. A. and Vyn, T. J. (1991), 'Crop rotation and tillage effects on corn growth and soil structural stability', *Agron. J.*, 83, 979–85.

Rankin, M. (2015), 'Choosing between Alfalfa and Corn Silage … or when do you trade *Medicago* for *Maize*?' Forage Resources & UW Extension Forage Resource. http://cdp.wisc.edu/jenny/crop/choosing.pdf (accessed 13 May 2016).

Rao, M. R. and Mathuva, M. N. (2000), 'Legumes for improving maize yields and income in semi–arid Kenya', *Agric. Ecosyst. Environ.*, 78, 123–37.

Reid, L. M., Zhu, X. and Ma, B. L. (2001), 'Crop rotation and nitrogen effects on maize susceptibility to gibberella (*Fusarium graminearum*) ear rot', *Plant Soil*, 237, 1–14.

Richards, R. (2000), 'Selectable traits to increase crop photosynthesis and yield of grain crops', *J. Exp. Bot.*, 51, 447–58.

Riedell, W. E., Pikul, J. L., Jaradat, A. A. and Schumacher, T. E. (2009), 'Crop rotation and nitrogen effects on soil fertility, maize mineral nutrition, yield, and seed composition', *Agron. J.*, 101, 970–9.

Robertson, G. P., Paul, E. A. and Harwood, R. R. (2000), 'Greenhouse gases in intensive agriculture: contributions of individual gases to the radiative forcing of the atmosphere', *Science*, 289, 1922–5.

Robinson, C. A., Cruse, R. M. and Kohler, K. A. (1994), 'Soil management', in I. L. Hatfield and D. L. Karlen (eds), *Sustainable Agricultural Systems*. Lewis Publ., CRC Press, Boca Raton, Florida, pp. 109–34.

Robertson, G. P., Gross, K. L., Hamilton, S. K., Landis, D. A., Schmidt, T. M., Snapp, S. S. and Swinton, S. M. (2014), 'Farming for ecosystem services: An ecological approach to production agriculture', *BioScience*, 64, 404–15.

Rodríguez-Lizana, A., Carbonell, R., González, P. and Ordóñez, R. (2010), 'N, P and K released by the field decomposition of residues of a pea-wheat-sunflower rotation', *Nutr. Cycl. Agroecosyst.*, 87, 199–208.

Russell, A. E. (2002), 'Relationship between crop–species diversity and soil characteristics in southwest Indian agroecosystems', *Agric. Ecosyst. Environ.*, 92, 235–49.

Schreiber, M. M. (1992), 'Influence of tillage, crop rotation, and weed management on giant foxtail (*Seraria faberi*) population dynamics and corn yield', *Weed Sci.*, 40, 645–53.

Shaner, D. L. (2014), 'Lessons learned from the history of herbicide resistance', *Weed Sci.*, 62, 427–31.

Sinclair, T. R. and Vadez, V. (2012), 'The future of grain legumes in cropping systems', *Crop Pasture Sci.*, 63, 501–12.

Smith, H. A. and McSorley, R. (2000), 'Intercropping and pest management: A review of major concepts', *Am. Entomol.*, 46, 154–61.

Stonehouse, D. B., Kay, B. D., Baffoe, J. K. and Johnston-Drury, D. L. (1998), 'Economic choices of crop sequences on cash cropping farms with alternative crop yield trends', *J. Soil Water Conserv.*, 43, 266–70.

Stoner, K. A. (2007), 'Management of insect pests with crop rotation and field layout', Available: http://www.sare.org/Learning-Center/Books/Crop-Rotation-on-Organic-Farms/Text-Version/Physical-and-Biological-Processes-In-Crop-Production/Management-of-Insect-Pests-with-Crop-Rotation-and-Field-Layout (accessed 13 May 2016).

Thierfelder, C., Cheesman, S. and Rusinamhodzi, L. (2012), 'A comparative analysis of conservation agriculture systems: Benefits and challenges of rotations and intercropping in Zimbabwe', *Field Crops Res.*, 137, 237–50.

Thierfelder, C., Cheesman, S. and Rusinamhodzi, L. (2013), 'Benefits and challenges of crop rotations in maize-based conservation agriculture (CA) cropping systems of southern Africa', *Int. J. Agric. Sustain.*, 11, 108–24.

Thom, M. D., Eberle, C. A., Forcella, F., Gesch, R., Weyers, S. and Lundgren, J. G. (2016), 'Nectar production in oilseeds: Food for pollinators in an agricultural landscape', *Crop Sci.*, 56, 727–39.

Tiemann, L. K., Grandy, A. S., Atkinson, E. E., Marin-Spiotta, E. and McDaniel, M. D. (2015), 'Crop rotational diversity enhances belowground communities and functions in an agroecosystem', *Ecol. Lett.*, 18, 761–71. doi: 10.1111/ele.12453.

Tilman, D., Balzer, C., Hill, J. and Befort, B. L. (2011), 'Global food demand and the sustainable intensification of agriculture', *Proc. Natl Acad. Sci. USA*, 108, 20260–4.

Tilman, D., Fargione, J., Wolff, B., D'Antonio, C., Dobson, A., Howarth, R., Schindler, D., Schlesinger, W. H., Simberloff, D. and Swackhamer, D. (2001), 'Forecasting agriculturally driven global environmental change', *Science*, 292(5515), 281–4.

Tollenaar, M. and Lee, E. A. (2004), 'Genetic yield improvement and stress tolerance in maize', in H. Nguyen and A. Blum (eds), *Physiology and Biotechnology Integration for Plant Breeding*. Marcel Dekker Inc., New York, NY, pp. 51–81.

Van Doren, D. M., Moldenhauer, W. C. and Triplet, G. B. (1984), 'Influence of long term tillage and crop rotation on water erosion', *Soil Sci. Soc. Am. J.*, 48, 636–40.

Van Kessel, C. and Hartley, C. (2000), 'Agricultural management of grain legumes: has it led to an increase in nitrogen fixation', *Field Crop Res.*, 65, 165–81.

Vanotti, M. B. and Bundy, L. G. (1995), 'Soybean effects on soil nitrogen availability in crop rotations', *Agron. J.*, 87, 676–80.

Varvel, G. E. and Wilhelm, W. W. (2003), 'Soybean nitrogen contribution to corn and sorghum in western corn belt rotations', *Agron. J.*, 95, 1220–5.

Wang, C. J., Pan, G. X., Tian, Y. G., Li, L. Q., Xu, H. Z., and Xiao, J. H. (2010), 'Changes in cropland topsoil organic carbon with different fertilizations under long-term agro-ecosystem experiments across mainland China', *Sci. China Life Sci.*, 53, 858–67.

Wang, Z., Gao, R. and Ma, B. L. (2014), 'Concurrent improvement in maize yield and nitrogen use efficiency with integrated agronomic management strategies', *Agron. J.*, 106, 1243–50.

Wang, Z., Ma, B. L., Gao, R. and Sun, J. (2015), 'Effects of different management systems on root distribution of maize', *Can. J. Plant Sci.*, 95, 21–8.

West, T. O. and Post, W. M. (2002), 'Soil organic carbon sequestration rates by tillage and crop rotation: A global data analysis', *Soil Sci. Soc. Am. J.*, 66, 1930–46.

Wu, T., Ma, B. L. and Liang, B. C. (2008), 'Quantification of seasonal soil nitrogen mineralization for corn production in eastern Canada', *Nutr. Cycl. Agroecosyst.*, 81, 279–90.

Wu, W. and Ma, B. L. (2015), 'Integrated nutrient management (INM) for sustaining crop productivity and reducing environmental impact: A review', *Sci. Tot. Environ.*, 512, 415–27.

Yu, Q. and Qiao, Y. (2016), 'Crop rotation and cover crop in pest and disease management in sustainable agriculture', in B. L. Ma (ed.), *Crop Rotations: Farming Practices, Monitoring and Environmental Benefits*. Nova Science Publisher, Inc., New York, NY, pp. 157–75.

Zentner, R. P., Lafond, G. P., Derksen, D. A., Nagy, C. N., Wall, D. D. and May, W. E. (2004), 'Effects of tillage method and crop rotations on non-renewable energy use efficiency for a thin Black Chernozem in the Canadian Prairies', *Soil Till. Res.*, 77, 125–36.

Chapter 10

Intercropping in sustainable maize cultivation

Abeya Temesgen, Shu Fukai and Daniel Rodriguez,
The University of Queensland, Australia

1 Introduction

Sustainable intensification of agriculture, defined as the capacity to continuously produce more food from less land (Bryan et al., 2014), has been proposed as one of the strategies to mitigate food insecurity in smallholder farming. Intercropping, the simultaneous growth of two or more crops on the same area of land, is widely practised in Africa and elsewhere in tropics as an intensified system. Maize-legume intercropping systems have the potential to improve soil fertility through N fixation (Chowdhury and Rosario, 1992; Ghaley et al., 2005; Hauggaard-Nielsen et al., 2001b), to conserve soil moisture by providing additional ground cover (Jahansooz et al., 2004) and to outcompete weeds (Hauggaard-Nielsen et al., 2001a; Workayehu and Wortmann, 2011). Intercropping is believed to provide insurance in an unpredictable environment, poor soil fertility and unstable market prices.

There is evidence that intercrops have higher productivity per unit area of land than sole crops (Belay et al., 2008; Ouda et al., 2007; Vahdettin et al., 2006; Ogindo and Walker, 2005; Tsubo et al., 2004). This might be as a result of improved resources capture (RC), and/or higher resource-use efficiencies (RUEs) in intercrops than the sole crops (Yang et al., 2011; Belay et al., 2008; Szumigalski and Van Acker, 2006; Kanton and Dennett, 2004; Zhang and Li, 2003; Hauggaard-Nielsen et al., 2001; Pilbeam et al., 1994; Morris and Garrity, 1993; Willey, 1990; Natarajan and Willey, 1986). However, the question of

http://dx.doi.org/10.19103/AS.2016.0002.11

whether intercropping should be favoured over sole cropping as the level of resource input increases due to maize intensification is unclear.

The interactive effects between component crops of intercropping systems can be highly complex. Component crops compete for belowground resources, such as water and nutrients. The aboveground competition (for solar radiation) is equally as important, particularly as the taller maize canopy shades legume crops. Reported advantages in RC and use efficiencies between sole crops and intercrops are still not well understood. It has been proposed that the increased performance of intercrops compared to sole crops might be related to the differences in the spatial and temporal use of available resources by the component crops (Tsubo et al., 2005). Differences between intercrops and sole crops in terms of resource use and RUEs in response to environmental factors (i.e. water, nitrogen and canopy stresses) under maize-dominated farming systems have yet to be properly analysed.

In this chapter, the performance of intercropping under various rainfall and N nutrition environments in maize-dominated farming systems is reviewed. Interactions of interest include potential synergisms, and competition between the component crops, in relation to the capture and use of water, N and solar radiation. In addition, socio-economic and biophysical factors influencing adoption of a maize–legume intercropping system as well as the performance of intercropping compared to sole crops in expected climate change is discussed in this chapter.

2 Intercropping under different conditions: moisture and nitrogen levels

2.1 Intercropping under contrasting moisture conditions

Intercropping systems have been identified as being more productive per unit land area than sole crops (Li et al., 2011; Yang et al., 2011; Zhang et al., 2008; Belay et al., 2008; Kanton and Dennett, 2004; Tsubo et al., 2004; Hauggaard-Nielsen et al., 2001; Pilbeam et al., 1994; Morris and Garrity, 1993; Willey, 1990; Natarajan and Willey, 1986). However, as technological improvements and farmers' investments increase the level of crop productivity, factors affecting the yield response in the intercropping systems as well as whether intercropping or sole cropping systems remain the best option is not known.

Published reports from experiments conducted in semi-arid Africa (Belay et al., 2008; Pilbeam et al., 1994) show the benefits of maize–bean intercrops in dryer environments, as compared to more humid environments. The advantages of intercrops over sole crops are usually quantified by using productivity measurements such as the land equivalent ratio (LER), that is, the relative land area required as sole crops to produce the same yields as intercrops (Mead and Willey, 1980). The LER is also considered as relative yield (Willey, 1979). The relative yield performance of intercrops is compared with sole crops by calculating the LER for productivity assessment as in Yang et al. (2009, 2011) and Kutu et al. (2009). The total LER is the sum of the partial LER. It is estimated by using the following formula:

$$LER = \left(\frac{Y_{IA}}{Y_{SA}}\right) + \left(\frac{Y_{IB}}{Y_{SB}}\right) \tag{1}$$

Table 1 Land equivalent ratio (LER) and rainfall in maize-legume intercrops across various environments

Reference	Crop species	LER	RF (mm)	Concurrency	Crop proportion	N/P2O5 K(i)	Year	Country
Pilbeam et al., 1994	Bean + maize	1.12	179.5	70	Additive series	0/40	1	Kenya
Pilbeam et al., 1994	Bean + maize	0.84	257.6	70	Additive series	0/40	1	Kenya
Belay, 2008	Bean + maize	1.1	763.0	Simultaneously[1]	50:100	64/46	1	Ethiopia
Belay, 2008	Bean + maize	1.4	541.1	Simultaneously[1]	50:100	64/46	1	Ethiopia
Niringiye et al., 2005	Climbing bean + maize	1	406.3	64	Mean[2]	Not mentioned	1	Uganda
Niringiye et al., 2005	Bean + maize	1.26	519.9	64	Mean[2]	Not mentioned	1	Uganda
Santalla et al., 1999	Bean + maize	0.93	974	67	57:43	127/75/122	1	Spain
Santalla et al., 1999	Bean + maize	1.01	1608	67	57:43	127/75/122	1	Spain
Ouda et al., 2007	Bean + maize	1.72[a]	1.2*	80	50:100	107/19.5/58.58	2	Egypt
Ouda et al., 2007	Soybean + maize	1.71[a]	1.0*	80	50:100	107/19.5/58.58	2	Egypt
Ouda et al., 2007	Soybean + maize	1.76[a]	0.8*	80	50:100	107/19.5/58.58	2	Egypt
Morales et al., 2009	Bean + maize	1.72	558	Simultaneously[1]	Not indicated	80/90/00	1	Mexico
Morales et al., 2009	Bean + maize	1.78	800	Simultaneously[1]	Not indicated	80/90/00	1	Mexico

[1]The crops were planted simultaneously at both sites (no date of sowing and harvest is indicated).
[2]Mean for plant population ranges from 25000 to 40000 for maize and 57152 to 95238 for beans, while 44000 and 111111 is for the pure stands, respectively.
[a]Estimated from mean yield data.
*Irrigation using 1.2 pan evaporation coefficient (PEC; control), irrigation using 1.0 PEC (about 7% reductions in irrigation water than the control) and irrigation using 0.8 PEC (about 14% reductions in irrigation water than the control).

where Y_{IA} = yield of intercrop 'A'; Y_{IB}= yield of intercrop 'B'; Y_{SA}= yield of sole crop 'A' and Y_{SB}= yield of sole crop 'B'.

Values of total LER greater than 1 indicate that the intercrop has an advantage over the sole crop when grown on the same land area (Yang et al., 2010, 2011; Hauggaard-Nielsen and Jensen, 2005).

Table 1 lists the LER values for maize-legume intercropping under different conditions. For example, intercrops produced a 40% and 10% relative yield advantage per unit area land than sole crops under 541 mm and 763 mm annual rainfall, respectively, in Ethiopia (Belay et al., 2008). In another report in semi-arid Kenya, intercropping under 122 mm of rainfall produced a 22% total yield advantage (LER = 1.22), while it was disadvantageous to grow an intercrop (LER = 0.88) under higher rainfall (258 mm). According to the investigators, the low values in LER were related to the crop failure to nodulate and fix nitrogen.

In contrast, Niringiye et al. (2005), for a maize–climbing bean intercropping system, found 1.26 and 1.00 values of LER for the first (519.9 mm) and second rainy seasons (406.3 mm), respectively. Climbing bean is known for its aggressive and taller growth behaviour, which could explain these results. With the exception of these results, all the reports accessed indicated greater benefits of intercropping in lower rainfall seasons or semi-arid zones. These results appear to indicate greater potential benefits from intercropping in dryer than wetter environments. However, the reasons behind this conclusion are not well understood.

Benefits of intercropping in high rainfall conditions have also been reported. For example, in Spain, under similar experimental trials evaluated at two locations (Santalla et al., 1999), a 10% yield advantage (LER = 1.1) was found in a location receiving 1608 mm annual rainfall, compared to a trial site receiving 978 mm (LER = 0.93). In Mexico under rain-fed conditions across different environments, Morales-Rosales (2009) reported almost similar values of LER (1.78) under both high (800 mm) and low rainfall-growing environments (558 mm and LER = 1.72).

Ouda et al. (2007) found similar LER values in soybean–maize intercropping under different moisture conditions in Egypt. They reported LER values of 1.72 and 1.71 by changing the irrigation level to achieve water at pan evaporation coefficients (PEC) of 1 and 0.8 (7% and 14% less irrigation water than for PEC = 1.2), respectively. At higher moisture levels (i.e. at PEC = 1.2), they reported a LER of 1.76.

From these studies, the differences in the values of LER in sub-humid and humid environments are not important; however, the yield advantage per unit land area between intercrop and sole crops can be very large (i.e. values of LER can be much greater than 1). This might reflect differences in the level of inputs, that is, high levels of rainfall/irrigation and fertilizers in high rainfall-growing environments. In terms of LER values, the intercropping systems in both high (humid) and low (sub-humid) rainfall environments appear to be similar (Table 1).

In summary, in relation to the relative advantage of intercropping in different rainfall environments, it appears that the variation across locations might not be significant between high rainfall and lower rainfall-growing environments. However, whether the different responses observed under contrasting rainfall regimes reflect other factors, such as the level of nutrients supply, could be important to understand.

2.2 Intercropping under low and high levels of nitrogen supply

A report from empirical research in China (Li et al., 2011) indicate no advantage on LERs at low or high levels of N supply when maize was intercropped with wheat and barley under

0, 75, 150 and 225 kg N ha^{-1}. However, higher LER values were obtained under lower N supply (Li et al., 2011) when maize was intercropped with legumes. In another experiment (Ofori and Stern, 1987), the LER of maize/cowpea intercrop decreased with increasing levels of N supply. A similar trend was reported in barley/pea intercrops (Hauggaard-Nielsen, 2001).

The above studies suggest that cereal/legume intercropping systems are advantageous under conditions of inadequate nitrogen fertilizer application.

3 Resource capture and use efficiency in maize-based intercropping: water, nitrogen and light

In most cases, the advantages of intercropping are associated with greater RC (water, nutrient and light) and utilization efficiencies. However, there have been few studies under water-limiting conditions.

According to Morris and Garrity (1993) and Caviglia et al. (2004), the percentage of relative change in RC (evapotranspiration (ET), nitrogen uptake and photosynthetically active radiation interception) and RUE (water-use efficiency (WUE), nitrogen-use efficiency (NUE) and Photosynthetically active radiation use efficiency (PARUE)) between the sole cropping and intercropping can be calculated as

$$\Delta RC = \left[\left(\frac{RC_{ic}}{P_a RC_{sa} + P_b RC_{Sb}}\right) - 1\right] \times 100 \tag{2}$$

$$\Delta RUE = \left[\left(\frac{RCE_{ic}}{P_a RCE_{sa} + P_b RCE_{Sb}}\right) - 1\right] \times 100 \tag{3}$$

where

$$P_a = \frac{A}{(A+B)} \text{ and } P_b = \frac{B}{(A+B)}$$

In these equations, *ic* is the intercropping of crops 'a' and 'b'; *sa* and *sb* are sole crops of crop 'a' and 'b'; P_a is the proportion of crop 'a' in intercrops 'a' and 'b' where *A* is the density of crop 'a' in the intercrop in relation to the sole crop; the same definition holds for P_b.

3.1 Water capture and utilization efficiency

In field crops research, water captured by the crop is usually quantified by the crop ET calculated from the soil–water balance (Yang et al., 2010, 2011; Zhang et al., 2007; Morris and Garrity, 1993) as

$$ET = (P + I + U) - \Delta SW - R - D \tag{4}$$

where ET is the evapotranspiration (mm); P is the effective precipitation (mm); I is the irrigation applied (mm); U is the upward capillary flow into the root zone (mm); D is the downward drainage out of the root zone (mm); R is runoff and ΔSW is the change of soil water stored in the rooted zone between sowing and maturity.

In field crops research, ET is generally assumed to represent the water taken up by the crop and transpired by the leaves (T), and the soil evaporative losses (Es). WUE can then be calculated as the crop dry weight or yield per unit of water used (Zhang et al., 2007; Morris and Garrity, 1993). WUE_{ET} has also been estimated from the slope of the linear regression between shoot dry matter or grain yield and cumulative ET (Sadras and Rodriguez, 2010; Sadras, 2004; Caviglia and Sadras, 2001) as

$$WUE(kg / mm) = \frac{YIELD(kg / ha)}{ET(mm)} \tag{5}$$

where yield can be either dry matter or grain.

In a review, Morris and Garrity (1993) they reported that intercropped systems have slightly higher values of ET and higher WUEs than sole crops (Table 2). The changes in ET and water utilization efficiency in the intercrops were as high as 4% and 99%, and as low as −5% and 4%, respectively.

In Northwest China, Yang et al. (2011) evaluated a wheat–maize intercropping system under applications of 30%, 40% and 50% of irrigation levels below the conventional application in the Oasis area. They reported significant increases in WUE of intercrops over the sole wheat and maize crops, at all three levels of irrigation. However, a slight increase in water capture was recorded for sole crops relative to the intercrops for all three irrigation levels.

The results of a study by Ouda et al. (2007) on maize–soybean intercrops in Egypt showed that the water utilization efficiencies under water stress were slightly higher in intercrops than in both sole crops of maize and soybean. A field experiment on maize–legume intercropping under irrigated conditions in India (Khan, 1996) showed that the intercrops had higher water utilization efficiencies over the sole maize crops. Kanton and Dennett (2004) also reported consistently higher WUEs in intercrops of maize–legumes than with sole crops in England.

Table 2 Water use (WU) measured as evapotranspiration, water-use efficiency (WUE) and relative change in water capture and use efficiency (ΔWC and ΔWUE) of intercrops relative to WC and WUE by sole crops

Reference	WU (mm)	ΔWU (%)	WUE (kg/mm)	ΔWUE (%)	Species	Concurrency[1] (%)	Crop proportion[2]	Yield measure
Hulugalle and Lal (1986)[3]	248 to 554	−5 to 0	1.6 to 4.2	4 to 99	Maize + cowpea	100	50:50,50:30	Grain
Suwanarit et al. (1984)[4]	362	4	20.5	78	Maize + mung bean	74	100:100	Dry matter

[1]Percent of time that the earlier harvested species was in the field with the later-harvest species.
[2]Crop proportion relative to respective sole crop plant densities.
[3]Nigeria, shallow gravelly soil; 1 dry and 1 wet season.
[4]Thailand; wet season with abundant rainfall; 1 year – Adapted from Morris and Garrity (1993).

According to Willey (1990), the advantage of intercropping over sole cropping on grain yields increased very markedly from 14% to 93% from no stress to severe water-stressed conditions. In contrast, studies by Ogindo and Walker (2005) indicated that intercrops did not have a significant difference in total soil water capture relative to sole crops.

Overall, intercrops, in general, have higher water capture and WUEs than the sole crops; however, it is still not clear how the levels of water supply/availability or nitrogen supply affect this relationship.

3.2 Nitrogen capture and utilization efficiency

Mineral N is taken up by the roots of a crop in the form of nitrate (NO_3^-) N and ammonium (NH_4^+) N. Their dynamics in the soil depend on soil organic matter, applied inorganic fertilizer, soil pH, soil temperature, soil moisture, C:N ratio and soil microorganisms. Soil microorganisms compete for mineral N with the crop (Barabasz et al., 2002). The fate of N in the soil and its availability to the crop are affected by biological N transformation processes. These processes are mineralization, microbial immobilization, nitrification, denitrification, leaching of NO_3^-, volatilization of NH_3 and soil erosion. Hence, the uptake of N by the crop is a function of many interactive processes in the soil.

Values of N use efficiency are, in general, lower in the tropics (<50%) and Mediterranean (<50%) environments than in temperate (<70%) climates (Luis et al., 2006; Baligar and Bennett, 1986). For example, in dry environments, much of the N applied to the crops remains in the soil (Luis et al., 2006; Pilbeam, 1994) or gets lost through denitrification, leaching and volatilization (Cassman et al., 2002; Zhang et al., 1994). On the contrary, in humid environments, more N is recovered in the crop (Pilbeam, 1994). However, the N that remains in the soil can be easily leached to below the rooting depth by precipitation. The recovery of the applied N usually ranges from 40% to 60% for many crops, while the world average is about 50% (Mosier, 2002). When looking at these figures, there is no doubt that the efficiency of N use is related to soil moisture or rainfall.

NUE is usually expressed in terms of yield/agronomic efficiency (YE), N recovery efficiency (NRE) and N physiological efficiency (NPE) (Tolessa et al., 2009) as reflected in equations 6, 7 and 8. However, the nitrogen recovery efficiency is estimated with the assumptions that mineralization and immobilization processes are similar in both fertilized and unfertilized conditions. In addition, equal amounts of mineral N are supposed to be captured in both growing environments

$$YE = \frac{(Yf - Yo)}{N} \tag{6}$$

$$NRE = \left[\frac{(NRf - NRo)}{N}\right] \times 100 \tag{7}$$

$$NPE = \left[\frac{(Yf - Yo)}{NRf - NRo}\right] \times 100 \tag{8}$$

where YE = yield/agronomic efficiency; Yf = yield of fertilized crop; Yo = yield of unfertilized crop; N = rate of applied N; NRE = N recovery efficiency; NRf = N recovered by fertilized crop; *NRo* = N recovered by unfertilized crop and *NPE* = N physiological efficiency.

Assessment of N use efficiency using the above methods can be easily affected by soil factors. In other words, different values of total N can be estimated with the same values of applied N from different plots (Hirel and Lemaire, 2005). This necessitates estimating the NUE (López-Bellido and López-Bellido, 2001; Delogu et al., 1998) in terms of yield (grain/dry matter) to nitrogen accumulation. Nitrogen uptake/accumulation can be determined under laboratory conditions from the total plant tissue N uptake. NUE is calculated as a yield (grain/dry matter) to crop nitrogen accumulation.

$$NUE = \frac{Yield\ (kg\,/\,ha)}{N\ uptake\ by\ the\ crop\ (kg\,/\,ha)} \qquad (9)$$

There are a large number of studies that have shown complementary uses of N sources in cereal–legume intercrops (Hauggaard-Nielsen, 2009; Anil et al., 1998; Jensen, 1996). A LER greater than unity (NLER = 1.24) was reported, based on crop grain N, in maize–faba bean intercropping by Li et al. (2011). Bedoussac and Justes (2010) reported an average LER value of 1.15 for shoot N. They further reported that the advantage of intercropping is greater under low N than high N.

In the reviewed papers summarized by Vandermeer (1989) on legume and non-legume intercropping trials (Table 3), a greater mass of N was taken up by intercrops than by their component sole crops. This makes it clear that legumes compete less for the available soil N when intercropped with cereals. However, the high N accumulation in legumes during the growing season is related to atmospheric N_2 fixation (Ghaley et al., 2005).

Cereals are more competitive for soil inorganic N, when compared to legumes in cereal–legume intercropping (Li et al., 2011; Hauggaard-Nielsen, 2009; Ghaley et al., 2005; Jensen, 1996).

Chowdhury and Rosario (1992) evaluated the performance of maize–mung bean intercrops and their component sole crops with *Rhizobia* inoculation and N fertilizer applied at 0, 30, 60 and 90 kg/ha in the Philippines. They found an increase in N uptake in maize with increased N application, but a decline in mung bean. However, inoculation improved the N uptake of both mung bean and maize in late growth stages. In general, they reported higher total N absorption in intercrops than in pure stands. However, the N conversion efficiency declined in maize intercrops, with slight increases in mung bean.

Table 3 Experiments involving legumes and nonlegumes

System	Experimental results	Reference
Maize–cowpea	Mixture contained more N than monocultures	Lipman, 1912
Maize–cowpea	Mixture contained more N (203 Kg/ha than monocultures (maize = 108 Kg/ha; cowpea = 101 Kg/ha)	Wahua, 1983
Maize–bean	Mixture contained more N than monocultures	Clark and Francis, 1985
Maize–faba bean	Mixture contained more N (493-Kg/ha than monocultures (faba bean = 212 Kg/ha; maize = 182.5 Kg/ha)	Li et al., 2011

Source: Adapted from Vandermeer (1989), with modifications.

Table 4 Changes in nitrogen capture by intercrops relative to capture by sole crops (ΔNU%)

	Maize intercropped with					
	2007			2008		
Rate of applied N (Kg/ha)	Barely	Wheat	Faba bean	Barely	Wheat	Faba bean
0	−17	−2	25	4	9	15
75	−7	−4	9			
150	−24	−8	27			
225	15	2	32			
300	−12	−2	28	29	16	30

Source: Adapted from Li et al. (2011).

Singh et al. (2000) reported significantly more dry matter and N accumulation in intercropped systems (maize + pea, maize + lentils) than in monocrops. The component crops showed an increasing yield and N uptake with increasing level of N. Several studies have shown decreases in N_2 fixation as a result of increased inorganic fertilizer N supply (Parsons et al., 1993; Vessey and Waterer, 1992; Kanayama et al., 1990).

Relative changes in N capture by an intercrop when compared to sole crops are presented in Table 4. N capture in maize/faba bean intercrops relative to sole crops was positive and greater than for N captured by intercropping of maize/barley and maize/wheat.

$$\Delta NU\% = \left[\left(\frac{NU_{i_a} + NU_{i_b}}{P_a\, NU_{Sa} + P_b\, NU_{Sb}}\right) - 1\right] \times 100 \qquad (10)$$

where i = intercrop; S = sole crop; a = crop 'a' in intercrop; b = crop 'b' in intercrop and p = proportion of crop 'a' and 'b' in intercrop.

In another study, faba bean also promoted N fixation when intercropped with maize (Zhang and Li, 2003). On the other hand, Li et al. (2011) indicated that maize gave high yields when intercropped with faba bean, but lower yields when intercropped with barley and wheat.

In general, the contribution of legume N is one of the advantages in intercrops in cereal–legume intercropping systems. Weak competition for soil nitrogen is discussed as one of the reasons for such advantages, because legumes rely more on symbiotic N_2 fixation as a source of N. This might add yield value to intercrops when compared to sole crops under nitrogen-limiting environments in Africa.

The level of N nutrition in crops can be determined by using the nitrogen nutrition index (NNI) (Dodras, 2011). The NNI can be used to indicate the level of N stress during crop growth (Ziadi et al., 2008). This index can be used to improve N management efficiency and decrease the risk of N loss. The concept was first developed by Lemaire and Salette (1984). Now it is widely used to estimate N deficits in maize (Ziadi et al., 2008, 2009, 2010; Plénet and Lemaire, 2000).

NNI is estimated by measuring the actual N concentration (Nc) in shoots in relation to a critical Nc. The critical Nc is defined as the minimum Nc necessary to achieve the maximum growth of a crop (Ziadi et al., 2008, 2010; Zhang et al., 2008). It is estimated from a critical N dilution curve for each sampling point.

$$NNI = \frac{N\%}{Nc\%} \tag{11}$$

$$N_C = aW^{-b} \tag{12}$$

where *NNI* = nitrogen nutrition index; *N%* = percentage of actual Nc; *Nc%* = percentage of total N concentration in shoots relative to the amount of biomass produced, expressed in g kg^{-1} DM; W = shoot biomass (M g ha^{-1}); *a* = the N concentration in the total shoot biomass of 1 Mg DM ha^{-1} and *b* = the coefficient of dilution describing the relationship between N concentration and shoot biomass.

The Nc curve (Nc = $34.0W^{-0.37}$) was developed for maize in France (Plénet and Lemaire, 2000) and validated in Australia (Lemaire et al., 2007), Canada (Ziadi et al., 2008) and Germany (Herrmann and Taube, 2004). Greenwood et al. (1990) proposed Nc = 5–7 W = 0.5 for C3 species.

When the value of NNI is greater than 1, nitrogen is not limiting for crop growth. However, when it is less than 1, it indicates N deficiency. A NNI equal to 1 is the ideal for crop growth. Ziadi et al. (2008) reported NNI values of 0.30–1.35 in Canada, while values higher than 1.7 were reported in France by Plenet and Lemaire (2000) for maize. These figures indicate that maize accumulated more N than required for growth and development. The N dilution curve was developed for relay intercropping systems of cotton and wheat by Zhang et al. (2008) in China. They reported that the dilution for the sole crops showed a slight decline when compared to the intercropped components. However, it appears that no work has been reported on NNI in maize–bean intercropping to determine the N stress condition under different environmental conditions.

3.3 Intercepted photosynthetically active radiation and PARUE

The fraction of photosynthetically active radiation intercepted (*f*PARI) by plants is calculated as the ratio of radiation flux density intercepted by the plants (I) to incident radiation flux density (Io) above the canopy (Gao et al., 2009; Tsubo et al., 2001; Sinclair and Muchow, 1999).

$$fPARI = 1 - \left(\frac{I_T}{I_0}\right) \tag{13}$$

where *f*PARI = fraction of flux density of intercepted radiation; I_T= flux density of transmitted radiation and I_0= flux density of incident radiation.

The total cumulative PAR is calculated from daytime integrated *f*PARI (F_{daily}) and cumulated incident PAR ($PAR_{cumulative}$) as defined by Liu et al. (2012).

$$PAR_{cumulative,t} = \sum_{d=emergance}^{d=t} \left(F_{daily} \times IPAR_{cumulative,d}\right) \tag{14}$$

The daily *f*PAR (F_{daily}) is estimated from *f*PARI equation 12.

$$F_{\text{daily}} = \frac{2fPARI}{(1+fPARI)} \quad (15)$$

PARUE is defined as the crop biomass accumulated per unit of solar radiation intercepted by the canopy of the crop (Gao et al., 2010; Tsubo and Walker, 2002; Hammer and Wright, 1994). It can also be defined by the slope of a regression equation between the produced biomass and intercepted solar radiation for a given period of time.

$$PARUE = \frac{\left(\text{Dry matter}_{(n)} - \text{Dry matter}_{(n-1)}\right)\left(g\,m^{-2}\right)}{\left(PAR_{\text{cumulative},\,(n)} - PAR_{\text{cumulative}(n-1)}\right)\left(MJ\,m^{-2}\right)} \quad (16)$$

where 'n' and '$n - 1$' represent days when dry matter is harvested.

In some studies (in intercropping, for example), PARUE is also based on a percentage of energy values of crops per unit solar radiation energy captured by the crop (Tsubo et al., 2001).

Photosynthetically active radiation is the most important resource of crop growth in intercropping when other resources (water and N, for example) are not limiting factors (Watiki et al., 1993). Several studies have indicated the advantages of photosynthetically active radiation interception and utilization in intercrops compared to alternative sole crops (Abeya et al., 2015a; Gao et al., 2010; Keating and Carberry, 1993; Szumigalski and Van Acker, 2006; Tsubo and Walker, 2002; Pilbeam et al., 1994; Muchow and Coates, 1986; Natarajan and Willey, 1985; Marshall and Willey, 1983; Reddy and Willey, 1981).

However, there have been some contradictory findings reported in intercropping studies. For example, from a field experiment on row orientation in intercropping, Tsubo et al. (2001) found that sole maize had a higher PARUE than a maize–bean intercrop, though the intercrop intercepted more PAR energy than the sole maize crop. The authors attributed the higher radiation interception in the intercrop to be a result of a higher leaf area index. On the other hand, the PARUE of the maize intercropped with peanut (3.03 g MJ^{-1}) was higher than for sole peanut (1.44 g MJ^{-1}), but slightly lower than that of sole maize (3.27 g MJ^{-1}) (Awal et al., 2006). However, little has been studied on the interrelationships among resources use and use efficiencies (light, water and nutrients, for example) to determine yield per unit area of land in intercropping.

4 Competition and complementary interactions in maize–legume intercropping

Many studies have shown yield advantages in intercrops (Belay, 2008; Ouda et al., 2007; Vahdettin et al., 2006; Ogindo and Walker, 2005; Tsubo et al., 2004). However, very little has been reported about the eco-physiological mechanisms driving changes in resource capture (RC) or increased RUE in intercropping systems.

Natarajan and Willey (1986) showed different responses of intercrops to water stress in response to replacement and additive arrangements. Kanton and Dennett (2004) found greater WUE in intercrops compared to sole crops in additive series arrangements.

According to Willey (1990), this might be explained by the different root distributions from the two crops in the soil profile or by different root densities at different soil depths.

It has been proposed that intercrops may use available resources more efficiently than sole crops, because the two component crops do not compete for exactly the same resource niche (Bedoussac and Justes 2010). Root distribution plays a major role in the interactions between crops (Gao et al., 2010). For instance, a plant could improve water capture through growing roots into deeper soil layers and taking up more water from lower soil depths (Brenda and Robert, 1997), when compared to a shallow rooted crop. This type of interaction is considered a 'spatial complementarity', where the advantage arises from resources use at different depths of the soil profile (Willey, 1990).

Water extraction depends on soil water content in the explored soil depth, root length density, the size of the transpiring canopy and the atmospheric demand (Roberston et al., 1993; Meyers et al., 1990). Similarly, Meyers et al. (1990) reported that the root length density limits water uptake at increased soil depths in wheat. Banik and Sharma (2009) found a higher root length density in legumes when intercropped with maize than when grown as sole crops. Compared to maize-legumes intercropping, Gregory and Reddy (1982) reported a 10–15% increase in root length (in terms of LER) in a sorghum–groundnut intercrop compared to their sole crops in the rainy season under annual rainfall of 982 mm in India. This could explain the higher LER of the intercrop.

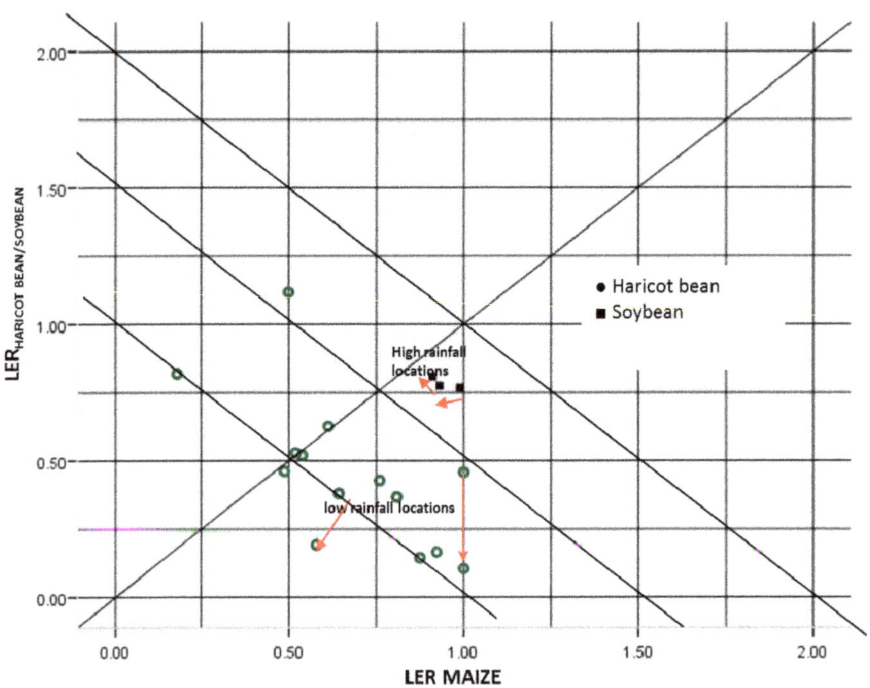

Figure 1 Partial LER of the component crops for maize intercropped with haricot bean and soybean (data from Table 5). The small arrows indicate locations differences in rainfall, with the direction of the arrows indicating the higher rainfall locations relative to the lower rainfall locations.

Table 5 Yield and LER of maize-legume intercrops in various environments

		Yield (t/ha)				LER			Concurrency	Remark
Author_year	Intercrop species	crop_1 inter	crop_2 inter	crop_1	crop_2	Total	crop_1	crop_2	70	
Pilbeam et al., 1994	Bean_1/maize_2	0.32	1.42	0.84	2.21	1.12	0.38	0.64	70	
Pilbeam et al., 1994	Bean_1/maize_2	0.18	1.11	0.93	1.92	0.84	0.19	0.58	Simultaneously[1]	
Belay, 2008	Bean_1/maize_2	0.90	16.40	8.50	16.40	1.10	0.11	1.00	Simultaneously[1]	
Belay, 2008	Bean_1/maize_2	1.10	8.50	2.40	8.50	1.40	0.46	1.00		
Ouda et al., 2007	Soybean_1/maize_2	1.61	7.10	1.99	7.80	1.72	0.81	0.91	80	
Ouda et al., 2007	Soybean_1/maize_2	1.38	6.75	1.78	7.25	1.71	0.78	0.93	80	
Ouda et al., 2007	Soybean_1/maize_2	0.93	4.60	1.21	4.65	1.76	0.77	0.99	80	
Oliaca et al., 2000	Bean_1/maize_2	0.94	6.44	1.50	10.52	1.21	0.63	0.61		
Oliaca et al., 2000	Bean_1/maize_2	0.61	5.13	1.32	10.52	0.95	0.46	0.49		
Tsubo et al., 2004	Bean_1/maize_2	1.20	6.00	2.80	7.90	1.17	0.43	0.76		Low plant density
Tsubo et al., 2004	Bean_1/maize_2	1.50	8.30	9.00	9.00	1.26	0.17	0.92		Medium
Tsubo et al., 2004	Bean_1/maize_2	1.50	9.00	10.30	10.30	1.15	0.15	0.87		High
Tsubo et al., 2004	Bean_1/maize_2	0.90	1.30	1.70	2.50	1.06	0.53	0.52		First sowing
Tsubo et al., 2004	Bean_1/maize_2	1.90	1.70	1.70	3.40	1.58	1.12	0.50		Second sowing
Ogindo and Walker, 2005	Bean_1/maize_2	0.88	1.34	1.69	2.49	1.06	0.52	0.54		
Ogindo and Walker, 2005	Bean_1/maize_2	1.06	4.01	2.88	4.96	1.18	0.37	0.81		
Vahdettin et al., 2006	Bean_1/maize_2	0.82	0.18	1.00	1.00	1.00	0.82	0.18		

[1]The crops were planted simultaneously at both sites (no date of sowing and harvest is indicated), Crop_1 inter = navy bean/soybean intercrop; crop_2 = maize intercrop.

It is also important to understand the relative yields of the component crops involved in an intercrop to determine dominance relationships between the two crop species in terms of partial LER (Mutsaers et al., 1993). The graph in Fig. 1 and the data in Table 5 compare the relative yields of component intercrops under different water conditions. The data were obtained from published reports on maize–navy bean/soybean intercrop trials. The overall trends show that maize is the dominant crop when intercropped with navy bean and soybean.

In general, it is clear from the graph that the partial LER of the component crops was below the bisectrix, indicating that maize was the dominant crop. Furthermore, the overall total LER values were greater than 1 (Table 5), which signifies the importance of intercropping over the sole crops in all environments. The graph also shows that the intercrops gave higher LER under more severe moisture stress conditions. However, in high rainfall areas, the performance of the intercrops was relatively similar.

Relative competition intensity (RCI) is applied to compare the competitive ability of different crops within an intercropping system. This is calculated as in Sammuel et al. (2000) and Bing-Cheng et al. (2013).

$$RCI_{ab} = \left(\frac{(Y_{aa} \times Z_{ab}) - Y_{ab}}{Y_{ab} \times Z_{ab}} \right) \quad (17)$$

where Y_{ab} is yield of 'a' intercropped with 'b'; Y_{aa} is yield of 'a' in sole crop and Z_{ab} is a proportion of 'a' in combination with 'b'. If the value of RCI is zero, then the inter-specific competition equals the intra-specific competition, and if the value is positive, then inter-specific competition is higher. But if the value is negative, intra-specific competition is higher.

The values of RCI for maize and navy bean, on biomass, grain yield and N accumulation basis are presented in Tables 6 and 7. Table 6 shows that maize RCI values, during both seasons, were negative. This indicates that maize in an intercropping system was not

Table 6 Relative competition intensity (RCI) of maize

	RCI in aboveground dry matter and grain yield		RCI in aboveground dry matter and grain N yield	
Experiments	Dry matter*	Grain	Dry matter†	Grain
In 2011/12				
Low W/N	−0.58 ± 0.16	−0.49 ± 0.19	−0.85 ± 0.30	−0.64 ± 0.37
Medium W/N	−0.38 ± 0.06	−0.32 ± 0.04	−0.68 ± 0.26	−0.30 ± 0.09
High W/N	−0.32 ± 0.08	−0.26 ± 0.09	−0.29 ± 0.17	−0.16 ± 0.09
In 2012/13				
Low W/N	−0.27 ± 0.03	−0.31 ± 0.12	−0.25 ± 0.18	−0.24 ± 0.22
Medium W/N	−0.40 ± 0.21	−0.36 ± 0.09	−0.45 ± 0.37	−0.25 ± 0.02
High W/N	−0.28 ± 0.07	−0.43 ± 0.06	−0.25 ± 0.05	−0.32 ± 0.04

Mean ± SE, *n* = 3.
* aboveground dry matter including grain yield; † aboveground dry matter excluding grain N.

Table 7 Relative competition intensity (RCI) of navy bean

	RCI in aboveground dry matter and grain yield		RCI in aboveground dry matter and grain N yield	
Experiments	Dry matter*	Grain	Dry matter†	Grain
In 2011/12				
Low W/N	0.15 ± 0.11	−0.01 ± 0.03	−0.12 ± 0.27	−0.06 ± 0.05
Medium W/N	0.39 ± 0.01	0.35 ± 0.04	0.25 ± 0.20	0.26 ± 0.05
High W/N	0.71 ± 0.05	0.70 ± 0.07	0.55 ± 0.15	0.68 ± 0.05
In 2012/13				
Low W/N	−0.06 ± 0.05	−0.48 ± 0.03	0.26 ± 0.06	−0.57 ± 0.04
Medium W/N	−0.15 ± 0.09	−0.15 ± 0.12	−0.23 ± 0.06	−0.14 ± 0.05
High W/N	0.26 ± 0.05	0.34 ± 0.27	−0.17 ± 0.06	0.19 ± 0.21

Mean ± SE, $n = 3$.
* aboveground dry matter including grain yield; † aboveground dry matter excluding grain N.

Source: Tables 6 and 7 are adapted from Abeya Temesgen, 2015. The role of intercropping and sole cropping across environments of contrasting level of productivity in Ethiopia. A thesis submitted for the degree of Doctor of Philosophy at The University of Queensland, Queensland Alliance for Agriculture and Food Innovations, Australia.

affected by inter-specific competition. Intra-specific competition on an intercropping maize crop was stronger than inter-specific competition when maize was intercropped with navy bean crop. The inter-specific competition effect on maize was even very much weaker under high water and nitrogen (W/N) than under low-W/N input (Table 6). The values of RCI for navy bean were consistently positive under high W/N input during the first cropping season (Table 7). In general, the inter-specific competition effect on maize crop was small, but navy bean was severely affected, particularly at high levels of resource input.

The inter-specific competition between crop species in the intercropping system was relatively stronger on navy bean, mainly under high resource availability. Across increasing levels of water and nitrogen availability, the more maize was aggressive, the less was the advantage of the intercropping system. From this finding, there is no doubt that competition between component crops limited navy bean crop to capture enough resource (e.g. nitrogen) under high water and nitrogen supply. In contrast, Monzon et al. (2014) reported a yield reduction in a maize intercrop due to an increased soybean competitive ability in maize/soybean intercropping system when short growing cycle maize was used in the intercropping system. In addition to maize growth cycle, a soybean plant population used by Monzon et al. (2014) could contribute to this effect.

5 Maize–legume intercropping evaluation

The results of the household survey and empirical modelling studies were discussed to increase our understanding of the performance of maize–legume intercrops compared to sole crops under different levels of resource availability.

Based on the biophysical and socio-economic data collected from two contrasting agro-ecologies in Ethiopia (Abeya et al., 2015b), smallholder resource-poor farmers who use low input and crop in areas of low rainfall tend to intercrop; while resource-rich farmers in better environments are more likely to grow sole crops. The main differences between adopters and non-adopters of intercropping systems were found to be related to biophysical factors (application of fertilizer, soil fertility, land slope and livestock ownership), farm resources and/or capital (farm size, availability of labour, household size, household income) and experience or education (household age, household experience, household member education).

Other studies related the adoption of less expensive technologies in smaller farm size (Bidogeza et al., 2009; Doss and Morris, 2001). Herath and Takeya (2002) and Iqbal et al. (2006) also reported a positive association between the probability of adoption of intercropping by smallholder farmers and their experience in farming.

Farmland is a scarce resource in smallholding farming systems (Vanlauwe et al., 2014). Reflecting the increasing population (Godfray and Garnett, 2014), a decline in the land to labour ratio has been reported previously (Austin, 2007; Heady and Jayne, 2014; Yan et al., 2014). Therefore, improved food production in smallholding farmers, for example, will primarily need to be based on increased productivity mostly from the same area of farmland, though increases in land productivity should occur in such a way that the future production potential of the land is not undermined and that the environment is not affected (Garnett et al., 2013). Maize-legume intercropping system practices under such conditions are important to improve soil fertility through N fixation (Chowdhury and Rosario, 1992; Ghaley et al., 2005; Hauggaard-Nielsen et al., 2001b), conserve soil moisture by providing additional ground cover (Jahansooz et al., 2004) and to outcompete weeds (Hauggaard-Nielsen et al., 2001a; Workayehu and Wortmann, 2011).

The advantages of intercropping, as per LER values, have been widely used to evaluate the productivity of the system. Experimental and modelling results showed that the advantage of maize–bean intercropping in terms of LER was larger under conditions of lower resource availability. For example, up to 41% of land would be required for sole crops to produce the same yield as an intercrop when the level of productivity (inputs and investment) is low (Abeya et al., 2015a). Interestingly, the intercropping advantage in LER decreased as the level of resource availability, that is, inputs and investment, increased. Results from the modelling exercise using 30 years of climate data for the semi-arid and sub-humid regions produced similar results (Abeya et al., unpublished). Intercropping systems in terms of grain LER were more advantageous under conditions of lower N in semi-arid regions than in sub-humid regions. The overall result of the long-term simulation modelling showed that the advantages of intercropping in terms of LER deceased as the amount of rainfall increased.

In agreement with these results, the household survey confirmed the advantages of intercropping in the lower yield potential environments. This conclusion was also supported by an analysis of farmers' perception about the benefits of intercropping over sole cropping practices (Abeya et al., 2015b). This was more evident in the semi-arid region compared to the sub-humid region. Intercropping systems are, therefore, potentially more productive in resource-limiting or low-productivity environments, and it can potentially be used as a strategy to achieve the current targets for poverty reduction (UN, 2014).

The advantage of intercropping in terms of LER was related to improved N capture and WUE (Abeya et al., 2015a). The reasons for this might be associated to the effects of N on leaf area expansion from the taller maize crop under conditions of high resource availability (Runkulatile et al., 1998), which limited N uptake by the legume component of the intercrop system (Chowdhury and Rosario, 1992, 1994). The effects of shading by the corn crop could possibly explain the relationship between LER, N capture and WUE, as the level of resource availability of the growing environment increases.

Crop simulation modelling using long-term weather and crops prices data confirmed that sole maize cropping appeared to be less profitable when compared to the intercropping system and the sole bean crop in the semi-arid regions (Abeya et al., unpublished). This finding is consistent with the results of the household survey conducted in semi-arid and sub-humid regions (Abeya et al., 2015b). For example, the results from the household survey (Fig. 2) showed that maize in the sub-humid region and bean in the semi-arid region accounted for 81% and 50% of the total household annual income, respectively.

The higher income achieved from navy bean in the semi-arid region could be the result of improved productivity of the bean crop in this environment. The yield prediction using 30 years of climate data indicated higher sole crop navy bean grain yields in the semi-arid than in the sub-humid regions. Similar results were reported from an empirical study by Setegn and Habtu (2003) in the same regions.

In general, the predicted monetary returns using long-term climate records were highest for the intercropping system in sub-humid regions and for sole navy bean in the semi-arid regions.

Climate impact on cropping systems was also evaluated by using sensitivity analyses for rainfall and temperature across increasing levels of resource availability (Abeya et al., unpublished). Changes in rainfall from the baseline by −10%, −20% and −30% had little or no effect on both maize and bean crop yields. On average, increases in daily temperature of 1°C, 2°C and 3°C caused yield reductions in the performance of navy bean and intercropping systems, both in the semi-arid and sub-humid regions. Grain yields in the intercropping system and sole maize cropping increased as temperatures increased by up to 3°C from the baseline temperature.

In general, the future trends in intercropping studies should address the intercropping research questions from multidisciplinary approaches rather than only from systems productivity. The intercropping research should focus on environmental, ecological and

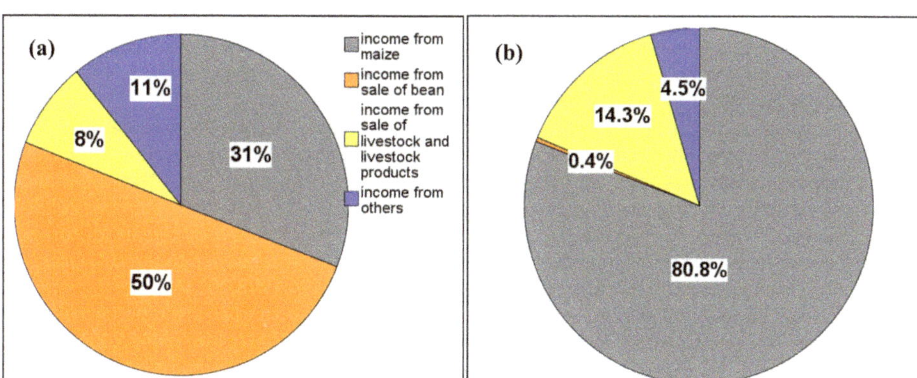

Figure 2 Annual household income sources in: (a) semi-arid regions; and (b) sub-humid regions of Ethiopia.

economical benefits along with system productivity. Cropping systems modelling and participatory research together with empirical experimentations are very powerful tools for deepening our understanding of the intercropping systems' productivity, profitability and resilience through a multidisciplinary lens

6 Conclusions

The information collated in this chapter indicates that the system productivity of intercropping is more important at lower than higher levels of productivity.

Maize-legume intercropping systems are more efficient and productive in relation to sole crops, under low N and water conditions. Lower competitive ability by the legume component as compared to the taller maize crop is proposed as one of the possible reasons for a decrease in system productivity of intercropping as the level of resource input increases.

In addition, studies on climate risk, farmers' drivers for adoption of intercrops or sole crops, household endowment, input use and farm size are important to fully understand the advantages of one cropping system over the other across a range of levels of resource inputs.

7 Where to look for further information

- John Vandermeer, 1989. *The Ecology of Intercropping.* Cambridge university press, Cambridge, Great Britain. 237 p.
- *Field Crop Research Journal* articles published in September 1993, 34(3–4): 239–472.
- International Maize and Wheat Improvement Centre (CIMMYT), website: SIMLESA. cimmyt.org

8 References

Abeya Temesgen, Shu Fukai and Daniel Rodriguez, 2015a. As the level of crop productivity increases: is there a role for intercropping in smallholder agriculture. *Field Crops Research*, 180: 155–66. http://dx.doi:10.1016/j.fcr.2015.06.003.

Abeya Temesgen, Shu Fukai, Moti Jaleta and Daniel Rodriguez, 2015b. Bio-physical and Socio-economic factors influence farmer's decisions on whether to intercrop or sole crops maize in Ethiopia. In Pathways for sustainable intensification of African agriculture?. 5th International symposium for farming systems design, Montpellier, France, 497–8, 7–10 September 2015.

Anil, L., Park, J., Phipps, R. H., Miller., F. A. 1998. Temperate intercropping of cereals for forage: a review of the potential for growth and utilization with particular reference to the UK. *Grass and Forage Science* 53: 301–17. http://dx.doi:10.1046/j.1365-2494.1998.00144.x.

Austin, G. 2007. Labour and land in Ghana, 1874–1939: a shifting ratio and an institutional revolution. *Australian Economic History Review* 47:95–120. http://dx.doi:10.1111/j.1467-8446.2006.00198.x.

Awal, M. A., Koshi, H., Ikeda, T. 2006. Radiation interception and use by maize/peanut intercrop canopy. *Agricultural and Forest Meteorology* 139 (1): 74–83. http://dx.doi:10.1016/j.agrformet.2006.06.001.

Baligar, V. C. and Bennett, O. L. 1986. NPK fertilizer efficiency – a situation analysis for the tropics. *Fertilizer Research* 10(2): 147–64. http://dx.doi:10.1007/BF01074369.

Banik, P. and Sharma, R. C. 2009. Yield and resource utilization efficiency in baby corn-legume-intercropping system in the Eastern Plateau of India. *Journal of Sustainable Agriculture* 33: 379–95. http://dx.doi:10.1080/10440040902834970.

Barabasz, W., Albińska, D., Jaśkowska, M. and Lipiec, J. 2002. Biological effects of mineral Nitrogen fertilization on soil microorganisms. *Polish Journal of Environmental Studies* 11 (3): 193–8.

Bedoussac, L. and Justes, E. 2010. Dynamic analysis of competition and complementarity for light and N use to understand the yield and the protein content of a durum wheat–winter pea intercrop. *Plant and Soil* 330: 37–54. http://dx.doi:10.1007/s11104-010-0303-8.

Belay, D., Schulthess, F. and Omwega, C. 2008. The profitability of maize–navy bean intercropping techniques to control maize stem borers under low pest densities in Ethiopia. *Phytoparasitica* 37: 43–50.

Bidogeza, J. C., Berentsen, P. B. M., De Graaff, J. and Oude Lansink, A. G. J. M. 2009. A typology of fram households for the Umutara Province in Rwanda. *Food Security* 1(3): 321–35. http://dx.doi:10.1007/s12571-009-0029-8.

Bing-Cheng, X., Wei-Zhou, X., Zhi-Juan, G., Jing, W. and Jin, H. 2013. Biomass production, relative competetive ability and water use efficiency of two dominant species in semiarid Loess Plateau under different water supply and fertilization treatments. *Ecological Research* 28(5): 781–92. http://dx.doi:10.1007/s11284-013-1061.

Brenda, B. C. and Robert, B. J. 1997. Plant competition underground. *Annual Review of Ecology and Systematics* 28: 545–70.

Bryan, B. A., King, D. and Zhao, G. 2014. Influence of management and environment on Australian wheat: information for sustainable intensification and closing yield gaps. *Environmental Research* 9 (4): 1–12. http://dx.doi:10.1088/1748-9326/9/4/044005.

Cassman, K. G., Dobermann, A. and Walters., D. T. 2002. Agroecosystems, nitrogen-use efficiency and nitrogen management. *Ambio* 31: 132–40. http://dx.doi:10.1579/0044-7447-31.2.132.

Caviglia, O. P. and Sadras, V. O. 2001. Effect of nitrogen supply on crop conductance, water- and radiation-use efficiency of wheat. *Field Crops Research* 69: 259–66. http://dx.doi:10.1016/S0378-4290(00)00149-0.

Caviglia, O. P., Sadras, V. O. and Andrade, F. H. 2004. Intensification of agriculture in the south-eastern Pampas: I. Capture and efficiency in the use of water and radiation in double-cropped wheat-soybean. *Field Crops Research* 87: 117–29. http://dx.doi:10.1016/j.fcr.2003.10.002.

Chowdhury, M. K. and Rosario, E. L. 1992. Utilization efficiency of applied nitrogen as related to yield advantage in maize/mungbean intercropping. *Field Crops Research* 30: 41–51. http://dx.doi:10.1016/0378-4290(92)90055-E.

Chowdhury, M. K. and Rosario, E. L. 1994. Comparison of nitrogen, phosphorus and potassium utilization efficiency in maize/mungbean intercropping. *Journal of Agricultural Science* 122: 193–9. http://dx.doi:10.1017/S0021859600087360.

Delogu, G., Cattivelli, L., Pecchioni, N., De Falcis, D., Maggiore, T. and Stanca, A. M. 1998. Uptake and agronomic efficiency of nitrogen in winter barley and winter wheat. *European Journal of Agronomy* 9: 11–20. http://dx.doi:10.1016/S1161-0301(98)00019-7.

Dordas, C. A. 2011. Nitrogen nutrition index and its relationship to N use efficiency in linseed. *European Journal of Agronomy* 34(2): 124–32. http://dx.doi:10.1016/j.eja.2010.11.005.

Doss, C. R. and Morris, M. L. 2001. How does gender affect the adoption of agricultural innovations? The case of improved maize technology in Ghana. *Agricultural Economics* 25: 27–39.

Gao, Y., Duan, A., Qiu, X., Sun, J., Zhang, J., Liu, H. and Wang, H. 2010. Distribution and use efficiency of photosynthetically active radiation in strip intercropping of maize and soybean. *Agronomy Journal* 102: 1149–57. http://dx.doi:10.2134/agronj2009.0409.

Gao, Y., Duan, A., Sun, J., Li, F., Liu, Z., Liu, Z. and Liu, H . 2009. Crop coefficient and water-use efficiency of winter wheat/spring maize strip intercropping. *Field Crops Research* 111: 65–73. http://dx.doi:10.1016/j.fcr.2008.10.007.

Garnett, T., Appleby, M. C., Balmford, A., Bateman, I. J., Benton, T. G., Bloomer, P., Burlingame, B., Dawkins, M., Dolan, L., Fraser, D., Herrero, M., Hoffmann, I., Smith, P., Thornton, P. K., Toulmin, C., Vermeulen, S. J. and Godfray, H. C. J. 2013. Sustainable intensification in agriculture: premises and policies. *Science* 341: 33–4. http://dx.doi:10.1126/science.1234485.

Godfray, H. C. and Garnett, T. 2014. Food security and sustainable intensification. Philosophical transactions of the Royal Society of London. *Series B: Biological Sciences* 369: 20120273. http://dx.doi:10.1098/rstb.2012.0273.

Ghaley, B. B., Hauggaard-Nielsen, H., Høgh-Jensen, H. and Jensen, E. S. 2005. Intercropping of wheat and pea as influenced by nitrogen fertilization. *Nutrient Cycling in Agroecosystems* 73: 201–12. http://dx.doi:10.1016/j.eja.2010.05.001.

Greenwood, D. J., Lemaire, G., Gosse, G., Cruz, P., Draycott, A. and Neeteson, J. J. 1990. Decline in percentage N of C3 and C4 crops with increasing plant mass. *Annals of Botany* 66: 425–36.

Gregory, P. J. and Reddy, M. S. 1982. Growth and resource use studies in an intercrop of pearl millet/groundnut. *Field Crops Research* 4: 13–24. http://dx.doi:10.1016/0378-4290(81)90050-2.

Hammer, G. L. and Wright, G. C. 1994. A theoretical analysis of nitrogen and radiation effects on radiation use efficiency in peanut. *Australian Journal of Agricultural Research* 45: 575–89. http://dx.doi:10.1071/AR9940575.

Hauggaard-Nielsen, H., Ambus, P. and Jensen, E. S. 2001a. Interspecific competition, N use and interference with weeds in pea–barley intercropping. *Field Crops Research* 70: 101–9. http://dx.doi:10.1016/S0378-4290(01)00126-5.

Hauggaard-Nielsen, H., Ambus, P. and Jensen, E. S. 2001b. Temporal and spatial distribution of roots and competition for nitrogen in pea-barley intercrops – a field study employing 32P technique. *Plant and Soil* 236: 63–74. http://dx.doi:10.1023/A:1011909414400.

Hauggaard-Nielsen, H., Gooding, M., Ambus, P., Corre-Hellou, G., Crozat, Y., Dahlmann, C., Dibet, A., von Fragstein, P., Pristeri, A., Monti, M. and Jensen, E. S. 2009. Pea-barley intercropping and short-term subsequent crop effects across European organic cropping conditions. *Nutrient Cycling in Agroecosystems* 85: 141–55. http://dx.doi:10.1007/s10705-009-9254-y.

Hauggaard-Nielsen, H. and Jensen, E. S. 2005. Facilitative root interactions in intercrops. *Plant and Soil* 274: 237–50. http://dx.doi:10.1007/s11104-004-1305-1.

Headey, D. D. and Jayne, T. S. 2014. Adaptation to land constraints: is Africa different? *Food Policy* 48: 18–33. doi:10.1016/j.foodpol.2014.05.005.

Herath, P. H. and Takeya, H. (2002). Factors determining intercropping by rubber smallholders in Sri Lanka: a logit analysis. *Agricultural Economics* 29: 159–68.

Herrmann, A. and Taube, F. 2004. The range of the critical nitrogen dilution curve for maize (Zea mays L.) can be extended until silage maturity. *Agronomy Journal* 96: 1131–8.

Hirel, B. and Lemaire, G. 2005. From agronomy and ecophysiology to molecular genetics for improving nitrogen use efficiency in crops. *Journal of Crop Improvement* 15: 213–57.

Iqbal, S. M. M., Ireland, C. R. and Rodrigo, V. H. L. 2006. A logistic of the factors determining the decisions of smallholder farmers to intercrop: a case study involving rubber-tea intercropping in Sri Lanka. *Agricultural Systems* 87: 296–312.

Jahansooz, M. R., Yunusa, I. and Coventry, D. 2004. Capture and use of water by wheat and chickpea in sole crops and intercrops, under dryland conditions of South Australia In: Proceedings of the Australian Agronomy Conference 2004, 12th AAC, 4th ICSC.

Jensen, E. 1996. Grain yield, symbiotic N_2 fixation and interspecific competition for inorganic N in pea-barley intercrops. *Plant and Soil* 182: 25–38. http://dx.doi:10.1007/BF00010992.

Kanayama, Y., Watanabe, I. and Yamamoto, Y. 1990. Inhibition of nitrogen fixation in soybean plants supplied with nitrate I. Nitrite accumulation and formation of nitrosylleghemoglobin in nodules. *Plant Cell Physiology* 31: 341–6.

Kanton, R. A. L. and Dennett, M. D. 2004. Water uptake and use by morphologically contrasting Maize/Pea cultivars in sole and intercrops in temperate conditions. *Experimental Agriculture* 40: 201–14. http://dx.doi:10.1017/S0014479703001571.

Keating, B. A. and Carberry, P. S. 1993. Resource capture and use in intercropping: solar radiation. *Field Crops Research* 34: 273–301. http://dx.doi:10.1016/0378-4290(93)90118-7.

Khan, G. M. 1996. Response of intercropping system involving maize (*Zea mays*) to irrigation level in Kashmir valley. *The Indian Journal of Agricultural Sciences* 66(5): 272–5.

Kutu, F. R., Deale, W. and Asiwe, J. A. N. 2009. Assessment of maize and dry bean productivity under different intercropping systems and fertilization regimes. In *Proceedings of African Crop Science Conference of Africa*, 267–70, Uganda.

Lemaire, G., van Oosterom, E., Sheehy, J., Jeuffroy, M. H., Massignam, A. and Rossato, L. 2007. Is crop demand more closely related to dry matter accumulation of leaf area expansion during vegetative growth? *Field Crop Research* 100: 91–106.

Lemaire, G. and Salette, J. 1984. Relationship between growth and nitrogen uptake in a pure gras stand. 1. Environmental effects (nitrogen nutrition, nitrogen content, allometric relation, variability between years, tall fescue). *Agronomy* 4(5): 423–30.

Li, C. J., Li, Y. Y., Yu, C. B., Sun, J. H., Christie, P., An, M., Zhang, F. S. and Li, L. 2011a. Crop nitrogen use and soil mineral nitrogen accumulation under different crop combinations and patterns of strip intercropping in northwest China. *Plant and Soil* 342: 221–31. http://dx.doi:10.1007/s11104-010-0686-6.

Li, Q. Z., Sun, J. H., Wei, X. J., Christie, P., Zhang, F. S., Li, L. 2011. Overyielding and interspecific interactions mediated by nitrogen fertilization in strip intercrop-ping of maize with faba bean, wheat and barley. *Plant Soil* 339: 147–61. http://dx.doi.org/10.1007/s11104-010-0561-5.

Liu, T., Song, T. F., Liu, S. and Zhu, X. 2012. Light interception and radiation use efficiency response to narrow-wide row planting patterns in maize. *Australian Journal of Crop Science* 6: 506–13.

López-Bellido, R. J. and López-Bellido, L. 2001. Efficiency of nitrogen in wheat under Mediterranean conditions: effect of tillage, crop rotation and N fertilization. *Field Crops Research* 71: 31–46. http://dx.doi:10.1016/S0378-4290(01)00146-0.

Luis, L. B., Rafael, J. L. B. and Francisco, J. L. B. 2006. Fertilizer nitrogen efficiency in Durum Wheat under rainfed Mediterranean conditions: Effect of split application. *Agronomy Journal* 98(1): 55–62. http://dx.doi:10.2134/agronj2005.0017.

Marshall, B. and Willey, R. W. 1983. Radiation interception and growth in an intercrop of pearl millet/groundnut. *Field Crops Research* 7: 141–60. http://dx.doi:10.1016/0378-4290(83)90018-7.

Mead, R. and Willey, W. 1980. The concept of 'land equivalent ratio' and advantages in yield from intercropping. *Experimental Agriculture* 16: 217–28.

Meyers, W. S., Tan, C. S., Barrs, H. D. and Smith R. C. G. 1990. Root growth and water uptake by wheat during drying of undisturbed and repacked soil in drainage lysimeters. *Australian Journal of Agricultural Research* 41: 253–65.

Monzon, J. P., Mercau, J. L., Andrade, J. F., Caviglia, O. P., Cerrudo, A. G., Cirilo, A. G., Vega, C. R. C., Andrade, F. H. and Calvino, P. A. 2014. Maize – soybean intensification alternatives for Pampas. *Field Crop Research* 162: 48–59. http://dx.doi:10.1016/j.fcr.2014.03.012.

Morris, R. A. and Garrity, D. P. 1993. Resource capture and utilization in intercropping: water. *Field Crops Research* 34: 303–17. http://dx.doi:10.1016/0378-4290(93)90119-8.

Mosier, A. R. 2002. Environmental challenges associated with needed increases in global nitrogen fixation. *Nutrient Cycling in Agroecosystems* 63: 101–16. http://dx.doi:10.1023/A:1021101423341.

Muchow, R. C. and Coates, D. B. 1986. An analysis of the environmental limitation to yield of irrigated grain sorghum during the *dry* season in tropical Australia using a radiation interception model. *Australian Journal of Agricultural Research* 37: 135–48.

Mutsaers, H. J. W., Ezumah, H. C. and Osiru, D. S. O. 1993. Cassava-based intercropping: a review. *Field Crops Research* 34: 431–57. http://dx.doi:10.1016/0378-4290(93)90125-7.

Natarajan, M. and Willey, R. W. 1985. Effect of row arrangement on light interception and yield in sorghum-pigeonpea intercropping. *The Journal of Agricultural Science* 104: 263–70. http://dx.doi:10.1017/S0021859600043902.

Natarajan, M. and Willey, R. W. 1986. The effects of water stress on yield advantages of intercropping systems. *Field Crops Research* 13: 117–31. http://dx.doi:10.1016/0378-4290(86)90015-8.

Niringiye, C. S., Ssekabembe, C. S. and Kyamanywa, S. 2005. Effects of plant population on yield of maize and climbing beans grown in an intercropping *African Crop Science Journal* 13 (1): 83–93.

Ofori, F. and Stern, W. R. 1987. Cereal-legume intercropping systems. *Advances in Agronomy* 41: 41–90.

Ogindo, H. O. and Walker, S. 2005. Comparison of measured changes in seasonal soil water content by rainfed maize-bean intercrop and component cropping systems in a semi-arid region of southern Africa. *Physics and Chemistry of the Earth* 30: 799–808.

Ouda, S. A., Mesiry, T. El., Abdallah, E. F. and Gaballah, M. S. 2007. Effect of water stress on the yield of Soybean and maize grown under different intercropping patterns. *Australian Journal of Basic and Applied Sciences* 1 (4): 578–85.

Parsons, R., Stanforth, A., Raven, J. A. and Sprent, J. I. 1993. Nodule growth and activity may be regulated by a feedback mechanism involving phloem nitrogen. 16: 125–36. http://dx.doi:10.1111/j.1365-3040.1993.tb00854.x.

Pilbeam, C. J., Okalebo, J. R., Simmonds, L. P. and Gathua, K. W. 1994. Analysis of maize-common bean intercrops in semi-arid Kenya. *Journal of Agricultural Science* (Cambridge) 123: 191–8.

Plénet, D. and Lemaire, G. 2000. Relationships between dynamics of nitrogen uptake and dry matter accumulation in maize crops. Determination of critical N concentration. *Plant Soil* 216: 65–82.

Reddy, M. S. and Willey, R. W. 1981. Growth and resource use studies in an intercrop of pearl millet/groundnut. *Field Crops Research* 4: 13–24. http://dx.doi:10.1016/0378-4290(81)90050-2.

Robertson, M. J., Fukai, S., Ludlow, M. M. and Hammer, G. L. 1993. Water extraction by grain sorghum in a sub-humid environment. II. Extraction in relation to root growth. *Field Crop Research* 33(1): 99–112. http://dx.doi:10.1016/0378-4290(93)90096-6.

Runkulatile, K., Homma, K., Horie, T., Kurusu, T. and Inamura, T. 1998. Land equivalent ratio of groundnut-fingermillet intercrops as affected by plant combination ratio, and nitrogen and water availability. *Plant Production Science* 1: 39–46.

Sadras, V. O. 2004. Yield and water-use efficiency of water- and nitrogen-stressed wheat crops increase with degree of co-limitation. *European Journal of Agronomy* 21: 455–64. http://dx.doi:10.1016/j.eja.2004.07.007.

Sadras, V. O. and Rodriguez, D. 2010. Modelling the nitrogen-driven trade-off between nitrogen utilisation efficiency and water use efficiency of wheat in eastern Australia. *Field Crops Research* 118: 297–305. http://dx.doi:10.1016/j.fcr.2010.06.010.

Santalla, M., Casquero, P. A. and De Ron, A. M. 1999. Yield and yield components from intercropping improved bush bean cultivars with maize. *Journal of Agronomy and Crop Science* 183: 263–9. http://dx.doi:10.1046/j.1439-037x.1999.00348.x.

Setegn, G. and Habtu, A. 2003. Genotype X Environment interaction and stability analysis of seed yield in navy bean genotypes. *African Crop Science Journal* 11(1): 1–7.

Sinclair, T. R. and Muchow, R. C. 1999. Radiation use efficiency. *Advance Agronomy* 65: 215–65.

Singh, H. P., Venkateswarlu, B., Vittal, K. P. R. and Ramachandran, K. 2000. In J. S. P. Yadav and G. B. Singh (eds), *Natural resource Management for Agricultural Production in India*, International Conference on Managing Natural Resources for Sustainable Agricultural Production in the 21st Century, 14–18 February 2000, New Delhi, India, 669.

Szumigalski, A. R. and Van Acker, R. C. 2006. Nitrogen yield and land use efficiency in annual sole crops and intercrops. *Agronomy Journal* 98: 1030–40.

Tolessa, D., Du Preez, C. C. and Ceronio, G. M. 2009. Effects of tillage system and nitrogen fertilization on officacy of applied nitrogen by maize in Western Ethiopia. *South African Journal of Plant and Soil* 26 (1): 36–48.

Tsubo, M. and Walker, S. 2002. A model of radiation interception and use by a maize–bean intercrop canopy. *Agricultural and Forest Meteorology* 110: 203–15. http://dx.doi:10.1016/S0168-1923(01)00287-8.

Tsubo, M., Walker, S. and Mukhala, E. 2001. Comparisons of radiation use efficiency of mono-/inter-cropping systems with different row orientations. *Field Crops Research* 71: 17–29. http://dx.doi:10.1016/S0378-4290(01)00142-3.

Tsubo, M., Walker, S. and Ogindo, H. O. 2005. A simulation model of cereal-legume intercropping systems for semi-arid regions: II. *Model Application. Field Crops Research* 93: 23–33. http://dx.doi:10.1016/j.fcr.2004.09.003.

Tsubo, M., Ogindo, H. O. and Walker, S. 2004. Yield evaluation of maize-bean intercropping in a semi-arid region of South Africa. *African Crop Science* Journal 12 (4): 351–8. http://dx.doi:10.4314/acsj.v12i4.27897.

United Nation (UN). 2014. Millennium development goals reports. http://www.un.org/millenniumgoals/reports.shtml.

Vahdettin, C., Togay, N. and Dogan, Y. 2006. The effect of intercropping sowing system with dry bean and maize on yield and some yield components. *Asian Network for Scientific Information* 5 (1): 53–6.

Vandermeer, J. H. 1989. *The Ecology of Intercropping*. Cambridge, UK: Cambridge University Press.

Vanlauwe, B., Coyne, D., Gockowski, J., Hauser, S., Huising, J., Masso, C., Nziguheba, G., Schut, M. and Van Asten, P. 2014. Sustainable intensification and the African smallholder farmer. *Current Opinions on Environmental Sustainability* 8: 15–22. http://dx.doi:10.1016/j.cosust.2014.06.001.

Vessey, J. K. and Waterer, J. 1992. In search of the mechanism of nitrate inhibition of nitrogenase activity in legume nodules: Recent developments. *Physiologia Plantarum* 84: 171–6. http://dx.doi:10.1111/j.1399-3054.1992.tb08780.x.

Wahua, T. A. T. 1983. Nutrient uptake by intercropped maize and cowpeas and a concept of nutrient supplementation index (NSI). *Experimental Agriculture* 19: 163–75.

Watiki, J. M., Fukai, S., Banda, J. A. and Keating, B. A. 1993. Radiation interception and growth of maize/cowpea intercrops as affected by maize plant density and cowpea cultivar. *Field Crops Research* 35: 123–33. http://dx.doi:10.1016/0378-4290(93)90145-D.

Willey, R. W. 1979. Intercropping: Its importance and research needs. Part 1, Competition and yield advantages. *Field Crops Abstract* 32: 1–10.

Willey, R. W. 1990. Resource use in intercropping systems. *Agricultural Water Management* 17: 215–31.

Workayehu, T. and Wortmann, C. S. 2011. Maize–bean intercrop weed suppression and profitability in Southern Ethiopia. *Agronomy Journal* 103 (4): 1058–63. http://dx.doi:10.2134/agronj2010.0493.

Yan, X., Bauer, S. and Huo, X. 2014. Farm size, land reallocation, and labour migration in rural China. *Population, Space and Place* 20: 303–15. http://dx.doi:10.1002/psp.1831.

Yang, C., Huang, G., Chai, Q. and Luo, Z. 2011. Water use and yield of wheat/maize intercropping under alternative irrigation in the Oasis field of Northern China. *Field Crops Research* 124: 426–32.

Yang, G., Aiwang, D., Xinqiang, Q., Zugui, L., Jingsheng, S., Junpeng, Z. and Hezhou, W. 2010. Distribution of roots and root length density in a maize/soybean strip intercropping system. *Agricultural Water Management* 98: 199–212. http://dx.doi:10.1016/j.agwat.2010.08.021.

Zhang, B., Huang, G. and Li, F. 2007. Effect of limited single irrigation on yield of winter wheat and spring maize relay intercropping. *PEDOSPHERE* 17: 529–37. http://dx.doi:10.1016/S1002-0160(07)60063-0.

Zhang, F. and Li, L. 2003. Using competitive and facilitative interactions in intercropping systems enhances crop productivity and nutrient-use efficiency. *Plant and Soil* 248: 305–12. http://dx.doi:10.1023/A:1022352229863.

Zhang, L., Spiertz, J. H. J., Zhang, S., Van der Werf, B. and Li, W. 2008. Nitrogen economy in relay intercropping systems of wheat and cotton. *Plant and Soil* 303: 55–68. http://dx.doi:10.1007/s11104-007-9442-y.

Zhang, N., Zhang, W. L., Tian, Z. X. and Li, X. Q. 1996. Nitrate pollution of groundwater in northern China. *Agriculture, Ecosystems and Environment* 59: 223–31. http://dx.doi:10.1016/0167-8809(96)01052-3.

Ziadi, N., Bélanger, G., Claessens, A., Lefebvre, L., Cambouris, A. N., Tremblay, N., Nolin, M. C. and Parent, L.-É. 2010. Determination of a critical nitrogen dilution curve for spring wheat. *Agronomy Journal* 102: 241–50. http://dx.doi:10.2134/agronj2009.0266.

Ziadi, N., Bélanger, G., Gastal, F., Claessens, A., Lemaire, G. and Nicolas Tremblay, N. 2009. Leaf nitrogen concentration as an indicator of corn nitrogen status. *Agronomy Journal* 101: 947–57. http://dx.doi:10.2134/agronj2008.0172x.

Ziadi, N., Brassard, G., Belanger, A. N., Cambouris, N., Tremblay, M. C., Nolin, A., Claessens and Parent, L. E. 2008. Critical nitrogen curve and nitrogen nutrition index for corn in eastern Canada. *Agronomy Journal* 100: 271–6. http://dx.doi:10.2134/agronj2007.0059.

Chapter 11

Climate risk management in maize cropping systems

Daniel Rodriguez, Caspar Roxburgh, Claire Farnsworth, Ariel Ferrante, Joseph Eyre, Stuart Irvine-Brown, James McLean, Martin Bielich, Queensland Alliance for Agriculture and Food Innovation (QAAFI), The University of Queensland, Australia

1 Introduction

Climate risk means different things to different people depending on their circumstance. To smallholder farmers from sub-Saharan Africa (SSA), drought might mean food insecurity, temporary migration or famine; to a large-scale commercial farmer from the West, drought might mean opportunity – as per higher commodity prices, or a time to mind investments, or to seek government support. In this chapter, we aim to make a 'two tail' analysis of how risk affects farmers operating under contrasting socio-economic circumstances, and discuss likely options and opportunities for the sustainable intensification of maize-based farming systems.

Maize is the cereal with the largest global production (>1 billion tonnes/year) (Fig. 1a). This compares to about 0.73 and 0.5 billion tonnes for wheat and rice, respectively (http://www.fao.org/giews/english/fo/index.htm). The world's maize production shows signs of exponential increase, both driven by changes in area planted (Fig. 1b) and a relatively constant rate of global yield increase (1.5%/year) (Fig. 2a). However, rates of yield change have been rather contrasting across regions. For example, yield increases

http://dx.doi.org/10.19103/AS.2016.0002.26

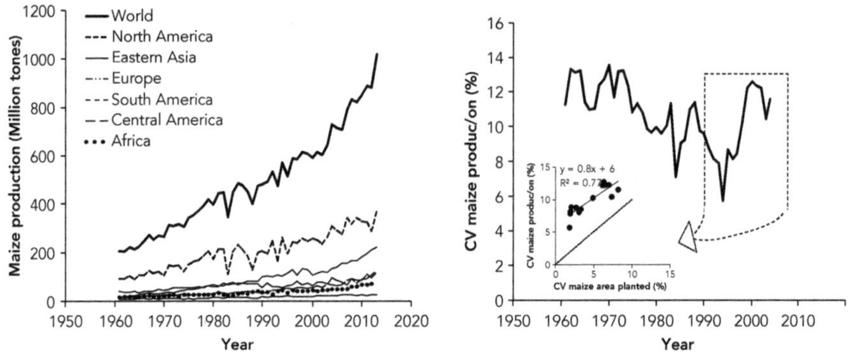

Figure 1 Maize production across the globe (a), and coefficient of variability (%) on world maize production (b). The inset in (b) shows the relationship between the coefficient of variability in world maize production and maize area planted.

Source: FAO Stat.

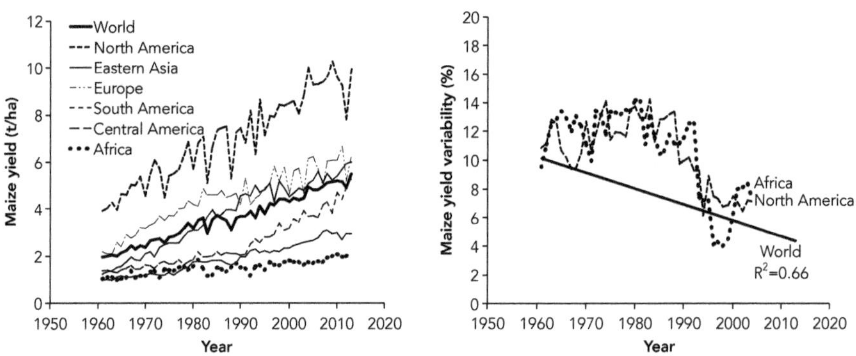

Figure 2 Maize yields across the globe (a), and coefficient of variation (%) on world (fitted line), Africa and North America maize yields (b).

Source: FAO Stat.

in South America, particularly in Argentina and Brazil, showed a rapid increase after the 1990s (ca. 2.6%/year – from 2.4 to 5.2 t/ha between 1990 and 2013), probably associated with the adoption of stubble retention practices, earlier planting and the use of higher plant populations with crowding tolerant hybrids, and a low baseline yield. On the contrary, yields in Africa remain low (about 1.8t/ha), growing at a slower pace than in the rest of the world (1.4%/year), with the exception of Europe 1.2% (i.e. from 4.2t/ha in 1990 to 6.2t/ha in 2013) – when the Russian Federation and Ukraine are also included in the calculation. Interestingly, yield variability appears to be decreasing (Fig. 2b), particularly after the 1990s.

Globally, about 41% of the yield variability in maize can be explained by year-to-year climate variability, though large variations exist around the globe (Ray et al., 2015) (Fig. 3).

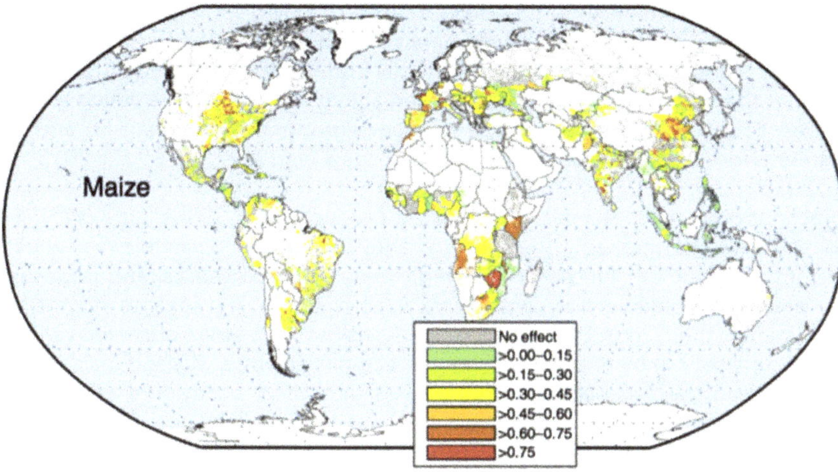

Figure 3 Total crop yield variability explained due to climate variability over the last three decades. A value of 1 indicates that all the yield variability is explained by climate variability.

Source: Ray et al., 2015.

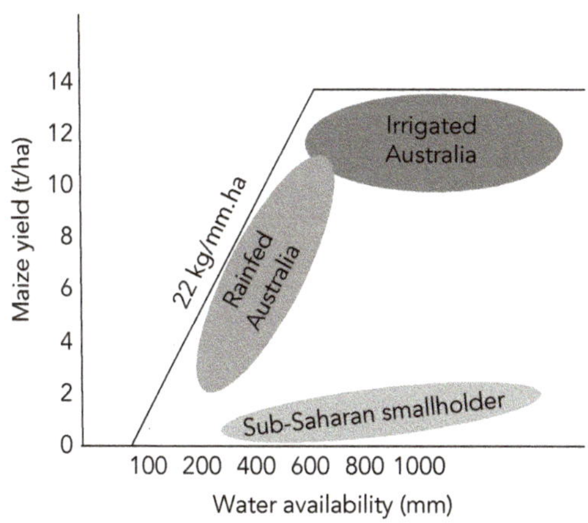

Figure 4 Schematic representation of actual farmers' yields plotted against water availability in large-scale commercial irrigated and rainfed cropping systems from Australia, and smallholder farmers from sub-Saharan Africa. The solid line represents the boundary function for cereal crops water productivity.

Source: Sadras and Angus, 2006.

It is difficult to attribute the changes after the 1990s (Fig. 2b) in Africa and North America to a single or common factor. Ray et al. (2015) identified that for many African countries rainfall variability related to El Niño Southern Oscillation (ENSO) best explained variability in maize yields. Eastern and southern Africa, Ethiopia, Zimbabwe, Malawi and Mozambique show the strongest relationships.

In Fig. 4, the performances of SSA rainfed maize cropping systems (Dimes et al., 2015) are compared to irrigated and rainfed maize production systems from Queensland in Australia. Assuming the boundary function from Sadras and Angus (2006), Fig. 4 shows that there is a significant gap between actual and achievable yields across SSA countries, ca. 80%. Most likely driven by poor agronomic management, and lack of investments on crop nutrition, and improved seed (van Ittersum et al., 2013; Dimes et al., 2015).

So even though the potential for improving productivity in SSA is large, Why success has been so evasive? Probably, the predominantly unimodal and variable rainfall patterns of eastern and southern Africa, farmers' high vulnerability to shocks, and poor terms of trade, provide formidable disincentives for farmers to reinvest meagre returns on the improvement of their farms. Here we then discuss risk not only as the observed variability in rainfall and yields, but try to build up from the eco-physiology of dryland maize cropping, connecting with the socio-economic conditions that affect behaviours and technology adoption.

2 Sensitivity of rainfed maize cultivation

Maize is a rather sensitive crop to environmental stresses particularly during a window around flowering, that is, the critical stage for grain number definition, and during early grain filling.

Figure 5 shows that the critical stage for grain number definition in maize is narrowly concentrated around anthesis, and that the impact of a stress during this window can

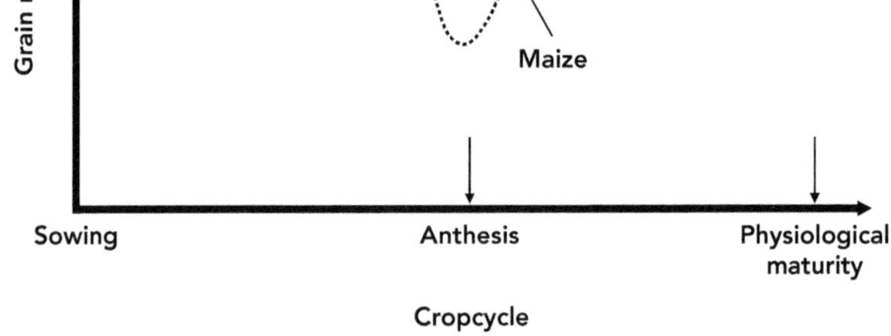

Figure 5 Comparative timing and severity effects of stresses around flowering on grain number for wheat, maize, sunflower and soybean. Grain number is presented as an arbitrary scale where the line is 100%.

Source: Calviño and Monzon, 2009.

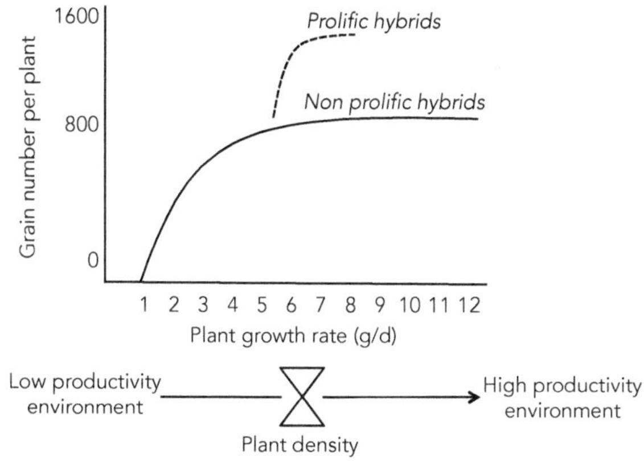

Figure 6 Relationship between grain number and plant growth rate during the critical window (−230–100°Cd) around silking for non-prolific and prolific maize hybrids. The level of environmental productivity and plant density are the main factors driving plant growth rate.

dramatically reduce the most important yield component in maize, that is, grain number. This contrasts with wheat and sunflower that have earlier and wider critical windows, and soybean, which has a later one, that is, during seed setting and grain formation (Calviño and Monzon, 2009). The sensitivity of maize to changes in the rate of growth of the crop around flowering is shown in Fig. 6. In maize, grain yield is highly related to grain number per unit area (Tollenaar et al., 2000; Andrade, 1995; Echarte et al., 2000), while grain number per unit of area appears to be related to the rate of crop growth during a critical window defined as −230 to +100°Cday around silking (Otegui and Bonhomme, 1998) (Fig. 6).

Figure 6 also shows the distinctive behaviour for so-called non-prolific and prolific materials. Prolificity in maize is the capacity of the crop to develop multiple cobs. There are several types of prolificity (or multi-cobbing), single cobbing hybrids, hybrids that develop multiple cobs on the main stem and hybrids that develop multiple cobs on multiple stems (tiller prolificity). We will discuss further the role of prolificity for rainfed maize cropping and its relation to management and environment factors in Section 3. Fig. 5 also indicates that in rainfed maize cropping stresses taking place at different stages of crop development are likely to have different effects on crop yield and that understanding the frequency of occurrence and intensity of water stresses around flowering across regions might help define 'stress environments'. Understanding the types of stress and their frequency of occurrence is likely to (i) guide breeders to identify representative sites for multi-environment networks of trials; (ii) identify physiological traits likely to introduce site-specific drought tolerance; and (iii) match hybrids (G) and managements (M), to environments (E) in GxExM approaches.

Chauhan et al. (2013) used a simulation model (APSIM, www.apsim.info) to simulate the yield variability of maize across north-eastern Australia and its relation to the frequencies of different water stress patterns around anthesis (Fig. 7 and 8). By clustering the simulated water supply demand ratio around anthesis, they defined five major water stress patterns; a no-stress environment (E1), a mild-early stress environment (E2), a mild terminal stress (E3), a moderate terminal stress (E4) and a severe terminal stress (E5).

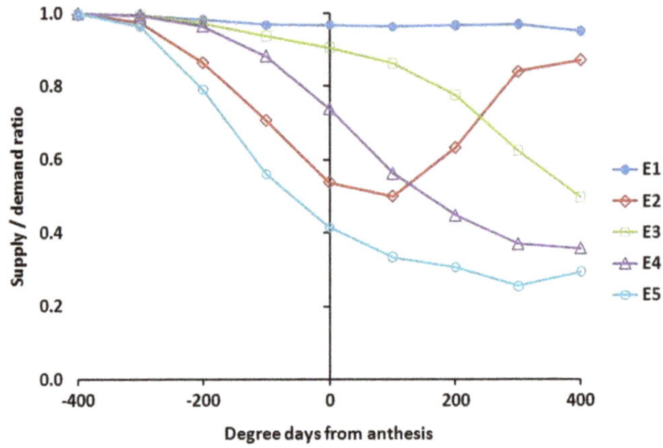

Figure 7 Five major water stress patterns common in the rainfed maize-growing areas of north-eastern Australia based on clustering of the water supply demand ratio around anthesis calculated with APSIM. E1 = no stress; E2 = mild-early stress; E3 = mild terminal stress; E4 = moderate terminal stress and E5 = severe terminal stress.

Source: Chauhan et al., 2013.

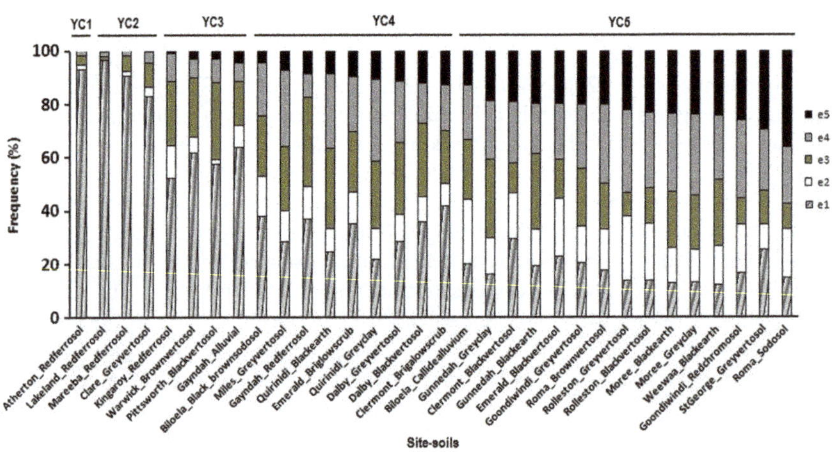

Figure 8 Frequencies of water stress environment types found in the rainfed maize-growing areas of north-eastern Australia.

Source: Chauhan et al., 2013.

Making adjustments in planting time is one of the main tools farmers have to fit the crop critical stage in Fig. 6, so that it takes place during the most favourable conditions. Thus, understanding the implications of planting date on potential yield is imperative to manage the risk of stress around flowering. As planting date progresses throughout the season, factors such as mean daily temperature, photoperiod, radiation and rainfall vary.

Successful cultivation of maize requires a minimum soil temperature of 12°C at planting and a mean daily temperature of at least 15°C throughout the growing period (Tollenaar and Dwyer, 1999). Maize will tolerate hot and dry conditions as long as sufficient water is available and mean daily temperatures do not exceed 45°C. Grain set can be disrupted if temperatures drop below 15°C during the critical period for grain number definition (Araus et al., 2012). Planting early in the season may see low temperatures delaying early growth stages and increasing the crop's exposure to frost damage, while planting late in the season will reduce the length of crop and yield potential. Maize genotypes developed for tropical regions are sensitive to day length and temperature interactions. Short days and warm temperatures, for example, may delay the onset of silking (Jones and Kiniry, 1985). The greatest sensitivity of maize to water deficits are observed when water is limited during the flowering period (Calviño and Monzon, 2009; Fig. 5 and 7). It follows that planting should occur to allow for the critical period around flowering and the period with the highest likelihood of rainfall to coincide. Biotic stresses, such as weeds and pests, can impact both growth and yield. The *Heliothine* moths, for example, can impact yield when its presence coincides with the period of cob development (Ortega, 1987). Damage by *Heliothis* can also increase the susceptibility of plants to fungal infections, particularly in high-humidity environments (Aguirre-Uribe et al., 2015).

In many instances, successful maize cultivation depends on the appropriate choice of hybrids to allow the optimal growing season to match the length of the growing period. The selection of hybrids tailored to the growing environment is crucial to maximizing yield. Under adequate soil moisture conditions, more slowly maturing hybrids are likely to give increased grain yields, whereas under dryland conditions, short-season hybrids have the ability to escape the high temperatures and terminal water stresses (Araus et al., 2012). There may be benefit in planting a variety of hybrids having different maturity times, or planting at a range of planting dates to overcome the unpredictability of the environment.

3 Assessing the risk of rainfall variability

In dryland cropping environments (E), where the amount of available water may limit crop growth, final grain yield is generally determined by the interactions between crop genetic

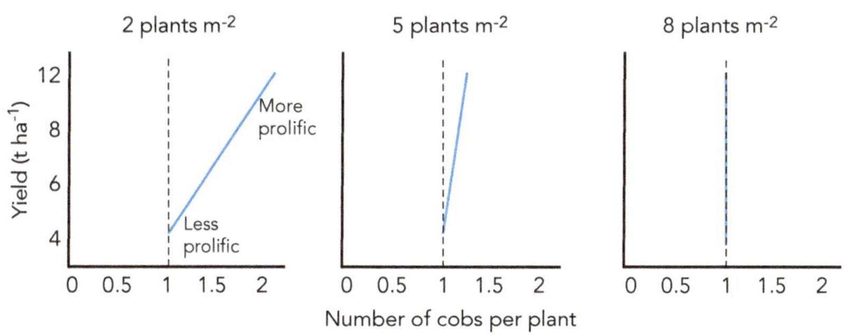

Figure 9 Relationship between rainfed maize yield for a range of hybrids showing a contrasting degree of multi-cobbing (prolificity) growing at different plant populations.

Figure 10 Types of multi-cobbing (prolificity) observed in modern maize hybrids when sown at low plant populations.

traits (G) and agronomic management (M) manipulations available to improve crop growth and yield (McCown et al., 1996).

Australia's high climate variability rainfed cropping provides a good example of how GxExM interactions can be used to manage the risks associated with water stress.

Farmers' adaptations are observed at both the field and cropping system levels. At the cropping system level, Australian rainfed farmers use dynamic or opportunistic cropping systems. For example, farmers shift from conservative strategies in likely drier-than-average years to more intensive and potentially more profitable strategies in wetter-than-average years (Sadras et al., 2003; Rodriguez et al., 2011, 2014).

At the field or crop level, climate risk adaptations may involve the early sowing (on stored soil moisture) of prolific hybrids (i.e. multi-cobbing) at low plant populations to reduce crop water use early in the season, that is, to transfer the use of water from the vegetative to reproductive stages (Ferrante et al., 2015), while retaining the higher potential of the prolific materials in case the season turns out to be good (Fig. 6 and 9).

Results shown in Fig. 9 indicate that prolific genotypes planted early at low plant populations are able to compensate for yield losses due to the lower plant population and be more resilient to drier conditions. When planted at higher plant populations, most plants tend to produce one cob per plant. The strategy allows for savings in seed cost, while reducing the likelihood of stress during critical stages (Fig. 5), and retaining the potential for yield in the wetter-than-average seasons.

Different types of maize prolificity in commercial hybrids in Australia have been reported (Eyre and Rodriguez, 2016), that is, one single cob on the main stem (non-prolific), multiple cobs on the main stem (main stem prolificacy), multiple cobs on multiple stems (tiller prolificacy) and a single cob with multiple stems (infertile tillers) (Fig. 10). Moulia et al. (1999) already indicated that at very low densities, modern hybrids produce a large number of mostly unproductive tillers with a low contribution to grain yield, that is, a low-tiller harvest index (0 to 20%). We also found that in materials that produce fertile tiller cobs, the likelihood that the secondary cob on the main stem would develop was much reduced. Further research is required to answer whether by restricting tillering more resources would be available for developing a secondary cob on the main stem (main stem prolificity) and increasing the harvest index of main stem of prolific materials (Vega et al., 2001).

4 Risk management in smallholder maize cropping

The particular eco-physiology of the maize crop sets the scene for a relatively high sensitivity to water stress of maize-dominant cropping systems (see previous sections). This conspires to produce an inherently risky enterprise in agro-ecologies where climate variability and farmers' vulnerability are high. This is the particular case of eastern and southern Africa (Fig. 11).

While climate and price variability are associated with variability in farmers profits in professionalized maize production systems (i.e. that of the United States, Argentina, Brazil and Australia), in vulnerable smallholder farming in Africa, the implications of drought and crop failure need to be considered as sources for food insecurity, malnutrition, migration and poverty traps (Wossen and Berger, 2015).

The yields of smallholder farmers from SSA show large productivity gaps that reach up to 80% of the achievable water-limited yields (Fig. 3). This shows the potential for improvement when compared to other parts of the world, for example, United States (32%), Brazil (65%) or Argentina (17%) (Table 1), but this also reflects on the complexity and difficulty of the problem.

In principle, the productivity gaps in smallholder African agriculture exist because known and effective technologies and innovations are yet to be adopted (Giller et al., 2015). Mechanization, use of fertilizers, manures, amendments, herbicides, improved germplasm, use of mulches, reduced tillage systems, crop mix and rotations are all cited as technologies with potential to increase productivity (Fischer et al., 2014).

The question is why have we found it difficult to ensure the widespread adoption of solutions when these are known and might even be available? In most cases this is the result of socio-economic, infrastructure and market constraints, and probably to a great extent due to farmers' perceived risk involved in adopting unknown or costly technologies

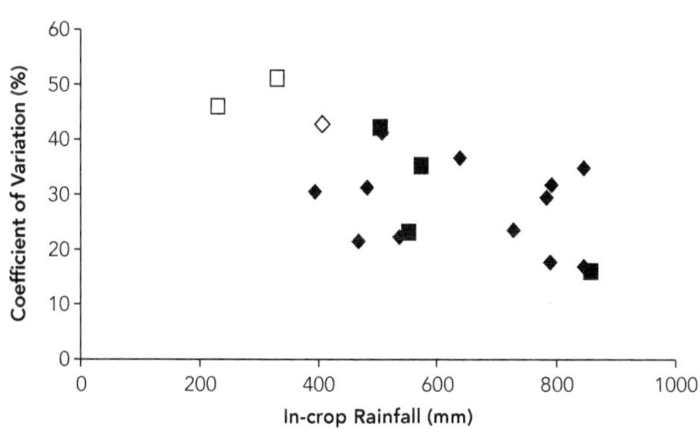

Figure 11 Average in-crop rainfall and its variability (coefficient of variation, %) for 16 site-seasons (solid symbols) across Kenya, Tanzania and Mozambique. Bimodal (squares) and unimodal (diamonds) cropping seasons are represented, along with rainfall in semi-arid environments (open symbols) where rainfed maize production is commonly practised.

Source: Dimes et al., 2015.

Table 1 Reported estimates for yield gaps in different maize production regions around the world (Fischer et al., 2014)

Region/country	Farm yield (Yf)	Water-limited yield potential (Yw)	Potential yield (Yp)	Rainfed yield gap	Rainfed gap
		(t/ha)			(% of Yf)
Higher rainfall – eastern Africa	1.3–2.0	1.2–5.7	6.1–7.1	0–3.7	20–80
Lower rainfall coastal tropics of eastern and western Africa	0.8–2.0	0.9–4.2	4.4–5.3	0.1–2.2	20–80
Midwest USA	11.4	–	15.0	3.6*	32*
Brazilian Cerrado	4.0	6.6	11.0	2.6	65
Argentinian Pampas	7.2	8.4	14.0	1.2	17

*value calculated from yield potential which was not specified as irrigated or non-irrigated.

(Monjardino et al., 2013). So, this means farmers' perceived risk act as a disincentive for investment (Rao et al., 2011). But in what manner does risk and climate variability affect smallholders' ability to adopt improved technologies? How does this differ from our understanding of risk in more developed agricultural systems? And what (if anything) can be done to overcome the 'risk perception' barrier?

We think the problem is that the lifetime experience in agriculture of a smallholder farmer from eastern or southern Africa might lead him/her to conclude that many events are unpredictable and unexpected and that every season brings the same challenge (Meze-Hausken, 2004), the same unmanageable uncertainty and the same lack of opportunity. Uncertainty is one of the key constraints with which farmers, NGOs and development agencies struggle when promoting the adoption of technological innovations among vulnerable, risk-averse and poorly resourced smallholder farmers. The protracted low productivity and a lack of personal success in mitigating this seasonal risk affect the African farmer profoundly, more so than do similar risks to farmers in more developed economies.

To a point, in highly uncertain and vulnerable environments, risk aversion makes sense, as we cannot manage what we do not measure or know, and while embedded in the puzzlement of the unknown-unknowns, lack of investment remains a justification for the present low-input low-output production systems. Yet, this uncertainty is arguably much greater in Africa. The wealth of historical climate data, years of relatively well-funded local research and large extension networks available in other maize-producing regions are not available to African smallholders. This makes the uncertainty and its effects more severe. When seasons are poor, or if a management change leads to a lower yield, the result for the farmer can be inability to pay for children to attend school, malnourishment of children and family or even starvation (Richmond et al., 2015; Grimm, 2011). This is in contrast to advanced economies where the results of poor seasons tend to revolve around additional farm business debt and government assistance (Stone, 2014). In the African scenario, climate risk acts as a much more powerful disincentive for farmers to invest in needed technologies and markets, reducing the chances of increasing yields through higher adoption rates.

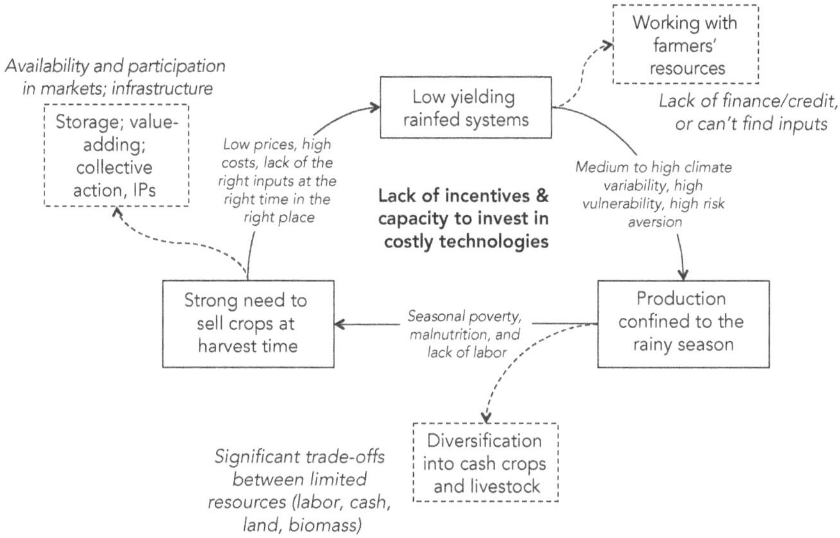

Figure 12 Poverty traps, commonly tested interventions in African agriculture development projects, possible solutions (dashed boxes) and most commonly found obstacles faced (italic fonts).

Source: Adapted from Naylor, 2014.

The short-term nature of many international research and development efforts (Kristjanson et al., 2009), much of them the primary driver of adoption efforts, combined with the climate variability during those years, has also meant that farmers involved in these projects often had negative experiences. This leaves them with less faith in the possibilities of managing change and improving productivity levels.

Further aggravating these challenges, poor socio-economic conditions in African countries act as major barrier to the feasibility of better management. A lack of efficient infrastructure means transport of primary products is much more expensive and time-consuming (Gollin and Rogerson, 2014). Low levels of education in rural populations mean that the required professionalized farm workforce is difficult to find (Glewwe et al., 2014). Complex management decisions requiring economic and environmental cost-benefit analyses can be difficult to make, as can the communication of their potential benefit to others (Pannell et al., 2014). A lack of adequate local markets for new crops or products, as well as local processing facilities, can reduce options for maize farmers who often participate in non-maize activities simultaneously (Jayne et al., 2010). Finally, without credit availability, the opportunities for farmers to improve their systems are limited (Juana et al., 2013). These factors negate many options for intensification or improvement, which are feasible in other maize-growing regions – particularly when considering the financial investment required for input use or mechanization.

As a result, high rainfall variability across SSA countries compounds the multiple constraints that the smallholder faces, creating protracted poverty traps (Fig. 12).

According to Rao et al. (2011), what matters is not climate variability or climate risk, but how poorly resourced and highly vulnerable smallholder farmers act upon their perception of high climate variability and risk. Across Africa, highly vulnerable risk-averse farmers

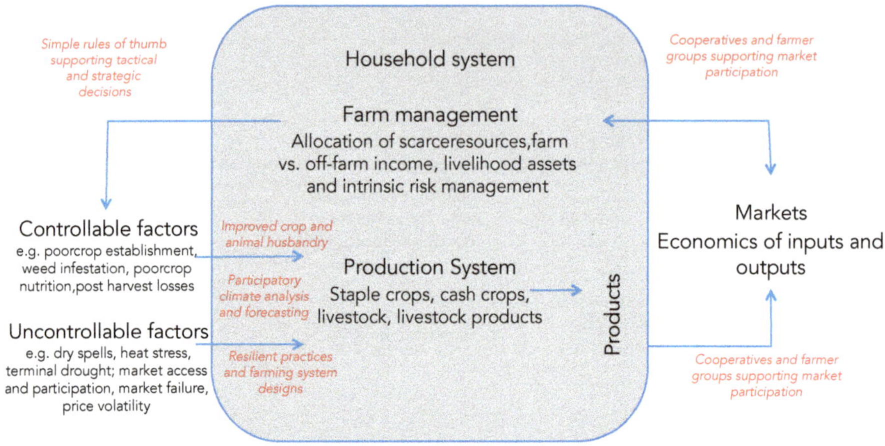

Figure 13 Sources of risk and risk-mitigation options in managing farms and value chains.

will tend to favour precautionary strategies that buffer against climatic extremes over activities that might be more profitable on average (Hansen et al., 2009, 2011; Rao et al., 2011). Farmers' perception of risk and its consequences are paramount here. How farmers' perceptions relate to the actual variability in yields driven by climate variability and to what extent existing yield gaps driven by poor agronomic practice or lack of use of agricultural inputs are responsible for farmer's risk-averse attitudes? Innovations that could increase productivity on average are avoided as they are seen as highly variable and therefore too risky, particularly if farmers have to pay expensive agricultural inputs in advance. Present perceptions on climate variability and risk among farmers from Kenya show that they give greater weight to negative experiences (Rao et al., 2011) and this is no different from findings in the Western world (Kahneman, 2011). The consequence is that these highly vulnerable and poorly resourced smallholder farmers consistently miss the good opportunities, and the cycle of low input and low output becomes difficult to break.

5 Climate risk at the whole farm level

Farms are complex systems (Fig. 13). Irrespective of their level of endowment, farmers use incomplete or imperfect knowledge to make rather technical and financial decisions, in a context of risk and uncertainty. In addition, farmers and their families have their own production objectives, aspirations and preferences. Thus, when decision making analyses are limited to the field or the crop, this is removed from the whole farm context, the disconnect between the more technical issues, for example, choosing a crop or cultivar, or whether to fertilize or not, and the final decision made conspires against understanding why an individual piece of technology is not adopted or why an apparently 'suboptimal' decision was made. So, after all issues surrounding feasibility (whether technical, economic or climatic) are resolved, and after farmers' risk preference is addressed, the decision then becomes one of providing incentives for the adoptions of the most feasible or 'best-fit' technologies (Giller et al., 2015).

6 A case study: the sustainable intensification of rainfed maize cropping in Mozambique

Mozambique provides a good example of the imperative for the sustainable intensification of maize farming systems. The country has a fast-growing population (currently 28 million, up 15% since 2010) with a significant proportion of it – 64% – rural (up 11% since 2010) (FAOSTAT 2015). As of 2011, the agricultural sector contributed 30.9% of the country's gross domestic product, with cropping being the largest sub-sector and maize being the dominant staple crop (FAO, 2013). However, actual maize productivity in the country is low by African standards, with mean production per hectare only 905 kg/ha as of 2013. By contrast, the same mean value for SSA in the same year was 1868 kg/ha (FAOSTAT, 2013). One of the most significant maize-producing regions is the Manica Province located close to the Zimbabwean border in the centre of the country. Studies have shown that land-use changes between 1990 and 2004 have led to reductions in permanent forest areas and expansion of settled/permanent production areas (Jansen et al., 2008). However, it has been argued that land availability is not the primary constraint for the province (Leonardo et al., 2015). Water is also not a major constraint on productivity, with the provincial capital city of Chimoio receiving an average of 921 mm of in-crop rainfall, that is, between October and April. Despite this, rainfall can vary widely (coefficient of variation = 31%). Soils in the province are mostly red sandy loam Ferralsols (FAO system) (Maria and Yost, 2006). Despite the challenges, the Manica Province could be considered as an area of high potential for intensification, given the relatively abundant land and water, as well as the comparatively easy market access (a large provincial capital – Chimoio – two hours by highway to an international port – Beira).

6.1 Household description and main constraints

Households in Manica are poor in knowledge and are constrained primarily by limited access to agricultural support services and inputs, key infrastructure and mechanization.

Figure 14 A group of Mozambican maize farmers learns how to take a soil sample from its field.

More than two thirds of the population (72.4%) have no complete primary education. Illiteracy in the province is 43%, with this figure rising to 60% for women (INE, 2010). The low level of education and literacy in the province means that individuals are poorly equipped to manage their farms in a business-like manner – recording expenses, productivity and profits. It also increases farmers' reliance on limited extension services for basic management advice, and market and climate information. In terms of infrastructure, most families live in houses constructed from natural materials, most are not connected to the electrical grid, and have no running water (INE, 2010). As an indication of the relative lack of basic infrastructure, 60% of households have no latrines. These conditions affect not only the health and well-being of the population, but also their productive capacity. Lack of electricity reduces access to information and communication technologies, though mobile technologies are widespread, and most farmers have mobile phones. In addition, without power, more household tasks (i.e. cooking, washing) must be done by hand – adding to labour demand. Lack of running water means more household labour must be used to collect water from communal bores. An almost complete lack of mechanization for smallholders means labour is critically limiting (Cunguara and Darnhofer, 2011). Finally, cash availability relative to local prices continues to act as a barrier to the use of agricultural inputs. Previous studies in the province have recorded mean farm income levels as US $335 (Uaiene, 2004), or more recently $297 (Roxburgh and Rodriguez, 2015). This means that the cost of inputs is a significant proportion of the annual income of farms. For example, fertilizing just one hectare of maize with locally recommended rates of basal and nitrogenous fertilizers would cost 63% of the annual farm income (Roxburgh and Rodriguez, 2015). Furthermore, this figure ignores the costs of transporting the fertilizer to the farm (both in time and money) and the labour needed for its application.

A large variability can be found in Manica in terms of levels of household endowment (Leonardo et al., 2015). This indicates that R4D projects and development agencies trying to increase the adoption of technologies should design their intervention targeting such diversity. Typically, this involves the classification of households into more homogenous groups with regard to key constraints like knowledge, labour availability, family size, availability of livestock, level of productivity and the level and source of the household income. For Manica Province, households may be classified into:

1 Subsistence farming

These farmers are largely, if not entirely, focused on subsistence activities and severely limited by available labour. Many farmers in this group require to have off-farm activities to achieve food security. Their income and productivity are below the average for Manica. These are households without any capacity to invest cash or even additional labour into their farming system. Gaps in knowledge are likely the most constraining to intensification. Here, we propose that for such households, innovations should aim primarily at increasing the efficiency with which existing resources are used (land, labour, cash), rather than proposing complex technologies that would require increases in investment or skilled operators; we call this 'doing better with the same' (resources). Previous work in Manica was able to identify simple, cost-neutral and low-risk interventions that would improve productivity for the lowest quartile of farmers through the improvement of agronomic management (Roxburgh and Rodriguez, 2015). These included reducing maize plant populations, in conjunction with better timing of planting and weeding activities, plant-to-plant spacing and row configurations. These simple agronomic changes could increase levels of productivity by up 120% without the need for any increases in labour or cash investment.

2 Moderately endowed farmers

These farmers sit around the averages discussed earlier. They have some cash cropping and livestock activities in addition to their subsistence farming. They might also be less labour constraint than the previous group, though off-farm income is still an important source of their livelihoods. In general terms, they are still restricted in cash availability and have poor participation in poorly developed or inexistent markets. For this group of households, we found that basic agronomy could still be improved. In addition, options for sustainable intensification would revolve around improving the efficiency in the use of inputs, given the limited capacity for investment. Modelling by Dimes et al. (2015) in a higher-producing area of Manica Province (Rotanda) found that for this type of household, investments in herbicides prior to investments in improved seed was critical to ensure seasonally stable returns on investment that would allow the reinvestment of profits on the farm, and free labour tide up to weeding activities. Many of these farms also combine cropping with livestock activities as well as cash crops like sesame, sweet potatoes and leafy vegetables and a range of bean crops. Supporting livestock activities by improving the feed supply during the dry season, as well as using manure on cash crops, is also likely to increase profits and farmers' capacity to invest in more productive, though costly, technologies.

3 Better endowed farmers

These farmers have mastered the management of rainfed maize cropping within the conventional system that exists in Manica. They have sufficient capacity to invest cash into their maize and are already effectively utilizing inputs to improve productivity. For these farmers, knowledge and cash are not primary barriers. Instead, these farmers are limited by their ability to access technologies, which shift their system closer to the efficiency frontier (Fig. 1). Interventions that would support this type of household would still include adjustments in the agronomy and use of inputs. In addition, more transformational measures like land acquisition, along with machinery purchases, and investments in irrigation would allow these farmers to specialize into the production of higher-value crops. In Manica Province, the limited number of studies undertaken suggests that few, if any, of these farmers currently exist among the cohort of smallholders being engaged by international research groups. Neither Dimes et al. (2015), Leonardo et al. (2015) nor Roxburgh and Rodriguez (2015) found farmers who appeared to be ready for marked change.

7 Future trends

Effective climate risk management in rainfed maize cropping systems need to develop from:

i. clear understanding of the existing production system and its diversity across targeted regions;
ii. quantitative evaluation of the sources of variability (climate, markets, both?), their frequency of occurrence, impacts and trends;

iii. ex-ante evaluations of options that are derived from participatory modelling exercises, accounting for benefits but also trade-offs in view of the high vulnerabilities; and followed by
iv. design and implementation phase that needs to involve not only the farm, but also key actors in the inputs, produce, and information value chain.

We stress the importance of three technologies to support the adaptation process (1) well-calibrated modelling tools able to integrate the climate-soil-crop-management continuum, for the quantification of benefits and trade-offs from alternative actions and technologies; (2) the use of participatory methods, so that interventions and changes are designed around local conditions, norms and values; and (3) the provision of simple good systems agronomy recommendations. It is clear that after five years of work across eastern and southern Africa in the scaling out of conservation agriculture practices, that the best climate risk management results we obtained originated just from the adoption of simple good agronomic practice.

8 Acknowledgements

The authors thank the Australian Centre for International Agriculture Research (ACIAR) and the Grains Research and Development Corporation for their support with many of the studies reported within.

9 Where to look for further information

- The report for the 2015 El Niño Conference at the International Research Institute for Climate and Society at Columbia University, can be found at: http://iri.columbia.edu/news/el-nino-2015-conference-report/
- Seasonal climate forecasts can be obtained from the following agencies and sites: International Research Institute for Climate and Society (IRI). http://iri.columbia.edu/our-expertise/climate/forecasts/
- Link to Elsevier's *Climate Risk Management Journal*. http://www.journals.elsevier.com/climate-risk-management/
- Link to the *World Meteorological Organization Climate Risk Management* web page. https://www.wmo.int/pages/themes/climate/risk_management.php
- Link to the *Weather and Climate Risk Management* web site from the Australian Bureau of Meteorology. http://www.bom.gov.au/watl/about-weather-and-climate/risk/
- Link to Climate Kelpie, a list of climate risk management tools from the Managing Climate Variability Program in Australia. http://www.climatekelpie.com.au/

10 References

Aguirre-Uribe, L. A., Hernandez, A., Flores, M., Frias, G., Cerna, E., Landeros, J. and Harris, M. 2015. Genetically modified maize resistance to corn earworm (Lepidoptera; Noctuidae) in Sinaloa, Mexico. *Florida Entomologist*. 98(3), 821–6.

Andrade, F. H. 1995. Analysis of growth and yield of maize, sunflower and soybean grown at Balcarce, Argentina. *Field Crops Res.* 41, 1–12.
Annual population (2015). Food and Agricultural Organization. http://faostat3.fao.org/download/O/OA/E (accessed 2 November 2015).
Araus, J. L., Serret, M. P. and Edmeades, G. O. 2012. Phenotyping maize for adaptation to drought. *Frontiers in Physiology* 3, 20–39.
Calviño, P. and Monzon, J. P. 2009. Farming systems of Argentina: yield constraints and risk management. In Victor O. Sadras and Daniel F. Calderini (Eds), *Crop Physiology*, pp. 55–97. Applications for Genetic Improvement and Agronomy, Ac Press Elsevier, Amsterdam.
Chauhan, Y. S., Solomon, K. F. and Rodriguez, D. 2013. Characterization of north-eastern Australian environments using APSIM for increasing rainfed maize production. *Field Crops Res.* 144, 245–55.
Cunguara, B. and Darnhofer, I. 2011. Assessing the impact of improved agricultural technologies on household income in rural Mozambique. *Food Policy* 36, 378–90.
Dimes, J., Rodriguez, D. and Potgieter, A. 2015. Raising productivity of maize based cropping systems in eastern and southern Africa: step-wise intensification options. In V. O. Sadras and D. Calderini (Eds), *Crop Physiology*, pp. 94–110. Applications for Genetic Improvement and Agronomy Second Edition.
Echarte, L., Luque, S., Andrade, F. H., Sadras, V. O., Cirilo, A., Otegui, M. E. and Vega, C. R. C. 2000. Response of maize kernel number to plant density in Argentinean hybrids released between 1965 and 1995. *Field Crops Res.* 68, 1–8.
Eyre, J. and Rodriguez, D. 2016. Conference Presentation. Australian Summer Grains Conference. Gold Coast.
FAO. 2013. FAO country programming framework: Mozambique 2012–15. Maputo: FAO.
FAOSTAT. 2013. FAO. http://faostat3.fao.org/faostat-gateway/go/to/home/E (accessed 3 April 2014).
FAOSTAT. 2015. FAO. http://faostat3.fao.org/home/E (accessed 3 April 2014).
Ferrante, A., Eyre, J., George-Jaeggli, B., McLean, J., Chenu, K., deVoil, P., McLean, G. and Rodriguez, D. 2015. Maize yield determination in the Northern Region: Hybrid by environment by management interactions Proceedings of the 17th ASA Conference, 20–24 September 2015, Hobart, Australia. Website www.agronomy2015.com.au.
Fischer, T., Byerlee, D. and Edmeades, G. O. 2014. Crop yields and global food security: will yield increase continue to feed the world. *ACIAR Monograph*. Canberra, Australian Centre for International Agricultural Research.
Giller, K. E., Andersson, J. A., Corbeels, M., Kirkegaard, J., Mortensen, D., Erenstein, O. and Vanlauwe, B. 2015. Beyond conservation agriculture. *Front. Plant Sci.* 6, 37–14. doi:10.3389/fpls.2015.00870.
Glewwe, P., Maïga, E. and Zheng, H. 2014. The contribution of education to economic growth: a review of the evidence, with special attention and an application to sub-saharan Africa. *World Development* 59, 379–93. doi:http://dx.doi.org/10.1016/j.worlddev.2014.01.021.
Gollin, D. and Rogerson, R. 2014. Productivity, transport costs and subsistence agriculture. *J. Dev. Econom.* 107, 38–48. doi:http://dx.doi.org/10.1016/j.jdeveco.2013.10.007.
Grimm, M. 2011. Does household income matter for children's schooling? Evidence for rural sub-saharan Africa. *Economics of Education Review* 30(4), 740–54. doi:http://dx.doi.org/10.1016/j.econedurev.2011.03.002.
Hansen, J. W., Mishra, A., Rao, K. P. C., Indeje, M., Ngugi, R. K. and Robinson, K. 2009. Potential value of GCM-based seasonal rainfall forecasts for maize management in semi-arid Kenya. *Ag Systems* 101, 80–90.
Hansen, J. W., Mason, S. J., Sun, L. Q. and Tall, A. 2011. Review of seasonal climate forecasting for agriculture in Sub-Saharan Africa. *Exp Agric* 47, 205–40
INE. 2010. INE. http://www.ine.gov.mz/ (accessed 3 April 2014).
Jansen, L. J. M., Bagnoli, M. and Focacci, M. 2008. Analysis of land-cover/use change dynamics in Manica Province in Mozambique in a period of transition (1990–2004). *For. Ecol. Manage.* 254(2), 308–26. doi:http://dx.doi.org/10.1016/j.foreco.2007.08.017.

Jayne, T. S., Mather, D. and Mghenyi, E. 2010. Principal challenges confronting smallholder agriculture in sub-saharan Africa. *World Development* 38(10), 1384–98. doi:http://dx.doi.org/10.1016/j.worlddev.2010.06.002.

Jones, C. A. and Kiniry, J. R. 1985. CERES-Maize: a simulation model of maize growth and development. Texas A&M Press. College Station, p. 194.

Juana, J. S., Kahaka, Z. and Okurut, F. N. 2013. Farmers' perceptions and adaptations to climate change in sub-sahara Africa: a synthesis of empirical studies and implications for public policy in African agriculture. *J. Agric. Sci* 5(4), 121–35.

Kahneman, D. 2011. Thinking Fast and Slow. Farrar, Straus and Giroux, 18 West 18th Street, New York 10011

Kristjanson, P., Reid, R. S., Dickson, N., Clark, W. C., Romney, D., Puskur, R., et al. 2009. Linking international agricultural research knowledge with action for sustainable development. *Proc. Nat. Acad. Sci. U.S.A.* 106(13), 5047–52. doi:10.1073/pnas.0807414106.

Leonardo, W., van de Ven, G. J., Udo, H., Kanellopoulos, A., Sitoe, A. and Giller, K. 2015. Labour not land constrains agricultural production and food self-sufficiency in maize-based smallholder farming systems in Mozambique. *Food Security* 7(4), 857–74. doi:10.1007/s12571–015-0480-7.

Maria, R. M. and Yost, R. 2006. A survey of soil fertility status of four agroecological zones of Mozambique. *Soil Sci.* 171(11), 902–14.

McCown, R. L., Hammer, G. L., Hargreaves, J. N. G., Holzworth, D. P. and Freebarirn, D. M. 1996. APSIM: a novel software system for model development, model testing and simulation in agricultural systems research. *Ag Systems* 50, 255–71.

Meze-Hausken, E. 2004. Contrasting climate variability and meteorological drought with perceived drought and climate change in northern Ethiopia. *Climate Res.* 27(1), 19–31. doi:10.3354/cr027019.

Monjardino, M., McBeath, T. M., Brennan, L. and Llewellyn, R. S. 2013. Are farmers in low-rainfall cropping regions under-fertilising with nitrogen? A risk analysis. *Agric. Syst.* 116, 37–51. doi:10.1016/j.agsy.2012.12.007.

Naylor, R. L. 2014. *The Evolving Sphere of Food Security*. Oxford Press Madison, NY, p. 394.

Moulia, B., Loup, C., Chartier, M., Allirand, J. M. and Edelin, C. 1999. Dynamics of architectural development of isolated plants of maize (*Zea mays* L.), in a non-limiting environment: the branching potential of modern maize. *Annals of Botany* 84, 645–56.

Ortega, C. A. 1987. Insect pests of maize: a guide for field identification. International maize and wheat improvement centre. Mexico, p. 106.

Otegui, M. E. and Bonhomme, R. 1998. Grain yield components in maize. I Ear growth and kernel set. *Field Crops Res.* 56, 247–56.

Pannell, D. J., Llewellyn, R. S. and Corbeels, M. 2014. The farm-level economics of conservation agriculture for resource-poor farmers. *Agric. Ecosyst. Environ.* 187, 52–64. doi:http://dx.doi.org/10.1016/j.agee.2013.10.014.

Rao, K. P. C., Ndegwa, W. G., Kizito, K. and Oyoo, A. 2011. Climate variability and change: farmer perceptions and understanding of intra seasonal variability in rainfall and associated risk in semi arid Kenya. *Ex. Agric.* 47, 267–91. doi:10.1017/S0014479710000918.

Ray, D. K.,Gerber, J. S., MacDonald, G. K. and West, P. C. 2015. Climate variation explains a third of global crop yield variability. *Nat. Commun.* 6, 5989. doi: 10.1038/ncomms6989.

Richmond, A. K., Malcomb, D. and Ringler, K. 2015. Household vulnerability mapping in Africa's Rift Valley. *J. Appl. Geogr.* 63, 380–95. doi:http://dx.doi.org/10.1016/j.apgeog.2015.07.013.

Rodriguez, D., deVoil, P., Power, B., Cox, H., Crimp, S. and Meinke, H. 2011. The intrinsic plasticity of farm businesses and their resilience to change. An Australian example. *Field Crops Res.* 124, 157–70. doi:10.1016/j.fcr.2011.02.012.

Rodriguez, D., Cox, H., deVoil, P. and Power, B. 2014. A participatory whole farm modelling approach to understand impacts and increase preparedness to climate change in Australia. *Agric. Syst.* 126, 50–61.

Roxburgh, C. W. and Rodriguez, D. 2015. Ex-ante analysis of opportunities for the sustainable intensification of maize production in Mozambique. *Agric. Syst.* 142, 9–22.

Sadras, V. O. and Angus, J. F. 2006. Benchmarking water use efficiency of rainfed wheat in dry environments. *Aust. J. Ag. Res.* 57, 847–56.

Sadras, V., Roget, D. and Krause, M. 2003. Dynamic cropping strategies for risk management in dry-land farming systems. *Agric. Syst.* 76, 929–48.

Stone, R. C. 2014. Constructing a framework for national drought policy: the way forward – The way Australia developed and implemented the national drought policy. *Weather and Climate Extremes* 3, 117–25. doi:http://dx.doi.org/10.1016/j.wace.2014.02.001.

Tollenaar, M. and Dwyer, L. M. 1999. Physiology of maize. In D. L. Smith and C. Hamel (Eds), *Crop yield, Physiology and Processes*, pp. 169–204. Springer Verlag, Berlin.

Tollenaar, M., Ying, J. and Duvick, D. N. 2000. Genetic gain in corn hybrids from the Northern and Central Corn Belt. In Proceedings of 55th Corn Sorghum Research Conference, 5–8 December, Chicago, IL. ASTA, Washington, DC, pp. 53–62.

Uaiene, R. N. 2004. Maize and sorghum technologies and the effects of marketing strategies on farmers' income in Mozambique. Masters, Purdue University, Layfayette.

van Ittersum, M. K., Cassman, K. G., Grassini, P., Wolf, J., Tittonell, P. and Hochman, Z. 2013. Yield gap analysis with local to global relevance – A review. *Field Crops Res.* 143, 4–17. doi:http://dx.doi.org/10.1016/j.fcr.2012.09.009.

Vega, C. R. C., Andrade, F. H., Sadras, V. O., Uhart, S. A. and Valentinuz, O. R. 2001. Seed number as a function of growth. A Comparative Study in Soybean, Sunflower, and Maize. *Crop Sci.* 41, 748–54.

Wossen, T. and Berger, T. 2015. Climate variability, food security and poverty: Agent-based assessment of policy options for farm households in Northern Ghana. *Environ. Sci. Policy* 47, 95–107. doi:10.1016/j.envsci.2014.11.009.

Chapter 12

Advances in maize post-harvest management

Tadele Tefera, International Center of Insect Physiology & Ecology (icipe), Ethiopia

1 Introduction

One of the key constraints to improving food and nutritional security in developing world is poor post-harvest management that leads to loss of grains. In Africa, annual post-harvest losses equate to an estimated US$ 2 billion. Maize is the most important food staple for resource-poor smallholders in developing countries, providing food and income to millions; however, poor post-harvest management leads to between 14% and 36% loss of maize grains in Africa, and 8–50% in Latin America. Post-harvest losses contribute to high food prices by removing part of the supply from the market. The lack of access to improved post-harvest management often forces smallholder farmers to sell their produce immediately after harvest to avoid post-harvest losses from pests and pathogens. Consequently, farmers receive low market prices for any surplus grain they may produce, cannot use their harvest as collateral to access credit, and ultimately their food security is undermined. Poor post-harvest management is one of the factors that cause mycotoxin contamination. Improved post-harvest management at farm level is crucial, as it directly impacts on poverty alleviation, food, nutrition and income security and prosperity for the smallholder farmers. Biological and socio-economic causes of post-harvest losses are discussed in relation to food security, and technological interventions to reduce the post-harvest losses are suggested.

http://dx.doi.org/10.19103/AS.2016.0002.13

2 Post-harvest losses in maize: an overview

2.1 Incidence and significance

Grains constitute the bulk of production in the developing world (The World Bank 2011). One of the key constraints to improving food and nutritional security in the developing world, however, is poor post-harvest management, which leads to losses of 20–30%, with an estimated monetary value of more than US$4 billion annually in sub-Saharan Africa (SSA) (FAO, 2010). Among grains, maize accounts for the major share of staple crops in SSA. A recent report of a joint FAO/World Bank (Suleiman and Rosentrater, 2015; Zorya et al., 2011) shows that post-harvest losses (PHLs) of cereals in Eastern and Southern Africa account for over 40% of the total PHL in SSA countries. This represents losses of about US$1.6 billion each year (Suleiman and Rosentrater, 2015). Such losses are equivalent to the annual caloric requirement for at least 20 million people (FAO, 2014) or more than half of the value of total food aid received by SSA in a decade (Zorya et al., 2011). Furthermore, Tefera (2012) reported that PHLs of maize in various chains of post-harvest activities in SSA are between 14% and 36%, while in Latin America it is between 8% and 50% (IICA, 2013).

Post-harvest and marketing systems are a chain of interconnected activities from the time of harvest to the delivery of food to the consumers (Abass et al., 2014). Agricultural commodities produced on the farm have to undergo several procedures such as harvesting, drying, threshing, winnowing, processing, bagging, storage, transportation and exchange before reaching the final consumer (Abass et al., 2014; De Groote et al., 2013). The primary role of an effective post-harvest system is to ensure that the harvested food reaches the consumer, while fulfilling customer satisfaction in terms of quality, volume and safety. PHLs in the developed countries are lower than in the developing countries because of more efficient farming systems, better transport infrastructure, better farm management, and effective storage and processing facilities that ensure a larger proportion of the harvested foods is delivered to the market in the most desired quality and safety (World Bank et al., 2011). For low-income countries, pre-harvesting management, processing, storage infrastructure and market facilities are either not available or inadequate (World Bank et al., 2011). In developing countries, significant volumes of grain are lost after harvest, aggravating hunger and resulting in the wasting of expensive inputs such as fertilizer, irrigation water and human labour.

2.2 Types and causes of losses

Causes of PHL are diverse, encompassing biological, physical, technological, environmental and socio-economic factors (Tefera, 2012). These include mechanical damage; excessive exposure to high ambient temperature, relative humidity and rain; contamination by spoilage fungi and bacteria; invasion by birds, rodents, insects and other pests; and inappropriate handling, storage and processing techniques (Abbas et al., 2014; World Bank, 2011). Losses may be aggravated by poor infrastructure, harvesting methods, post-harvest handling procedures, distribution, sales and marketing policies (World Bank, 2011). According to Tyler (1982), the economic importance of the factors leading to high PHLs varies from commodity to commodity, from season to season and with the enormous diversity of circumstances under which commodities are grown, harvested, stored, processed and marketed. During post-harvest operations, there may be losses of both cereal quantity and quality. Quantitative loss indicates the reduction in physical weight, for example, a portion of grain damaged by pests or lost during transportation. A qualitative loss is contamination of grain by moulds and includes loss in nutritional quality, edibility, consumer acceptability of the products and the caloric value (Suleiman and Rosentrater, 2015; Zorya et al., 2011). Economic loss is the reduction in monetary value of the product due to a reduction in quality and quantity of food (Tefera, 2012). These losses also have an impact on environmental and climate change, as land, water and non-renewable resources such as fertilizer and energy are used to produce, process, handle and transport food that no one consumes (World Bank, 2011).

Weight loss (WL) is the standard international measure expressing PHL in grains loss (De Lima, 1979), generally regarded as a loss of food. WL is expressed as a loss in the dry matter or dry weight basis (Tefera, 2012), and is estimated in two ways: first, scattering of grain due to poor post-harvesting handling practices including harvesting, threshing, drying, poor packaging and transport and, second, from bio-deterioration brought by pests and other organisms such as insects, moulds and fungi, rodents and birds (Hodges, 2013). WL can be calculated by the count and weight method developed by Adams and Schulten (1978). Other methods include standard volume and weight method (Reed, 1987) and the thousand grain mass and converted percentage damage method (Dick, 1988).

3 Major storage insects and fungi

Post-harvest operations for maize follow a chain of activities starting from harvesting, threshing, transport to storage, farm storage, marketing and consumption (Tefera, 2012). PHL occurs along all these chains of activities. Maize insect pests are a major constraint to PHL reduction and income generation in SSA because of significant yield losses and grain quality degradation. The most economically important post-harvest pests of maize include the maize weevil (*Sitophilus zeamais*), the larger grain borer (LGB) (*Prostephanus truncatus*), the Angoumois grain moth (*Sitotroga cerealella*) and the lesser grain weevil (*S. oryzae*). These pests are widespread in Africa and co-occur in the field and in storage (Boxall, 2002). The LGB was introduced to Africa from Mesoamerica in the early 1980s, and has currently established itself in most parts of the continent, threatening maize production due to its aggressive nature and the considerable damage it causes within a short period of time (Tefera, 2012).

Poor post-harvest management also affects human health. Mycotoxin contamination (especially aflatoxin and fumonisin production by *Aspergillus flavus* and *Fusarium verticillioides* Sacc Nirenberg, respectively) makes grain unsafe for food and animal feed. Mycotoxin contamination in post-harvest occurs due to inadequate drying and storage of grains. The situation might be aggravated in the presence of insect infestations or when grains are exposed to high humidity or actual wetting due to poor storage management. Fungal infection results in rots and development of aflatoxins, which are poisonous compounds to livestock and cause cancer in human. Mycotoxins are influenced by environmental factors such as geography and agricultural practices. Mycotoxins are toxic secondary metabolites of fungi that contaminate the maize in the field and/or during storage (Smith et al., 2012); this type of contamination poses a health risk to humans and domesticated animals (Suleiman and Rosentrater, 2015; Mboya et al., 2012; Suleiman et al., 2013). The most important mycotoxins in maize are the aflatoxins, fumonisins, deoxynivalenol (DON) and ochratoxin (Kimanya et al., 2012). Aflatoxin is a group of mycotoxin produced by two fungi species, *A. flavus* and *A. parasiticus,* as secondary metabolites during spoilage of food materials (Marin et al., 2013; Feng et al., 2011). Fumonisins are mycotoxins synthesized mainly by *F. verticillioides* and *F. proliferatum* (Garrido et al., 2012). DON is a common type of mycotoxins produced by pink mould *F. graminearum* (Garrido et al., 2012). Ochratoxin is another type of mycotoxins mostly produced by *Penicillium verrucosum, A. ochraceus, A. niger* and other associated species (Lai et al., 2014).

4 Traditional storage structures and food security

The most common traditional maize storage facilities used by smallholder farmers in Africa include traditional cribs or granaries, baskets, and large pots. Although many smallholder farmers in developing countries still rely on traditional storage methods, storing unshelled maize in traditional cribs has proved difficult with the advent of the LGB and increased grain theft (De Groote et al., 2013). Traditional storage structures expose grain to rodents and insect attack, and provide favourable climatic conditions for their proliferation, as well as for that of microorganisms, thus leading to substantial PHLs (Ngamo et al., 2007). In Africa, farmers more commonly store the shelled grain in polypropylene bags (De Groote et al., 2013). Many smallholder farmers sell off their grain immediately after harvest to avoid damage by storage pests and consequently receive low prices (Boxall, 1998; Proctor, 1994). The LGB is an invasive storage pest accidentally introduced into Africa from Central America, and is now recognized as the most destructive maize pest (Boxall, 2002).

One of the reasons we are dealing with PHL is its contribution to food security; reducing PHL means increasing food availability. Food security exists when all people have, at all times, access to sufficient, safe and nutritious food to meet their dietary needs and food preferences in order to lead an active and healthy life (Pinstrup-Andersen, 2009). In Eastern and Southern Africa, maize is the most important food crop, and its consumption as a basic staple is relatively constant over the year. Unfortunately, maize production is highly seasonal, resulting in strongly fluctuating supplies (De Groote et al., 2013). Grain storage is therefore essential to food security by bridging the period between two harvests, as well as by stabilizing prices by taking the produce off the market during the peak season, and releasing it back when the grain is in short supply (Proctor, 1994). Good storage methods therefore become an important aspect of household food security and rural livelihoods as they ensure a continuous stable supply of food and better farm incomes (Tefera, 2012). Although farmers' primary objective in storing grain is household food security, some store it for market speculation purposes and others use it as seed in the following season (Proctor, 1994). However, farmers often still sell their produce immediately after harvest at low prices. When their stocks run out, they have to buy food later in the market at high prices (De Groote et al., 2013).

Apart from on-farm storage, most governments in SSA also engage in establishing strategic food reserves to provide for market failures, to stabilize food prices and to prepare for emergencies. However, the reduction of losses before and after harvest at the farm level is probably more efficient, but has not received as much attention (Tefera et al., 2011; De Groote et al., 2013). Food loss is a subset of PHL and represents the part of the edible share of food that is available for consumption at either the retail or the consumer levels, but is not consumed for whatever reason (De Groote et al., 2013; Hodges et al., 2010). Reducing PHL could increase food availability by both reducing physical losses and increasing income from the improved market opportunities that could be used to buy food (FAO and World Bank, 2010). The total PHL in Africa from maize are estimated to range from 14 to 36%, and the on-farm storage losses from 4 to 10% (Tefera, 2012). In another analysis, based on an extensive review of the literature, the storage WLs for maize in Eastern and Southern Africa for 2007 was estimated at 17.5%, valued at US$920 million (Rembold et al., 2011).

5 Institutional factors impacting post-harvest losses

Post-harvest policy include poor sales, procurement, storage, marketing and distribution policies or practices; absence of mechanisms for dealing with cash flow needs (such as warehouse receipts systems, or WRS); mismanagement or malfeasance in handling grain stocks and associated financing; or difficulty in dealing with the ownership, control, and payment aspects of grain storage and price stabilization programmes (World Bank, 2011). Overall, food losses contribute to high food prices by removing part of the food supply from the market.

Private sector involvement in market development for post-harvest technologies has been limited, compounding the inaccessibility of the technologies to farmers, which further reduces the demand for the technologies (Tefera, 2012; Ferris and Wheatley, 2001; FAO, 2010). In turn, this keeps the supply low, which consequently keeps the cost of production (and prices) relatively high (Mylene et al., 2002). These barriers are exacerbated by the lack of policy-based incentives for both research and private sector involvement in post-harvest technology development and dissemination, and adoption of post-harvest management practices. Although many countries are increasing investment in grain production, none of them has explicit policy-based incentives to support the adoption of better post-harvest technologies. This is in spite of recognition of their multiple benefits. Indeed, there are several disincentives to the adoption of post-harvest management technologies, such as high taxes on materials and consequently high prices of the technologies. In the case of metal silos, the cost of the metal sheets, which are used in the construction of metal silos, constitutes about 60% of the total cost of the silos (De Groote et al., 2013). Access to market, trade information and communication between producers and consumers is essential for grain marketing, post-harvest handling and food security (Tefera, 2012). However, the majority of smallholder farmers and private sector input dealers, especially in rural areas, have limited access to such information (Ferris and Robbins, 2004). Most handlers involved directly in harvesting, threshing/shelling, transporting and marketing in Africa have limited knowledge of how to maintain grain quality. An effective and far-reaching extension programme on these aspects is critically needed, and will continue to be essential in the future.

6 Reducing post-harvest losses through technological interventions

Losses are a combination of those which occur on the field, in storage, during processing and other marketing activities. After harvesting, grain moisture content plays a major

Figure 1 Metal silos filled in with maize grains in Kenya.

role. Wet grains attract insects and mould. Therefore, the grain must be dried as soon as possible after harvesting. To minimize the risk of mould growth and mycotoxin contamination, maize grain should have a moisture content between 12% and 13% for storage in bags, or below 12.5% for bulk storage in bins or silos (Tefera, 2012). It is one of the key post-harvest operations as all downstream operations depend on it. Sometimes, the next season starts early when the crops are still in the field. Drying permits the escape of moisture from the grain to an acceptable level, which can sustain very low metabolism. The enzyme activities and grain tissue respiration are reduced to a very low level, thus inhibiting sprouting/germination. During drying, the dry air rapidly takes moisture away from the grain, especially if the air is moving and has low humidity. Grain can be dried in a crib before shelling and on tarpaulins after it has been shelled. However, drying below 12% may lead to increased breakage of grains during handling. Maize grain on the cob can be stored, safe from mould growth, in open-sided cribs. In this case, the moisture content should not be less than 13.5%. Cobs can be loaded into the crib at high moisture content and allowed to dry by natural aeration (Tefera, 2012).

In response to challenges posed by the maize weevil and LGB, improved storage technologies have been developed, including the insecticide, actellic super (a mixture of pirimiphos-methyl and permethrin), super grain bags and the metal silo (De Groote et al., 2013). Metals silos are made from galvanized flat iron sheets. They are cylindrical containers with grain holding capacity from 50 to 3000 kg (Fig. 1). Metal silos provide a strong barrier against insect pests and rodents and are hermetically sealed, killing any remaining insect through oxygen depletion (Tefera et al., 2011). They are durable, allow for long storage periods and are therefore an important storage technology in the fight against hunger and food insecurity in developing countries.

A metal silo with a capacity of 990 kg can conserve enough grain to feed a household of five members for one year (FAO, 2010). Metal silos were first developed and disseminated in Central America through the Postcosecha (meaning 'post-harvest') Program in 1983 by the Swiss Development Cooperation (SDC) (Bokusheva et al., 2012). Farmers used phostoxin pellets to kill the storage pests and the silos were made airtight to prevent leaking of the fumigant (Fischler et al., 2011). Phostoxin is a fumigant insecticide, contains aluminium phosphide, is volatile and is highly toxic; therefore it should be used only by trained operators. In Kenya, metal silos were introduced by the Catholic Relief Services (CRS) in 2000, mainly in the western region (Nguyo, 2007). Following the success story in Central America, an SDC-funded Effective Grain Storage Project (EGSP) was launched in Eastern and Southern Africa in 2008 to promote the fabrication and dissemination of metal silos to smallholder farmers.

Metal silos were tested and found effective in Kenya (De Groote et al., 2013). On-station trials in different maize agro-ecological zones showed that weevils or LGBs had not damaged grains stored in metal silos after six months in storage, even without insecticides, while for maize stored in polypropylene bags with insecticides (Actellic Super) good control was only

Figure 2 PICS bags (http://modernfarmer.com/2014/06/simple-bag-help-millions-african-farmers-bill-gates-thinks/).

provided for four months. Although metal silos are technically effective against the main grain storage pests, their distribution and adoption by smallholder farmers remain limited. The cost of a metal silo with a capacity of holding half a ton is about KSh15,000 (US$200), while the cost for a 2.7-ton silo is about KSh35,000 (US$460). A survey of metal silo artisans in 2011 showed that over 2300 silos have been fabricated and distributed in Kenya. Half of those fabricated have a capacity of 1800 kg each, and 15% having a capacity of 540 kg each. The relatively high cost of the metal silos and lack of awareness dampen the demand and widespread adoption by farmers (Gitonga et al., 2015).

Hermetic bagging and insect-resistant maize variety are an extension of the 'choice' of technologies available to farmers to explore to lever food security and sustain maize-dependent livelihoods (Likhayo et al., 2015, 2016). Hermetic bags are airtight containers which prevent oxygen and water movement between the outside atmosphere and the stored grain or product. The respiration by the grain and insect inside the bags changes inter-granular atmosphere consuming oxygen and producing carbon dioxide (Baoua et al., 2012). Depending on the type of the container and insect population, oxygen levels is reduced (hypoxia) from 21 to below 10% and carbon dioxide levels increased (hypercarbia) within a short period of time (Mutungi et al., 2014). However, the attainment of high enough carbon dioxide level to kill all the insects in hermetic bags is a challenge. Whereas several studies have examined grain storage infestation under hermetic conditions, little work, however, has been reported on the combined effect of hermetic bagging and insect-resistant maize variety. The hermetic bags are placed inside polypropylene bags to provide support and handling convenience. PICS® (Fig. 2) and SuperGrain II™ bags (Fig. 3) are a type of multi-layer co-extruded tougher plastics with low permeability in which oxygen is depleted fast. Big storage facilities such as cocoons (Fig. 4) can be used for long-term safe storage for large-scale agricultural community or community food reserve and are available from GrainPro for most farmers in the developing areas of the world (http://grainpro.com/gpi/index.php?option=com_content&view=article&id=231&Itemid=1644; accessed on 19 June 2016).

Figure 3 SuperGrain bags filled with grains (http://www.everychina.com/buy/c-zbe6914/p-38447242/showimage.html).

Figure 4 Cocoons for outdoor storage (http://www.israel21c.org/giant-storage-bags-save-crops-from-mold-pests/).

Purdue Improved Crop Storage (PICS) bags (formerly Purdue Improved Cowpea Storage bags) provide a simple, low-cost method of reducing post-harvest cowpea (*Vigna unguiculata*) losses due to bruchid infestations in west and central Africa. A PICS bag consists of two layers of polyethylene liners and a third layer made from woven polypropylene. When each layer is tied and closed separately, it creates a hermetically sealed environment for storing harvested grain. This oxygen-deprived environment proves fatal for *Bruchidius atrolineatus* and *Callosobruchus maculatus* and other post-harvest insects (http://www.entm.purdue.edu/PICS2/project_overview.php; accessed on 29 June 2016).

The SuperGrain II™ are manufactured by GrainPro Inc. (http://www.grainpro.com/grainpro-supergrainbag. php) through their agents in many countries in Asia, Africa and Latin America. Their dimensions were the same size as the polypropylene bags so they can be used as liners. The super grain bags are manufactured from high-density polyethylene that reduces gas exchange. Most agricultural commodities stored in these bags will develop a modified atmosphere of low oxygen and high carbon dioxide content, created by respiration of living organisms such as insects and fungi (Villers et al., 2008). After filling the super grain bag with grain, the free plastic portion (above the grain) is squeezed in order to remove excess air. The opening is then closed by tightly twisting the free portion and sealing it with a special strap fastener provided by the manufacturer (Fig. 2). The top end of the bag is twisted once more, folded back and sealed with another fastener. As recommended, the super grain bags are used as a liner bag inside polypropylene bags, which provide support and ease in handling. The outer bag is also closed (Likhayo et al., 2016).

Storability of maize grain is constrained by two post-harvest insect pests namely the maize weevil (*S. zeamais*) and LGB (*P. truncatus*). Host plant resistance is the most economical way to manage such pests especially for resource-constrained farmers as a component of integrated pest management. Host plant resistance is a valuable component of integrated pest management. The use of high-yielding maize varieties with resistance to post-harvest insect pests, particularly, the maize weevil and LGB, has been reported as a potential option to minimize storage losses. The resistant varieties possess chemical compounds that deter insect feeding or inhibit their growth, or some of these varieties have tight husk cover and ear tips, creating inaccessible conditions for insect infestation (Abebe et al., 2009). These varieties may also reduce the potential risk associated with consumption of treated maize with insecticides and they are generally compatible with other insect-control methods (Tefera et al., 2011, 2013). Breeding for resistance to these pests in maize is dependent on good understanding of genetic mechanisms underlying the resistance. The insect-damage traits and protein content are correlated. The CIMMYT-derived resistant genotypes to maize weevil and LGB include CKPH08003, BRAZ 2451, CKSPL10028, MTPEH0701, MTPEH0702 and EMB0701 (Tefera et al., 2016). Protein and oil contents were high in grain of resistant compared to susceptible genotypes. The resistant hybrids and inbred lines had flint texture, whereas the susceptible ones had dent kernel texture (Mwololo, 2013).

The insecticide Actellic Super has been adopted by small-scale farmers for grain storage in Kenya as well as in other African countries (Tefera, 2012). Actellic Super is a cocktail of 1.6% pirimiphos-methyl and 0.3% permethrin. It has been promoted as a chemical which is effective against the LGB in combination with practices such as immediate shelling and treating (Farrell and Schulten, 2002). In Tanzania, it was the most common control method of treating maize before storage; its use having been reported by more than 93% of farmers in both high rainfall and low rainfall zones (Kaliba et al., 1998).

Niger seed oil at the rates of 10%, 20%, 30%, 40%, 50% and 100% of the recommended application rate, 5 ml kg^{1}, combined with malathion at the respective rates of 100%, 50%, 40%, 30%, 20% and 10% of the recommended application rate, 0.5 g kg^{1}, provided complete protection to maize seed from the maize weevil up to 90 days after infestation (Yuya et al., 2008). Therefore, these combinations could be considered as a potential component in an effort to establish integrated maize weevil management. The effectiveness of the diatomaceous earth SilicoSec, a mineral industrial filter cake and domestic wood ash, applied at three different rates for the control of the maize weevil, *S. zeamais*, on three maize genotypes was determined by Demissie et al. (2008). Treatment with SilicoSec was the most effective followed by filter cake and wood ash. The treatments reduced progeny emergence, percentage grain damage and grain WLs, but did not affect percentage seed germination. Grain treated with wood ash at all rates resulted in a relatively low mortality 3 days after infestation as compared to other treatments. However, all treatments caused high mortality (97–100%) after 15 days of exposure. Therefore, SilicoSec, filter cake and wood ash can be considered as potential components of an integrated pest management strategy against the maize weevil (Demissie et al., 2008).

7 Conclusions

In summary, maize grain losses occur during post-harvest handling because of biological, socio-economic and matters of policy. The first critical step in reducing post-harvest grain losses is maintaining grain moisture content to acceptable levels. Use of hermetic grain storage technologies is ecologically sound and economically affordable. The cocoons, PICS, SuperGrain bags and the metal silo work hermetically, where the lack of oxygen inside the container causes the death of insect pests. These containers also exclude insects and rodents. Hermetic storage technologies can be integrated with resistant varieties wherever available to get optimum results. Utilization of post-harvest technologies is influenced by market and prevailing policy environment; hence, concerted efforts are required to develop institutional innovations and enabling policies to ensure adoption of effective post-harvest management practices, and bring the benefits to both producers and consumers.

8 Where to look for further information

Further information can be found in the following reports:

FAO/World Bank Workshop on Postharvest Losses, http://www.fao.org/3/a-au092e.pdf

UN calls for better grain storage to reduce post-harvest losses, http://www.un.org/apps/news/story.asp?NewsID=38548#.WlWlzH3MJlk

University of Greenwich, Natural resources Institute, postharvest loss reduction center, http://postharvest.nri.org/scenarios/grains

ADM Institute for the Prevention of Postharvest Loss | University of Illinois at Urbana-Champaign, http://postharvestinstitute.illinois.edu/literature.html

Postharvest Losses and Food Waste: The Key Contributing Factors to African Food Insecurity and Environmental Challenges, http://worldfoodscience.com/article/postharvest-losses-and-food-waste-key-contributing-factors-african-food-insecurity-and.

What do farmers say?, https://www.researchgate.net/publication/267816160_Post-harvest_loss_in_sub-Saharan_Africa-what_do_farmers_say

9 References

Abass, A. B., Ndunguru, G., Mamiro, P., Alenkhe, B., Mlingi, N. and Bekunda, M. (2014). Postharvest food losses in a maize-based farming system of semi-arid savannah area of Tanzania. *Journal of Stored Products Research* 57: 49–57.

Abebe, F., Tefera, T., Mugo, S., Beyene, Y. and Vidal, S. (2009). Resistance of maize varieties to the maize weevil *Sitophilus zeamais* (Motsch.) (Coleoptera: Curculionidae). *African Journal of Biotechnology* 8: 5937–43.

Adams, J. M. and Schulten, G. G. M. (1978). Loss caused by insects, mites and microorganisms. In Harris, K. L. and Lindbland, C. L. (Eds), *Post-Harvest Grain Loss Assessment Methods*. American Association of Cereal Chemists, USA, pp. 83–95.

Baoua, I. B., Amadou, L., Margam, V. and Murdock, L. L., 2012. Comparative evaluation of six storage methods for postharvest preservation of cowpea grain. *Journal of Stored Products Research* 49: 171–5.

Bokusheva, R., Finger, R., Fischler, M., Berlin, R., Marín, Y., Pérez, F. and Paiz, F. (2012). Factors determining the adoption and impact of a postharvest storage technology. *Food Security* 4(2): 279–93.

Boxall, R. A. (2002). Damage and loss caused by the Larger Grain Borer *Prostephanus truncatus*. *Integrated Pest Management Reviews* 7: 105–21.

Boxall, R. A. (1986). *A Critical Review of the Methodology for Assessing Farm-level Grain Losses after Harvest*. Tropical Development and Research Institute, Slough, UK.

Chapoto, A. and Jayne, T. S. (2010). Maize Price Instability in Eastern and Southern Africa: The Impact of Trade Barriers and Market Interventions. Prepared for the COMESAOMESA policy seminar on 'Variation in staple food prices: Causes, consequence, and policy options', Maputo, Mozambique, 25–26 January 2010 under the COMESAOMESA-MSU-IFPRI African Agricultural Markets Project (AAMP).

De Groote, H., Kimenju, S. C., Likhayo, P., Kanampiu, F., Tefera, T. and Hellin, J. (2013). Effectiveness of hermetic systems in controlling maize storage pests in Kenya. *Journal of Stored Products Research* 53: 27–36.

De Lima, C. P. F. (1979). Appropriate techniques for use in the assessment of country loss in stored produce in the tropics. *Tropical Stored Products Information* 38: 15–19.

Demissie, G., Tefera, T. and Abraham, T. (2008). Efficacy of Silicosec, filter cake and wood ash against the maize weevil, *Sitophilus zeamais* Motsch. (Coleoptera: Curculionidae) on three maize genotypes. *Journal of Stored Products Research* 40: 227–31

Dick, K. (1988). A review of insect infestation of maize in farm storage in Africa with special reference to the ecology and control of *Prostephanus truncatus* (NRI Bulletin No. 18).

FAO and World Bank (2010). FAO/World Bank Workshop on Reducing Post-Harvest Losses in Grain Supply Chains in Africa: Lessons learned and practical guidelines. FAO Headquarters, Rome, Italy, 18–19 March 2010, Rome.

FAO (2011). FAO Statistical Databases (FAOSTAT). Available by Food and Agriculture Organization (FAO) (verified 11 March 2013).

FAOSTAT. (2014). Africa maize production – 2012/13. http://faostat3.fao.org/browse/Q/QC/E.

Farrell, G. and Schulten, G. G. M. (2002). Larger grain borer in Africa; a history of efforts to limit its impact. *Integrated Pest Management Reviews* 7: 67–84.

Feng, Y., Tao, B., Pang, M., Liu, Y. and Dong, J. (2011). Occurrence of major mycotoxins in maize from Hebei Province, China. *Frontiers of Agriculture in China* 5(4): 497–503.

Ferris, S. and Robbins, P. (2004). Developing marketing information services in Eastern Africa: The FOODNET experience. International Institute of Tropical Agriculture (IITA), Ibadan, Nigeria. Available at http://old.iita.org/cms/articlefiles/736-ASARECA_9-Cassava_Marketing_in_Uganda.pdfhttp://old.iita.org/cms/articlefiles/736-ASARECA_9-Cassava_Marketing_in_Uganda.pdf.

Ferris, S. and Wheatley, C. (2001). FAO/GFAR Global Initiative on Post-harvest Technology, Phase 1. Report on the Regional Workshop for Africa, Held at Entebbe, Uganda, 17–19 September 2001. Available at http://www.foodnet.cgiar.org/Post%20Harvest/ Accessed on 10 November 2014.

Fischler, M., Berlin, R., Bokusheva, R., Finger, R., Marín, Y., País, F., Pavón, K. and Pérez, F. (2011). POSTCOSECHA Programme Central America. SDC.

Garrido, C. E., Hernández Pezzani, C. and Pacin, A. (2012). Mycotoxins occurrence in Argentina's maize (*Zea mays* L.), from 1999 to 2010. *Food Control* 25(2): 660–5.

Gitonga, G., De Groote, H. and Tefera, T. (2015). Metal silo grain storage technology and household food security in Kenya. *Journal of Development and Agricultural Economics* 7(6): 222–30.

Hodges, R. J. (2013). Rural 21. Focus. Tackling post-harvest cereal losses in sub-Saharan Africa. Available at: http://www.rural21.com/uploads/media/rural2013_01-S16-18.pdf. Accessed on 18 January 2015.

Hodges, R. J., Buzby, J. C. and Bennett, B. (2010). Post-harvest losses and waste in developed and less developed countries: opportunities to improve resource use. *Journal of Agricultural Science* 1; 9.

Kaliba, A. R. M., Verkuijl, H., Mwangi, W., Byamungu, D. A., Anandajayasekeram, P. and Moshi, A. J. (1998). *Adoption of Maize Production Technologies In Western Tanzania*. Mexico, CIMMYT.

Kimanya, M. E., De Meulenaer, B., Camp, J., Baert, K. and Kolsteren, P. (2012). Strategies to reduce exposure of fumonisins from complementary foods in rural Tanzania. *Maternal & Child Nutrition* 8(4): 503–11.

IICA (Inter-American Institute for Cooperation on Agriculture). (2013). IICA 1889 F Street, N.W. Suite 360, Washington, DC.

Lai, X., Liu, R., Ruan, C., Zhang, H. and Liu, C. (2014). *Occurrence of Aflatoxins and Ochratoxin a in Rice Samples from Six Provinces in China*. Food Control.

Likhayo, P., Bruce, A. Y., Mutambuki, K.,Tefera, T. and Mueke, J. (2016). On-farm evaluation of hermetic technology against maize storage pests in Kenya. *Journal of Economic Entomology* 109: 1–8, doi: 10.1093/jee/tow134.

Likhayo, P., Tefera, T., Mugo, S. and Mueke, J. (2015). Combined effect of hermetic bag and insect resistant variety for the control of larger grain borer and maize weevil in stored maize. *International Journal of Science and Research* 4: 2305–13.

Marin, S., Ramos, A. J., Cano-Sancho, G. and Sanchis, V. (2013). Mycotoxins: Occurrence, toxicology, and exposure assessment. *Food and Chemical Toxicology* 60: 218–37.

Mboya, R. and Bogale, A. (2012). Mycotoxin Contamination of Food and its Possible Implications on Sustainable Development in Rungwe district, Tanzania. *OIDA International Journal of Sustainable Development* 5: 8, p37.

Mutungi, C. M., Affognon, A., Njoronge, A. W., Baributsa, D and Murdock, L. L. (2014). Storage of mung bean (Vigna radiata [L.] Wilczek) and pigeopea grains (Cajanus cajan [L.] Millsp) in hermetic triple - layer bags stops losses caused by *Callosobruchus maculatus* (F.) (Coleoptera: Bostrichidae). *Journal of Stored Product Research* 58: 39–47.

Mwololo, J. K. (2013). *Resistance in Tropical Maize to the Maize Weevil and Larger Grain Borer*. PhD Thesis, Mekerer University. http://hdl.handle.net/10570/2880.

Mylene, K., Christopher, L. D., Gabre-Madhin, E. Z., Nicholas, M. and Michael, J. (2002). *Reforming Agricultural Markets in Africa*. Washington, International Food Policy Research Institute (IFPRI).

Ngamo, L. S. T., Ngassoum, M. B., Mapongmetsem, P. M., Maliasse, F., Hauburg, E., Lognay, G., et al. (2007). Current post-harvest practices to avoid insect attacks on stored grains in northern Cameroon. *Agricultural Journal* 2: 242–7.

Nguyo, R. (2007). Grain storage: keeping weevils at bay with metal silos. In Rootheart, R. and Taylor, P. (Eds), *Creating Sustainable Agricultural Impact for Smallholders: MATF Innovative Partnerships and Approaches 4th MATF Experience Sharing Workshop*. Mombasa, Kenya, pp. 19–23.

Pinstrup-Andersen, P. (2009). Food security: definition and measurement. *Food Security* 1: 5–7.

Proctor, D. L. (Ed.) (1994). Grain storage techniques. Evolution and Trends in Developing Countries. FAO Agricultural Services Bulletin No. 109. FAO, Rome.

Reed, C. (1987). The precision and accuracy of the standard volume weight method of estimating dry weight losses in wheat, grain sorghum and maize, and a comparison with the thousand grain mass method in wheat containing fine material. *Journal of Stored Products Research* 23 (4): 223–31.

Rembold, F., Hodges, R., Bernard, M., Knipschild, H. and Léo, O. (2011). *The African Postharvest Losses Information System (APHLIS)*. European Union, Luxembourg.

Smith, L. E., Stoltzfus, R. J. and Prendergast, A. (2012). Food chain mycotoxin exposure, gut health, and impaired growth: a conceptual framework. *Advances in Nutrition: An International Review Journal* 3(4): 526–31.

Suleiman, R., Rosentrater, K. and Bern, C. (2013). Effects of Deterioration Parameters on Storage of Maize: *A Review. Journal of Natural Sciences Research* 3(9): 147–65.

Suleiman, R. A. and Rosentrater, K. A. (2015). Current Maize Production, Postharvest Losses and the Risk of Mycotoxins Contamination in Tanzania. *Agricultural and Biosystems Engineering Conference Proceedings and Presentations*. Paper 442.

Tefera, T. (2012). Post-harvest losses in African maize in the face of increasing food shortage. *Food Security* 4(2): 267–77.

Tefera, T., Demissie, G., Mugo, S. and Beyene, Y. (2013). Yield and agronomic performance of maize hybrids resistant to the maize weevil *Sitophilus zeamais* Motschulsky (Coleoptera: Curculionidae). *Crop Protection* 46: 94–9.

Tefera, T., Kanampiu, F., De Groote, H., Hellin, J., Mugo, S., Kimenju, S., Beyene, Y., Boddupalli, P. M., Shuferaw, B. and Banziger, M. (2011). The metal silo: An effective grain storage technology for reducing post-harvest insect and pathogen losses in maize while improving smallholder farmers' food security in developing countries. *Crop Protection* 30(3): 240–5.

Tefera, T., Mugo, S. and Beyene, Y. (2016). Developing and deploying insect pest resistant maize in Africa. *Food Security* 8: 211–20.

Tefera, T., Mugo, S., Likhayo, P. and Beyene, Y. (2011). Resistance of three-way cross experimental maize hybrids to postharvest insect pests, the larger grain borer *(Prostephanus truncatus)* and maize weevil *(Sitophilus zeamais). International Journal of tropical Insect Sciences* 31: 3–12.

Villers, P., Navarro, S. and Bruin, T. D. (2008). Development of hermetic storage technology in sealed flexible storage structures. In Daolin, G. (Ed.), *Proceedings of the 8th International Conference on Controlled Atmosphere and Fumigation in Stored Products*, CAF 2008, Controlled Atmosphere and Fumigation, Green, Safe, Harmony and Development, Chengdu, China, September 21e26. Grain Pro Inc, Chengdu, China. Accessed on 16 April 2015. http://grainpro.com/pdf/PU2015PV0708-CHENGDU-PAPER-CAF-Presentation-Development-of-Hermetic-Storage-Technology.pdf.

World Bank, FAO, NRI (2011). Missing Food: the Case of Post-harvest Grain Losses in Sub-Saharan Africa. In: Economic Sector Work Report No. 60371-AFR. World Bank, Washington, DC.

Yuya, A., Tadesse, A., Azerefegne, F. and Tefera, T. (2009). Efficacy of combining Niger seed oil with malathion 5% dust formulation on maize against the maize weevil, *Sitophilus zeamais* (Coleoptera: Curculionidae). *Journal of Stored Products Research* 45: 67–70.

Zorya, S., Morgan, N. and Rios, L. D. (2011). Missing food: The Case of Postharvest Grain Losses in Sub-Saharan Africa. The International Bank for Reconstruction and Development/The World Bank. Report No. 60371-AFR. The World Bank, Washington, DC.

Part 2

Maize pests, diseases and weeds

Chapter 13

Economically important insect pests of maize

William D. Hutchison and Theresa M. Cira, University of Minnesota, USA

1 Introduction

As emphasized previously in this volume, maize continues to be the world's most abundant food crop, recently exceeding 1.04 billion tons produced on ~184.8 million ha (FAO 2014). Maize is also an essential staple crop in many developing countries, and thus critical for global food security. Maize is popular for its rapid processing time from dry grain to various flours and markets, and its nutritional value for both human and livestock consumption.

In addition to advances in hybrid technology in the Americas and in Europe, there have been an increasing number of hybrids adapted for use in African countries such as Ethiopia, Kenya and South Africa. These hybrids are bred for insect and pathogen resistance, and drought tolerance (e.g. CIMMYT, 2016). Although yields in many industrial countries range from 5 to 15 tons/ha, yields in most sub-Saharan African countries remain low, at 0.5–1.0 ton/ha (Worku et al. 2012). While some of this yield disparity can be attributed to the use of open-pollinated varieties (OPVs), much of the yield reduction is a result of numerous economically damaging insect pests. Insect pests can cause direct injury via feeding on vegetative and reproductive plant tissue. This can then lead to indirect injury by facilitating the introduction and establishment of maize pathogens. In this chapter we provide descriptions of the most important and representative insect pests for the primary maize production areas of the world and a summary of emerging pest management advancements, challenges and research needs.

http://dx.doi.org/10.19103/AS.2016.0002.14

2 Below-ground insect pests

2.1 Western corn rootworm *Diabrotica virgifera virgifera* LeConte (Coleoptera: Chrysomelidae) and Northern corn rootworm *Diabrotica barberi* R. Smith and Lawrence (Coleoptera: Chrysomelidae)

Injury from *D. virgifera virgifera* and *D. barberi*, commonly referred to as the rootworm complex, can lead to high economic losses. The primary injury and yield loss is due to larval feeding on the developing root system (Steffey et al. 1999, Gray et al. 2009). A compromised root system can result in a reduction in the plant's ability to take up water

Figure 1 Western corn rootworm adults, *Diabrotica virgifera virgifera* LeConte (Coleoptera: Chrysomelidae), credits: a) Jim Kalisch – University of Nebraska-Lincoln and b) Eric Burkness – University of Minnesota.

and nutrients and stalk lodging, particularly when exposed to high winds, which then makes mechanical harvesting problematic. A secondary, less consistent source of injury is adult feeding on silk and ear kernels during late-season beetle emergence. Both New World pests, *D. barberi* and *D. virgifera virgifera*, occur in North America. *Diabrotica virgifera virgifera*'s range extends further south, however, into Central America and is now found across Europe. In Europe, the western corn rootworm was first detected in Serbia in 1992, in a maize field near the Belgrade airport (Kiss et al. 2005), and quickly spread into Italy and Eastern Europe (Gray et al. 2009). There is increasing evidence that the current

Figure 2 Northern corn rootworm adults, *Diabrotica barberi* R. Smith and Lawrence (Coleoptera: Chrysomelidae), credits: a) Jim Kalisch – University of Nebraska-Lincoln and b) Eric Burkness – University of Minnesota.

European distribution of the pest is the result of multiple, independent introductions (Gray et al. 2009).

Both species lay eggs in the soil where the eggs go through an obligate winter diapause before hatching. Eggs are a flattened oval shape. Larvae have a yellowish brown head capsule and a pale yellow white body with a brown plate on the terminal abdominal segment. Larvae feed on the roots of maize and go through three instars before pupating in the soil (Fig. 1). Fifty per cent of *D. virgifera virgifera* first-instar larvae will emerge after ~265–290 degree days (base temperature 11°C) (Fisher 1989, Schaafsma et al. 1993); adults emerge after 415 degree days (base temperature 9°C) (Jackson and Elliott 1988). Elytra of *D. virgifera virgifera* adults are greenish yellow with varying degrees of dark brown or black markings, often in the form of longitudinal stripes (Fig. 2). Adults of *D. barberi are* uniformly greenish yellow (Fig. 3). Adults of both species will feed on maize silk, leaf tissue, young kernels and pollen.

In North America, the widespread use of genetically engineered (GE) maize, transformed to express various toxins from the bacterium, *Bacillus thuringiensis* (Bt) (James 2015), and cultural methods of crop rotation were initially quite successful against the rootworm complex (Gray et al. 2009). However, the rootworm complex has evolved both behavioural and physiological mechanisms against these management strategies. The first case of resistance to Bt was documented in 2009 in Iowa (Gassmann et al. 2011). In the past four years, there has been an increasing incidence of field-evolved resistance by *D. virgifera virgifera* to multiple Bt toxins, in the Midwestern U.S. (e.g. Zukoff et al. 2016). In addition to Bt resistance, the opportunistic *D. virgifera virgifera* has also adapted to the yearly rotation of maize and soybean. Previously, this rotation prevented the build-up of rootworm populations. Females laid eggs in a field of maize in the fall but in the following spring, hatching larvae would soon die when exposed to soybean (Gray et al. 2009). Adapting to this cultural management tactic, *D. virgifera virgifera* females have begun laying eggs

Figure 3 Corn rootworm larvae, credit: Jim Kalisch - University of Nebraska-Lincoln.

in soybean fields in the fall so that eggs hatching the following spring would be exposed to maize. This phenomenon was primarily limited to an area of eastern Illinois, western Indiana and southern Michigan. Additionally, in some populations of *D. barberi*, eggs can enter an extended diapause which lasts one or two additional winters so that they are more likely to emerge in a field that is planted with maize (French et al, 2014).

2.2 Seed corn maggot, *Delia platura* (Meigen) (Diptera: Anthomyiidae)

The seed corn maggot is a sporadic but damaging early season pest of maize and is distributed worldwide (Steffey et al. 1999). The fly overwinters in the pupal stage, with adults emerging in early spring. Females lay eggs in newly planted maize fields, and once eggs hatch, the larvae are well positioned to attack maize seed and subsequently cause stand loss. Depending on location, the species can go through 4–5 generations per year. Adult flies are attracted to the volatiles released from animal manure applied to fields (e.g. turkey manure). Although these risk factors can often be avoided in conventional maize systems, the propensity for damage can be higher for organic maize production. *Delia platura* has a wide host range of up to 50 species, including several edible and common bean crops where economic loss can also occur.

As with other seed-feeding maggot species, adult *D. platura* are mostly grey and slightly smaller than the common housefly (*Musca domestica*); unlike the housefly, however, the adults will hold their wings over, and parallel with the abdomen when at rest. Larvae (maggots) are white, tapered at the front end with no visible head or legs (Fig. 4). Their colour and size

Figure 4 Seed corn maggot larvae, *Delia platura* (Meigen) (Diptera: Anthomyiidae), credit: Jim Kalisch – University of Nebraska-Lincoln.

are also not unique to this species, and thus visual identification to species, as well as field monitoring is difficult, until damaged seeds are found. Understanding the degree to which *D. platura* is present in or near maize fields is best accomplished with the use of yellow stick cards or traps baited with specific lures (e.g. Kuhar et al. 2006). Where there is a history of *D. platura* infestations within a farm or region, many growers have begun using insecticidal seed treatments (e.g. imidacloprid, or thiamethoxam) to protect against stand loss.

2.3 Wireworms: *Aeolus, Agriotes, Conoderus, Limonius, Melanotus* spp. (Coleoptera: Elateridae)

Wireworm species are found globally throughout the major maize growing regions, including North America, Europe, Africa and Asia (e.g. Steffey et al. 1999, Zhang et al. 2017). Over 9000 species have been documented worldwide; fortunately, only a small proportion are considered pests of agricultural crops (Barsics et al. 2013). Depending on the species, adult beetles, commonly called click beetles, can have a solid brown to black colouration, or an orange with prominent black markings. When placed on their back the beetles will move in a characteristic way to right themselves, accompanied by a clicking sound. Adults have a distinct body shape; elongated oval with parallel sides and with the posterior corners of the pronotum elongated backwards into points or spines. Adults are phytophagous but are not considered the damaging stage. Most species prefer to oviposit in grassy areas but will also lay eggs in newly tilled fields. The larvae of this beetle family damage plants by feeding on seeds and roots in the soil. The larvae are slender, shiny and hard-bodied, and generally are orangish brown in colour but with slight variations by species (Fig. 5). They prefer porous soils, and can occur at several depths

Figure 5 Wireworm larvae, photo credit: Jim Kalisch – University of Nebraska-Lincoln.

below the seedbed (often within 12 cm). Depending on the species, larvae require 1–5 years to complete development. Wireworm damage to maize can be substantial in fields planted in recently tilled pasture or forages (Steffey et al. 1999).

Wireworm injury results from direct feeding on seeds prior to germination, or young seedlings, resulting in stand loss. Wireworm damage can also be highly aggregated throughout a field, making sampling and monitoring a challenge. Moreover, there is a narrow window of time to monitor for wireworms prior to planting. In the United States, 'bait station' traps have been developed and used with some success (Steffey et al. 1999) and in Europe (Barsics et al. 2013). Bait stations have been recommended at approximately 2–3 per ha, and placed in fields 2–3 weeks before planting. Stations should be ~5–8 cm deep and 15–23 cm wide at the soil surface, filled with ½ cup of maize and wheat seeds, then filled with soil creating a mounded top. The mound is then covered with a 45-cm square piece of black plastic, and a larger sheet of clear plastic – both of which accelerate the spring soil temperatures and seed germination to attract wireworm larvae. An average of one wireworm per bait trap is indicative of an economic infestation, and appropriate liquid, or granular insecticides should be considered at planting (Steffey et al. 1999). Alternatively, and because of the aggregated and sporadic nature of wireworm infestations, insecticidal seed treatments (e.g. imidacloprid) have become commonplace in several countries for maize and other crops (Zhang et al. 2017).

3 Above-ground insect pests

3.1 Corn leaf aphid *Rhopalosiphum maidis* (Fitch) (Hemiptera: Aphididae)

Rhopalosiphum maidis reproduces parthenogenetically and gives live birth to nymphs which pass through four instars before adulthood. From 15°C to 30°C development rate increases more than double; at the optimum 30°C, the total lifespan averages 16.8 ± 1.3 d (El-Ibrashy et al. 1972). From a base temperature of 4.44°C, 119.05 degree days are required to progress through all four instars. Maximum adult longevity is seen at 15°C where they survive for 26.7 d (Kuo et al. 2006). Nymphs have a lighter colouration than adults, but both stages have a greyish green body and bluish grey head and legs (Fig. 6). They are unable to survive cold winters (e.g. Canada, Northern Asia, Europe and the United States) but can migrate to these areas in spring from warmer areas. This cosmopolitan insect species does not often reach severe pest status on maize. Feeding in the whorl, deposition of excessive amounts of honeydew during the tassel and ear development, and vectoring maize dwarf mosaic virus (MDMV) (Steffey et al. 1999) are the main concerns. However, even with high aphid infestations, insecticidal treatments are rarely necessary.

3.2 Maize leafhoppers, *Cicadulina mbila* Naudé, *C. parazeae* Ghauri, *C. storeyi* China (Hemiptera: Cicadellidae)

The eggs, nymphs and adults of these species look nearly identical. Cream-coloured elliptical eggs are laid in leaf tissue and pale wingless nymphs are 1–3 mm smaller than adults. They pass through 5 instars; at 25°C nymphs require 14–20 days to become adults (Rose 1973). Adults are 2–4 mm, their wings are held tent-like over the body, and the head

Figure 6 Corn leaf aphid, *Rhopalosiphum maidis* (Fitch) (Hemiptera: Aphididae), photo credit a) Jim Kalisch - University of Nebraska-Lincoln and b) Eric Burkness - University of Minnesota.

is bluntly rounded at the tip. Adults and nymphs feed on maize, but the primary concern is their ability to transmit maize streak virus (van den Berg et al. 2014). Although some hybrids have been developed to confer resistance, such hybrids are not available in all countries.

3.3 European corn borer, *Ostrinia nubilalis* (Hübner) (Lepidoptera: Crambidae)

Throughout North America and much of Europe, *O. nubilalis* is the most widely distributed pest of maize, including ~36.4 million ha of field corn grown for livestock feed and ethanol use in the United States (Hutchison et al. 2010), or sweet corn grown for direct human consumption (Hutchison et al. 2004). Historically, *O. nubilalis* has been a persistent, yet cryptic pest, as it can damage all above-ground parts of the plant, including leaf/whorl feeding, boring into the stalk, feeding at the base (shank) of ears causing premature ear loss or feeding directly on ear kernels. Yields are subsequently reduced both indirectly, via feeding within the stalk that inhibits normal nutrient flow throughout the plant, and directly, via kernel feeding within the ear (Mason et al. 1996, Steffey et al. 1999, ISU 2015). In addition, stalk and ear feeding also predispose the plant to various pathogens causing stalk and ear rots, including the continued concern with aflatoxin, a known carcinogen and particularly problematic in more humid maize production regions such as the southern U.S. Consequently, there are many ways that economic losses can occur in any given year. The *O. nubilalis* distribution includes central and eastern North America, Eastern and Western Europe, North Africa and selected Middle East countries (CABI 2016a). Depending on local climate, between one and six generations can be completed in a calendar year. Ecotypes related to voltinism, or the number of generations per year, have been characterized (e.g. Coates et al. 2013). In northern latitudes in North America (e.g. >48°N), populations have historically been univoltine, whereas south of this latitude, bivoltine and multi-voltine strains exist that permit the species to take advantage of longer growing seasons (e.g. Calvin and Song 1994). Also, in North America there exist two main ecotypes that differ in the pheromone type emitted by females. The E- and Z-ecotypes, which refer to the isomer ratio blends (i.e. Z:E) in the primary pheromone components (e.g. Bartels et al. 1997). For example, the Z-strain of *O. nubilalis* is dominant in the central U.S. 'Corn belt', where the pheromone Z:E ratio 97:3(%) is most effective. Previous research has shown that any deviation from this blend (e.g. 95Z:5E ratio) will greatly reduce male moth catch in pheromone traps (Bartels et al. 1997). Black light traps are therefore often used to monitor this species for integrated pest management (IPM) programmes, as a more consistent monitoring tool (Mason et al. 1996).

Eggs of *O. nubilalis* are a flattened oval shape, creamy white, darkening as they age and laid in clusters of ~15–20 eggs on the underside of leaves. Young larvae have a brownish black head capsule and their body is a creamy white. As larvae develop they take on a darker colouration and dark circles in each segment appear. Approximately 212 degree days (base temperature 10°C) are needed for hatch, after which larvae will mature through five or six instars over approximately 792 degree days (ISU 2015). If located in an area where overwintering occurs, late instars will overwinter in the stalk, corn stubble or plant debris. Regardless of overwintering, pupation occurs in the stalk, corn stubble or plant debris, and eclosion as an adult moth occurs after 1192 accumulated degree days from egg lay (ISU 2015). Adult wingspan is ~20–26 mm for males and ~25–34 mm for females. The colour pattern of males is slightly darker brown compared to the female but they both have zigzag lines running horizontally across the wings and yellowish gold patches (Fig. 7).

Since the commercial approval of Bt maize in the United States in 1996, the use of Bt maize technology has become the dominant IPM strategy for *O. nubilalis* in field (or dent) maize (Hellmich et al. 2008, Hutchison et al. 2010). Prior to 1996, growers relied on crop scouting and the use of economic thresholds to determine whether insecticide applications were necessary. Conventional breeding for corn borer resistance (tolerance) also improved considerably from 1980 to 1995, and many complementary agronomic traits continue to be incorporated into existing inbred and hybrid development programmes (e.g. Mason et al. 1996, Hellmich et al. 2008, CIMMYT 2016). In addition to the direct yield benefits of Bt hybrids, grain quality has improved with several studies demonstrating reductions in fungal infections associated with damage from *O. nubilalis*,

Figure 7 European corn borer, *Ostrinia nubilalis* (Hübner) (Lepidoptera: Crambidae) a) larva, credit: Eric Burkness – University of Minnesota and b) adult female, photo credit: Jim Kalisch – University of Nebraska-Lincoln.

such as *Fusarium* spp. (e.g. Munkvold et al. 1999). Historically, conservation biological control has been encouraged in North America, Europe and China (e.g. Mason et al. 1996, Zimmerman et al. 2016). Effective biological control agents include the microsporidium *Nosema pyrausta*, generalist predators and a suite of parasitic wasps that have provided additional suppression benefits (Mason et al. 1996, Steffey et al. 1999, Zimmerman et al. 2016). As a result of extensive and current use of Bt maize in North America (>28.3 million ha in the United States, as of 2015), the pest status of *O. nubilalis* has decreased significantly since 1997 (Hutchison et al. 2010, Bohnenblust et al. 2014, Milne et al. 2015). Due to landscape-level suppression, resulting economic benefits have spilled over to producers of non-Bt maize as well. For more information on *O. nubilalis* and Bt maize, see Section 5.

3.4 Asian corn borer *Ostrinia furnacalis* (Guenée) (Lepidoptera: Crambidae)

Ostrinia furnacalis has a similar life history and feeding habits as *O. nubilalis*, with multiple generations per year throughout China and tropical Asia (CABI 2016b). They feed on all above-ground parts of the plant and can cause serious economic injury. This stem borer is part of a complex that is difficult to identify to species (see Mutuura and Munroe 1970) (Fig. 8). The egg masses, larvae and adults look very similar to *O. nubilalis*. In many countries, particularly for small-scale farmers, insecticidal and biological control continue to be the primary control tactics (CABI 2016b). For over a decade, Bt maize (Cry1Ab) has been grown in the Philippines and substantially reduced the damage from this species; yields have increased and become more consistent (e.g. Alcantara et al. 2011). For more on the biology and management of *O. furnacalis*, see Nafus and Schreiner (1991).

Figure 8 Asian corn borer, *Ostrinia furnacalis* (Guenée) (Lepidoptera: Crambidae) a) eggs and b) larva, credit: Merdelyn C. Lit – University of the Philippines Los Baños.

3.5 South-western corn borer *Diatraea grandiosella* and *Diatraea* spp. (Lepidoptera: Crambidae)

The south-western corn borer is native to central and north-western Mexico, and began expanding its range into the United States in the early 1900s (Mason et al. 1996, Steffey et al. 1999). In the United States, it is now found from Arizona to Alabama and as far north as Missouri and Illinois. Larvae of *D. grandiosella* have distinctive black spots along the dorsal surface, but these are not present in overwintering larvae. As another borer in the family Crambidae, its life history and feeding damage is similar to *O. nubilalis*, with first-generation *D. grandiosella* feeding on whorl stage corn, but also damaging the growing point, which can result in 'dead-heart' (Steffey et al. 1999). However, an important distinction is the 'girdling' damage done by late-season larvae feeding within the stalk at the base of plants. This feeding behaviour, in late summer and fall in preparation for overwintering, results in an internal girdling of the stalk above the soil surface that predisposes plants to lodging. High winds prior to harvest will increase the risk of lodging and exacerbate the potential for yield loss. The advent of Bt maize (Cry1Ab) in the United States has thus far been effective against this species (Qureshi et al. 2006).

Another pest species in the *Diatraea* genus in North America is the sugarcane borer, *D. saccharalis*. As with the other borer species in the family Crambidae, feeding behaviour is cryptic, with larvae quickly boring into plant stalks after hatching, making control difficult. In addition to sugarcane, this species will attack maize, sorghum, rice and wheat (Steffey et al. 1999). Bt maize has been effective against this species, but resistance has been discovered in Louisiana populations (Huang et al. 2012). A related species that has also recently reached pest status in the United States is the Mexican rice borer, *Eoreuma loftini* (Steffey et al. 1999). This species causes substantial losses in sugarcane but also attacks maize, rice and sorghum.

3.6 Corn earworm *Helicoverpa zea* (Boddie) (Lepidoptera: Noctuidae)

Although this species is highly polyphagous, and as consequence, known by numerous common names (e.g. corn earworm, tomato fruitworm, soybean pod borer), maize is often a preferred host throughout the Americas. *Helicoverpa zea* most commonly causes injury when larvae feed on silk and ear tissue. This New World pest is currently known in North, South and Central America. *Helicoverpa zea* is seasonally migratory, with adult dispersal triggered by senescence of host plants (Lingren et al. 1993). For example, in the central U.S., *H. zea* adults depart from senescing cornfields in several south-central states, usually in late July; by early August significant increases in pheromone-baited trap captures are observed in northern states (Westbrook et al. 1997, Olmstead et al. 2016). This timing allows the pest to colonize northern corn that is silking and attractive for oviposition in August. The northern overwintering limit is approximately the 40th parallel (Hardwick 1965, Morey et al. 2012).

Eggs are typically laid singly. They are radially ribbed and begin as a light green colour changing to grey before hatching. Larvae have a light brown to orange head capsule, and usually display a broad dark lateral line above the spiracles and a light line below the spiracles. Colouration of the body is variable, pink, yellow, brown or green. A key diagnostic trait is the presence of small black spines on the body's cuticle. As summarized by Morey (2010), *H. zea* egg development requires 40–76 degree days (base temperature between 12.2 and 12.8°C) to hatch. Larvae typically progress through 6 instars (less commonly 5 or 7 instars), and require 200–290 degree days. Pupation occurs in the soil, typically 5–15 cm below the surface. In temperate areas, the pupae will enter a facultative diapause for overwintering (Morey et al. 2012). The following spring, adults emerge after 162–355 degree days. Male and female moths have green eyes and yellowish brown forewings that measure 32–45 mm across (Fig. 9).

Figure 9 Corn earworm, *Helicoverpa zea* (Boddie) (Lepidoptera: Noctuidae) a) larva, photo credit: Eric Burkness – University of Minnesota and b) adult male, credit: Silvana Paula-Moraes – University of Florida.

Although Bt maize in both field (Burkness et al. 2015) and sweet corn (Shelton et al. 2013) varieties is available for *H. zea* management in North America, and in selected countries such as Brazil (e.g. Bentivenha et al. 2016), growers in many areas continue to rely on foliar insecticides for management (e.g. Hutchison et al. 2004). Since 1996 the initial Cry1Ab event provided excellent efficacy against *H. zea*, but in recent years, efficacy has declined and resistance is now evident in several locations of the United States, including North Carolina (Reisig and Reay-Jones 2015) and Maryland (Dively et al. 2016). During the past decade, synthetic pyrethroids such as bifenthrin (Capture®) in the United States have gradually lost their effectiveness, as variable levels of resistance have now been documented in several U.S. states (Jacobson et al. 2009).

3.7 Old World bollworm, cotton bollworm *Helicoverpa armigera* (Hübner) (Lepidoptera: Noctuidae)

Helicoverpa armigera causes a great deal of damage when the larvae feed on maize ears and is considered a major pest where it occurs. Historically, *H. armigera* has been limited to Africa, Asia, Australia and Europe, but in 2013, it was detected for the first time in the Americas, in Brazil (Czepak et al. 2013). Since then, it has been found in Argentina and Paraguay (Murúa et al. 2014, EPPO 2015). Most recently, male moths were detected in pheromone traps in Florida in 2015 (Smith 2015) and in 2014 and 2015 males were trapped in Puerto Rico (USDA-APHIS 2015). The spread of *H. armigera* into Argentina, and throughout South and Central America, as well as the United States is being monitored closely (e.g. Kriticos et al. 2015). Eggs are laid on the silk of maize and are ridged, starting as yellowish white when laid and changing to a dark brown before hatching. Larvae go through five to six instars with their colouration darkening as they age and varying from pale green, grey, yellow or pink with dark setae. Larvae require 215 degree days to mature (base temperature 10.5°C) and pupae 151.8 degree days (base temperature 11.3°C) (Jallow and Matsumura 2001). Adults have a 3.5–4 cm wingspan, their coloration and patterning is difficult to differentiate from that of the corn ear worm, *H. zea* (Fig. 10). Twine (1978) found that 475 degree days (base temperature

Figure 10 Old World bollworm, cotton bollworm *Helicoverpa armigera* (Hübner) (Lepidoptera: Noctuidae) a) larva and b) adult male, photo credit: Silvana Paula-Moraes – University of Florida.

11°C) are needed to complete development from larvae to adult. As this species continues to expand its range in the Americas, it is important to note that the larvae cannot be distinguished visually from *H. zea* larvae (Kriticos et al. 2015). In addition, the only reliable method for separating adult moths between the two species is examination of the male genitalia (Czepak et al. 2013). Although molecular markers are being investigated for rapid identification of each species in the field, this approach has not yet been validated (e.g. Bentivenha et al. 2016).

Throughout Asia, *H. armigera* has developed resistance to many foliar insecticides, for example in India (Kranthi et al. 2002), primarily because of insecticide use on cotton. However, this can carry over to populations feeding on maize. Recently, in Brazil, there was some question as to the utility of maize as a major host for *H. armigera*, particularly in areas where cotton or soybean are grown. However, a recent study by Bentivenha et al. (2016) showed that in the western Bahia region of Brazil the second crop of maize in the year supported nearly equal populations of *H. armigera* and *H. zea*. Thus, maize may play a more significant role in the seasonal dynamics of *H. armigera* than initially thought, following its introduction into Brazil.

3.8 Western bean cutworm *Striacosta albicosta* (Smith) (Lepidoptera: Noctuidae)

Striacosta albicosta is native to North America, and historically attacked maize and bean crops in the western U.S., from Arizona and Idaho to Nebraska (Steffey et al. 1999). Beginning in 1999, however, researchers in the Midwestern U.S. began detecting moth flights in Iowa, Minnesota and Missouri, as well as larval populations causing significant damage in South Dakota and Iowa (e.g. Michel et al. 2010). Beginning in 2001, the pest was found in Wisconsin, Michigan, eastward to New York, Pennsylvania, and Ontario and Quebec, Canada (e.g. Michel et al. 2010).

The egg masses of this species are usually laid on the undersides of maize leaves, but are relatively easy to detect due to their characteristic purple colour. Despite the use of the term 'cutworm' in the common name, larvae do not feed like typical cutworms. Most cutworms attack plants in the seedling stage, damaging only the foliage or cutting the stem near the soil surface. Larvae of *S. albicosta* on the other hand, are usually hatching during the early tassel stage of maize development, and feed in both the tassel and developing ears. Ear and kernel damage can be quite extensive; because the larvae are not cannibalistic, each ear may support five or more larvae. The larvae are of typical size and length for many species in the Noctuidae family, typically progressing through 5–7 instars. The adult moths are light brown with distinctive black and white lines on the dorsal surface, slightly resembling the dingy cutworm (*Feltia jaculifera*) in North America (Fig. 11). Colour photographs of all life stages are available in Michel et al. (2010).

Compared to many lepidopteran pests in North America, *S. albicosta* is somewhat unique with only one generation per year. Moth flights typically peak in mid-late July (Michel et al. 2010). Historically, this species has been managed well with Bt maize, but in recent years the often used Cry1F event has not been providing consistent control (C. DiFonzo, Michigan State University, pers. comm.). Many growers therefore continue to rely on foliar applications of pyrethroids (e.g. Michel et al. 2010), and can use rapid sampling plans for egg masses (Paula-Moraes et al. 2011) and degree-day models to forecast moth flight phenology and control activities (Hanson et al. 2015).

Figure 11 Western bean cutworm, *Striacosta albicosta* (Smith) (Lepidoptera: Noctuidae) a) eggs, photo credit: Samantha R. Daniel – University of Nebraska-Lincoln, b) larva and c) adult, photo credit: Julie Peterson – University of Nebraska-Lincoln.

3.9 Fall armyworm *Spodoptera frugiperda* (J. E. Smith) (Lepidoptera: Noctuidae)

This highly mobile pest of maize is widely distributed throughout the Americas, with damaging infestations most common in South and Central America. Populations in the United States are limited to the south-central, south eastern and north eastern U.S. (Steffey et al. 1999). Summer infestations in the north eastern and north-central U.S. are facilitated by migratory flights from southern states each summer, although such migrations are most consistently observed in the northeast. Most recently, *S. frugiperda* was detected for the first time in West and Central Africa (Goergen et al. 2016), and it is currently a concern in South Africa (J. van den Berg, Northwest Univ., SA, pers. comm.). It is unclear to what extent this species can overwinter in selected countries, or maintain itself in tropical regions; considerable research is needed to address the potential impacts of this new pest of African maize. As a polyphagous species, *S. frugiperda* feeds on many non-crop grass hosts; however, the primary crop hosts of concern are typically maize and cotton.

Eggs are laid in masses which appear fuzzy from the light-coloured scales deposited by the female. Larvae are tan and yellow with broad longitudinal stripes and a characteristic

Figure 12 Fall armyworm, *Spodoptera frugiperda* (J. E. Smith) (Lepidoptera: Noctuidae) a) larva and b) adult, credit: Carla Dutra – Monsanto Brazil .

'Y' shape which is light-coloured on the front of the head and usually go through six instars (Fig. 12). Pupation occurs in the soil after which adult moths emerge. They have a wingspan of 32–40 mm, and forewing colouration of greyish brown with white and dark brown markings (Steffey et al. 1999). Beyond traditional insecticidal control, the current Vip3A Bt event has provided the most consistent control, whether it be in Bt maize or Bt cotton. Beginning in 2008, *S. frugiperda* developed resistance to Cry1F, initially in Puerto Rico, and most recently in Brazil (e.g. Zhu et al. 2015). Moreover, these authors also provided evidence of cross resistance to an organophosphate insecticide. In brief, this species continues to be one of the most challenging pests to manage throughout the Americas.

3.10 Spotted stalk borer *Chilo partellus* (Swinhoe) (Lepidoptera: Crambidae)

Historically, *C. partellus* has been one of the most widespread and damaging maize pests in Africa and India (Polaszek 1998, CABI 2016c). In several African countries, there can be overlapping generations, and as with most stem borers, the cryptic life stage of the larvae makes it nearly impossible to control once it has bored into maize stalks or ears. *Chilo partellus* occurs in Asia and the Middle East, as well as Eastern and Southern Africa (Yonow et al. 2017). The pest was recently detected in Turkey, suggesting its range expansion remains active (CABI 2016c).

Eggs of *C. partellus* are laid on the underside of leaves overlapping each other, eggs are creamy white and scale-like. Larvae are yellowish brown to cream coloured with four purplish brown stripes running longitudinally down the body and usually with dark dorsal spots. At first, larvae feed on the leaf whorl then older instars tunnel in and feed on the stalk and pupate there. Depending on the local climate, five or more generations may occur in a year and a facultative larval diapause in winter or the dry season may occur in dry corn stalks. From egg to adult, ~588.3 degree days (base temperature 17.6°C) are needed (Jalali and Singh 2001). Adult moths have a wingspan of 7–17 mm, and

Figure 13 Spotted stalk borer, *Chilo partellus* (Swinhoe) (Lepidoptera: Crambidae) a) eggs, b) larva and c) adult, credit: Annemie Erasmus – Agricultural Research Council, South Africa.

forewings are light yellowish brown in colour with darker coloured longitudinal striations. Examination of genitalia is needed to definitively distinguish from other *Chilo* species (Fig. 13).

Interestingly, for the past five years, populations in South Africa are at historic lows, and likely due to the use of Bt maize (Cry1Ab) (J. van den Berg, Northwest Univ., pers. comm., see also van den Berg and Campagne 2015). Although Bt maize has been found to be highly effective against *C. partellus* in Kenya, the technology has not been approved for use in this country, or East Africa (Tefera et al. 2016). One of the most innovative IPM solutions for stem borers in East Africa is the 'push-pull' system. This involved the strategic intercropping of legumes (e.g. *Desmodium* spp.) that effectively repel (*push*) *C. partellus* moths from the maize crop, coupled with various grass species (e.g. Napier grass) planted on the perimeter of the maize crop that attract (*pull*) the moths for oviposition (Khan et al. 1997, Kfir et al. 2005, Hassanali et al. 2008). Volatiles produced from both the legume and grass species have been identified to understand the underlying mechanisms for the 'diversionary tactics' against the moths. Moreover, some of the volatiles (e.g. (*E*)- 4,8-dimethyl-1,3,7-nonatriene) have the added benefit of attracting beneficial parasitoids of *C. partellus* (Khan et al. 1997). In addition to providing stem borer suppression, the push-pull system has also been successful in controlling *Striga* spp. ('witchweed'), and the additional legume/grass plantings provide increased yields and diversity of fodder for livestock (Hassanali et al. 2008). By 2000, the push-pull system had been adopted by thousands of farmers in Kenya, since then adoption has greatly expanded to Uganda and Ethiopia as well (Hassanali et al. 2008).

3.11 Maize stalk borer *Busseola fusca* (Fuller) (Lepidoptera: Noctuidae)

Busseola fusca is widespread in sub-Saharan Africa and is not known to occur elsewhere. It is considered the most destructive lepidopteran species in Africa (Kfir et al. 2005, van den Berg and Campagne 2015, CABI 2016d). Young larvae will feed on leaf tissue in the whorl, and older larvae bore into and tunnel in the stalk and ears.

Females lay 30–100 flat round creamy coloured eggs in a cluster on the underside of leaves. Larvae lack markings, their head capsule is dark brown in colour and body is creamy white and have between six and eight instars. Calatayud et al. (2014) summarizes the durations of different life stages. It takes around seven days for eggs to hatch, 27–19 days from neonate to fifth instar, seven days until pupation and finally 9–11 days before emerging as an adult. Late instar larvae diapause in tunnels in dry maize stalks. Adult moths have brownish grey forewings with dark markings and are 25–35 mm across (Polaszek 1998) (Fig. 14).

Figure 14 Maize stalk borer, *Busseola fusca* (Fuller) (Lepidoptera: Noctuidae) a) larva and b) adult, credit Annemie Erasmus – Agricultural Research Council, South Africa.

Historically, *B. fusca*, similar to other stem borers, was difficult to control with insecticides, particularly when feeding on maize ears (Polaszek 1998). Although this is a fairly aggressive pest of maize there is also some evidence of displacement by *C. partellus* in some areas of East Africa (e.g. Kriticos et al. 2015). Given the need to control both of these species in many African countries, the approval of Bt maize (Cry1Ab) for South Africa in 1998 was welcomed by maize farmers, and quickly adopted. Although the initial years of Bt maize use were quite successful, *B. fusca* developed resistance within approximately five years, and resistance is now evident in most maize growing regions of South Africa (van den Berg and Campagne 2015). The evolution of resistance in this case was likely due to several factors, including the lack of a high dose, the use of a single toxin (Cry1Ab) and the inherent ability of many species within the Noctuidae family to develop resistance to insecticidal toxins. However, an extensive survey of maize farmers indicated a surprisingly high proportion that did not follow the necessary refuge requirements (Kruger et al. 2012). As noted with *C. partellus*, the push-pull system has been evaluated for *B. fusca* as well (e.g. Khan et al. 1997, Kfir et al. 2005, Hassanali et al. 2008).

3.12 African pink stem borer *Sesamia calamistis* (Hampson) and *Sesamia* spp. (Lepidoptera: Noctuidae)

Sesamia calamistis is common throughout west and southern Africa, exploiting several wild grass species as well as maize, sorghum and sugarcane (Polaszek 1998). Although *S. calamistis* is often suppressed by early maize planting and biological control, this species can inflict damage to maize with early stage stalk feeding. The species is somewhat unique from other maize borers in that the larvae do not spend much time feeding on leaf/whorl tissue; larvae will penetrate the stem directly from behind leaf sheaths, where eggs are laid (Fig. 15). In addition, late-season boring within the stalks and ear feeding contribute to

Figure 15 African pink stem borer, *Sesamia calamistis* (Hampson), and *Sesamia* spp. (Lepidoptera: Noctuidae) a) eggs, b) larva and c) adult, credit Annemie Erasmus – Agricultural Research Council, South Africa.

reductions in yield and grain quality (e.g. Polaszek 1998). The species is also susceptible to Bt maize (Cry1Ab) in South Africa (van den Berg and Van Wyk 2007).

Two other related species, *S. nanogrioides* and *S. cretica*, are common in the Mediterranean region, including Spain (Velasco et al. 2007). *Sesamia cretica* occurs primarily in southern Spain, while *S. nonagrioides* is distributed throughout the maize growing regions of the country, and occurs sympatrically with *O. nubilalis*. In some years, the damage caused by *S. nonagrioides* was greater than that of *O. nubilalis* in Spain (Velasco et al. 2007). In recent years, Bt maize (Cry1Ab) has been effective in controlling both *Sesamia* spp. and *O. nubilalis* in Spain (Soberon et al. 2015).

4 Stored products pests

Insects can continue to damage and reduce maize production post-harvest. Stored product insect pests present an increasing challenge, especially in developing countries where two or more of the species may be present, including several countries in East Africa (Tefera et al. 2011). We briefly mention three cosmopolitan species which are considered major pests: *Sitophilus zeamais* Motschulsky, *Prostephanus truncatus* (Horn) and *Rhyzopertha dominica* (Fabricius) are small (3–4.5 mm) brown to black beetles. As a weevil, *S. zeamais* has an elongated snout shaped head. As typical to beetles in the family Bostrichidae, the heads of *P. truncates* and *R. dominica* are tucked under their prothorax, facing downward. All three species lay eggs in stored maize as well as other stored grains, though *S. zeamais* prefers maize. Larvae of all three species are white and grub shaped. Adults and larvae feed on grain and produce frass, both of which decreases the value of the final product.

5 Case study: management of European corn borer in the United States

Historically, this species was difficult to control in maize because of its cryptic life history, including the tendency, upon larval hatch, to relatively quickly bore into the whorl, main stalk or ears of plants. Once in the stalk or ear, plant injury was underway, and control options were limited (Mason et al. 1996, Steffey et al. 1999). Effective monitoring plans for larvae and economic thresholds were developed and validated. However, due to the necessary sampling time, many growers and crop consultants would focus primarily on whorl feeding only (first-generation generation damage), and often avoid sampling ear stage corn (second- or late generations) due to the difficulty and logistics of locating egg masses and larvae on tall, late-season maize plants. Further contributing to later-season damage, growers could only expect about 75% control from insecticides and it was often cost prohibitive to apply two or more sprays per year. Consequently, very little maize in the U.S. Corn Belt was regularly scouted, particularly for late-season infestations of *O. nubilalis*, and farmers reluctantly accepted the associated yield loss. With the commercialization of Bt maize, farmers now had a tool of convenience and high efficacy which targeted infestations season-long. Neonate larvae would be exposed to Bt proteins as soon as feeding commenced, larvae would stop feeding after taking a few bites, and usually die within 24 hours. This successfully prevented larvae from boring into the plant and causing significant injury.

As noted previously, the deployment of Bt maize for the European corn borer, *O. nubilalis*, in the United States, has thus far been considered a success (e.g. Huang et al. 2011), with both Bt maize and non-Bt maize growers reaping the economic benefits of long-term, consistent efficacy against this species, coupled with a lack of field-evolved resistance (Hutchison et al. 2010, Huang et al. 2011, Siegfried and Hellmich 2012). Compared to several case studies of field-evolved Bt maize resistance, such as *Spodoptera frugiperda* and *Busseola fusca* (Tabashnik et al. 2013, van den Berg and Campagne 2015), the success of Bt maize for *O. nubilalis* can be attributed to several critical assumptions being met for long-term insecticide resistance management IRM (Tabashnik et al. 2013, Soberon et al. 2015). For example, the high-dose refuge (HDR) strategy for resistance management, as established by the U.S. Environmental Protection Agency (EPA), includes the following five primary assumptions: 1) Bt transgenic plants express one or more toxins at a high dose that effectively kills both heterozygote resistant and susceptible larvae (operationally, a dose that is 25X greater than the known LC_{99} for a susceptible population, Fitt et al. 2004); 2) non-Bt maize refuge must be planted on each farm within ½ mile of the Bt maize (initially, 20–50% of the crop area, for northern and southern states, respectively; 3) random mating of moths emerging from non-Bt maize (and/or other non-Bt host plants), with any rare moths that might emerge from a nearby Bt field occurs; 4) that the initial gene frequency for resistant alleles be quite low (e.g. $<10^{-4}$); and 5) that inheritance for resistance would be recessive. Despite the substantial selection pressure underway (multiple generations per year for 20 years), annual resistance monitoring reported by Siegfried and Hellmich (2012) indicated only two reports of unexpected larval survival on Bt maize in the Midwestern U.S., including events expressing Cry1Ab (Minnesota, 2001) and Cry1F (Iowa, 2004). However, ongoing monitoring indicated a lack of sustained field populations in either area, possibly reflecting the influence of various fitness costs (Siegfried and Hellmich 2012). Several additional Bt toxins have been approved with unique modes of action, making it more difficult for *O. nubilalis* to develop resistance (Hellmich et al. 2008, Burkness et al. 2010, Siegfried and Hellmich 2012).

During the early years of Bt adoption, there were several million hectares of non-Bt maize available to serve as refugia for *O. nubilalis*, and grower compliance with non-Bt maize refuge requirements set by the EPA was high (e.g. Hutchison et al. 2010). These factors may have contributed to a lack of resistance evolution. In recent years, however, grower compliance with refuge requirements has declined (e.g. Goldberger et al. 2005). To address this IRM concern, Bt maize seed companies developed 'refuge in a bag' (RIB) seed mixtures. In the United States, Bt seed bags contained Bt/non-Bt seed blends, where typically 5% of the seed was non-Bt maize (e.g. Onstad et al. 2011). This approach became popular for addressing IRM for several pests. It was a convenient way for companies to automatically assure grower compliance, and more easily document compliance. For lepidopteran pests feeding on ears, however, concerns of cross-pollination of non-Bt ears with nearby Bt pollen challenge this strategy. Larvae feeding on such 'refuge' ears which express a mosaic of kernels with and without Bt proteins could compromise IRM for *O. nubilalis*, or any of the other ear feeding pests such as *Helicoverpa zea, H. armigera, Striacosta albicosta* and *Spodoptera frugiperda* (e.g. Yang et al. 2014, Huang et al. 2014, Burkness et al. 2015, Dively et al. 2016). Insect resistance management is a perpetual challenge for all chemical and genetic control methods.

Finally, and germane to IRM, most researchers believe that Bt maize is compatible with biological control, following extensive risk assessment research and meta-analysis reviews indicating that Bt proteins are not active on most generalist predators and parasitoids (e.g. Romeis et al. 2008a, Naranjo 2009). In addition, the ubiquitous corn borer pathogen, *Nosema pyrausta* (Zimmermann et al. 2016), has maintained activity in several Midwestern U.S. states, despite low *O. nubilalis* populations (White et al. 2014). Predators, parasitoids and pathogens are therefore important control methods that are complementary to Bt maize. In countries where Bt maize has not been approved, *O. nubilalis* remains a concern and biological control plays a dominant role in managing this species. For example, in Germany, biological control is encouraged, including the use of augmentative releases of *Trichogramma* spp., as egg parasitoids of *O. nubilalis* (Meissle et al. 2011).

6 Future trends and conclusion

After contrasting maize insect pest challenges in both industrialized and developing countries, it is clear that the following research needs will remain a priority over the next 5–10 years. The following suggestions will require considerable infrastructure support for cooperating universities (e.g. Juma 2011); international research centres (e.g. CGIAR system); and local Ministers of Agriculture, invasive species experts and extension/outreach agencies. The following areas of research should be viewed as complementary objectives:

1) Landscape ecology and pest biology – A greater understanding of habitat management, within a landscape ecology framework, is needed. A few challenges that this research could address include better management of stem boring pest species via expanded use of push-pull systems (e.g. Kfir et al. 2005, Hassanali et al. 2008), including the use of climate-smart push-pull, where more drought-tolerant and water-efficient legume and grass species are used. Diverse cropping systems are also needed, to maintain alternate hosts for the primary target pests, to help mitigate the risk of resistance to insecticides and GE technologies.

2) Genomics, transgenic, RNAi and CRISPR/Cas9 innovation – Novel genomics and population genetic research and development should receive increasing support to more rapidly identify plant genes for pest resistance, which can be coupled with conventional breeding to produce improved insect-maize resistant hybrids (Ricroch 2014, NAS 2016). Notably, the numerous GE maize hybrids expressing various Bt proteins (via *Bacillus thuringiensis*) available for stem borer management should receive serious consideration by regulatory authorities in more African and Asian countries, to improve maize yields on an area-wide basis (e.g. Fitt et al. 2004, Romeis et al. 2008b, van den Berg and Campagne 2015). For example, even where Bt maize may be limited to larger farms, potential area-wide suppression effects of stemborers in a region could benefit smallholder farmers.

3) Invasive species – Early detection and proper identification of invasive arthropod species, and associated plant pathogens, is not only an ongoing need in industrialized countries but a tremendous challenge for developing countries, where regulatory staff and pest detection systems are often limited. Pest species may go undetected for several years because of inadequate monitoring systems, resulting in unnecessary maize losses and delays in timely research and sharing of knowledge (e.g. Kriticos et al. 2015, Venette 2015).

4) Climate change – Effects of climate change must continue to be monitored. Using appropriately downscaled data coupled with insect population models, range expansion of endemic and invasive maize insect pests can be studied in relation to changes in ambient temperature and rainfall frequency. This research is essential for addressing

climate adaptation strategies and understanding the changing plant-insect interaction impacts on crop yields that can be tailored to specific countries (e.g. Both et al. 2009).

5) Biological control – Continued support for the exploration and conservation of new biological control agents should be encouraged for compatibility with enhanced pest-resistant hybrids, and reduced reliance on insecticidal inputs which are often prohibitively expensive and dangerous to human and environmental health in developing countries (e.g. Polaszek 1998, Norton et al. 2005).

6) Post-harvest insect control – A renewed commitment must be made for innovative, cost-effect management solutions for insect pests of stored grain, particularly for maize in developing countries. Greater yield and quality gains at harvest will only enhance food security to a certain extent. Post-harvest grain protection remains a challenge and must be addressed to give farmers the opportunity for improved incomes via market timing, as well as increased odds of adequate family nutrition throughout the year. Recent studies have documented a wide range in post-harvest grain losses (20–60%) in several countries (e.g. Abass et al. 2014, Affognon et al. 2015).

In summary, despite the many challenges to sustainable maize IPM, the outlook for improved maize productivity is promising, with diverse research opportunities and encouraging new technologies to address insect pests of maize. The future also bodes well for career opportunities for students interested in all aspects of innovative IPM, for both private and public sector positions (see also Juma 2011).

7 Where to look for further information

For a comprehensive review of primary and secondary arthropod pests and beneficial insects of maize in North America, including colour photographs, see Steffey et al. (1999). For a review of the African stem borers, their taxonomy, biological control and IPM, see Polaszek (1998). van den Berg et al. (2014) also provided a comprehensive review of the maize pests of South Africa. In addition, a global up-to-date resource on the distribution and literature for insect pests of maize, given that many are invasive species, is available online at CABI (2017). This also includes arthropod and pathogen pests for most agricultural crops. The international research centre, CIMMYT (*Centro Internacional de Mejoramiento de Maíz y Trigo*), continues to be active in developing pest and disease-resistant germplasm, as well as drought-tolerant maize hybrids (see CIMMYT 2016). An excellent resource for understanding the advantages and challenges of GE maize and integrating GE crops in IPM programmes was compiled by Hellmich et al. (2008), Romeis et al. (2008b) and Ricroch et al. (2014). For an overview of current and emerging transgenic crop technology and strategies, including an overview of RNAi, readers should consider *Emerging Technologies to Benefit Farmers in Sub-Saharan Africa and South Asia* (NAS 2009). Annual updates on transgenic maize and crop adoption globally is provided by the *International Service for the Acquisition of Agri-Biotech Applications* (ISAAA) (see James 2015). The most recent global review of the benefits and risks of GE crops was published by the National Academy of Sciences, titled *Genetically Engineered Crops: Experiences and Prospects* (NAS 2016). For all insect-resistant transgenic crops, and particularly for insect pests of Bt maize, the texts by Onstad (2014) and Soberon et al. (2015) provide comprehensive reviews of the science and practical implications necessary for effective IRM. Three general resources for a review of numerous IPM tactics and strategies, including maize examples, are provided by Koul et al. (2008), Radcliffe et al. (2009), and

Pedigo and Rice (2009). Finally, Norton et al. (2005), following many years of international development experience, review the importance of considering socio-economic and gender roles in successful IPM implementation (maize and other crops), including IPM impact evaluation in Africa and South Asia.

8 References

Abass, A. B., G. Ndunguru, P. Mamiro, B. Alenkhe, N. Mlingi and M. Bekunda. 2014. Post-harvest food losses in a maize-based farming system of semi-arid savannah area of Tanzania. *J. Stored Prod. Res.* 57: 49–57.

Alcantara, E., A. Estrada, V. Alpuerto and G. Head. 2011. Monitoring Cry1Ab susceptibility in Asian corn borer (Lepidoptera: Crambidae) on Bt corn in the Philippines. *Crop Prot.* 30(5): 554–9.

Affognon, H., C. Mutungi, P. Sanginga and C. Borgemeister. 2015. Unpacking postharvest losses in Sub-Saharan Africa: A meta-analysis. *World Dev.* 66: 49–68.

Barsics, F., E. Haubruge and F. J. Verheggen. 2013. Wireworm management: An overview of the existing methods, with particular regards to *Agriotes* spp. (Coleoptera: Elateridae). *Insects.* 4:117–52. doi:10.3390/insects4010117.

Bartels, D. W., W. D. Hutchison and S. Udayagiri. 1997. Pheromone trap monitoring of Z-strain European corn borer (Lepidoptera: Pyralidae): Optimum pheromone blend, comparison with light traps, and trap number requirements. *J. Econ. Entomol.* 90: 449–57.

Bentivenha J. P.F., S. V. Paula-Moraes, E. L. L. Baldin, A. Specht, I. F. da Silva and T. E. Hunt. 2016. Battle in the New World: *Helicoverpa armigera* versus *Helicoverpa zea*(Lepidoptera: Noctuidae). *PLoS One* 11(12): e0167182. doi:10.1371/journal.pone.0167182.

Bohnenblust E. W., J. A. Breining, J. A. Shaffer, S. J. Fleischer and et al. 2014. Current European corn borer, *Ostrinia nubilalis*, injury levels in the north eastern United States and the value of Bt field corn. *Pest Manag Sci.* 70: 1711–19.

Both, C., M. van Asch, R. G. Bijlsma, A. B. van den Burg and M. E. Visser. 2009. Climate change and unequal phenological changes across four trophic levels: Constraints or adaptations? *J. Anim. Ecol.* 78: 73–83.

Burkness, E. C., G. Dively, T. Patton, A. C. Morey and W. D. Hutchison. 2010. Novel Vip3A *Bacillus thuringiensis* (Bt) maize approaches high-dose efficacy against *Helicoverpa zea* (Lepidoptera: Noctuidae) under field conditions: Implications for resistance management. *GM Crops.* 1(5): 337–343.

Burkness, E. C., T. M. Cira, S. E. Moser and W. D. Hutchison. 2015. Bt Maize seed mixtures for *Helicoverpa zea* (Lepidoptera: Noctuidae): Larval movement, development, and survival on non-transgenic maize. *J. Econ. Entomol.* 108(6): 2761–9.

CABI. 2016a. Invasive Species Compendium: European corn borer. On-line: http://www.cabi.org/isc/datasheet/46129 (accessed 12 December 2016).

CABI. 2016b. Invasive Species Compendium: Asian corn borer: On-line: http://www.cabi.org/cpc/search/?q=Asian+corn+borer (accessed 15 December 2016).

CABI. 2016c. Invasive Species Compendium: Spotted stem borer. On-line: http://www.cabi.org/isc/datasheet/12859 (accessed 12 December 16).

CABI. 2016d. Invasive Species Compendium: *Busseola fusca*. On-line: http://www.cabi.org/isc/search/?q=Busseola+fusca&types=26.

CABI. 2017. Invasive Species Compendium. http://www.cabi.org/isc/ (accessed 15 January 2017).

Calatayud, P.-A., B. Le Ru, J. van den Berg and F. Schulthess. 2014. Ecology of the African maize stalk borer, *Busseola fusca* (Lepidoptera: Noctuidae), with special reference to insect-plant interactions. *Insects.* 5: 539–63.

Calvin, D. D. and P. Z. Song. 1994. Variability in postdiapause development periods of geographically separate *Ostrinia nubilalis* (Lepidoptera: Pyralidae) populations in Pennsylvania. *Environ. Entomol.* 23: 431–6.

CIMMYT. 1989. Toward insect resistant maize for the third world. In: Proceedings of the international symposium on methodologies for developing host plant resistance to maize insects, p. 175. CIMMYT (International Maize and Wheat Improvement Center), Mexico.

CIMMYT. 2016. Global Maize Research. http://www.cimmyt.org/global-maize-research/ CIMMYT, Mexico. (accessed 15 December 16).

Coates B. S., H. Johnson, K-S. Kim, et al. 2013. Frequency of hybridization between *Ostrinia nubilalis* E-and Z-pheromone races in regions of sympatry within the United States. *Ecol.Evol.* 3: 2459–70.

Czepak, C., K. C. Albernaz, L. M. Vivan, H. O. Guimarães and T. Carvalhais. 2013. First reported occurrence of *Helicoverpa armigera* (Hübner) (Lepidoptera: Noctuidae) in Brazil. *Pesqui. Agropecu. Trop.* 43: 110–13.

Dively G. P., P. D. Venugopal and C. Finkenbinder. 2016. Field-evolved resistance in corn earworm to Cry proteins expressed by transgenic sweet corn. *PLoS One* 11(12): e0169115. doi:10.1371/journal.pone.0169115.

El-Ibrashy, M. T., S. El-Zaidy and A. A. Riad. 1972. Laboratory studies on the biology of the corn leaf aphid, *Rhopalosiphum maidis* (Homoptera: Aphididae). *Ent. Exp. Appl.* 15:166–74.

EPPO Global Database. 2017. Helicoverpa armigera. (on-line: https://gd.eppo.int/taxon/HELIAR/distribution (accessed 21 April 2017).

FAO. 2014. Global crop statistics (see maize), on-line: http://www.fao.org/faostat/en/#home (accessed, 18 December 2016).

Fisher, J. R. 1989. Hatch of *Diabrotica virgifera virgifera* (Coleoptera: Chrysomelidae) eggs exposed to two different overwintering and hatch regimes. *J. Kansas Entomol. Soc.* 62: 607–10.

Fitt, G. P., D. A. Andow, Y. Carriere, W. J. Moar, T. H. Schuler, C. Omoto, et al. 2004. Resistance risks and management associated with Bt maize in Kenya, pp. 209–50. In: A. Hillbeck and D. A. Andow (Eds), *Environmental Risk Assessment of Genetically Modified Organisms: A Case Study of Bt Maize in Kenya (Vol. 1)*, p. 281. CABI: UK.

French, B. W., B. S. Coates and T. W. Sappington. 2014. Inheritance of an extended diapause trait in the Northern corn rootworm, *Diabrotica barberi* (Coleoptera: Chrysomelidae). *J. Appl. Entomol.* 138: 213–21.

Gassmann A. J., J. L. Petzold-Maxwell, R. S. Keweshan and M. W. Dunbar. 2011. Field-evolved resistance to Bt maize by Western corn rootworm. *PLoS One* 6(7): e22629. doi:10.1371/journal.pone.0022629.

Goergen, G., P. L. Kumar, S. B. Sankung, A. Togola and M. Tamò. 2016. First report of outbreaks of the Fall Armyworm, *Spodoptera frugiperda* (J E Smith) (Lepidoptera, Noctuidae), a new alien invasive pest in West and Central Africa. *PLoS One* 11(10): e0165632. doi:10.1371/journal.pone.0165632.

Goldberger, J., J. Merrill and T. Hurley. 2005. Bt corn farmer compliance with insect resistance management requirements in Minnesota and Wisconsin. *AgBioForum* 8(2&3): 151–60.

Gómez-Barbero, M., J. Berbel and E. Rodríguez-Cerezo. 2008. Bt corn in Spain—the performance of the EU's first crop. *Nat. Biotechnol.* 26: 384–6.

Gray, M. E., T. W. Sappington, N. J. Miller, J. Moeser, and M. O. Bohn. 2009. Adaptation and invasiveness of Western corn rootworm: Intensifying research on a worsening pest. *Ann. Rev. Entomol.* 54: 303–21.

Hanson, A. A., R. D. Moon, R. J. Wright, T. E. Hunt and W. D. Hutchison. 2015. Degree-day prediction models for the flight phenology of western bean cutworm (Lepidoptera: Noctuidae) assessed with the concordance correlation coefficient. *J. Econ. Entomol.* 108(4):1728–38.

Hardwick, D. F. 1965. The Corn Earworm Complex. *Mem. Entomol. Soc. Canada.* 97: 1–247.

Hassanali, A., H. Herren, Z. R. Khan, J. A. Pickett and C. M. Woodcock. 2008. Integrated pest management: The push–pull approach for controlling insect pests and weeds of cereals, and its potential for other agricultural systems including animal husbandry. *Phil. Trans. R Soc. Lond. B Biol. Sci.* 363(1491): 611–21. doi:10.1098/rstb.2007.2173.

Hellmich, R. L., R. Albajes, D. Bergvinson, J. R. Prasifka, Z-Y. Wang and M. J. Weiss. 2008. The present and future role of insect-resistant genetically modified maize in IPM. In J. Romeis, A. M. Shelton and G. G. Kennedy (Eds), *Integration of Insect-Resistant Genetically Modified Crops within IPM Programs*. Springer: Germany.

Huang, F., D. A. Andow and L. L. Buschman. 2011. Success of the high dose/refuge resistance management strategy after 15 years of Bt crop use in North America. *Entom. Exp. App.* 140: 1–16.

Huang, F., M. N. Ghimire, B. R. Leonard, C. Daves, R. Levy and J. Baldwin. 2012. Extended monitoring of resistance to *Bacillus thuringiensis* Cry1Ab maize in *Diatraea saccharalis* (Lepidoptera: Crambidae). *GM Crops*. 3: 245–54.

Huang, F., J. A. Qureshi, R. L. Meagher Jr., D. D. Reisig, G. P. Head and et al. 2014. Cry1F resistance in fall armyworm *Spodoptera frugiperda*: Single gene versus pyramided Bt maize. *PLoS One* 9(11): e112958. doi:10.1371/journal.pone.0112958.

Hutchison, W. D., B. Flood and J. A. Wyman. 2004. Advances in United States Sweet Corn and Snap Bean Insect Pest Management, pp. 247–78. In A. R. Horowitz and I. Ishaaya (Eds), *Insect Pest Management*. Springer-Verlag: Heidelberg, Germany.

Hutchison, W. D., E. C. Burkness, P. D. Mitchell, R. D. Moon, T. W. Leslie, S. J. Fleischer, M. Abrahamson, K. L. Hamilton, K. L. Steffey, M. E. Gray, R. L. Hellmich, L. V Kaster, T. E. Hunt, R. J. Wright and et al. 2010. Area wide suppression of European corn borer with Bt maize reaps savings to non-Bt maize growers. *Science*. 330: 222–5.

Hutchison, W. D. 2015. Insect resistance management and integrated pest management: Prospects for an area-wide view, pp. 186–201. In M. Soberon, Y. Gao and A. Bravo (Eds), *Bt Resistance: Characterization and Strategies for GM Crops Producing Bacillus thuringiensis toxins*. CABI: Wallingford, UK.

Iowa State University (ISU), Department of Entomology. 2015. The European corn borer, on-line: http://www.ent.iastate.edu/pest/cornborer/insect/lifecycle (accessed 15 December 2016).

Jackson, J. J. and N. C. Elliott. 1988. Temperature-dependent development of immature stages of the western corn rootworm, *Diabrotica virgifera virgifera* (Coleoptera: Chrysomelidae). *Environ. Entomol.* 17: 166–71.

Jacobson, A., R. Foster, C. Krupke, W. Hutchison, B. Pittendrigh and R. Weinzierl. 2009. Resistance to pyrethroid insecticides in Helicoverpa zea (Lepidoptera: Noctuidae) in Indiana and Illinois. *J. Econ. Entomol.* 102(6): 2289–95.

Jalali, S. and S. Singh. 2001. Studies on thermal requirements for estimating the number of generations of *Chilo partellus* (Swinhoe) and its natural enemies in the field. *Ann. Plant Prot. Sci.* 9: 213–19.

Jallow, M. F. A. and M. Matsumura. 2001. Influence of temperature on the rate of development of *Helicoverpa armigera* (Hübner) (Lepidoptera: Noctuidae). *Appl. Entomol. Zool.* 36: 427–30.

James, C. 2015. *20th Anniversary (1996 to 2015) of the Global Commercialization of Biotech Crops and Biotech Crop Highlights in 2015. ISAAA Brief* No. 51. ISAAA: Ithaca, NY.

Juma, C. 2011. *The New Harvest: Agricultural Innovation in Africa, 268*. Oxford Univ. Press: UK.

Khan, Z. R., K. Ampong-Nyarko, P. Chiliswa, A. Hassanali, S. Kimani, W. Lwande, W. A. Overholt, J. A. Pickett, et al. 1997. Intercropping increases parasitism of pests. *Nature*. 388: 631–2.

Kfir, R., W. A. Overholt, Z. R. Khan, and A. Polaszek. 2005. Biology and management of economically important Lepidopteran cereal stem borers in Africa. *Ann. Rev. Entomol.* 47: 701–31.

Kiss, J., C. Edwards, H. Berger, P. Cate, M. Cean, S. Cheek, J. Derron, H. Festic, L. Furlan and et al. 2005. Monitoring of western corn rootworm (*Diabrotica virgifera virgifera* LeConte) in Europe 1992–2003, pp. 29–39. In Vidal, S., Kuhlmann, U. and Edwards, C. (Eds), *West. Corn Rootworm Ecol. Manag.* CAB International: Wallingford, UK.

Koul, O., G. W. Cuperus and N. Elliott (Eds). 2008. *Areawide Pest Management: Theory and Practice*. CABI: Wallingford, UK, p. 590.

Kranthi, K. R., D. Russell, R. Wanjari, M. Kherde, S. Munje, N. Lavhe and N. Armes. 2002. In-season changes in resistance to insecticides in *Helicoverpa armigera* (Lepidoptera: Noctuidae) in India. *J. Econ. Entomol.* 95: 134–42.

Kriticos, D. J., N. Ota, W. D. Hutchison, J. Beddow, T. Walsh and et al. 2015. The potential distribution of invading *Helicoverpa armigera* in North America: Is it just a matter of time? PLoS One 10(3): e0119618. doi:10.1371/journal.pone.0119618.

Kruger, M., J. B. J. Van Rensburg and J. van den Berg. 2012, Transgenic Bt maize: farmers' perceptions, refuge compliance and reports of stem borer resistance in South Africa. *J. Appl. Entomol.* 136: 38–50.

Kuhar, T. P., W. D. Hutchison, J. Whalen, D. G. Riley, J. C. Meneley, H. B. Doughty, E. C. Burkness and S. J. Wold-Burkness. 2006. Field evaluation of a novel lure for trapping seed corn maggot adults. *Plant Health Progress*. doi:10.1094/PHP-2006–0606-01-BR. (on-line: https://www.plantmanagementnetwork.org/pub/php/brief/2006/lure/).

Kuo, M. H., M. C. Chiu and J. J. Perng. 2006. Temperature effects on life history traits of the corn leaf aphid, *Rhopalosiphum maidis* (Homoptera: Aphididae), on corn in Taiwan. *Appl. Entomol. Zool.* 41: 171–7.

Lingren, P. D., V. M. Bryant, J. R. Raulston, M. Pendleton, J. Westbrook and G. D. Jones. 1993. Adult feeding host range and migratory activities of corn earworm, cabbage looper, and celery looper (Lepidoptera: Noctuidae) moths as evidenced by attached pollen. *J. Econ. Entomol.* 86: 1429–39.

Mason, C. E., M. E. Rice, D. D. Calvin, J. W. Van Duyn, W. B. Showers, W. D. Hutchison, J. F. Witkowski, R. A. Higgins, D. W. Onstad and G. P. Dively. 1996. *European Corn Borer Ecology and Management*. North Central Regional Extension Publication No. 327. Iowa State University: Ames.

Meissle, M., J. Romeis and F. Bigler. 2011. Bt maize and integrated pest management – a European perspective. *Pest Manag. Sci.* 67: 1049–58.

Michel, A. P., C. H. Krupke, T. S. Baute and C. D. Difonzo. 2010. Ecology and management of the Western Bean Cutworm (Lepidoptera: Noctuidae) in corn and dry Beans. *J. Integr. Pest Manag.* 1(1): A1–A10. doi:https://doi.org/10.1603/IPM10003.

Milne, A. E., J. R. Bell, W. D. Hutchison, F. van den Bosch, P. D. Mitchell, D. Crowder and et al. 2015. The effect of farmers' decisions on pest control with Bt crops: A billion dollar game of strategy. *PLoS Comput. Biol.* 11(12): e1004483. doi:0.1371/journal. pcbi.1004483.

Morey, A. C. 2010. Corn earworm (*Helicoverpa zea* Boddie), cold hardiness, and climate change: Implications for future distributions and IPM. M.S. Dissertation, University of Minnesota, St. Paul, USA.

Morey, A. C., W. D. Hutchison, R. C. Venette and E. C. Burkness. 2012. Cold hardiness of corn earworm (Lepidoptera: Noctuidae) pupae. *Environ. Entomol.* 41(1): 172–9.

Munkvold, G. P., R. L. Hellmich and L. G. Rice. 1999. Comparison of fumonisin concentrations in kernels of transgenic Bt maize hybrids and nontransgenic hybrids. *Plant Dis.* 83(2): 130–8.

Murúa, M. G., F. S. Scalora, F. R. Navarro, L. E. Cazado, A. Casmuz, M. E. Villagran, E. Lobos and G. Gastaminza. 2014. First record of *Helicoverpa armigera* (Lepidoptera: Noctuidae) in Argentina. *Florida Entomol.* 97: 854–6.

Mutuura, A. and E. Munroe. 1970. Taxonomy and distribution of the European corn borer and allied species: Genus *Ostrinia* (Lepidoptera: Pyralidae). *Mem. Entomol. Soc. of Canada.* 71:1–112.

Nafus, D. M. and I. H. Schreiner. 1991. Review of the biology and control of the Asian corn borer, *Ostrinia furnacalis* (Lep: Pyralidae). *Trop. Pest Manag.* 37:41–56.

Naranjo, S. E. 2009. Impacts of Bt crops on non-target organisms and insecticide use patterns. *CAB Reviews: Perspectives in Agriculture, Veterinary Science, Nutrition and Natural Resources* 4, No. 011 (doi: 10.1079/PAVSNNR20094011) (on-line: http://www.cabi.org/cabreviews/FullTextPDF/2009/20093131779.pdf).

NAS (National Academy of Sciences). 2009. *Emerging Technologies to Benefit Farmers in Sub-Saharan Africa and South Asia*. The National Academies Press: Washington, D.C., p. 269.

NAS (National Academy of Sciences, Engineering and Medicine). 2016. *Genetically Engineered Crops: Experiences and Prospects*. The National Academies Press: Washington, D.C., p. 584. doi:10.17226/23395.

Norton, G. W., E. A. Heinrichs, G. C. Luther and M. E. Irwin. 2005. *Globalizing Integrated Pest Management: A Participatory Research Process*. Blackwell Publishing Ltd.: Oxford UK, p. 338.

Olmstead, D. L., B. A. Nault and A. M. Shelton. 2016. Biology, ecology, and evolving management of *Helicoverpa zea* (Lepidoptera: Noctuidae) in sweet corn in the United States. *J. Econ. Entomol.* 109(4): 1667–76.

Onstad, D. W., P. D. Mitchell, T. M. Hurley, J. G. Lundgren, R. P. Porter, C. H. Krupke, J. L. Spencer, C. D. DiFonzo, T. S. Baute, R. L. Hellmich, L. L. Buschman, W. D. Hutchison and J. F. Tooker. 2011. Seeds of change: Corn seed mixtures for resistance management and IPM. *J. Econ. Entomol.* 104(2): 343–52.

Paula-Moraes, S., E. C. Burkness, T. E. Hunt, R. J. Wright, G. L. Hein and W. D. Hutchison. 2011. Cost-effective binomial sequential sampling of Western bean cutworm, *Striacosta albicosta* (Lepidoptera: Noctuidae), egg masses in corn. *J. Econ. Entomol.* 104(6): 1900–8.

Pedigo, L. P. and M. E. Rice. 2009. *Entomology and Pest Management.* Pearson-Prentice Hall: London, p. 784.

Polaszek, A. (Ed.). 1998. *African Cereal Stem Borers: Economic Importance, Taxonomy, Natural Enemies and Control.* CABI: Wallingford, UK, p. 530.

Qureshi, J. A., L. L. Buschman, J. E. Throne and S. B. Ramaswamy. 2006. Dispersal of adult *Diatraea grandiosella* (Lepidoptera: Crambidae) and its implications for corn borer resistance management in *Bacillus thuringiensis* maize. Ann. Entomol. Soc. Am. 99: 279–91.

Radcliffe, E. B., W. D. Hutchison and R. Cancelado (Eds). 2009. *Integrated Pest Management: Concepts, Tactics, Strategies and Case Studies.* Cambridge Univ. Press: UK, p. 450.

Reisig, D. and F. P. F. Reay-Jones, 2015. Inhibition of *Helicoverpa zea* (Lepidoptera: Noctuidae) growth by transgenic corn expressing Bt toxins and development of resistance to Cry1Ab. *Environ. Entomol.* 44(4): 1275–85.

Ricroch, A., S. Chopra and S. J. Fleischer (Eds). 2014. *Plant Biotechnology: Experience and Future Prospects.* Springer: Germany, p. 291.

Romeis, J., D. Bartsch, F. Bigler, M. P. Candolfi and et al. 2008a. Assessment of risk of insect-resistant transgenic crops to nontarget arthropods. *Nat. Biotech.* 26: 203–8.

Romeis, J., A. M. Shelton and G. G. Kennedy (Eds), 2008b. *Integration of Insect-Resistant Genetically Modified Crops within IPM Programs.* Springer, Germany, p. 441.

Rose, D. J. W. 1973. Laboratory observations on the biology of *Cicadulina* spp. (Hom., Cicadellidae), with particular reference to the effects of temperature. *Bull Entomol Res.* 62: 471–6.

Schaafsma, A. W., J. D. Fuentes, T. J. Gillespie, G. H. Whitfield and C. R. Ellis. 1993. Performance of a model for egg hatching of the western corn rootworm, *Diabrotica virgifera virgifera* LeConte, using measured and modelled soil temperatures as input. *Int. J. Biometeorol.* 37: 11–18.

Shelton, A. M., D. L. Olmstead, E. C. Burkness, W. D. Hutchison, G. Dively, C. Welty and A. N. Sparks. 2013. Multi-state trials of Bt sweet corn varieties for control of the corn earworm, *Helicoverpa zea* (Lepidoptera: Noctuidae). *J. Econ. Entomol.* 106(5): 2151–9.

Siegfried, B. D. and R. L. Hellmich. 2012. Understanding successful resistance management: The European corn borer and Bt corn in the United States. *GM Crops.* 3:184–93.

Smith, E. 2015. *Helicoverpa armigera* (Old World Bollworm) - Detection in Florida. NAPPO. (http://www.pestalert.org/oprDetail.cfm?oprID=629&keyword=armigera).

Soberon, M., Y. Gao and A. Bravo (Eds). 2015. *Bt resistance: Characterization and Strategies for GM Crops Producing Bacillus thuringiensis toxins.* CABI: Wallingford, UK.

Steffey, K. L., M. E. Rice, J. All, D. A. Andow, M. E. Gray and J. W. Van Duyn (Eds). 1999. *Handbook of Corn Insects.* Entomological Society of America: Lanham, MD, USA, p. 164.

Tabashnik, B. E., T. Brévault and Y. Carrière. 2013. Insect resistance to Bt crops: Lessons from the first billion acres. *Nat. Biotech.* 31: 510–21. doi:10.1038/nbt.2597.

Tefera, T., S. Mugo and P. Likhayo. 2011. Effects of insect population density and storage time on grain damage and weight loss in maize due to the maize weevil *Sitophilus zeamais* and the larger grain borer *Prostephanus truncates. Afr. J. Agr. Res.* 6: 2249–54.

Tefera, T., S. Mugo, M. Mwimali, B. Anani, R. Tende, Y. Beyene, S. Gichuki, S. O. Oikeh, et al. 2016. Resistance of Bt-maize (MON810) against the stem borers *Busseola fusca* (Fuller) and *Chilo partellus* (Swinhoe) and its yield performance in Kenya. *Crop Prot.* 89: 202–8.

Twine, P. 1978. Effect of temperature on development of larvae and pupae of the corn earworm, *Heliothis armigera* (Hübner) (Lepidoptera: Noctuidae). *Queensl. J. Agric. Anim. Sci.* 35: 23–8.

USDA-APHIS. 2015. Old world bollworm Puerto Rico 2014–2015. On-line: (https://www.aphis.usda.gov/plant_health/plant_pest_info/owb/downloads/program-summary.pdf).

van den Berg, J., A. Erasmus and M. Van Rooyen. 2014. Maize. In Prinsloo, G. L. and Uys, V. M. (Eds), *Insects of Cultivated Plants and Natural Pastures in Southern Africa*, pp. 88–118. Entomological Society of Southern Africa: Pretoria.

van den Berg, J. and A. Van Wyk. 2007. The effect of Bt maize on *Sesamia calamistis* in South Africa. *Entomol. Exp. et Appl.* 122: 45–51.

van den Berg, J. and P. Campagne. 2015. Resistance of *Busseola fusca* to Cry 1Ab Bt maize plants in South Africa and challenges to insect resistance management in Africa, pp. 36–48. In M. Soberon, Y. Gao and A. Bravo (Eds), *Bt Resistance: Characterization and Strategies for GM Crops Producing Bacillus Thuringiensis Toxins*. CABI: Wallingford, UK.

Velasco, P., P. Revilla, L. Monetti, A. Butron, A. Ordas and R. A. Malvar. 2007. Corn borers (Lepidoptera: Noctuidae; Crambidae) in Northwestern Spain: Population dynamics and distribution. *Maydica*. 52: 195–203.

Venette, R. C. (Ed.) 2015. *Pest Risk Modelling and Mapping for Invasive Alien Species*. CABI: Wallingford, UK, p. 256.

Westbrook, J. K., W. W. Wolf, P. D. Lingren, J. R. Raulston, J. D. Lopez, J. H. Matis, R. S. Eyster, J. F. Esquivel and P. G. Schleider. 1997. Early-season migratory flights of corn earworm (Lepidoptera: Noctuidae). *Environ. Entomol.* 26: 12–20.

White, J. A., E. C. Burkness and W. D. Hutchison. 2014. Biased sex ratios, mating frequency and *Nosema* prevalence in European corn borer, at low population densities. *J. Appl. Entomol.* 138:195–201.

Worku, M., S. Twumasi-Afriyie, L. Wolde, B. Tadesse, G. Demisie and et al. (Eds). 2012. Meeting the Challenge of Global Climate Change and Food Security through Innovative Maize Research. Proceedings of the Third National Maize Workshop in Ethiopia, April 18–20 2011. Addis Ababa, Ethiopia. CIMMYT, Mexico City.

Yang, F., D. L. Kerns, G. P. Head, B. R. Leonard, R. Levy, Y. Niu and F. Huang. 2014. A challenge for the seed mixture refuge strategy in Bt maize: Impact of cross-pollination on an ear-feeding pest, corn earworm. *PLoS One* 9(11): e112962. doi:10.1371/journal.pone.0112962.

Yonow, T., D. J. Kriticos, N. Ota, J. van Den Berg and W. D. Hutchison. 2017. The potential global distribution of *Chilo partellus*, including consideration of irrigation and cropping patterns. *J. Pest Sci.* 90(2): 459–77. doi:10.1007/s10340-016-0801-4.

Zhang, Z., X. Zhang, Y. Zhao, W. Mu and F. Liu. 2017. Efficacy of insecticidal seed treatments against the wireworm *Pleonomus canaliculatus* (Coleoptera: Elateridae) in China. *Crop Prot.* 92: 134–42.

Zhu, Y-C., C. A. Blanco, M. Portilla, J. Adamczyk, R. Luttrell and F. Huang. 2015. Evidence of multiple/cross resistance development to Bt and organophosphate insecticides in Puerto Rico population of the fall armyworm, *Spodoptera frugiperda*. *Pesti. Biochem. Physiol.* 122:15–21.

Zimmermann, G., A. M. Huger, G. A. Langenbruch, et al. 2016. Pathogens of the European corn borer, *Ostrinia nubilalis*, with special regard to the microsporidium *Nosema pyrausta*. *J. Pest. Sci.* 89: 329–46. doi:10.1007/s10340-016-0749-4.

Zukoff, S. N., K. R. Ostlie, B. Potter, L. N. Meihls, A. L. Zukoff, L. French, M. R. Ellersieck, B. W. French, and B. E. Hibbard. 2016. Multiple assays indicate varying levels of cross resistance in Cry3Bb1-selected field populations of the Western corn rootworm to Cry3A, eCry3.1Ab, and Cry34/35Ab1. *J. Econ. Entomol.* 109(3): 1387–1398.

Chapter 14

Nematodes associated with maize

T. L. Niblack, The Ohio State University, USA

1 Introduction

Maize is one of the most important crops in the world, regardless of how one measures 'importance'. As a graminaceous plant that produces grain and silage in amounts and nutritive value competitive with other crop plants in the Poaceae, which can be grown under an amazing variety of climates and variations within climates, maize deserves the attention it has received from scientists dedicated to maximizing yields under current and future conditions. The term 'cultivar' is used in this chapter as a general label for maize hybrids, inbreds, open-pollinated varieties and tropical accessions per Baldwin and Barker (1970).

Abiotic and biotic limitations to maize yield come in many forms, and one of the least-understood categories is that of plant-feeding nematodes, referred to somewhat misleadingly as 'herbivores' in some of the literature of the past 50 years or so. Plant-feeding, or plant-parasitic, nematodes are ubiquitous in soils that support plant growth, and it is a very rare occurrence that one does not observe at least one species in association with the roots or shoots of maize. However, the macro-effects of phytophagous nematodes are notoriously difficult to test and measure (Norton, 1984), and those that have been investigated have shown that such effects are dependent on the multitude of variations in the host plant and its milieu, including the species and even subspecific variation in the associated nematodes. It is for that reason that the title of this chapter was chosen.

http://dx.doi.org/10.19103/AS.2016.0002.17

Some excellent and informative reviews of the literature related to maize–nematode associations have been published (Norton, 1983, 1984; Todd, 2016; Rivoal and Cook, 1993; Swarup and Sosa-Moss, 1990; Windham, 1998) and these are instructive both for the amount of information they contain and for indirectly highlighting the relative lack of information we have that clearly addresses nematode management in maize, or even the need for such. Indeed, the assumption that nematodes may be the cause of maize yield loss in many papers was based on the observation of increased yields following treatment with nematicides (Norton, 1984). Maize is an interesting plant in that it seems to have the ability, barring other stresses, to support population densities of nematodes that we would otherwise consider to be highly injurious to the host. For example, Norton (1983) noted that maize would tolerate 3000 unspecified lesion nematodes/g dry root, a level well above the commonly agreed upon (at the time) 'damage threshold' of 100 lesion nematodes/100 cm^3 soil (D. Edwards, *unpublished*). We have moved away from the 'damage threshold' concept as a trigger for recommendations for nematode management in maize in the Midwestern US.

2 Soil and plant nematodes: characteristics, identification and effects on maize

2.1 General characteristics of soil and plant nematodes

Nematodes are the major constituents of the soil meiofauna, invertebrates so categorized because they are larger than microfauna (visible only with a microscope) and smaller than macrofauna (visible to the naked eye). In general, a standard set of protocols is required to extract nematodes efficiently from soil and plant tissue (e.g. Barker et al., 1985; Perry and Moens, 2006; Shurtleff and Averre, 2000) and dedicated training and practice are necessary to distinguish their identifying characters.

Nematodes have six life stages: the egg stage, in which the embryos are surrounded by a pliant eggshell and progress to either the first or the second juvenile stage (J1 and J2, respectively) before emergence; two successive juvenile stages after the J2, namely J3 and J4; and the adult stage. Each life stage is punctuated by a moult of the cuticle, the outer covering of the worm. Nematodes do not undergo metamorphosis during development to the adult and therefore the first four stages are more correctly termed 'juveniles' rather than 'larvae'. Detailed information on the physiology of nematodes is available in a number of sources, for example, Perry and Wright (1998), and on plant-parasitic nematodes, for example, Perry and Moens (2006).

Plant-parasitic nematodes feed on plants by means of the stylet (Fig. 1), which is connected to the anterior cephalic (head) structure by muscles that allow the nematode to thrust the stylet into plant cell walls and thereby both inject secretions (Vieira et al., 2011) and ingest plant material. Nematode feeding may be characterized as ectoparasitic, in which the nematode remains outside the plant tissue; endoparasitic, in which the nematode feeds from within plant tissue; or some combination of the two. Nematodes are also categorized as migratory or sedentary feeders. Migratory nematodes retain a vermiform (wormlike) shape throughout their life histories, and may migrate within root tissue or between root tissue and soil, whereas sedentary feeders may remain vermiform or become enlarged. Knowledge of general plant-parasitic nematode life histories is

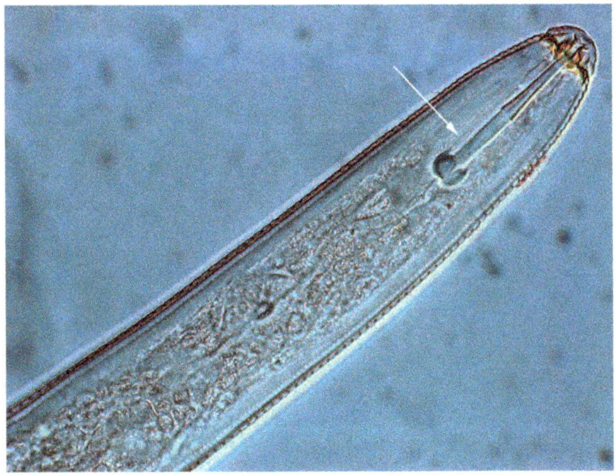

Figure 1 Head of a lance nematode with arrow pointing to the stylet (photo by Tesfamariam Mekete).

important because the methods used for extraction should match the likely locations of the nematodes, for example, in soil and/or roots, or in shoot tissue.

Nematode population densities in cultivated (monocultured) soils range from a few million to a few billion per 'acre plough layer', the equivalent of about .405 hectares to a depth of about 15–20 cm, although plant-feeding nematodes may be found in significant numbers much deeper than 20 cm (Pudasaini et al., 2006). Assemblages of nematode species in cultivated soils are among the least diverse, compared with plant-covered, non-disturbed soils; nonetheless, they contain representatives of all of the trophic group (food source) categories often used in nematode assemblage analysis: microbivores (~bacterivores), fungivores, predators, omnivores and plant parasites. Actual delegation of a nematode to one of these groups is often dependent on oesophageal morphology rather than direct knowledge of the nematode's feeding preferences (Niblack, 1989), but all plant-parasitic nematodes have feeding structures known as stylets (although not all stylet-bearing nematodes are phytophagous). Most, but not all, plant-parasitic nematodes found in cultivated soils are also obligate plant feeders, meaning that they require a plant host in order to survive and reproduce. Plant parasitism is not necessarily associated with plant pathogenicity; some cultivated plant hosts exhibit remarkable tolerance to nematode feeding and can support significant population densities without suffering measurable harm.

2.2 Identification

For purposes of identification, one generally needs to observe the morphological features of female plant-parasitic nematodes, distinguished by the presence of a vulval opening somewhere along the ventral surface of the nematode (Fig. 2). The position of the vulva is an important diagnostic characteristic (Mai et al., 1996). The important exception to this is for sedentary endoparasitic or semi-endoparasitic nematodes, such as cyst and root-knot nematodes, which are massively swollen and attached to or embedded in root tissue, respectively. Because the protocols for extracting female cyst and root-knot nematodes

Figure 2 Body of a spiral nematode with arrow pointing to the vulva, evidence that this is a female nematode (photo by Tesfamariam Mekete).

are different from those used for extracting vermiform nematodes, the presence of cyst and root-knot juveniles or males in soil extracts signals that additional steps must be taken to extract and identify the females.

Other important diagnostic characters that can be more or less well discerned under a stereoscope are the shape of the head, length and morphology of the stylet, morphology of oesophageal features, prominence of annules, sclerotization (particularly of the cephalic region), and tail length and shape. Few of these characters are easy to distinguish for those not trained in nematode identification, and inspection of many other characters discernable only with a compound microscope, and often only with an oil immersion objective. I highly recommend the use of a pictorial diagnostic key, such as Mai et al. (1996) or Powers' (2016) interactive diagnostic key for initial generic diagnosis.

2.3 Plant-parasitic nematodes on maize

Some plant-parasitic nematodes have high probabilities of causing yield losses and other damage to maize (such as predisposition to other infectious agents); however, the effects of nematode parasitism of maize have not been well-studied throughout the wide geographic or genetic range of the crop, nor the wide distributions and host adaptation of the nematodes. This may seem like an enormous gap in our knowledge, given the importance of maize as a crop, but is a consequence of two factors. The first is that damage to maize by nematodes is usually difficult to identify and measure (Fig. 3), and that characteristic affects the status of nematodes as a target for research and research funding. The second factor is that there are few scientists with the training in plant nematology who

Figure 3 Uneven canopy (left) and close-up (right) of plant damage of maize associated with very high population densities of spiral nematodes.

can focus on unproven pathosystems, given the high profile and fundability of research on highly pathogenic nematodes of high-value crops. In addition, very few studies have included genetically modified maize cultivars. However, the recent development of several commercial nematode management products has increased interest in nematodes and maize (Tylka et al., 2011).

In addition, for example, a typical maize sample from the Midwestern US 'Corn Belt' is highly likely to exhibit five to seven or more species of plant-parasitic nematodes, and it may well be that a combination of species is responsible for more effects than one species alone (Niblack, *unpublished*; Norton, 1983; Norton et al., 1978). McSorley and Dickson (1984) likewise observed that maize fields in the southeastern United States usually contain several plant-parasitic nematode species. Anecdotal evidence from nematologists on other continents suggests that it is typical for maize to host multispecific plant-parasitic nematode assemblages without a single predominant known pathogenic species. Higher diversity in plant-parasitic nematode assemblages, especially when combined with good representation from nematodes in other trophic groups, is a reflection of relative health of a soil community. Lower diversity, or high total numbers with predomination of the assemblage by one or two plant-parasitic species, is characteristic of situations in which nematodes may be responsible for maize yield loss.

Between 1889 and 1965, only about 35 species of plant-parasitic nematodes had been reported on maize (Goodey et al., 1965), very few of which had been demonstrated as pathogenic. Norton (1983) referred to reports of 120 nematode species associated with maize, of which 60 were known from North America. Todd (2016) noted that a survey of plant-parasitic nematodes associated with maize yield loss in the United States included eight genera (Koenning et a*l*., 1999), which likely include most of the 60 species mentioned by Norton.

3 Lesion nematodes, *Pratylenchus* spp.

Lesion, or root lesion, nematodes are so named because most species induce necrotic lesions on or within roots as a result of their migratory feeding. Generally regarded as

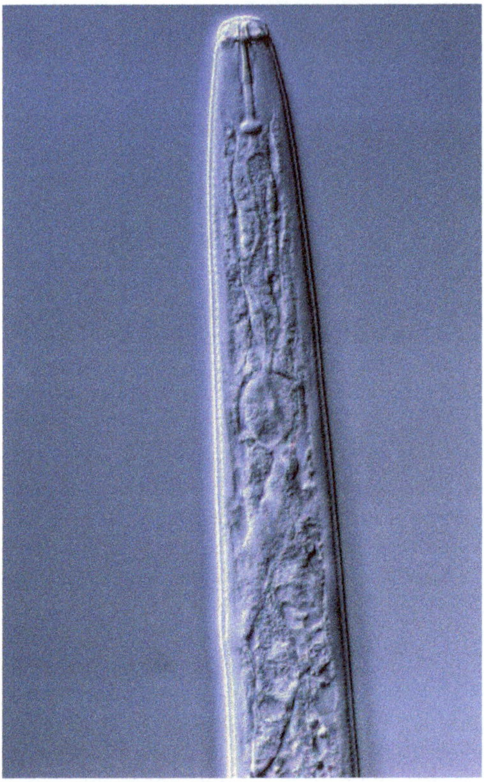

Figure 4 Oesophageal region of a lesion nematode, *Pratylenchus* spp. (photo by Ursula Reuter-Carlson).

endoparasites, lesion nematodes have been observed feeding on epidermal as well as cortical cells of maize (Rebois and Huettel, 1986); lesion nematodes do not typically feed in the vascular tissue. Lesion nematodes (Fig. 4) are tylenchoid, with strong, short stylets, overlapping oesophageal glands and single ovaries, that is, the position of the vulva is about 60–80% of the total body length (Castillo and Vovlas, 2007). Males are common for some species and absent or rare for others.

Three issues should be considered when lesion nematodes are suspected as the cause of injury to maize (Fig. 5). The first is that lesion nematodes will be found both in soil and within root tissues (MacGuidwin, 1989), even varying between seminal and adventitious roots (Todd and Oakley, 1996), and the relative distribution between root and soil tissues may vary depending on many factors, including the nematode species involved (MacGuidwin and Stanger, 1991, MacGuidwin and Bender, 2016). The second is that population densities of lesion nematodes, both in soil and in roots, vary with depth (MacGuidwin and Stanger, 1991). The typical depth of soil sampling (~15 cm) may lead to underestimation of nematode numbers; in addition, although the usual recommendation for nematode diagnosis is to sample soil at planting, pre-plant population densities of lesion nematodes are poor predictors of population development or yield effects (Todd and Oakley, 1996). The third, and most important issue, is that lesion nematodes

Figure 5 Greenhouse-grown maize roots of the same ages, inoculated with *Pratylenchus* (left) and non-inoculated (right).

are exceptionally difficult to identify to species level (Castillo and Vovlas, 2007). Because such identification is required in order to establish pathogenicity (e.g. Pili et al., 2016), a nematologist trained in morphological and molecular identification of lesion nematodes must be included in diagnostic and pathogenicity studies. In the Midwestern United States, it is typical for individual soil samples from maize to contain more than one species of *Pratylenchus*; thus, distinguishing species is critical to diagnosis.

The probability of lesion nematode damage to maize is a function of the nematode species (Castillo and Vovlas, 2007), the maize cultivar (Georgi et al., 1983; Smolik and Wicks, 1987) and the growing conditions (Jordaan et al., 1989; McDonald and van den Berg, 1993; Smolik and Evenson, 1987; Zirakparvar et al., 1980). Most of the research on organism-level interactions between lesion nematodes and maize predates the wide use of genetically modified cultivars. A recent paper by MacGuidwin and Bender (2016) contains an excellent discussion of development of functions relating lesion nematode population densities to maize yield loss, that is, establishment of 'damage thresholds'. The reality is that, in the absence of good, relevant information about specific nematode–maize cultivar relationships, and weighing the difficulty of species identification, a diagnosis of lesion nematode injury to maize has to be based on measurement of lesion nematode population densities and relative abundance of other plant-parasitic nematodes.

The *Pratylenchus* species reported in association with maize by Goodey et al. (1965) included *P. brachyurus, P. crenatus, P. hexincisus, P. neglectus, P. penetrans, P. scribneri, P. thornei* and *P. zeae*. All of these have been the subject of study, rarely intensive, in North America when they were identified in association with maize, but only a few have been demonstrated as maize pathogens. Maize is listed as a main host of *P. brachyurus, P. coffeae, P. penetrans, P. vulnus* and *P. zeae* in the Invasive Species Compendium published by CABI (Anon., 2016), and numerous other *Pratylenchus* species are considered pathogens of maize. McDonald and van den Berg (1993) stated that *P. zeae* and *P. brachyurus* were the most common and injurious nematodes to maize in South Africa, as is the case in the southeastern United States (Norton, 1984; Todd, 2016) and probably in maize-production regions in South America.

Pratylenchus spp. are well-documented as the nematode partners in several disease interactions involving fungi, which can result in additive or synergistic effects on the plant hosts (Back et al., 2002, da Silva et al., 2016). More severe leaf blight caused by *Colletotrichum graminicola* occurred in maize infected with *P. hexincinsus* (Nicholson et al., 1985). da Silva et al. (2016) observed a measurable interaction on maize root development when seedlings were co-infected by *P. penetrans* and *Fusarium verticillioides*, but not with *F. graminearum*. Likewise, *P. scribneri* and *F. verticillioides* showed a synergistic disease interaction on maize (Palmer and MacDonald, 1974), and Jordaan et al. (1987) showed a synergistic interaction among a combination of *P. brachyurus, P. zeae* and *F. moniliforme* (=*verticillioides*). da Silva et al. (2016) expressed some doubt as to whether blanket statements can be made about such interactions, given the wide variability expressed among isolates of *Fusarium* within a species.

4 Root-knot nematodes, *Meloidogyne* spp.

The root-knot nematodes are, as a group, the best-studied plant-pathogenic nematodes because of their wide host range, distribution and pathogenicity. However, until relatively recently, maize was often recommended as a rotation crop to control populations of root-knot nematodes. It is possible that pathogenic relationships between various isolates or host races of *Meloidogyne* species have simply been overlooked because the usual reaction of susceptible maize cultivars to infection does not involve the typical galling that characterizes root-knot disease. In addition, root-knot J2 (the stage commonly seen in soil extracts) are quite small compared with other plant-parasitic nematodes, and their stylets are difficult to see even at the highest magnification of a dissecting microscope; one requires familiarity with the overall characteristics of the J2 in order to identify them. Root staining was recommended to confirm root-knot infection of maize (Swarup and Sosa-Moss, 1990; Thies et al., 2002).

Meloidogyne species are sedentary endoparasitic obligate plant parasites (Perry and Moens, 2006). In general, those that infect maize reproduce by parthenogenesis, although there are variations within that general category. Females produce eggs within a gelatinous matrix, which can be visualized on the root surfaces after staining (Thies et al., 2002). The embryos proceed through development to the vermiform J2 stage, during which stage they can hatch and infect roots and may be seen in soil extracts. The infection site is often just behind the root tip, from which they migrate intercellularly until they induce the formation of a feeding site in or adjacent to living vascular cells (phloem or parenchyma). Successful establishment of a feeding site is accompanied by a loss of lateral musculature in the nematode and resultant swelling to a 'sausage stage' J2, after which the nematode is unable to move from the feeding site. Eventually, due to the effects of nematode secretions and host responses to them, several cells are recruited into the 'giant cell' complex that becomes a metabolic sink that provides nourishment for and is maintained by the nematode throughout its progression through the next two juvenile stages and the adult female's reproductive phase. The giant cell complexes formed by many root-knot nematodes infecting the same root system are formidable competitors for host plant resources even if galls are not produced, and are the means of the root-knot nematodes' pathogenicity. The mechanisms of susceptibility and resistance in maize to root-knot nematodes are not well understood but are receiving some attention (e.g. Gao et al., 2008).

Even in species that do not require males for reproduction, males may be produced. Males regain a vermiform shape and exit the roots; they have not been observed to feed on roots, although they have well-formed stylets which presumably aid them in escaping the J4 cuticle and the root. Males may be observed in extracts from soil samples; they are quite large relative to other plant-parasitic nematodes, and are excellent indicators of root-knot nematode infection when they appear.

Only a few *Meloidogyne* species are addressed in the following sections, included because of their wide distributions. It should be kept in mind that additional maize-parasitic species may be economically important in more localized areas (Elling, 2013), and to keep in mind that root-knot disease may be difficult to diagnose on maize due to the lack of obvious galling. About 100 species of *Meloidogyne* are known, and differentiating species is a difficult task (Perry et al., 2009). In addition, the traditional separation of the major *Meloidogyne* species into narrowly defined host races by means of a bioassay (Barker et al., 1985) is insufficient to describe the possible variations in reproduction and pathogenicity of *Meloidogyne* populations (Robertson et al., 2009). *Meloidogyne* species are also known for their roles in disease interactions involving other plant-pathogenic organisms (Back et al., 2002).

Meloidogyne incognita, 'southern' root-knot nematode, is limited by its inability to withstand freezing, but it is found throughout the tropical and subtropical crop production regions of the world. *M. incognita* isolates vary in pathogenicity to maize depending on whether they have had the opportunity to adapt to maize (Baldwin and Barker, 1970), and race designation does not predict isolates' abilities to parasitize other hosts (Robertson et al., 2009). Although interpretation of the literature is somewhat hampered by identifications of isolates according to race, which does not address adaptation to maize, several studies showed that resistance may be present in maize germplasm (McSorley and Gallaher, 1997; Williams and Windham, 1998; Windham, 1998; Windham and Williams, 1994). Resistant hybrids were used to reduce *M. incognita* population densities in one study (McSorley and Dickson, 1995) and failed to do so in another (Johnson et al., 1999). Davis and Timper (2000) screened several transgenic hybrids and found that all were good hosts of *M. incognita*. Despite the availability of resistance in maize germplasm, resistance has not been deployed in commercial cultivars and is not currently used to manage this nematode. Consideration of maize as tolerant to *M. incognita* and therefore possibly used in crop rotations to manage root-knot disease may be counterproductive; tolerance to nematodes simply means that the host is able to support large populations without harm compared with a susceptible host. The large nematode population densities remaining after a maize crop would be detrimental to a succeeding susceptible crop (Johnson et al., 1999; Windham and Williams, 1994).

Meloidogyne arenaria is commonly known as the peanut root-knot nematode, and it, like *M. incognita*, has a wide distribution in regions where the lowest annual temperature extreme is above freezing. Also like *M. incognita*, *M. arenaria* isolates are highly variable in their adaptation to maize cultivars. Davis and Timper (2000) identified variable responses to *M. arenaria* in transgenic and non-transgenic hybrids, similar to earlier results reported by Windham and Williams (1987, 1994) on non-transgenic and tropical cultivars. Windham and Williams (1987) also reported that some maize inbreds were resistant to both *M. arenaria* and *M. javanica*, and resistance to these two species was correlated

(Windham and Williams, 1990). Johnson et al. (1999) recommended maize, as a poor host of *M. arenaria*, for a combination of rotation and chemical treatment to manage pests and pathogens of peanut (*Arachis hypogaea*).

Meloidogyne chitwoodi (Santo and O'Bannon, 1981), an economically important pathogen of potato (*Solanum tuberosum*), is also able to reproduce on graminaceous hosts, including maize. Maize was a poor host in their test, producing small galls in response to infection, but as with other *Meloidogyne* species, isolates given the opportunity to adapt to maize probably will do so.

5 Vermiform nematodes restricted to sandy soils: *Belonolaimus*, *Longidorus*, *Paratrichodorus* and *Trichodorus* spp.

Belonolaimus spp., commonly known as sting nematodes, and *Longidorus* spp., known as needle nematodes, are members of two distinct nematode orders (Tylenchida and Dorylaimida, respectively) and only distantly related. As pathogens of maize, however, they have some interesting similarities.

First, these are rare among nematodes associated with maize in that they can cause seedling death (Todd, 2016). Damage to maize is usually dramatic and highly aggregated in the field (Fig. 6).

Second, they are large for plant-parasitic nematodes; adult *Belonolaimus* spp. can exceed 2 mm in length, and adult *Longidorus* spp. approach 5 mm in length, and both have exceptionally long stylets, approximately 100 and 85 μm long (Fig. 7), respectively, relative to overall body length. They are large enough that standard passive nematode extraction procedures might result in failure to detect or severe underestimation of the population densities.

Figure 6 Damage to maize caused by needle nematodes, *Longidorus* spp. (photo by Tamra Jackson).

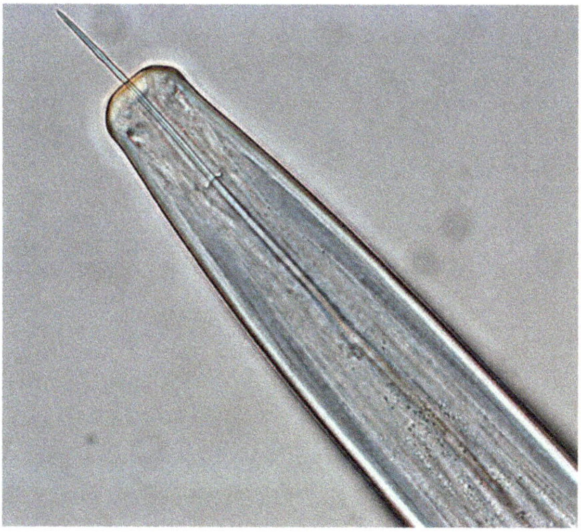

Figure 7 Head region of a needle nematode, *Longidorus* spp. (photo by Tesfamariam Mekete).

Third, both are restricted in distribution to soils with very high sand content: 50% or more for *L. breviannulatus* (Norton and Hoffman, 1975), with greater damage occurring in soils that exceed 90% sand (Todd, 2016), and 80% or more for *Belonolaimus* spp. (Todd, 1989, 2016).

Fourth, marked seasonal variation in the vertical distribution of population densities of these nematodes complicates sampling and diagnosis; detectable levels may occur deeper than those of other plant-parasitic nematodes (MacGuidwin, 1989; Todd, 1989), and the standard 15-cm depth of sampling may result in failure to detect (MacGuidwin and Stanger, 1991).

Fifth, both nematodes have host ranges including dicots and monocots, which feed on, and severely damage or kill root tips. The result of their feeding is severely restricted root systems, but the symptoms are not clearly diagnostic (Fig. 8).

Distinguishing these two nematodes is relatively easy once the proper extraction procedures have been applied. *Belonolaimus* spp. are known from only North America and a few central Asian countries. They are tylenchoid, with very long stylets (~100 μm) and overlapping oesophageal glands. On the other hand, *Longidorus* spp. have a worldwide distribution and are dorylaimoid, with long stylets (~85 μm) lacking basal knobs or swellings, and the oesophagus is non-overlapping. Maize is highly susceptible to *L. breviannulatus* and moderately susceptible to *L. elongatus* (Yassin, 1969).

In addition to sting and needle nematodes, potentially maize-pathogenic genera that are more damaging in sandy soils are the stubby-root nematodes, *Trichodorus* and *Paratrichodorus*, particularly *P. minor*. Stubby-root nematodes are dorylaimoid, with short, curved stylets, and relatively wide compared with body length. Stubby-root nematodes feed on root tips, as do sting and needle nematodes, but the root symptoms are more distinctive, causing a root morphology similar to those damaged by herbicides (Fig. 9). Some resistance in maize to *P. minor* was reported (Timper et al., 2007). These nematodes are extremely sensitive to manipulation and may not survive the handling required in

Figure 8 Field-grown maize plants at 4 weeks after planting, with severe symptoms of needle nematode injury (left) compared with less severely damaged plants (right) (photo by Tamra Jackson).

Figure 9 Maize root injury caused by stubby root nematodes (photo by Ursula Reuter-Carlson).

taking and extracting soil samples; their population densities may be underestimated as a result. In addition, both *Trichodorus* and *Longidorus* were found to travel relatively long distances rapidly, compared with the spiral nematodes *Rotylenchus* and *Helicotylenchus*, under controlled circumstances (Rossner, 1972).

6 Other nematodes

Very few other nematodes have been investigated as the cause of maize disease; however, as is the case with some of those which received attention, it may be that only certain selected populations of widely distributed nematodes are pathogenic, or that some nematodes are so limited in distribution that resources have not been available to work on them.

6.1 Awl nematodes, *Dolichodorus* spp.

The awl nematode, *Dolichodorus heterocephalus*, is comparable with the previously mentioned sting nematode for its ability to damage maize severely in Florida (Rhoades, 1985), but it appears to be much more restricted in distribution. This nematode is similar in appearance to the sting nematode as well, but can be distinguished at low magnification by the tail shape (CIH Descriptions of Plant-Parasitic Nematodes, set 4).

6.2 Burrowing nematodes, *Radopholus* spp.

Radopholus similis, the burrowing nematode, is a significant pathogen of banana/plantain and also parasitizes maize (Price, 1994). Its wide host range and widespread distribution make it a species of interest for quarantine organizations. A member of the family Pratylenchidae, its morphology is very similar to *Pratylenchus* spp., with which it can be easily confused.

6.3 Cyst nematodes

Cyst nematodes are obligate plant parasites so named because the female nematode is enlarged and serves as a protective cyst in which eggs survive in soil. The J2 which hatches from the egg is the infective stage, therefore the number of eggs per volume of soil represents the number of potential infectious juveniles. The number of cysts per volume of soil is sometimes used to assess damage potential because cysts are relatively easy to extract from soil and root samples and require only slight magnification (x40) for identification and counting. If cyst nematodes are suspected as the cause of maize injury or yield loss, extensive information on soil sampling recommendations and cyst, egg, and juvenile extraction and enumeration is available in the literature on the soya bean (Schmitt et al., 2004) and potato cyst nematodes (Marks and Brodie, 1998).

Heterodera avenae, the cereal cyst nematode, was reported to cause yield losses in maize of up to 50% and severe losses in northern Europe when maize cultivation was extended into that region (Rivoal and Cook, 1993). Although maize is considered resistant to *H. avenae*, serious injury to young plants may apparently be caused by a hypersensitive response when large numbers of J2 penetrate young roots.

Heterodera zeae, commonly known as the corn cyst nematode, was described first from India and later reported from Egypt, Pakistan and the United States (Golden and Mulvey, 1982). Isolates from various locations showed differences in host ranges (reviewed in Ringer et al., 1987), but all hosts were graminaceous. *H. zeae* has a relatively high temperature optimum for reproduction (36°C), and is therefore more likely to be of economic importance in tropical and subtropical maize production (Hashmi et al., 1993). There is evidence of moderate resistance in some Indian maize cultivars, as well as evidence of pathotypes or ecotypes among Egyptian isolates, so the relationship between *H. zeae* and maize in tropical and subtropical conditions is likely to be complicated by cultivar-pathotype interactions.

Punctodera chalcoensis, a cyst nematode associated with teosinte and with severe damage to maize in Mexico (Stone et al., 1976), was shown to induce syncytia anatomically similar to those induced by cyst nematode pathogens of potato (*Globodera* spp.) and soya bean (*Heterodera glycines*) (Suarez et al., 1985), but there is very little information available on its distribution and pathogenicity. It has been referred to as the 'Mexican corn cyst nematode' (Swarup and Sosa-Moss, 1990).

Several other cyst nematode species were reported as associates of maize; species in this category, according to Swarup and Sosa-Moss (1990), included *H. cajani, H. delvii, H. gambiensis, H. oryzae* and *H. sorghi. Vittatidera zeaphila*, a new genus and species of cyst nematode, was described as a parasite of maize in the United States (Bernard et al., 2010). *Heterodera goldeni*, a recently described species from Egypt, reduced shoot and dry weights of maize in a greenhouse test (Ibrahim et al., 2012). The potential for these species as economically important pathogens of maize is yet to be determined.

6.4 Dagger nematodes, *Xiphinema* spp.

The dagger nematodes are dorylaimoid, with long stylets characterized by basal swellings, unlike other dorylaimoid plant parasites (Fig. 10). *Xiphinema americanum, X. chambersi* and *X. rivesi* were noted as potential pathogens of maize by Todd (2016), and *X. americanum* population densities were negatively correlated with maize yield (Norton et al., 1978). In contrast, Evans *et al.* (2007) found that *X. americanum* declined on maize in field and greenhouse studies, and suggested the use of maize as a rotation crop to reduce the occurrence of soya bean severe stunt virus, for which the nematode is a putative vector. Because dagger nematodes are important pathogens of other plant species, by themselves and as plant virus vectors, they should be noted if they are observed in soil extracts as they may affect rotation recommendations.

6.5 Lance nematodes, *Hoplolaimus* spp.

The lance nematodes, in the genus *Hoplolaimus*, have been observed to feed both ecto- and endoparasitically on maize roots. *Hoplolaimus* species are large for tylenchoid

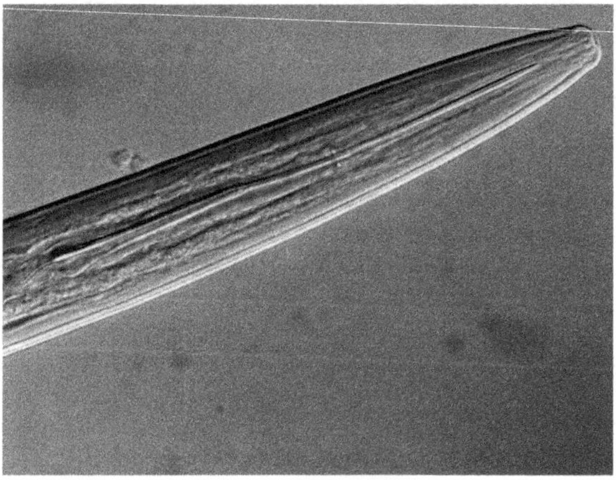

Figure 10 Head region of a dagger nematode, *Xiphinema* spp., with the characteristic flanges at the base of the stylet (photo by Tesfamariam Mekete).

nematodes, ranging between 1 and 2 mm in length and appearing stout. Females have well-developed sclerotization in the cephalic region, and strong, short stylets about 40–50 μm, some species with anterior projections on the knobs. *H. galeatus* caused yield loss in maize in Iowa when population densities were high and when combined with *Pratylenchus hexincisus* (Norton and Hinz, 1976). *H. magnistylus* was associated with stunted maize plants (Donald et *al*., 2013). Although maize was reported as an excellent host of the serious cotton and soya bean pathogen *H. columbus* (Fassuliotis et al., 1974), maize is considered to be tolerant of this species (Todd, 2016). *H. aegypti* reduced maize growth in greenhouse tests (Koura and Oteifa, 1990).

6.6 Pin nematodes

Species of pin nematodes in the genera *Pratylenchus* and *Gracilacus* are migratory ectoparasites, common associates of maize in the Midwestern United States (*personal observation*). Pin nematode females have very long stylets relative to body length, little cephalic sclerotization, non-overlapping oesophageal glands and single ovaries (Fig. 11). These nematodes are quite small and easily overlooked in extracts from soil samples, a situation that is exacerbated for *Pratylenchus* by the frequent appearance in soil extracts of J4 stage nematodes (and sometimes males) that do not have stylets and are not easily distinguishable. It was thought that these stylet-less nematodes are in an arrested or dauer stage that indicates they are not feeding on whatever host was present when they were collected; however, this may not be the case. For example, *P. microdorus* occurred more frequently in the driest (summit) position in loess toposequences (Norton and Oard, 1981) and the dauer stage may be an indicator of poor conditions for nematode activity. Maize was found to be a moderately susceptible host of *P. projectus* (Townshend and Davidson, 1989).

6.7 Reniform nematodes, *Rotylenchulus* spp.

Maize was reported as a poor host of *Rotylenchulus reniformis* (Windham and Lawrence, 1992), and could be useful in rotations as this nematode is a pathogen of cotton, soya bean

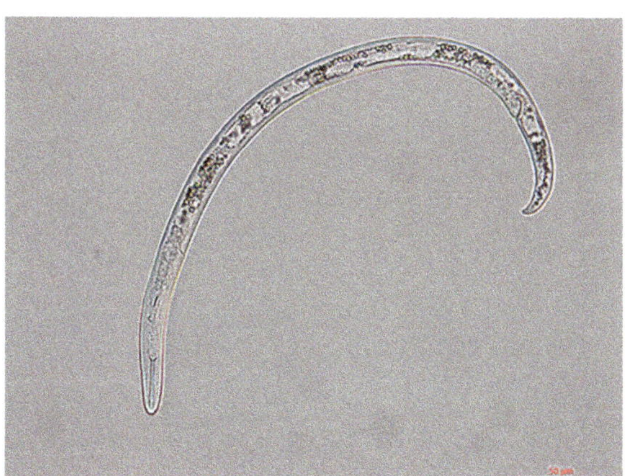

Figure 11 Full body of a female pin nematode, *Pratylenchus* spp. (photo by Katharine Ankrom).

and a number of other crops. Maize was considered a good host for one isolate of *R. reniformis* in India (Dasgupta and Seshadri, 1971), which again emphasizes the need to consider subspecific variation in nematode–maize relationships. Maize was reported to be a good host of *R. parvus* in South Africa (Jordaan et al., 1989). *R. borealis* was reported on maize in eastern Europe (Norton, 1984).

6.8 Ring nematodes

Ring nematodes, including those identified as *Criconemella ornata* and *C. sphaerocephala*, have been associated with maize. Although considered weak pathogens (Ayers et al., 1989), they have both been suspected of causing loss when they occur in multispecific assemblages. The nematodes' common name is based on the obvious strongly annulated cuticle (Fig. 12). It should be noted that the generic names of ring nematodes have changed several times, and many of those identified as associates of maize are now classified as *Mesocriconema*. Ring nematodes are short, less than 1 mm in length, stout and have very strong stylets, measuring about 50 to 100 μm. Maize was observed to be a moderate host for a *C. ornata* population isolated originally from peanut (Barker et al., 1982).

6.9 Spiral nematodes

Spiral nematodes are extremely common in maize fields. Spiral nematodes, including species in the genera *Helicotylenchus*, *Rotylenchus* and *Scutellonema*, are so named because they assume a spiral shape at rest (Fig. 2). They have well-developed cephalic sclerotization, strong stylets, overlapping oesophageal glands and are didelphic. There are many species, and within *Helicotylenchus* there are at least 200, with considerable variation in morphological characters even within species (Subbotin et al., 2011). *H. pseudorobustus* can reach very high population densities on maize in the United States, but there is little evidence that this common prairie species by itself is pathogenic (Norton, 1977). Vovlas

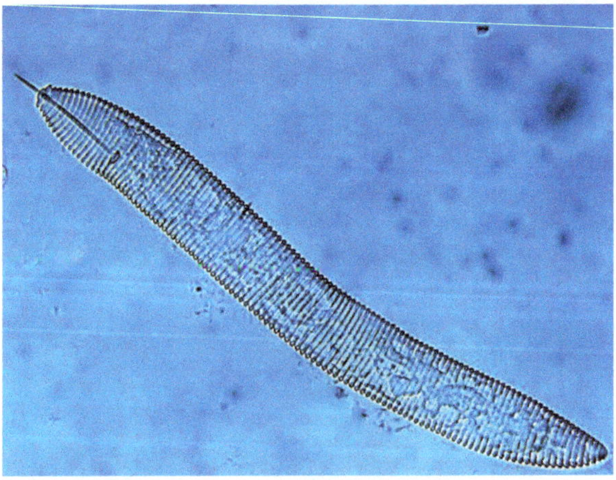

Figure 12 Full body of a ring nematode, *Mesocriconema* spp. (photo by Tesfamariam Mekete).

and Inserra (1985) observed that *H. pseudorobustus* induced modified food cells in the cortex of maize roots similar to those induced by pathogenic species of *Tylenchulus* (citrus nematode) and *Rotylenchulus* (reniform nematodes). *H. digonicus* damaged maize roots in a greenhouse test (Griffin, 1964) and *H. dihystera* has an established damage threshold of 200 nematodes/100 cm^3 soil (Todd, 2016).

6.10 Stem and bulb nematode, *Ditylenchus dipsaci*

Ditylenchus dipsaci, the stem and bulb nematode, may damage maize by causing it to lodge (Norton, 1984). Under cool, wet conditions, the nematode infects the stalk, causing swelling and necrosis at the base and resulting in a reduction in secondary roots. This nematode species has a widespread distribution and host range, is a documented pathogen of maize (Caubel and Rivoal, 1972) and should be assessed as a potential cause of disease when lodging is the primary symptom.

6.11 Stunt nematodes

Stunt nematodes, in the genera *Tylenchorhynchus*, *Quinisulcius* and *Merlinius*, are commonly identified from soil samples taken from maize fields in the United States and frequently occur in plant-parasitic nematode assemblages. Stunt nematodes are relatively easy to identify, having a tylenchoid oesophagus with a strong stylet, well-developed basal bulb and obvious annulation (Fig. 13). The three genera are distinguished by the number of longitudinal incisures in the lateral field, four for *Tylenchorhynchus* spp., five for *Quinisulcius* spp. and six for *Merlinius* spp.; these are visible only under high magnification (e.g. × 600 to 1,000). Stunt nematodes feed ectoparasitically on root epidermal tissue and

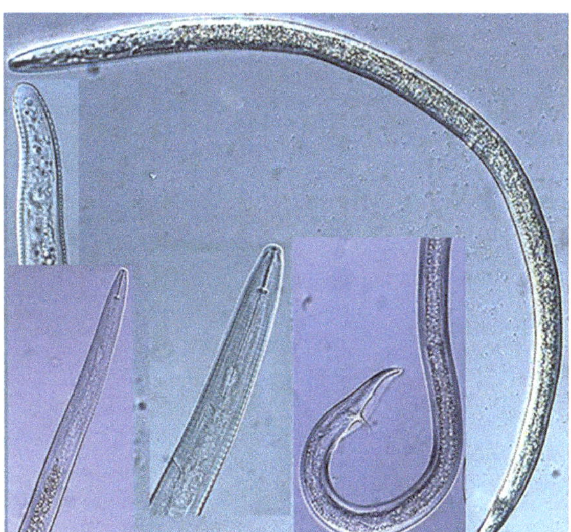

Figure 13 Full and partial bodies of a stunt nematode, *Tylenchorhynchus* spp. (photo by Tesfamariam Mekete).

may cause enough damage to serve as a point of infection by opportunistic organisms in the root zone. Except for *T. claytoni*, which can significantly reduce maize root systems (Krusberg, 1959) and was observed to cause changes in the morphology of nuclei of maize cells (Deubert *et al.*, 1967), there is little information available on the effects of stunt nematodes alone on maize growth. *T. zambiensis* was considered to be a weak pathogen of maize by itself, based on greenhouse tests (Venditti and Noel, 1995). Norton's (1984) review includes a table of references on six *Tylenchorhynchus* and one *Quinisulcius* species and their associations with maize. *T. maximus* was associated with stunting of maize in a greenhouse test (Griffin, 1964), noted here because this species is common in the United States and was observed in soil samples from symptomatic maize in Illinois (Niblack, *unpublished*).

7 Conclusions

As is often noted in reviews of plant-parasitic nematodes, they are ubiquitous, and maize fields are no exception. Their presence alone does not indicate that a pathogenic relationship exists. Visible damage caused by nematodes in maize is often subtle and symptoms are rarely diagnostic. It maybe that in field situations, an assemblage of plant-parasitic nematodes may cause additive injuries (Fig. 14); such assemblages are highly variable both in terms of species composition and population densities. In addition, local adaptation of certain nematodes as pathogens of maize appears to be highly likely but has received little attention. Investigations on the whole-organism or assemblage levels have been remarkably sparse in the past couple of decades. These deserve considerably more attention, given the degree of change that has taken place in maize cultivars and production practices, both of which are likely to have affected nematode–maize interactions.

Figure 14 Damage to maize in a field infested with stunt and lance nematodes; not treated with an experimental nematode management product (left) and treated (right).

8 Where to look for further information

The reviews mentioned in the Introduction are excellent places to begin finding information about nematodes associated with maize (Norton, 1983, 1984; Todd, 2016; Rivoal and Cook, 1993; Swarup and Sosa-Moss, 1990; Windham, 1998). I also recommend consultation of Ferris' (2016) comprehensive nematode–host database, NEMABASE (in the list below), which contains references for 128 species of plant-parasitic nematodes associated with maize, along with a number of references for genera only. Norton (1984) considered the following nematodes to be possible maize parasites, but with little evidence to support the supposition: *Aglenchus agricola, Anguina tritici, Aphelenchoides arachidis, Scutellonema brachyurum* and *Subanguina radicicola.*

I also highly recommend the following currently available online resources for further information:

- 'Nemaplex, the Nematode-Plant Expert Information System, a Virtual Encyclopedia on Soil and Plant Nematodes', http://plpnemweb.ucdavis.edu/nemaplex/index.htm
- 'Nemabase, A Database of the Host Status of Plants to Nematodes', http://plpnemweb.ucdavis.edu/nemaplex/Nemabase2010/Nemabase%20Search%20Menu.htm
- The Invasive Species Compendium, http://www.cabi.org/isc/
- The Plantwise Knowledge Bank, http://www.plantwise.org/knowledgebank/home.aspx

9 References

Anon. (2016), 'Invasive species compendium,' CABI http://www.cabi.org/isc/ (accessed 23 February 2016).

Ayers, A. R., Duncan, H. E., Barker, K. R. and Beute, M. K. (1989), 'Effects of crop rotation and nonfumigant nematicides on peanut and corn yields in field infested with *Criconemella* species', *Journal of Nematology* 21(2), 268–75.

Back, M. A., Haycock, P. P. J. and Jenkinson, P. (2002), 'Disease complexes involving plant parasitic nematodes and soilborne pathogens', *Plant Pathology* 51, 683–97.

Baldwin, J. G., and Barker, K. R. (1970), 'Host suitability of selected hybrids, varieties and inbreds of corn to populations of *Meloidogyne* spp.', *Journal of Nematology* 2(4), 345–50.

Barker, K. R., Carter, C. C. and Sasser, J. N., eds. (1985), *An Advanced Treatise on Meloidogyne, Volume II: Methodology*, Department of Plant Pathology, North Carolina State University.

Barker, K. R., Schmitt, D. P. and V. P. Campos (1982), 'Response of peanut, corn, tobacco, and soybean to *Criconemella ornata*', *Journal of Nematology* 14(4), 576–81.

Bernard, E. C., Handoo, Z., Powers, T. O., Donald, P. A. and Heinz, R. D. (2010), '*Vittatidera zeaphila* (Nematoda: Heteroderidae), a new genus and species of cyst nematode parasitic on corn (Zea mays)', *Journal of Nematology* 42, 139–50.

Castillo, P., and Vovlas, N. (2007), *Pratylenchus (Nematoda:Pratylenchidae) Diagnosis, Biology, Pathogenicity and Management*, Koninklijke Brill NV, Leiden, the Netherlands.

Caubel, G., and Rivoal, R. (1972), 'Observations sur les attaques des deux nematodes nuisibles au mais en 1971', *Phytoma* 239, 15–18.

da Silva, M. P., Tylka, G. L. and Munkvold, G. P. (2016), 'Seed treatment effects on maize seedlings coinfected with *Fusarium* spp. and *Pratylenchus penetrans*', *Plant Disease* 100, 431–7.

Dasgupta, D. R., and Seshadri, A. R. (1971), 'Reproduction, hybridization, and host adaptation in physiological races of the reniform nematode, *Rotylenchulus reniformis*', *Indian Journal of Nematology* 1, 128–44.

Davis, R. F., and Timper, P. (2000), 'Resistance in selected corn hybrids to *Meloidogyne arenaria* and *M. incognita*', *Journal of Nematology* 32(4S), 633–40.

Deubert, K. H., Norgren, R. L., Paracer, S. M. and Zuckerman, B. M. (1967), 'The Influence of *Tylenchus agricola* and *Tylenchorhynchus claytoni* on corn roots under gnotobiotic conditions', *Nematologica*, 13(1), 56–62.

Dickson, D. W., and Hewlett, T. C. (1987), 'Effect of two nonfumigant nematicides on corn grown in two adjacent fields infested with different nematodes', *Annals of Applied Nematology* 1, 89–93.

Donald, P. A., Holguin, C. M. and Agudelo, P. A. (2013), 'First report of lance nematode (*Hoplolaimus magnistylus*) on corn, soybean, and cotton in Tennessee', *Plant Disease* 97(10), 1389.

Elling, A. (2013), 'Major emerging problems with minor *Meloidogyne* species', *Phytopathology* 103, 1092–102.

Evans, T. A., Miller, L. C., Vasilas, B. L., Taylor, R. W. and Mulrooney, R. P. (2007), 'Management of *Xiphinema americanum* and soybean severe stunt in soybean using crop rotation', *Plant Disease* 91(2), 216–19.

Gao, X., Starr, J., Gobel, C., Engelberth, J., Feussner, I., Tumlinson, J. and Kolomiets, M. (2008), 'Maize 9-lipoxygenase ZmLOX3 contols development, root-specific expression of defense genes, and resistance to root-knot nematodes', *Molecular Plant-Microbe Interactions* 21(1), 98–109.

Georgi, L., Ferris, J. M. and Ferris, V. R. (1983), 'Population development of *Pratylenchus hexincisus* in eight corn inbreds', *Journal of Nematology* 15(2), 243–52.

Golden, A. M., and Mulvey, R. H. (1982), 'Redescription of *Heterodera zeae*, the corn cyst nematode, with SEM observations', *Journal of Nematology* 15(1), 60–70.

Goodey, J. B., Franklin, M. T. and Hooper, D. J. (1965), *T. Goodey's The Nematode Parasites of Plants Catalogued under Their Hosts*, Commonwealth Agricultural Bureaux.

Griffin, G. D. (1964), 'Association of nematodes with corn in Wisconsin', *Plant Disease Reporter* 48(6), 458–9.

Hashmi, S., Krusberg, L. R. and Sardanelli, S. (1993), 'Reproduction of *Heterodera zeae* and its suppression of corn plant growth as affected by temperature', *Journal of Nematology* 25(1), 55–8.

Ibrahim, I. K. A., Awd-Allah, S. F. A. and Handoo, Z. A. (2012), 'Host suitability of some poaceous crop cultivars for *Heterodera goldeni*', *Nematropica* 42, 324–7.

Johnson, A. W., Minton, N. A., Brenneman, T. B., Burton, G. W., Culbreath, A. K., Gascho, G. J., and Baker, S. H. (1999a), 'Bahiagrass, corn, cotton rotations, and pesticides for managing nematodes, diseases, and insects on peanut', *Journal of Nematology* 31(2), 191–200.

Johnson, A. W., Sumner, D. R., Windham, G. L. and Williams, W. P. (1999b), 'Effects of a resistant corn hybrid and fenamiphos on *Meloidogyne incognita* in a corn-squash rotation', *Journal of Nematology* 31(2), 184–90.

Jordaan, E. M., Loots, G. C., Jooste, W. J. and De Waele, D. (1987), 'Effects of root-lesion nematodes (*Pratylenchus brachyurus* Godfrey and *P. zeae* Graham) and *Fusarium moniliforme* Sheldon alone or in combination, on maize', *Nematologica* 33, 213–19.

Jordaan, E. M., De Waele, D. and van Rooven, P. J. (1989), 'Endoparasitic nematodes in maize roots in the western Transvaal as related to soil texture and rainfall', *Journal of Nematology* 21(3), 356–60.

Koura, F. H., and Oteifa, B. A. (1990), 'Effect of *Hoplolaimus aegypti* population density on maize growth response and nematode reproduction', *Annals of Agricultural Science (Cairo)*, 35(1), 491–6.

Krusberg, L. R. (1959), 'Investigations on the life cycle, reproduction, feeding habits and host range of *Tylenchorhynchus claytoni* Steiner', *Nematologica* 4(3), 187–97.

MacGuidwin, A. E. (1989), 'Abundance and vertical distribution of *Longidorus breviannulatus* associated with corn and potato', *Journal of Nematology* 21, 404–8.

MacGuidwin, A. E. (1989), 'Distribution of *Pratylenchus scribneri* between root and soil habitats', *Journal of Nematology* 21, 409–15.

MacGuidwin, A. E., and Bender, B. E. (2016), 'Development of a damage function model for Pratylenchus penetrans on corn', *Plant Disease* 100, 764–9.

MacGuidwin, A. E., and Stanger (1991), 'Changes in vertical distribution of *Pratylenchus scribneri* under potato and corn', *Journal of Nematology* 23(1), 73–81.

Mai, W. F., Mullen, P. G., Lyon, H. H. and Loeffler, K. (1996), *Plant-Parasitic Nematodes A Pictorial Key to Genera,* Cornell University Press.

Marks, R. J., and Brodie, B. B., eds. (1998), *Potato Cyst Nematodes: Biology, Distribution, And Control*, New York, CAB International.

McDonald, A. H., and van den Berg, E. H. (1993), 'Effect of watering regimen on injury to corn and grain sorghum by *Pratylenchus* species', *Journal of Nematology* 25(4), 654–8.

McSorley, R., and Dickson, D. W. (1989), 'Effects and dynamics of a nematode community on miaze', *Journal of Nematology* 21(4), 462–71.

Niblack, T. L. (1989), 'Applications of nematode community structure research to agricultural production and habitat disturbance', *Journal of Nematology* 21(4), 437–43.

Nicholson, R. L., Bergeson, G. B., DeGennaro, F. P. and Viveiros, D. M. (1985), 'Single and combined effects of the lesion nematode and Colletotrichum graminicola on growth and anthracnose leaf blight on corn', *Phytopathology* 75, 654–61.

Norton, D. C. (1977), '*Helicotylenchus pseudorobustus* as a pathogen on corn, and its densities on corn and soybean', *Iowa State Journal of Research* 51, 279–85.

Norton, D. C. (1983), 'Maize nematode problems', *Plant Disease* 67(3), 253–6.

Norton, D. C. (1984), 'Nematode parasites of corn', in Nickle, W. R., ed., *Plant and Insect Nematodes,* pp. 61–94. Marcel Dekker.

Norton, D. C., and Hinz, P. (1976), 'Relationship of *Hoplolaimus galeatus* and *Pratylenchus hexincisus* to reduction of corn yields in sandy soils in Iowa', *Plant Disease Reporter* 60(3), 197–200.

Norton, D. C., and Hoffman, J. K. (1975), '*Longidorus breviannulatus* n. sp. (Nematoda: Longidoridae) associated with stunted corn in Iowa', *Journal of Nematology* 7, 168–71.

Norton, D. C., and Oard, M. (1981), 'Plant-parasitic nematodes in loess toposequences planted with corn', *Journal of Nematology* 13(3), 314–22.

Norton, D. C., Tollefson, J., Hinz, P. and Thomas, S. H. (1978), 'Corn yield increases relative to nonfumigant chemical control of nematodes', *Journal of Nematology* 10, 160–6.

McSorley, R., and Dickson, D. W. (1995), 'Effect of tropical rotation crops on *Meloidogyne incognita* and other plant-parasitic nematodes', *Journal of Nematology* 27(4S), 535–44.

McSorley, R., and Gallaher, R. N. (1997), 'Effect of compost and maize cultivars on plant-parasitic nematodes,' *Journal of Nematology* 29(4S), 731–6.

Perry, R. N., and Moens, M., eds. (2006), *Plant Nematology*, CAB International.

Perry, R. N., Moens, M. and Starr, J. L., eds. (2009), *Root-knot Nematodes*, CAB International.

Perry, R. N., and Wright, D. J., eds. (1998), *The Physiology and Biochemistry of Free-Living and Plant-Parasitic Nematodes*, CAB International.

Pili, N. N., Kyndt, T., Gheysen, G., Janssen, T., Couvrer, M., Bert, W. and Mibey, R. K. (2016), 'First report of *Pratylenchus zeae* on upland rice from Kwale County, Kenya', *Plant Disease* http://dx.doi.org/10.1094/PDIS-07–15-0743-PDN.

Powers, T. O., et al. 2016. *Interactive Diagnostic Key to Plant Parasitic, Freeliving and Predaceous Nematodes, http://nematode.unl.edu/key/nemakey.htm* (accessed 23 February 2016).

Price, N. S. (1994), 'Alternate cropping in the management of *Radopholus similis* and *Cosmopolites sordidus*, two important pests of banana and plantain', *International Journal of Pest Management* 40(3), 237–44, doi: 10.1080/09670879409371890.

Pudasaini, M. P., Schomaker, C. H., Been, T. H. and Moens, M. (2006), 'Vertical distribution of the plant-parasitic nematode, *Pratylenchus penetrans*, under four field crops', *Phytopathology* 96, 226–33.

Rebois, R. V., and Huettel, R. N. (1986), 'Population dynamics, root penetration, and feeding behavior or *Pratylenchus agilis* in monoxenic root cultures of corn, tomato, and soybean', *Journal of Nematology* 18(3), 392–7.

Rhoades, H. L. (1985), 'Effects of separate and concomitant populations of *Belonolaimus longicaudatus* and *Dolichodorus heterocephalus* on *Zea mays*', *Nematropica* 15, 171–4.

Ringer, C. E., Sardanelli, S. and Krusberg, L. R. (1987), 'Investigations of the host range of the corn cyst nematode, Heterodera zeae, from Maryland', *Annals of Applied Nematology* 1, 97–106.

Rivoal, R., and Cook, R. (1993), 'Nematode pests of cereals,' in Evans, K., Trudgill, D. L., and Webster, J. M. (eds.), *Plant Parasitic Nematodes in Temperate Agric*ulture, pp. 259–303. CAB International.

Robertson, L., Díez-Rojo, M. A., López-Pérez, J. A., Piedra Buena, A., Escuer, M., López Cepero, J., Martínez, C. and Bello, A. (2009), 'New host races of *Meloidogyne arenaria, M. incognita*, and *M. javanica* from horticultural regions of Spain', *Plant Disease* 93, 180–4.

Rossner, J. (1972), 'Vertikalverteilung wandernder wurzelnematoden im boden in abhangigkeit von wassergehalt und durchwurzelung', *Nematologica* 18, 360–72.

Santo, G. S., and J. H. O'Bannon (1981), 'Pathogenicity of the Columbia root-knot nematode (*Meloidogyne chitwoodi*) on wheat, corn, oat, and barley', *Journal of Nematology* 13(4), 548–50.

Schmitt, D. P., Wrather, J. A. and Riggs, R. D., eds. (2004), *Biology and Management of the Soybean Cyst Nematode*, Marceline, MO USA, Schmitt & Associates of Marceline.

Shurtleff, M. C., and Averre, C. W. (2000), *Diagnosing Plant Diseases Caused by Nematodes*, APS Press.

Smolik, J. D., and Evenson, P. D. (1987), 'Relationship of yields and *Pratylenchus* spp. population densities in dryland and irrigated corn', *Annals of Applied Nematology* 1, 71–3.

Smolik, J. D., and Wicks III, Z. W. (1987), 'Reproduction of Pratylenchus hexincisus and P. scribneri in corn inbreds', *Annals of Applied Nematology* 1, 29–31.

Stone, A. R., Sosa-Moss, C. and Mulvey, R. H. (1976), 'Punctodera chalcoensis n. sp. (Nematoda: Heteroderidae) a cyst nematode from Mexico parasitizing *Zea mays*', *Nematologica* 22, 381–9.

Suarez, Z., Sosa-Moss, C. and Inserra, R. N. (1985), 'Anatomical changes induced by Punctodera chalcoensis in corn roots', *Journal of Nematology* 17(2), 242–4.

Subbotin, S. A., Inserra, R. N., Marais, M., Mullin, P., Powers, T. O., Roberts, P. A., van den Berg, E., Yeates, G. W. and Baldwin, J. G. (2011), 'Diversity and phylogenetic relationships within the spiral nematodes of *Helicotylenchus* Steiner, 1945 (Tylenchida:Hoplolaimidae) as inferred from analysis of the D2-D3 expansion segments of 28S rRNA gene sequences', *Nematology* 13(3), 333–45.

Swarup, G., and Sosa-Moss, C. (1990), 'Nematode parasites of cereals', in Luc, M., Sikora, R. A., and Bridge, J. (eds.), *Plant Parasitic Nematodes in Subtropical and Tropical Agriculture*, pp. 109–36. CAB International.

Thies, J. A., Merrill, S. B. and Corley, E. L. (2002), 'Red food coloring stain: new, safer procedures for staining nematodes in roots and egg masses on root surfaces', *Journal of Nematology* 34(2), 179–81.

Timper, P., Krakowski, M. D. and Snook, M. E. (2007), 'Resistance in maize to *Paratrichodorus minor*', *Nematropica* 37, 9–20.

Todd, T. C. (1989), 'Population dynamics and damage potential of *Belonolaimus* sp. on corn', *Journal of Nematology* 21(4S), 697–702.

Todd, T. C. (2016), 'Diseases caused by nematodes', in Munkvold, G. P. and White, D. G. (eds.), *Compendium of Corn Diseases Fourth Edition*, pp. 117–19. The American Phytopathological Society.

Todd, T. C., and Oakley, T. R. (1996), 'Seasonal dynamics and yield relationships of *Pratylenchus* spp. in corn roots', *Journal of Nematology* 28(4S), 676–81.

Townshend, J. L., and Davidson, T. R. (1989), 'Grass and legume hosts of *Pratylenchus projectus*', *Nematologica* 35(1), 128–31.

Tylka, G. L., Todd, T. C., Niblack, T. L., MacGuidwin, A. E. and Jackson, T. (2011), 'Sampling for plant-parasitic nematodes in corn strip trials comparing nematode management products', *Plant Health Progress*. doi:10.1094/PHP-2011-0901-01-DG.

Venditti, M. E., and Noel, G. R. (1995), 'Pathogenicity of *Tylenchorhynchus zambiensis* to maize', *Nematropica* 25(1), 7–13.

Vieira, P., Danchin, E. G. J., Neveu, C., Crozat, C., Jaubert, S., Hussey, R. S., Engler, G., Abad, P., de Almeida-Engler, J., Castagnone-Sereno, P. and Ile Rosso, M.-N. (2011), 'The plant apoplasm is an important recipient compartment for nematode secreted proteins', *Journal of Experimental Botany* 62(3), 1241–53.
Vovlas, N., and Inserra, R. N. (1985), 'Single modified food cell induced by *Helicotylenchus pseudorobustus* in corn roots', *Journal of Nematology* 17(3), 371–3.
Williams, W. P., and Windham, G. L. (1998), 'Registration of root-knot nematode-resistant maize germplasm lines Mp 709, Mp710, Mp711, and Mp712', *Crop Science* 38, 563.
Windham, G. L. (1998), 'Corn', in Barker, K. R., Pederson, G. A., and Windham, G. L. (eds.), *Plant and Nematode Interactions*, pp. 335–57. American Society of Agronomy, Crop Science Society of America, and Soil Science Society of America.
Windham, G. L., and Williams, W. F. (1987), 'Host suitability of commercial corn hybrids to *Meloidogyne arenaria* and *M. incognita*', *Annals of Applied Nematology* 1, 13–16.
Windham, G. L., and Williams, W. F. (1990), 'Resistance of maize to *Meloidogyne arenaria* and *Meloidogyne javanica*', *Theoretical and Applied Genetics* 80, 810–12.
Windham, G. L., and Williams, W. F. (1994), 'Penetration and development of *Meloidogyne incognita* in roots of resistant and susceptible corn genotypes', *Journal of Nematology* 26(1), 80–5.
Windham, G. L., and Williams, W. F. (1988), 'Resistance of maize inbreds to *Meloidogyne incognita* and *M. arenaria*', *Journal of Nematology* 26(1), 80–5.
Windham, G. L., and Lawrence, G. W. (1992), Host status of commercial maize hybrids to *Rotylenchulus reniformis*, *Journal of Nematology* 24(4S), 745–8.
Yassin, A. M. (1969), 'Glass-house and laboratory studies on the biology of needle nematode, *Longidorus* elongatus', *Nematologica* 15, 169–78.
Zirakparvar, M. E., Norton, D. C. and Cox, C. P. (1980), 'Population increase of *Pratylenchus hexincisus* on corn as related to soil temperature and type', *Journal of Nematology* 12(4), 314–18.

Chapter 15

Control of rodent pests in maize cultivation: the case of Africa

Loth S. Mulungu, Sokoine University of Agriculture, Tanzania

1 Introduction

Rodents cause serious problems to human communities in Africa as a result of their involvement in the spread of diseases (Katakweba et al. 2012) and in the losses of crops through direct consumption (Mulungu et al. 2003; Bekele et al. 2003) and spoilage (Mdangi et al. 2013). For example, Taylor (1968) reported 20% damage to maize crop after the outbreak of rodents in Western Kenya. Earlier reports (Taylor 1968) on economic losses due to rodents in Kenya indicated 20–30% damage to maize crops, and a 34–100% loss during rodent outbreaks. In Ethiopia, it has been estimated that rodents consume up to 26.4% of maize crop in most years (Bekele et al. 2003). In Northern Ethiopia, surveyed farmers estimated 9–44% pre-harvest yield losses in annual production of cereal crops due to rodent attacks (Meheretu et al. 2010), while Central Ethiopia showed 26.4% loss of yield in maize (Bekele et al. 2003). In Tanzania, rodents are estimated to cause on average 15% yield loss (Makundi et al. 1991), which would mean the loss of around 382 673 tonnes per year of the actual yield (FAO statistics 2014). This amount of maize, with an estimated value of US$42.5 million (at US$11.1 per 100 kg bag of maize), would be enough to feed 2.1 million people for a whole year (at about 0.5 kg/day/person). However, in many parts of Africa, this figure has risen dramatically over the last few years, most noticeably in places where rodent outbreaks occur (Mwanjabe et al. 2002; Taylor and Green 1968, 1976). Today, it is not usual for smallholder maize farmers to report chronic rodent damage

http://dx.doi.org/10.19103/AS.2016.0002.18

of more than 80% in certain cropping seasons and localities (Mwanjabe and Leirs 1997; Mulungu et al. 2003). To overcome such problems, a sound understanding of rodent population dynamics and breeding patterns can help in the design and implementation of appropriate control strategies (Leirs et al. 1997; Singleton et al. 1999). According to Mulungu et al. (2005), there are about 31 rodent pest species involved in crop damage in Tanzania. However, the most important pest species are *Mastomys natalensis*, *Arvicanthis* spp. and *Gerbilliscus* spp. (Taylor and Green 1976; Fiedler 1994; Leirs et al. 1994), with *M. natalensis* being the most predominant rodent pest species (Leirs 1995; Mulungu et al. 2011; Massawe et al. 2012).

Outbreaks of *M. natalensis* cause significant impacts on food security at different scales, from the household through to the regional level (Leirs et al. 2010). Losses of up to 80% in maize crop have been reported in Tanzania (Mulungu et al. 2003). Irregular rodent outbreaks can result in extreme crop losses, with the potential for causing widespread famine (Mwanjabe et al. 2002). Acute effects of rodent outbreaks have been reported over large areas from many localities in Africa (Leirs et al. 2010), but chronic seasonal damage to agriculture can be even more important in terms of socio-economic impact (Skonhoft et al. 2006).

In typical farm-fallow mosaic habitats under maize cultivation, the population dynamics of *M. natalensis* generally show seasonal trends related to variations in rainfall, peaking towards the end of the rainy season when food resources are plenty (Leirs 1992; Leirs et al. 1996). Their breeding decreases during the drier months (Wube 2005), indicating that rainfall is the ultimate source of variation in rodent density (Taylor and Green 1976; Leirs 1995; Massawe et al. 2012). By contrast, in areas where farmers grow maize twice a year, rodent density increases (Mulungu et al. submitted). These two scenarios suggest that breeding is highly influenced by the presence of a staple crop and that the population dynamics and breeding patterns of *M. natalensis* are influenced by local cropping and environmental conditions. Thus, extended rainy seasons could result in longer breeding duration, with subsequent increases in population size attributed to the quality of food resources (Leirs et al. 1994).

Severe rodent damage to grains in stores contributes to the food shortage in rural communities of Africa, resulting in the immediate financial loss. Post-harvest losses in India, for example, has been reported to be 2.5% due to rodent (Haris and Linblad 1978) while in Tanzania, the extent of rodent damage on maize seeds under storage has been reported to be 35% (Mdangi et al. 2013; Makundi et al. 2006). This would mean a loss of around 1 280 650 tonnes per year of the actual yield (FAO statistics 2007), which could be higher in years with rodent population outbreaks in houses.

In recent years, in Tanzania, new developments in rodent pest management include a system designed to provide an early warning of damage, which includes an understanding of the population processes that give rise to rodent pest's problems and provides the framework for evaluating the causes and solutions. To implement control strategies, population models incorporating simulations and aspects of economics for predicting outbreaks have been developed (Skonhoft et al. 2006). Simulations based on bioeconomics models have shown that the most economically rewarding strategies (such as controlling rodents one month before planting) differ significantly from current practices of symptomatic treatment when severe rodent damage is noticed in the maize fields. Therefore, shifting from symptomatic practices and controlling rodents on a calendar basis can substantially improve the economic conditions for the majority of maize producers in Africa.

2 The impact of rodents on maize crops in Africa

2.1 Rodent damage in fields

The major vertebrate pests in Africa are rodents, although birds and other mammals may sometimes affect maize crop. Although rodents are often used as a protein supplement in some regions of Tanzania for example, they are, however, serious agricultural pests throughout the continent (Mulungu et al. 2003; Makundi et al. 1991). In fact, they do more damage than plant diseases and all other animal pests put together (Hoppe 1980). Annual crops such as maize are affected adversely by rodents. Several reports including the extent of damage caused by rodents to maize have been published (Myllymaki 1987; Leirs 1989; Key 1990; Mwanjabe and Sirima 1993; Fiedler 1994; Mwanjabe et al. 2002; Mulungu et al. 2003, 2005, 2008). These studies support the fact that rodents are a serious pest problem in maize fields.

Nature of the damage

The pattern and the extent of damage by pest rodents depend upon the species, the intensity of infestation, the type and the growth stage of the crop, and the nature of the surrounding habitat (Mulungu et al. 2005). Much of the estimated damage occurs in the pre-harvest stages of the crop, although damage occurs even at the maturity stage of the crop, at sowing stage and at seedling stage. Understanding the mechanism, the extent of the damage and the situations vulnerable to attack by rodents in maize crop and regions is important in planning management strategies (Parshad 1999). Maize damage by rodent pests is therefore not the same in time but follow crop growth stages with different rodent pest species. In Africa, maize seeds are usually sown directly into fields by hand. Rodent pest species commonly dig up and consume both seeds and seedlings (Fig. 1). Damages of this type are light and are often ignored by farmers, as it merely leads to a patchy crop, but heavily damaged fields may require re-sowing after implementing some form of rodent control programme. The soft stems of the young maize plants are also vulnerable to rodent attacks as they are highly attractive to rodents, although their loss does not necessarily have a direct effect on reducing crop yield. For example, Mulungu et al. (2012) reported that *Gerbilliscus vicinus* ate more plant material indicating that this species prefer seedling/soft stems of young plants.

The most commonly reported type of damage that is inflicted by rodents is to the developing maize cobs (Mulungu et al. 2005). Maize stems are relatively robust, and the

Figure 1 Rodents dig up sown maize seeds and seedlings in fields.

more agile rodents such as *M. natalensis* are able to climb them to attack the cobs in situ. Cobs damaged this way may be completely destroyed, but more often, they are only partially damaged. Mulungu et al. (2003) identified two types of partial damages to maize cobs. In one type, the damage is confined to the removal of kernels in strips down the longitudinal axis of the cobs, and in the other, kernels are taken in rings around the circumference.

Rodent pests consume both milky (Fig. 2) grains, when the maize stems are standing upright, and dry grains; therefore, the rodent damage is continuous and persists until the crop has been harvested. Few maize plants could fall and be attacked by rodents. It has been reported that rodent activity is increased, especially when the maize fields are surrounded by vegetation cover (Funmilayo and Akande 1977). Most of the damage is caused to ears on dislodged or leaning plants, starting with the outer rows and moving towards the centre of the field.

The spatial patterns of rodent individuals in maize fields vary according to habitat quality due to changes in land use, and it is expected that the individuals of resident animals will exhibit important spatio-temporal differences that can potentially affect crop damage patterns and severity (Mulungu et al. 2005; Mulungu et al. 2015a). Understanding the pest distribution patterns is an important prerequisite for developing an effective management programme for the pest in question. Massawe et al. (2003) reported a clustered distribution of rodent captures, and Mulungu et al. (2015a) reported that *M. natalensis* in irrigated rice fields generally exhibits an aggregated spatio-temporal distribution at different crop growth stages. The distribution of damage within maize fields, however, probably depends on local conditions; areas close to other infested crops or to land offering harborage to rodents are more prone to attack (Mulungu et al. 2005). The spatial distribution of rodent damage in the maize field however, is either random or regular, depending on the cropping patterns (Mulungu et al. 2005) suggests that crop damage depends on the individual's movements (Cheson 1981). Random damage distribution is observed in fields which are planted with maize and that particular field is surrounded by other maize cultivated fields (Mulungu et al. 2005). These observations suggest that the individual rodents present in these fields after ploughing are either residents or passerby animals indicating that no visitors from the fallow land to maize fields because the fallow land was far from the maize field. The distribution of crop loss over a wide area is related to the pest distribution in both time and space (Kumar 1984). However, populations of rodents not only vary in relation to time, but also spatially, with the spatial variation being influenced by habitat type (Leirs et al. 1997). With very few exceptions of some rodent pest species, majority of them favour habitats with a high degree of heterogeneity and are discouraged by intensively cultivated mono-cultures (Myllymäki 1987). Similarly, the timing

Figure 2 Rodents damage to maize cobs at maturity stage of the crop.

of rodent damage and distribution within the crop field vary considerably with the rodent species, the local environment and the age of the crop (Hampson 1984). In maize fields surrounded by fallow land, the rodent damage is reported to be regular, indicating that rodent pest species have more time to cause damage to maize seeds and plants, as fallow lands are the focal point for rodent and they act as a cover for rodent to some extent. The feeding behaviour of *M. natalensis* at the crop's maturity stage means more cob damage at the periphery of the maize fields. It is common for rodent to cause damage to maize at maturity; however, the damage is often limited, or negligible, unless the maize has fallen on the ground, which may be mainly due to termite attack on the roots.

Crop compensation/recovery after rodent damage

It has been reported that in many studies, non-uniform plant spacing within the row has little or no effect on plant growth and on grain yield of maize if the plant population is adequate for high yield (Edmeades and Daynard 1979). The same has been reported by Myllymäki (1987) who reported that compensation in maize crop is minimal after rodent damage has occurred. This in fact has been observed at low levels of rodent damage, when farmers may actually thin the seedlings to two per planting hole so as to reduce competition between plants. It is also important to note that different factors contribute to crop loss, and the yield response is usually variable at a given location and time (Walker 1990), which may be due to environmental factors, for example rainfall (Mulungu et al. submitted).

However, Abdulahi (1994) reported that when maize seeds have been planted in a space of 75 by 25 cm and 0, 15, 30 and 45% damage levels of the plant population were removed at the 6-leaf, 9-leaf and tasselling stages, no significant yield difference between the 0% plant removed and at 15% of the plant population at the 6-leaf and 9-leaf stages suggesting the compensation by the remaining plants. By contrast, yield difference was recorded between the 0% plant removed and the other plant removal levels such as 30 and 45. Yield losses of 10.9 and 26.4% were recorded when 30 and 45%, respectively, of the plant population was removed at the early, indicating that crop damage and yield loss are not the same, and maize plants could tolerate up to 15% loss of plant populations due to pest damage at the early stage of crop development without significant yield reductions (Abdulahi 1994). However, Pommel and Bonhomme (1998) reported that with plant populations of 130 000 ha$^{2\ 1}$, the ears lost corresponding to missing plants are poorly compensated by increased yield of surrounding plants because of additional light interception: when two or three adjacent plants were missing, compensation for missing plants was only 16% and 34%, respectively.

2.2 Rodent damage in stores

Rodents cause direct damage to maize grains by gnawing and feeding (Table 1) and indirect damage by spoilage (Mdangi et al. 2013). Many farmers in Africa store maize grains in sacks; however, storing in this way, rodent problems may increase although the problem may vary from just an occasional damaged grain sack to severe damage that results in the collapse of bag stacks depending on the type of houses. Mdangi et al. (2013) reported that damage of maize grains is greater in houses thatched with grasses as compared to those with corrugated iron. The author also found higher rodent population in the storage of houses which are thatched with grasses as compared to those with corrugated iron sheets. They contaminate (with their droppings, urine, hair, and other body parts),

Table 1 Stored maize grains at farm and village level in Asia, Africa and Latin America

Continent	Country	Type of storage	Percentage damage or loss
Asia	Nepal	Sack	3–5
	The Philippines	Cribs, sacks	2–3
	Thailand	Sacks in roof, maize in cribs	5
	Turkey	Farm houses, underground pits	5
Africa	Egypt	Houses and stores	50
	Ethiopia	Huts on stilts, underground, bags	5–15
	Ghana	Traditional barns or in rooms; drying in field, on platform, on ground, and use of mud silos.	2–3
	Malawi	Woven cane bins, grass baskets, Cribs Cob maize	0.5–15
	Sierra Leone	Roof and cribs	2–3
	Zaire	Bags in roof	3
	Zambia	Farm cribs	10
	Tanzania	Sacks	35
Latin America	Mexico	Cribs, sacks in roofs	5–10
	Brazil	Stacks, sacks, cribs	4–8

deteriorate, and make the crop susceptible to fungal and bacterial infections during pre- and post-harvesting stages (Gregory 2002). It has been reported that, within six months, one pair of mice can eat more than 2 kgs of food and may deposit about 18 000 droppings (Alberta Agriculture Food and Rural Development 1996). Maize seeds contaminated by mice are about 10 times more than what is eaten. Apart from the maize seedseaten, spoiled or contaminated, there are additional 'invisible' losses such as the replacement or repair of packaging materials and the cost of rebagging spilled food. Much of the spillage arises when rodents attack food packaging to obtain nesting material; stacks of heavily infested bagged foodstuffs may ultimately collapse. Rats and mice gnaw inedible materials including electrical wiring, so their presence in buildings can constitute a fire hazard.

2.3 Disease transmission to humans

Rodents are capable of transmitting diseases to people either directly – by bites, through the air or during the handling of rodent carcasses – or indirectly, through contact with food and water contaminated with rodent hair, droppings and urine, which also constitute filth in the grain (Mgode et al. 2014). Therefore, they play an important role as reservoirs and hosts for many pathogens of animal and humans (Gratz 1994). Agents of rodent-borne zoonoses include viruses, bacteria, rickettsia, protozoa and helminths (Gratz 1994, 1997). Infections with zoonotic haemoparasites are widespread in wild rodents (Korbawiak et al. 2005). They include borrelia, trypanosomes, bacilli, plasmodia and coccobacilli (Juha et al. 2003; Powelczyk et al. 2004). In humans, these pathogens are responsible for many rodent-borne diseases including plague, leptospirosis, toxoplasmosis, leishmaniasis and haemorrhagic fevers. In additional, they cause spoilage and contamination of food such as maize grains with hair, urine, and faeces. They also bite people, killing chicks and lead to storage structural damage (Meerburg et al. 2009).

3 Rodent pests affecting maize

3.1 Species

Africa has almost 400 rodent species, but only about 5% are crop pests. Of these *Mastomys natalensis* (multimammate rat) and, to a lesser extent, *Arvicanthis* species (grass rats) are the dominant rodent pests of sub-Saharan Africa, and they frequently involved in rodent population outbreaks. Both species damage maize crop (Taylor and Green 1976; Table 2) and are found in cereal-growing areas. They have a similar ecological requirement, being essentially animals of grasslands, and under natural conditions, they feed on grass or grass seeds, supplemented by insects (Leirs et al. 1994; Mulungu et al. 2012). All species have short life spans of one to two years and a high reproductive potential (Taylor and Green 1976).The African giant rat (*Cricestomys giambianus*), cane rat (*Thryonomys* spp), gerbils (*Gerbillicus* spp), spine mice (*Acomys* spp), stripped grass mouse (*Lemniscomys* spp), crested porcupine (*Hystrix cristata*), ground squirrels (*Xerus* spp), and mole rat (*Tachyoryctes* spp) and *Cryptomys* spp are found in different localities and countries (Fiedler 1994; Table 2). These species are also pests of maize, and they cause serious damage to crops before and after harvest (Mulungu et al. 2006), but are relatively slow breeding.

3.2 Distribution

Multimammate and grass rats are widely distributed. *Mastomys natelensis* is found throughout the continent, with other *Mastomys* species more locally present. *Arvicanthis* species live in the northern part of the continent. Both live in grasslands and wooded savannah, cultivated areas and villages. By contrast, the orange-toothed mole rat is found in the moist uplands of Ethiopia, Somalia, Kenya, Tanzania, Uganda and the eastern Democratic Republic of the Congo.

3.3 Ecology and breeding patterns

Rodents, for example mole rat, feed on sweet potato roots and can cause serious damage by digging through the mounds or ridges to eat them, or by attacking them when they are exposed above ground. Rats live above ground and nest on or in the ground or in trees, depending on the species. Species that cause outbreaks have high reproductive capabilities. The gestation period of the multimammate rat is only 21 days; as soon as they give birth they can conceive again after 24 hours. Mole rats burrow into the soil, eating storage roots from below ground and live and breed in underground burrows. Gestation is longer at about 7 weeks. It has been reported that rat outbreaks is caused by several factors, including (i) a long rainy season that provides more food and cover, allowing better survival (ii) a reduction in competition from other rat species, predators and disease, when there is a return to rains after consecutive dry years (iii) early breeding, when unusually heavy rains occur during the rainy season and the progeny join the main breeding season

Table 2 Rodent pest species in Africa

S/N	Rodent pest species	Crop attacked
1	*Mastomys natalensis* Smith (Shamba or multimammate rat)	Cereals, potatoes, root crops, legumes, cotton, sugar cane
2	*Mastomys erythroleucus*	Cereals, potatoes, root crops, legumes, cotton, sugar cane
3	*Rattus rattus* L. (Roof rat)	Cereals in stores, cassava, potatoes, beans, cotton, groundnuts
4	*Arvicanthis abyssinicus* Rupell (Unstriped grass mouse)	Thin-stemmed cereals, especially wheat, oats, barley, millets, root crops, planted seeds, sugar cane
5	*Arvicanthis niloticus* Desmarest (nile or grass rat)	Thin-stemmed cereals, especially wheat, oats, barley, millets, root crops, planted seeds, sugar cane
6	*Arvicanthis neumanni*	Thin-stemmed cereals, especially wheat, oats, barley, millets, root crops, planted seeds, sugar cane
7	*Arvicanthis nairobae*	Thin-stemmed cereals, especially wheat, oats, barley, millets, root crops, planted seeds, sugar cane
8	*Mus musculus*	Stored cereals, beans and other stored products
9	*Mus minutoides*	Cereals
10	*Mus mahomet*	Cereals
11	*Funisciurus spp* (Tree squirrels)	Maize, fruits, etc.
12	*Heliosciurus spp* (Sun squirrels)	Maize, fruits, etc.
13	*Cricetomys gambianus* (Giant African rat)	Cassava, coffee, cereals, etc., debarking of young trees and eating roots
14	*Gerbilliscus spp* (Gerbils)	Root crops (cassava, sweet potatoes, etc.), attack cereal seeds and seedlings in the field
15	*Heliophobius spp.* (Mole rats)	Root crops (cassava, sweet potatoes), roots of other crops, including cereals, cabbages, tomato and banana rhizomes
16	*Tachyoryctes spp.* (orange-toothed mole rats)	Root crops (sweet potatoes, Irish potatoes, cassava), cereals (damage to the shoots), vegetables (cabbages, tomato), pineapples, forage crops, banana rhizomes
17	*Lemniscomys striatus*	Cereals and grasses (minor damage to a variety of crops)
18	*Lemniscomys rosalia*	Cereals and grasses (minor damage to a variety of crops)
19	*Lemniscomys zebra*	Cereals and grasses (minor damage to a variety of crops)
20	*Lemniscomys barbarus*	Cereals and grasses (minor damage to a variety of crops)
21	*Otomys spp* (Swamp rats)	Young plantation trees and cereals
22	*Thryonomys spp.* (Cane rats)	Cereals
23	*Rhabdomys spp.* (Four-striped mouse)	Cereals and young trees
24	*Dasymys spp.* (short-tailed mouse)	Cereals, cassava
25	*Hystrix spp.* (porcupines)	Cereals, cassava

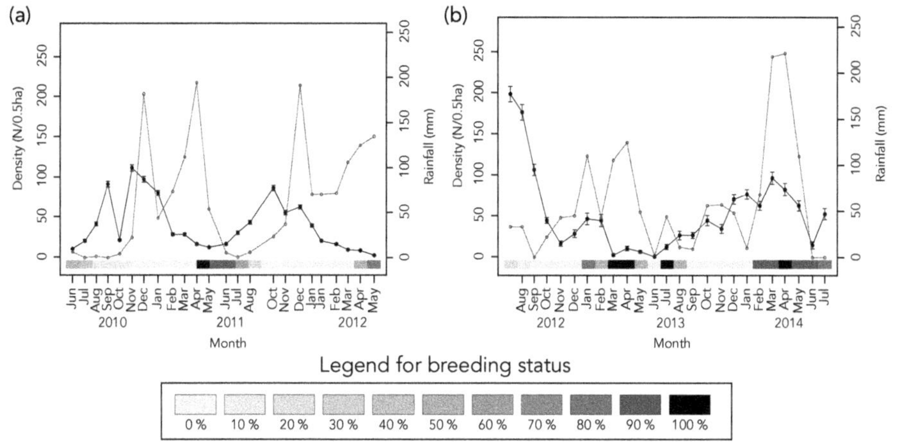

Figure 3 Population density (M(h)) of *Mastomys natalensis* ± standard error bars (full black line), rainfall (grey dotted line) and proportion of breeding females (greyscale gradient, from light = 0% to dark = 100% breeding) in three different cropping systems: (a) rainfed maize/single crop farm (3h area); (b) rainfed maize/double crop (0.5 ha).

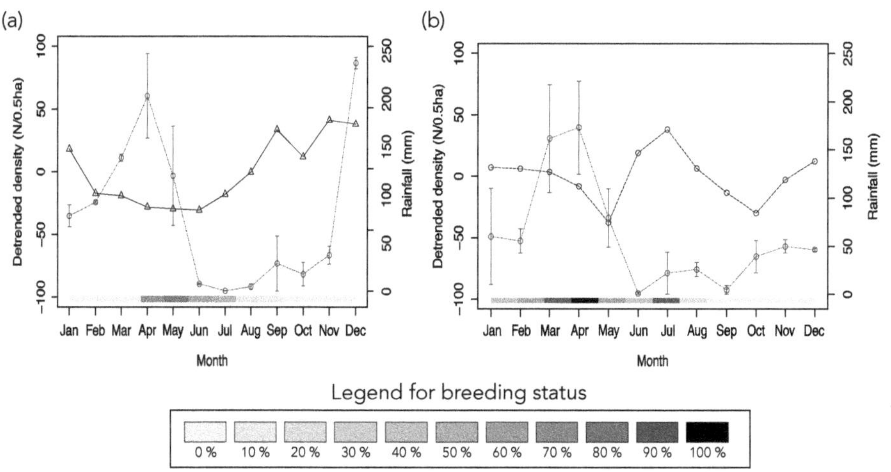

Figure 4 Seasonal density pattern after de-trending over the year at (a) rainfed maize/single crop, and (b) rainfed maize/double crop. Detrended population density bars (full black line), rainfall (grey dotted line) and mean monthly proportion of breeding females ± standard error (greyscale gradient, from light = 0% to dark = 100% breeding).

population later in the year. Breeding and population size of *M. natalensis* are reported to be linked to rainfall in the rainfed maize areas. In these areas, breeding, and hence population, are much more determined by rainfall and its population is low (Fig. 3a and b). Taylor and Green (1976) observed aseasonal breeding in an aseasonal crop in Kenya. Delany and Monro (1986) reported that African murid rodent populations generally respond to rainfall through breeding, which can be triggered by the appearance of new vegetation that may contain chemical substances that promote reproduction through reproductive hormones following rainfall (Linn 1991). Firquet et al. (1996) reported that the presence of germinating grasses induces sexual maturation in subadult female *M. natalensis*. The same authors also reported that, in October 1994, when no more reproductive activity was seen in *M. natalensis* in harvested rainfed maize fields in Morogoro, most adult females were still pregnant in old irrigated maize fields , just 50 km away; the pregnant females and young animals, showing that sexual maturation was still continuing.

In areas where farmers produce two maize crops per year, breeding duration is extended (Fig. 4a and b). Bekele and Leirs (1997) reported that breeding in mosaic crop-fallow fields is seasonal and is related to rainfall periods, but extended rainy seasons resulted in longer periods of breeding, higher litter sizes and an increase in population. Previous studies in Zimbabwe showed that breeding of *M. natalensis* was continuous and extended in irrigated and longer rainfall duration in wheat fields (Swanepoel 1980). In this situation, the link with rainfall is much weaker, although breeding is still most prominent in the rainy seasons as in single maize production areas. Evidently, in these areas, the rodents are not

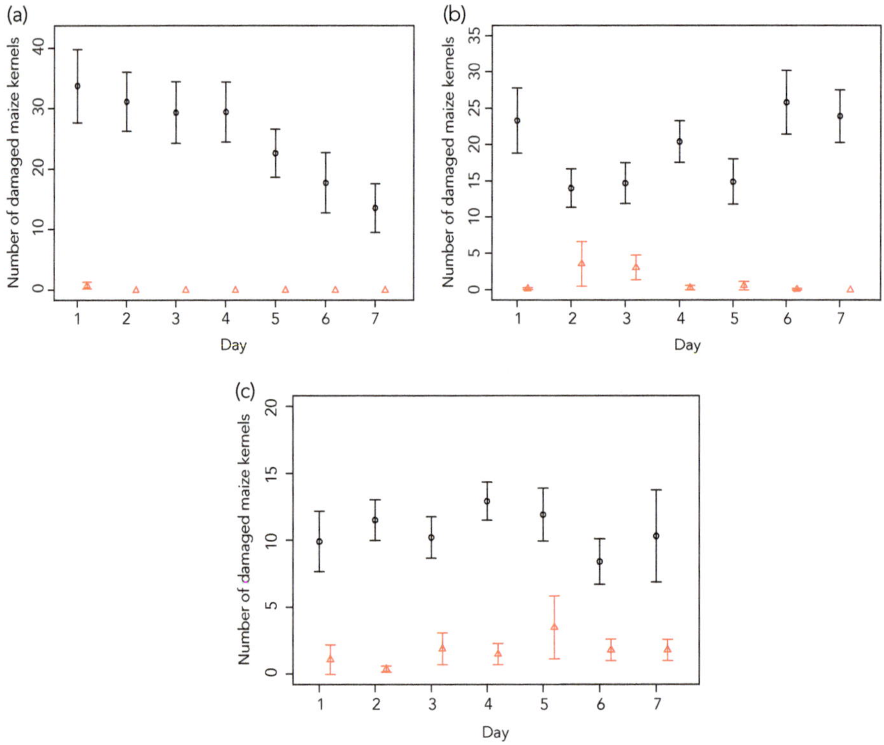

Figure 5 Mean proportion of damaged treated maize vs. the number of damaged maize seeds for each day in the control group (•) and the treated group (Δ) for different plant extracts of (a) *Ricinus communis*, (b) *Jatropha curcas* and (c) *Capscum chinense* fruit.

influenced by rainfall per se, but rather by the quantity and quality of their food, which is dependent on the phenology of crops and the surrounding vegetation.

Odhiambo et al. (2008) reported that *M. natalensis* is an opportunistic feeder consuming all types of food in different frequencies reflecting the availability of food items in its habitat. Variation in food supply of small mammals in different habitats has been reported (Bomford 1987). Odhiambo et al. (2008) reported that, although *M. natalensis* had a wide range of food items in its diet, there was a clear seasonal effect on the consumption of different food categories. They fed more on seeds, arthropods and grasses during the wet season and on other plant materials during the dry season in the south-west of Tanzania. Similar findings have been reported elsewhere in the species' range (Monadjem and Perrin 2003). The population dynamics and breeding patterns strongly suggest that food quality affects the timing and duration of breeding (Leirs et al. 1994). Elsewhere, it has been reported that the apparent seasonal reproductive cycles in tropical rats probably relate to the seasonal changes in the quantity and quality of cover and food (Boutin 1990; Christensen 1993; Pucek et al. 1993; Leirs et al. 1993; Leirs et al. 1994; Jędrzejewski and Jędrzejewska 1996; Wolff 1996).

Currently, there are 32 recorded outbreaks of the multimammate rat in Africa from 1925–2005. *Mastomys natalensis* is the most frequently involved, either alone or, in a few instances, with other *Mastomys* species, or with *Arvicanthis* species. Most outbreaks occurred in East Africa, although a few were in southern Africa and West Africa, with a major outbreak across the Sahel in 1975–6. Much attention has been paid to impacts on maize, rather than other crops such as root crops, so the loss of yield under normal circumstances and in outbreak years is not well documented.

4 Managing rodent pests in maize crops

4.1 Rodent management methods

From time immemorial, rodent pests have been causing alarming losses of both human properties and lives. Inevitably, rodent control is as old as the rodent problem itself. Indeed, this is evidenced by the presence of rat traps in archaeological digs from some parts of Asia (Jackson 1981), and by the fact that Egyptians domesticated cats as early as 2000 B.C. to protect their grains from rodent pest species (Boelter 1990). Rodent control such as hunting remained unscientific until after 1940 when principles underlying modern rodent control technologies such as using synthetic rodenticides as poison were founded following several research works conducted in Europe and USA (Brooks and Rowe 1979). Zinc phosphide was first synthesized in 1740 and first used a rodenticide in 1911 in Italy. In general, rodent control techniques such as natural enemies, cultural techniques and killing by rodenticides only cannot keep rodent outbreaks below damaging levels (Myllymäki 1987) because a prominent resilience capacity can be expected from the high local reproductive rate and the dispersal ability, immigration and survival (Leirs et al. 1994). Farmers in Africa, normally take action if pests are present in large numbers in the fields and storage areas (Makundi et al. 1999). Some farmers may respond to the risk of uncertain pest attack by scheduling prophylactic treatments in which the timing and amount of chemicals are quite independent of actual pest numbers (Myllymäki 1987). The decision is therefore normally taken irrespective of whether the pest will attack the crop (Myllymäki 1987).

4.2 Cultural approaches

A home remedy employs the legume shade tree, *Gliricidia sepium* ('rat killer'). Pounded young leaves or barks, and mixes with cooked rice, maize or other bait, or boils the Giricidia with rice or other cereal grains have been shown to control rats. Bacteria convert chemicals in the leaves to substances similar to brodifacoum (an anti-coagulant used as rat poison). These are less toxic than brodifacoum, so larger amounts must be eaten. Change the bait daily and protect children and pets by placing it in bamboo sections or tins. Planting the legume, for example *Tephrosia vogelii* (commonly known as fish bean), randomly throughout the crop cassava field and along the borders has been reported to control mole rat (Sichilima et al. 2003). This shrub contains rotenone, a fish poison and insecticide, so care should be taken when disposing of it. Mixture of cow dung and pepper placed in the burrows and burnt to smoke out the rodents, pouring one-week-old fermented cattle urine into the burrows to chase away mole rats, or digging deep ditches around crop plantings to stop rodents from tunnelling straight into the field also has been shown effect to control rats.

Mdangi et al. (Submitted) reported that extracts from *Ricinus communis* seed, *Jatropha curcas* and *Capscum chinense* reduce maize seed damage by *M. natalensis*, and its residual effectiveness was observed even after one week (Fig. 5). This result indicates that plant extract of these plant materials has some compounds associated with repellency to *M. natalensis*.

4.3 Chemical control

Generally, management of rodent pests in Africa relies on the ad hoc use of chemical rodenticides (Makundi et al. 1999; Myllymaki 1987). Rodenticides used in this way, however, are rarely economically and ecologically sustainable, and often they are applied only when damage has already occurred (Makundi et al. 1999; Mulungu 2003; Skonhoft et al. 2006). Rodenticide application is not widely practical on individual farms. The success in the use of rodenticides (whether acute or anticoagulants) depends on (i) availability of the required rodenticides (often influenced by available funds for their purchase) and (ii) acceptability of baits formulations to rodents (often influenced by palatability under field conditions. For example, the currently used bait formulation for *M. natalensis* is bromadiolone in cracked maize and sardines. Rice is more acceptable by rodents than maize, but since it is a highly valuable crop the government cannot afford to use it as bait), and (iii) the timing of bait application. This is critical for alleviating damage (Makundi et al. 1999, Mulungu 2003).

Chemical control measures against rodents in Africa continue to be implemented extensively. Most of the countries in Africa, unregistered compounds with acute vertebrate toxicity are commonly used to kill rodents. These poisons are used also as suicide, murder and killing livestock as non-targeted animals. However, acute poisons are often perceived to be more effective at killing rodents than chronic poisons, and changing these misconceptions is one of the major challenges facing experts working to improve rodent management in Africa, where acute poisons are widely used. Fast-acting, acute rodenticides, for example zinc phosphide, is used widely by smallholder farmers (Buckle, 1999), but, in Tanzania, they are supplied mostly free of charge by the government during outbreaks. Till date, however, zinc phosphide has not been registered in Tanzania, but it is strictly used under the supervision of Ministry of Agriculture, Food Security and Cooperative and is not allowed to be sold in open market. The decision to use this toxic compound has been made by the government, as currently there are no other cheap alternative rodenticides to be used during outbreaks. Farmers also favour this compound because its effect is immediately apparent after application. However, the proper application can be a problem under farmer's situation. The compound is supplied as a concentrate, which must be mixed with high-quality cereal baits if it is to be effective, and many smallholder farmers have insufficient resources and knowledge to be able to do this accurately and safely and, therefore, need a close supervision. In fact, a programme for rodent control which is solely a government responsibility would be difficult to implement due to a shortage of manpower, apart from the exorbitant costs involved (Mkondya 1977).

Rodents are well known for being suspicious of new objects (neophobic), a characteristic which is poorly understood or accounted for by smallholders (Myllymäki 1987). Rodents also develop bait shyness (toxiphobia), leading to avoidance of baits used against them following violent reactions to the chemical. In order to avoid bait shyness, more potent anticoagulants such as bromadiolone have been developed. Bromadiolone is effective against species 'naturally resistant' to the first generation anti-coagulant poison (Myllymäki 1987). Anticoagulant rodenticides are probably best for an Integrated Rodent Management (Buckle 1994). Even with an integrated rodent management strategy that uses monitoring systems to target the timing of rodenticides applications, it is felt that the emphasis of such an approach is on control rather than management. Similar concerns have arisen with insect pests (Flint and Bosch 1987) and weeds (Higley and Pedigo 1999).

4.4 Biological control

Biological control of rodents in Africa in general is an almost unexplored area (Vibe-Petersen 2003). The importance of predation in vertebrate prey population dynamics has been debated for many years, particularly in populations exhibiting regular density fluctuations in the northern temperate zones (Hansson and Henttonen 1988; Norrdahl 1995; Krebs 1996; Abrams 2000). Considerably less attention has been directed to the role of predation in tropical rodent population dynamics. Handwerk (1998) investigated the predator–prey relationship of rodents in Egypt, but monitored predators and rodents in areas distant from each other. Leirs et al. (1997) showed that survival of non-reproducing *M. natalensis* is an inverse density-dependent process, an effect that would be predicted if predation is an important mortality factor.

In Tanzania, Van Gulck et al. (1998) reported that raptor activity increased in areas with perch poles and an increased survival of *M. natalensis* in areas excluding avian predators. Vibe-Petersen (2003) investigated the effects of different levels of predation pressure on the population dynamics of *M. natalensis* in maize fields and its consequences on crop damage and maize yield production. The author showed that (i) population growth during the annual increase phase was faster in the absence of predators and peak population size increased, (ii) predators might be a strong driving force for rodent emigration, and (iii) manipulating predation pressure by perch poles and nest boxes did not affect rodent population dynamics directly, but might have an indirect beneficial effect on maize yield by changing the rodents' foraging behaviour.

4.5 Integrated rodent management

Very little work has been done on integrated rodent pest management (IPM) in Africa. In Tanzania, a rodent IPM package for rice crop in Morogoro region has been developed. This package includes: (i) population monitoring before land preparation, (ii) application of rodenticides when population is high, (iii) cleaning or preparation of the farms, (iv) population monitoring and application of rodenticides before planting, (v) reduction of rice bund size to minimize building of nests and burrows,(vi) cleaning of rice bunds, (vii) monitoring population before booting stage and application of rodenticides if the population is still high and weeding, (viii) permanent water buckets along bands for daily trapping of rodents, and (ix) timely harvesting without pilling of harvested rice in the farms (Ngowo 2003; unpublished data). An IPM programme for rodents is not applicable during outbreaks because the mortality may not be high enough to reduce damage while the remaining population often compensates due to better survival and breeding performance (Leirs et al. 1994; Myllymäki 1987).

4.6 Use of early warning signs

Rodent outbreaks, particularly of the multimammate rats *M. natalensis*, are experienced regularly in Africa (Mwanjabe et al. 2002, Bekele and Leirs 1997). A system designed to provide early warning of potential damage must therefore form an integral part of any control strategy (Mwanjabe and Leirs 1997). Research in Tanzania has found that cumulative rainfall of December and January could be used as a predictor for an outbreak 6 months in advance. Furthermore, calculations showed that if rainfall exceeded 366 mm for December and January then control actions would be cost-effective, taking into account the amount of damage that was likely. Davis et al. (2004) discussed how to find the economically optimal cut-off value in such a prediction model, depending on the cost of damage, the cost of control actions and the reduction of damage when control is undertaken.

Also a more mechanistic model was developed that allows simulating populations of *M. natalensis* in Tanzania (Leirs et al. 1997). This model uses recent rainfall and density dependence to predict, monthly wise, how a population of these rodents develops. By including time series of rainfall, it can simulate the population dynamics. Due to the stochastic nature of precipitation, however, it cannot be used for actual forecasting. It does however function well if it is used to find out the conditions to which the population is sensitive or the ways in which interventions could change the population dynamics (Leirs 1999). However, little has been done in other African countries on the predictive model.

5 Summary

Rodents cause serious problems to human communities in Africa as a result of their involvement in the losses of crops through direct consumption of maize grains (Mulungu et al. 2003) and through spoilage (Mdangi et al. 2013). In Africa, damage to maize is largely attributed to *M. natalensis* and *Arvicanthis* spp, and it has been estimated for these rodent species to cause damage up to 80% in maize field (Mulungu et al. 2003). Therefore, an outbreak of *M. natalensis* causes significant impacts on food security at different scales, from the household through to the regional level. Acute effects of rodent outbreaks have been reported over large areas from many localities in Africa, but chronic seasonal damage to agriculture can be even more important in terms of socio-economic impact (Skonhoft et al. 2006). However, in different localities, the Nile rat, *Arvicanthis* spp, African giant rat (*Cricestomys giambianus*), cane rat (*Thryonomys* spp), gerbils (*Gerbillicus* spp), spine mice (*Acomys* spp), stripped grass mouse (*Lemniscomys* spp), crested porcupine (*Hystrix cristata*), ground squirrels (*Xerus* spp), and mole rat (*Tachyoryctes* spp) and *Cryptomys* spp are found. These rodent species are found throughout the maize-growing areas and are also pests of maize, causing serious damage to crops before and after harvest.

Rodent management programmes in Africa have been reactive and have not considered the population ecology of the target species. The strategies used to manage rodents are probably most suited to managing low-density rodent populations and are selected to solve localized rodent problems in certain areas. These include bounty schemes, burning of homes and vegetation, trapping and poisoning. In recent years, new developments in rodent pest management include a system designed to provide early warning of potential damage. An understanding of the population processes that give rise to rodent pest problems provides the framework for evaluating the causes and solutions. Population models for predicting outbreaks have been developed. These have incorporated simulations and aspects of economics in the implementation of control strategies. Simulation based on bioeconomics models have shown that the most economically rewarding strategies differ significantly from current practices of symptomatic treatment when severe rodent damage is noticed in the fields (Skonhoft et al. 2006). Therefore, shifting from symptomatic practices and controlling rodents on a calendar basis can substantially improve the economic conditions for the majority of maize producers in the African context.

6 Future trends in rodent research

6.1 Disease

6.1.1 Haemoparasites

With haemoparasites and other pathogens causing schistosomiasis, borelia, leptospirosis, toxoplasmosis in humans, we need to establish knowledge on their prevalence, mode of transmission, and pathogenicity. More studies need to be conducted on their transmission ecology in Africa continent.

6.1.2 Behaviour of rodents

To establish the social behaviour of species which transmit disease and damage agricultural crops in order to develop sustainable technologies for their management. We need to establish how behaviour affects interaction with people in the transmission of diseases.

6.2 Taxonomy and systematics

We need to conduct research on the evolutionary relationship between species of rodents using molecular systematics. By establishing the phylogenetic relationships, future research will expand the number of species. Further, we need to establish the evolutionary relationships of rodents and the pathogens they carry. There is critical need to establish the phylogenetic relationship of the pathogenic and non-pathogenic arenaviruses found in rodent.

6.3 Population ecology

We need to conduct studies on the role of rodent population dynamics in the transmission of infectious agents (e.g. plague, lassa fever and leptospirosis). Further, critical studies on the transmission ecology of the infectious pathogens harboured by rodents are required.

6.4 Monitoring and modelling rodent populations

Monitoring rodent populations and modelling to develop ecologically based management systems are important research areas. Some species are serious pests, while others harbour and transmit infectious diseases.

6.5 Long-term effects of manipulation of environmental conditions in rodent pest management

In the face of climate change, we need to establish how environmental changes will impact on rodent populations and their consequences on crop damage and disease transmission. We need to answer questions such as: what will be the effects of climate change on rodent populations and rodent-borne disease outbreaks?

6.6 Rodent conservation

Not much is being researched on methods to conserve those rodents which are not pest species. There is a need to establish how habitat fragmentation impacts of species and its effects on human-rodent interactions.

6.7 Rodent management technologies

We need to shift from poison-based rodent management to sustainable non-chemical technologies including those which are ecologically based and environmentally sound.

7 Where to look for further information

Some key resources (web sites, book chapters etc.) are provided below.

Makundi, R. H. and Massawe, A. W. 2003. Review of recent advances in studies of the ecology of *Mastomys natalensis* (Smith 1834) (Rodentia: Muridae) in Tanzania, eastern Africa. In: Singleton G. R., Hinds L. A., Krebs, C. J. and Sprat, D. M. (eds), *Rats, Mice And People: Rodent Biology And Management*, 242–5. ACIAR, Canberra. Available at: http://aciar.gov.au/files/node/451/mn96rats_mice_and_people_rodent_biology_and_mana_29893.pdf#page=242

Makundi, R. H., Oguge, N. O. and Mwanjabe, P. S. 1999. Rodent pest management in East Africa – an ecological approach. In: Singleton, G., Hinds, L., Leirs, H. and Zhang, Z. (eds), *Ecologically-based Rodent Management*. ACIAR Monograph No. 59 Australian Centre for International Agricultural Research, Canberra. http://0-purl.umn.edu.innopac.up.ac.za/114821. Available at: http://aciar.gov.au/files/node/323/ecologically_based_rodent_management_part_1_19909.pdf

Mulungu, L. S., Makundi, R. H. and Leirs, H. 2003a. Robustness of techniques for estimating rat damage and yield loss in maize fields. In: Singleton, G. R., Hinds, L. A., Krebs, C. J. and Sprat, D. M. (eds), *Rats, Mice And People: Rodent Biology And Management*, 224–8.

ACIAR Monograph Series 96, Canberra. Available at:http://aciar.gov.au/files/node/451/mn96rats_mice_and_people_rodent_biology_and_mana_29893.pdf#page=295
Mulungu, L. S., Makundi, R. H., Leirs, H., Massawe, A. W., Vibe-Petersen, S. and Stenseth, N. C. 2003b. The rodent density-damage function in maize fields at an early growth stage. In: Singleton, G. R., Hinds, L. A., Krebs C. J. and Sprat, D. M. (eds), *Rats, Mice And People: Rodent Biology And Management*, 301–3. ACIAR Monograph Series 96, Canberra. Available at: http://aciar.gov.au/files/node/451/mn96rats_mice_and_people_rodent_biology_and_mana_29893.pdf#page=295
Buckle and Smith (2015): *Rodent Pests and Their Control* (2nd Edition) (Published by CAB International): This is the 2nd edition of the classic book. It has been updated and republished. Available at: http://www.cabi.org/bookshop/book/9781845938178
Mulungu, L. S. 2003. *Assessment of maize (zea mays l.) Damage and yield loss due to rodents in the field.* Sokoine University of Agriculture, Morogoro, Tanzania.

8 References

Abdulahi, A. 1994. *Simulation of Pest Damage on Maize.* http://agris.fao.org/agris-search/search.do?recordID=ET9500061
Abrams, P. A. 2000. The evolution of predator-prey interactions: theory and evidence. *Ann. Rev. Ecol. Syst.*, 31: 79–105.
Alberta Agricultural Food and Rural Development, 1996. *Mice and their Control.* http://www.agric.gov.ab.ca/agdex/600/8300000.html
Bekele, A. and Leirs, H. 1997. Population ecology of rodents of maize fields and grass-lands in central. *Ethiopia. Belg. J. Zool.*, 127: 39–48.
Bekele, A., Leirs, H. and Verhagen, R. 2003. Composition of rodents and damage estimates on maize farms at Ziway, Ethiopia. In: Singleton, G. R., Hinds, L. A., Krebs, C. J. and Spratt, D. M. (eds), *Rats, Mice and People: Rodent Biology and Management.* Australian Centre for International Agricultural Research, Canberra, pp. 262–3.
Boelter, W. R. 1909. *The Rat Problem.* John Bale, Sons and Danielsson Ltd., Oxford, pp. 17–28.
Bomford, C. A. (1987). Food and reproduction of wild house mice. II. A field experiment to examine the effect of food availability and food quality on breeding in spring. *Australian Wildlife Research* 14: 197–206.
Boutin, S. 1990. Food supplementation experiments with terrestrial vertebrates: patterns, problems, and the future. *Canad. J. Zool.*, 68: 203–20.
Brooks, J. E. and Rowe, F. P. 1979. *Commensal Rodent Control.* WHO/VBC/79.726, p. 109.
Buckle, A. P., 1999. Rodenticide – their role in rodent pest management in tropical agriculture. In: Singleton, G. R., Hinds, L. A., Leirs, H. and Zhang, Z. (eds), *Ecologically – Based management of Rodent Pests.* Australian Centre for International Agricultural Research Canberra, pp. 163–77.
Cheson, P. L. 1981. Models for spatially distributed populations: the effect of within-patch variability. *J. Theo. Population Biol.*, 19: 288–325.
Christensen, J. T. 1993. The seasonal variation in breeding and growth of Mastomys natalensis (Rodentia, Muridae): evidence for resource limitation. *Afr. J. Ecol.*, 31: 1–9.
Cleveland, R. B., Cleveland, W. S., McRae, J. E. and Terpenning, I. S. T. L. 1990. A seasonal-trend decomposition procedure based on loess. *J. Off. Stat.*, 6: 3–73. doi:citeulike-article-id:1435502.
Davis, S. A., Leirs, H., Pech, R., Zhang, Z. and Stenseth, N. C. 2004. On the economic benefit of predicting rodent outbreaks in agricultural systems. *Crop Prot.*, 23: 305–14.
Delany, J. M. and Monro, H. R. 1986. Population dynamics of *Arvicanthis niloticus* (Rodentia: Muridae) in Kenya. *J. Zool. Lond.*, 209: 85–103.
Duplantier, J. M., Granjon, L. and Bouganaly H. 1996. Reproductive characteristics of three sympatric species of Mastomys In Senegal, as observed In the field and In captivity. *Mammalia* 60: 629–38.

Edmeades, G. O. and Daynard, T. B. 1979. The development of plant-o-plant variability in maize at different planting densities. *Can. J. Plant Sci.*, 59: 561–76.
Fagerstone, K. A., Fall, M. W., Witmer, G. W. and Pitt, W. C. 2012.'Rats! Foiled Again: A History of Rodent Control Methods Development at the National Wildlife Research Center'. USDA National Wildlife Research Center Staff Publications Paper 1493. FAO Statistics, 2007. *Maize Production In Tanzania*. http://www.FAO.org.
FAO Statistics 2014. *Maize Production In Tanzania*. http://www.FAO.org.
Fiedler, L. A. 1994. *Rodent Pest Management In East Africa*. FAO Plant Production and Protection Paper 123, FAO, Rome, Italy.
Firquet, E., Leirs, H. and Bronner, G. 1996. Germinating grasses and reproductive seasonality of Mastomys species (Rodentia, Muridae). *Mammalia*, 60: 775–9.
Flint, M. L and van den, B. R. 1987. *Introduction to Integrated Pest Management*. Plenum Press, Nwe York and London, p. 240.
Flowerdew, J. R. 1985. The population dynamics of wood mice and yellow-necked mice. *Sym. Zoolog. Soc. London*, 55: 315–38.
Funmilayo, O. and Akande, M. 1977. Vertebrate pest of rice in Southwest Nigeria. *Pest Articles and News Summaries*, 23(1): 38–48.
Gratz, N. G. 1994. Rodents as carriers of diseases. In: Buckle, A. P. and Smith, R. H. (eds), *Rodent Pests and their Control*. CAB International, Cambridge, UK, pp. 85–108.
Gratz, N. G., 1997. The burden of rodent-borne diseases in Africa south of Sahara. *Belgian J. Zool.*, 127: 71–84.
Gregory, K. 2002. *Information On Common Pests*. http://www.statistics of common pests.html
Hampson, S. J. 1984. A review of rodent damage to sugar cane with criteria for the use of rodenticides. In: Dubock, A. C. (ed.), *Proceedings of a Conference on: The Organization and Practice of Vertebrate Pest Control*. 30 August – 3 September 1982, Elvetham Hall Hampshire England, pp. 227–51.
Handwerk, J. 1998. Zur Ökologie von Prädatoren (Mammalia, Aves, Serpentes) und deren Wirkung auf die Populationen von *Mus musculus praetextus* Brants, 1827, und *Avicanthis niloticus* Desmarest, 1822, im Niltal Ägyptens. PhD thesis, der Rheinischen Friedrich-Wilhelms-Universität zu Bonn.
Hansson, L. and Henttonen, H. 1988. Rodent dynamics as community processes. *Trends Ecol. Evol.*, 3: 195–200.
Higley, L. G. and Pedigo, L. P. 1999. Decision thresholds in pest management. In: Ruberson, J. R. (ed.), *Handbook of Pest Management*. New York, USA, pp. 741–63.
Hoppe, A. H. 1980. The control of rodent pests in grain stores. In Post harvest problems, documentation of a OAU/GTZ Seminar. Deutsche Gesellschaft fu Technische Zusammenarbeit (GTZ) GmbH, Germany, p. 291.
Jackson, W. B. 1981. *Of Men and Rats*. Professorial Lecturer Series, Bowling Green State University, p. 15.
Jędrzejewski, W. and Jędrzejewska, B. 1996. Rodent cycles in relation to biomass and productivity of ground vegetation and predation in the Palearctic. *Acta Theriologica*, 41: 1–34.
Juha, L., Jukka, T. L., Haingotiana, R. and Wright, P. C. 2003. Trypanosome parasites in the invading *Rattus rattus* and endemic rodents in Madagascar. In: Singleton, G. R., Hinds, L., Krebs, C. J. and Spratt, M. D. (eds), *Rats Mice and People: Rodent Biology and Management*. Australia Centre for International Agricultural Research, Canberra, pp. 37–9.
Katakweba, A. A. S., Mulungu, L. S., Eiseb, S., Mahlaba, T. A., Makundi, R., Massawe, A. and Belmain, S. R. 2012. Prevalence of haemoparasite, leptospires and cocobacilli with potential for human infection in the blood of rodents and shrews from selected localities in Tanzania, Namibia and Swaziland. *Afr. Zool.*, 47(1): 119–27.
Key, A. 1990. Pre-harvest crop losses to the African striped ground squirrel, *Xerus erythropus*, in Kenya. *Tropical Pest Manage.*, 36(3): 223–9.
Kilonzo, B., Mhina, J., Sabuni, C. and Mgode, G. 2005. The role of rodents and small carnivores in plague endemicity in Tanzania. *Belgian J. Zool.*, 135 (Suppl.): 119–25.
Korbawiak, G., Rychlik, L., Norwakowsk, W. and Wita, I. 2005. Natural infestation of small mammals with blood parasites on the borderland of boreal and temperate forest zones. *Acta Theriologica*, 50: 31–42.

Krebs, C. J. 1996. Population cycles revisited. *J. Mammalogy*, 77: 8–24.
Laudisoit, A., Neerinckx, S., Makundi, R. H., Leirs, H. and Krasnov, B. R. 2009. Are local plague endemicity and ecological characteristics of vectors and reservoirs related? A case study in north-east Tanzania. *Curr. Zool.*, 55(33): 200–11.
Leirs, H., Stenseth, N. C., Nichols, J. D., Hines, J. E., Verhagen, R. and Verheyen, W. N. 1997. Seasonality and non-linear density dependence in the dynamics of African Mastomys rats. *Nature*, 389: 176–80.
Leirs, H. 1992. Population ecology of *Mastomys natalensis* (Smith 1834) Multimmamate rats: possible implications for rodent control in Africa. PhD Thesis. University of Antwerp, Belgium.
Leirs, H. 1995. Population ecology of *Mastomys natalensis* (Smith 1834): implication for rodent control in East Africa. Agricultural Edition No. 35, Belgian Administration for Development Cooperation, Brussels, Belgium, pp. 1–268.
Leirs, H., Sluydts, V. and Makundi, R. 2010. Rodent outbreaks in sub-Saharan Africa. In: Singleton, G. R. (ed.), *Rodent Outbreaks: Ecology And Impacts*. International Rice Research Institute, Los Banos, pp. 269–80.
Leirs, H., Verhagen, R. and Verheyen, W. 1993. Productivity of different generations in a population of *Mastomys natalensis* rats in Tanzania. *Oikos*, 68: 53–60.
Leirs, H., Verhagen, R. and Verheyen, W. 1994. The basis of reproductive seasonality in Mastomys rats (Rodentia: Muridae). *J. Trop. Ecol.*, 10(1): 55–66.
Leirs, H., Verhagen, R., Verheyen, W., Mwanjabe, P. and Mbise, T. 1996. Forecasting rodent outbreaks in Africa: an ecological basis for Mastomys control in Tanzania. *J. Appl. Ecol.*, 33: 937–43.
Leirs, H. 1994. Population ecology of *Mastomys natalensis* (Smith, 1834): implications for rodent control in Africa. PhD. Thesis. Submitted at University of Antwerp, p. 268.
Linn, I. J. 1991. Influence of 6-methoxybenzoxazolinone and green vegetation on reproduction of Multimmamate rat (*Mastomys coucha*). *S. Afr. J. Wild. Res.*, 21: 33–7.
Machang'u, R. S., Mgode, G. and Mpanduji, D. 1997. Leptospirosis in animals and humans in selected areas of Tanzania. *Belgian J. Zool.*, 127: 97–104.
Makundi, R. H., Massawe, A. W. and Mulungu, L. S. 2006. Ecological considerations for management of rodent pests in Tanzania. In: Makundi, R. H. (ed.), *Management of Selected Crop Pests in Tanzania*, pp. 220–35. Tanzania Publishing House Limited, Dar Es Salaam, Tanzania (ISBN 9976 1 0210 0), 476pp.
Makundi, R. H., Massawe, A. W. and Mulungu, L. S. 2005. Rodent population fluctuations in three ecologically distinct locations in North-East, Central and South-West Tanzania. *Belg. J. Zool.*, 135(Suppl.): 159–65.
Makundi, R. H., Massawe, A. W., Mulungu, L. S., Katakweba, A., Mbise, T. J. and Mgode, G. 2008. Potential mammalian reservoir in bubonic plague outbreak focus in Mbulu District, northern Tanzania, in 2007. *Mammalia*, 72(3): 253–7.
Makundi, R. H., Mbise, T. J. and Kilonzo, B. S. 1991. Observations on the role of rodents in crop losses in Tanzania and control strategies. *Beitrage Tropical Landwirsch Veterinary Medicine*, 29(4): 465–74.
Massawe, A. W., Makundi, R. H., Mulungu, L. S., Katakweba, A. and Shayo, T. N. 2012. Breeding dynamics of rodent species inhabiting farm-fallow mosaic fields in Central Tanzania. *Afr. Zool.*, 47(1): 128–37.
Massawe, A. W., Leirs, H., Rwamugira, W. P. and Makundi, R. H., 2003. Effect of land preparation methods on spatial distribution of rodents in crop fields. In: Singleton, G. R., Hinds, L. A., Krebs, C. J. and Spratt, D. M. (eds), *Rats, Mice And People: Rodent Biology And Management*. ACIAR, Canberra, pp. 229–32.
Meerburg, B. G., Singleton, G. R. and Kijlstra, A. 2009. Rodent-borne diseases and their risks for public health. *Crit. Rev. Microbiol.*, 35(3): 221–70.
Mdangi, M., Mulungu, L. S., Massawe, A. W., Eiseb, S. J., Tutjavi, V., Kirsten, F., Mahlaba, T., Malebane, P., von Maltitz, E., Monadjem, A., Dlamini, N., Makundi, R. H. and Belmain, S. R. 2013. Assessment of rodent damage to stored maize (Zea mays L.) on smallholder farms in Tanzania, Intern. *J. Pest Manag.*, 59(1): 55–62.

Meheretu, Y., Kiros, W., Deckers, S., Raes, D., Makundi, R. and Leirs, H. 2010.Farmers' perspectives of rodent damage and management from the highlands of Tigray, Northern Ethiopia. *Crop Prot.*, 29 (6): 532e539.

Mkondya, C. B. 1977. Preliminary proposals and hints for approaches to outbreak evaluation and control strategies against heavy rodent infestations in the Shinyanga region out-break foci in Tanzania. Ministry of Agriculture, Crop Development Division, Dar es Salaam, Tanzania, p. 121.

Mgode, G. F., Katakweba, A. S., Mhamphi, G. G., Fwalo, F., Bahari, M., Mdangi, M., Kilonzo, B. S. and Mulungu, L. S., 2014. Prevalence of leptospirosis and toxoplasmosis: A study of rodents and shrews in cultivated and fallow land, Morogoro rural district, Tanzania. *Tanzania J. Health Res.*, 16(3). http://dx.doi.org/10.4314/thrb.v16i3.11

Monadjem, A. and Perrin, M.(2003). Population fluctuations and community structure in a Swaziland grassland over a 3-year period. *African Zoology,* 38: 127–137.

Monadjem, A., 1998. Reproduction biology, age structure and diet of *Mastomys natalensis* (Muridae: Rodentia) in a Swaziland grassland. *Z S¨augetierkunde,* 63: 347–56.

Mulungu, L. S., Borremans, B., Mdangi, M. E., Massawe, A. W., Lopa, H., Mgode, G. F., Makundi, R. H., Leirs, H. and Belmain, S. R. submitted. Population dynamics and breeding patterns of *Mastomys natalensis* (Smith 1834) in three different agricultural practices.

Mulungu, L. S., Mahlaba, T. A., Massawe, A. W., Kennis, J., Crauwels, D., Eiseb, S., Monadjem, A., Makundi, R. H., Katakweba, A. A. S., Leirs, H. and Belmain, S. R. 2011.Dietary differences of the multimammate mouse, *Mastomys natalensis* (Smith 1834), across different habitats and seasons in Tanzania and Swaziland. *Wildlife Res.*, 38(7): 640–6.

Mulungu, L. S., Makundi, R. H., Leirs, H., Massawe, A. W., Machangu, R. S. and Ngowo, V. 2005. Spatial pattern and distribution of rodent damage in maize fields in Tanzania. *Belg. J. Zool.*, 135 (Suppl.): 183–5.

Mulungu, L. S., Makundi, R. H., Leirs, H., Massawe, A. W., Vibe-Petersen, S. and Stenseth, N. C. 2003. The rodent density–damage function in maize fields at an early growth stage, In: Singleton, G. R., Hinds, L. A., Krebs, C. J. and Spratt, D. M. *Rats, Mice and People: Rodent Biology and Management* (ed.), Australia Centre for International Agricultural Research, Canberra, Australia, pp. 301–3.

Mulungu, L. S., Makundi, R. H., Massawe, A. W. and Leirs, H. 2007. The Relationship Between Sampling Intensity and Precision for Estimating Damage to Maize Caused by Rats. *Integrative Zool.*, 2(3): 131–5.

Mulungu, L. S., Ngowo, V., Mdangi, M., Katakweba, A. S., Tesha, P., Mrosso, F. P., Mchomvu, M., Sheyo, P. M., Kilonzo, B. S. 2013. Population dynamics and breeding patterns of Multimammate mouse, *Mastomys natalensis* (Smith 1834) in irrigated rice field in Eastern Tanzania. *Pest Manag. Sci.*, 69(3): 371–7.

Mulungu, L. S., Sixbert, V., Ngowo, V., Mdangi, M., Katakweba, A. S., Tesha, P., Mrosso, F. P., Mchomvu, M., Kilonzo, B. S. and Belmain, S. R. 2014a. Spatio-temporal patterns in the distribution of the multimammate mouse, *Mastomys natalensis*, in rice crop and fallow land habitats in Tanzania. *Mammalia*, 79(2): 177–84. doi: 10.1515/mammalia-2014-0006.

Mulungu, L. S., 2003. Assessment of maize (*Zea mays* l.) damage and yield loss due to rodents in the field. PhD Thesis Sokoine University of Agriculture, Morogoro, Tanzania, p. 178.

Mulungu, L. S., Mlyashimbi, E. C. M, Ngowo, V., Mdangi, M., Katakweba, A. S., Tesha, P., Mrosso, F. P., *Mchomvu*, M., Kilonzo, B. S. and Belmain, S. R. 2014b. Food preferences of the Multi-mammate mouse, *Mastomys natalensis* (Smith 1834) in irrigated rice field in eastern Tanzania. *Int. J. Pest Manage.*, 60(1): 1–8.

Mwanjabe, P. S. 1993. The role of weeds on population dynamics of *Mastomys natalensis* in Chunya (Lake Rukwa) valley. In: Machang'u, R. S. (ed.), *Proceeding Of Economic Importance And Control Of Rodents in Tanzania.* 6–8 July 1992, Morogoro, Tanzania, pp. 134–42.

Mwanjabe, P. S. and Leirs, H. 1997. An early warning system for IPM-based rodent control in 362 smallholder farming systems in Tanzania. *Belgian J. Zool.*, 127: 49–363.

Mwanjabe, P. S. and Sirima, F. B. 1993. Large scale control in Tanzania: present status. In: Machang'u R. S. (ed.), *Proceeding of Economic importance and control of rodents in Tanzania.* 6–8 July 1992, Morogoro, Tanzania, pp. 134–42.

Mwanjabe, P. S., Sirima, F. B. and Lusingu, J. 2002. Crop losses due to outbreaks of *Mastomys natalensis* (Smith 1834) Muridae, Rodentia, in the Lindi region of Tanzania. *Intern. Biodet. and Biodeg.*, 49: 133–7.

Myllymäki, A. 1987. Control of rodent problems by the use of rodenticides: rationale and constraints. In: Richards, C. G. J. and Ku, T. Y. (eds), *Control of Mammal Pests*. Taylor & Francis, London, pp. 83–111.

Norrdahl, K. 1995. Population cycles in northern small mammals. *Biol. Rev.*, 70: 621–37.

Odhiambo, R. O., Makundi, R. H., Leirs, H. and Verhagen, R. 2008. Dietary selection in *Mastomys natalensis* (Rodentia: Muridae) in the maize agro-ecosystems of Central and South-western Tanzania. *Mammalia*, 72: 169–77.

Pommel, B. and Bonhomme, R. 1998. Variations in the vegetative and reproductive systems in individual plants of an heterogeneous maize crop. *European J. Agron.*, 8(1–2): 39–49.

Powelczyk, A., Bejer, A., Behnke, J. M., Gilbert, F. S. and Sinski, E. 2004. Factors affecting the component community structure of haemoparasites in common voles (Microtus arvalis) from the Mazury Lake District region of Poland. *Parasitology Res. J.*, 92(4): 270–84.

Parshad, V. 1999. Rodent control in India. *Int. Pest Manage. Rev.*, 4: 97–126.

Pucek, Z., Jędrzejewski, W., Jędrzejewska, B. and Pucek, M. 1993. Rodent population dynamics in a primeval deciduous forest (Białowieża National Park in relation to weather, seed crop, and predation. *Acta Theriologica,* 38: 199–232.

Sichilima, A. M., Zulu, M. S. and Leirs, H. 2003. The effects of Tephrosia vogelii and land preparation methods on mole rat activity in cassava fields. In: Singleton, G. R., Hinds, L. A., Krebs, C. J. and Spratt, D. M. (eds), *Rats, Mice and People: Rodent Biology and Management*. Australia Centre for International Agricultural Research, Canberra, Australia, pp. 254–5.

Sikes, R. S. and Gannon, W. L. 2011. Guidelines of the American Society of Mammalogists for the use of wild mammals in research. *J. Mammal.*, 92(1): 235–53.

Singleton, G. R., Leirs, H., Hinds, L. A. and Zhang, Z. 1999. Ecologically-based management of rodent pests- re-evaluating our approach to an old problem. In: Singleton, G., Leirs, H., Zhang, Z. and Hinds, L. (eds), *Ecologically-based Management of Rodent Pests*. Australian Centre for International Agricultural Research Publications, Canberra, Monograph No. 59, pp. 17–29.

Skonhoft, A., Leirs, H., Andreassen, H. P., Mulungu, L. S. and Stenseth, N. C. 2006. The bioeconomics of controlling an African rodent pest species. *Environ. & Devel. Econ.*, 11: 453–75.

Swanepoel, C. M., 1980. Some factors influencing the breeding season of *Praomys natalensis. S. Afr. J. Zool.*, 15: 95–8.

Taylor, K. D. and Green, M. G. 1976. The influence of rainfall on diet and reproduction in four African rodent species. *J. Zool.*, 180: 367–89.

Telford, Jr. S. R. 1989. Population biology of Multimmamate rat, *Praomys* (*Mastomys*) *natalensis* at Morogoro, Tanzania 1981–1985. *Bull. Fla. State Mus Biol. Sci.*, 34: 249–88.

Van Gulck, T., Stoks, R., Verhagen, R., Sabuni, C. A., Mwanjabe, P. and Leirs, H. 1998. Short-term effects of avian predation variation on population size and local survival of the multimammate rat, *Mastomys natalensis* (Rodentia, Muridae). *Mammalia*, 62: 329–39.

Vibe-Petersen, S. 2003. Predation pressure and population dynamics of African Mastomys mice - possibilities for integrated pest management? Unpublished Thesis for Award of PhD Degree at Danish Pest Infestation Laboratory, Lyngby, Denmark and The Royal Veterinary & Agricultural University, Frederiksberg C, Denmark, p. 122.

Walker, P. T., 1990. Determining Pest-loss relationships and quantifying loss. In: Walker, P. T. (ed.), *Crop Loss Assessment in Rice*. International Rice Research Institute, Philippines, pp. 151–60.

Wolff, J. 1996. Population fluctuations of mast-eating rodents are correlated with production of acorns. *J. Mammal.*, 77: 850–6.

Wube, T. 2005. Reproductive rhythm of the grass rat, *Arvicanthis abyssinicus*, at the Entoto Mountain, Ethiopia. *Belg. J. Zool.*, 135: 53–6.

Chapter 16

Rapid response to disease outbreaks in maize cultivation: the case of maize lethal necrosis

George Mahuku, International Institute of Tropical Agriculture (IITA), Tanzania; and P. Lava Kumar, International Institute of Tropical Agriculture (IITA), Nigeria

1 Introduction

Plant diseases caused by insect pests and pathogens account for an estimated 23.1% of maize production loss annually, which translate to between 14.2 and 21 million metric tonnes (Oreke et al., 1995; Oreke, 2006). Emerging plant diseases (EPDs) are a significant constraint to agricultural productivity and profitability. They act as barriers to economic development, threaten ecosystems due to their impact on native biodiversity and control measures incur additional costs leading to indirect loses (Anderson et al., 2004; Fisher et al., 2012; Vurro et al., 2010). Disease outbreaks can spread rapidly and often result in enormous crop losses that can lead to considerable human suffering. For instance, this has been observed following the outbreak of maize lethal necrosis (MLN) in Kenya in 2011 (Fig. 1) (Wangai et al., 2012; Mahuku et al., 2015a). The best strategy to avoid losses is to stop disease from spreading at an early stage, or to prevent outbreaks altogether. This

http://dx.doi.org/10.19103/AS.2016.0002.20

Figure 1 A maize field in Tanzania severely affected by MLN. Production loss was 100% in this field. Photo: George Mahuku.

relies on rapid and accurate identification of the causal agent(s) and the deployment of appropriate control measures. This is particularly critical in outbreaks of new diseases or of those hitherto unknown in a region or on a crop. Such a situation often demands that new control measures must be found as quickly as possible to prevent the disease from spreading.

2 Emerging plant diseases

Diseases are categorized as 'emerging' if their incidence or virulence has recently increased, if they begin to infect a novel host, or have changed in pathogenesis, or have been discovered in new geographical locations (Daszak et al., 2000; Jones et al., 2008; Fisher et al., 2012; Morens et al., 2004). EPDs are those diseases caused by plant-infecting pathogens which develop into epidemics that may be both unexpected and devastating (Jones et al., 2008; Fisher et al., 2012). As illustrated (Fig. 2), disease results from a complex interplay of interactions among the pathogen, host and environment (Fletcher et al., 2010; Scholthof, 2007). Important catalysts for disease emergence include human activity, the weather and the emergence of new pathogen strains (Anderson et al., 2004; Daszak et al., 2000; Harvell et al., 2002; Josling et al., 2003). Although crop production is affected by established as well as emerging diseases, the epidemics caused by EPDs infecting hosts with little resistance or tolerance to the emerging pathogen can be particularly devastating.

One of the most notable EPD on maize was the southern corn leaf blight (SCLB) epidemic, caused by *Cochliobolus heterostrophus* that resulted in yield losses of more than 90% and an estimated loss of US$1.09 billion during the 1970–1 season in the US corn belt (Ullstrup, 1972). The epidemic resulted due to the emergence of a new fungus race, *C. heterostrophus* race T, that was able to overcome resistance gene *T-urf*13 conferring

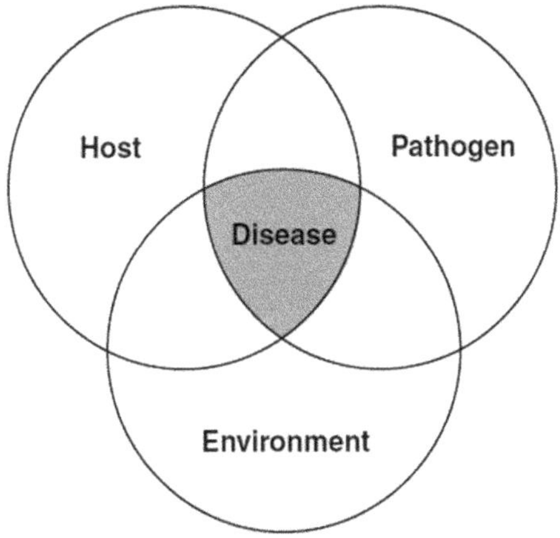

Figure 2 The disease triangle concept (host–pathogen–environment) illustrates the complex interplay of factors that result in disease at the individual and population levels. The presence of a pathogen is necessary but not sufficient to cause disease epidemics (*IOM, 2008*). Some pathogens (especially viruses) reach epidemic proportions with the introduction of suitable vectors.

Table 1 Economic losses from outbreaks of some notable disease epidemics on maize

Southern corn leaf blight, United States (1970)	Grey leaf spot (1996)	Mal del Río Cuarto virus, Argentina (1996–7 season)	Maize streak virus (Africa)	Tar spot complex, Guatemala (2008–9)	Maize lethal necrosis (Kenya 2012–14)
US$ 1.09 billion was lost in a single year (Ullstrup, 1972)	An estimated US$ 663 million was lost from a grey leaf spot (GLS) outbreak in 9.3 million hectares, (Iowa, Illinois, Indiana, Ohio and Kentucky) translating to a loss of 6.98 million tonnes (Ward et al., 1999)	An estimated 300000 hectares were affected. translating to losses of US$120 million (Lenardon et al., 1998)	US$ 120–480 million is lost annually in Africa (Karavina, 2014)	Estimated yield losses of up to 75% in the 2008/9 season reported in the northern provinces of Guatemala (ProMED-mail, 2009). The disease spread to United States (Ruhl et al., 2016) and Mexico (Rios-Herrera et al., 2016)	Losses of 0.5 million metric tonnes valued at US$ 188 million (de Groote et al., 2016)

cytoplasmic male sterility in the widely grown hybrids (Ullstrup, 1972). The detection of MLN in Kenya in 2011/2012 is the most recent EPD on maize that sent alarm bells ringing through the global community. In 2012, an estimated 126000 bags of maize, equivalent to US$ 52 million were lost to MLN in Kenya alone (MDRAT, 2012). Subsequent surveys revealed that MLN was widespread across most maize-growing areas of Kenya. Losses to production were estimated at 0.5 million metric tonnes (23%), valued at US$ 188 million

(De Groote et al., 2016). Such a loss has huge economic and social implications, especially in a country reliant on maize as a staple. The disease has spread rapidly to other countries in East Africa, threatening food security of millions of people (Mahuku et al., 2015a). Examples of diseases that have caused huge economic losses to maize production are listed in Table 1.

3 Factors influencing the emergence of maize diseases

The EPDs of maize are driven mainly by anthropogenic activity and environmental changes including (i) the introduction of a pathogen to a new geographical locations; (ii) the movement of new virulent strains, or the emergence of a new aggressive strain in an area where the pathogen already existed; (iii) the introduction of new vectors that can transmit a pathogen efficiently; (iv) changes in agronomic practices that favour one or more components of epidemics of a specific disease; (v) the intensification of agriculture to maximize productivity and profit; (vi) changes in cultivars; and (vii) a consistent change in weather in the short term. Some of these factors are intrinsic to crop husbandry, whereas others relate to extraneous forces such as trade, policy and international exchange (Anderson et al., 2004; Jones, 2009). Pathogen introduction is the most important driver of EPDs, and according to Anderson et al. (2004), more than 56% of outbreaks of diseases from 1996 to 2002 resulted from the introduction of new pathogens (Fig. 3).

Among the numerous pathways by which pathogens can spread, the international seed trade, worth an estimated US$40–80 billion, is an efficient means of introducing pathogens (Contreras, 2002; Condon, 1997). Over 2400 microorganisms occur in the seeds of 380 genera of plants (McGee, 1997) and it is estimated that up to one-third of

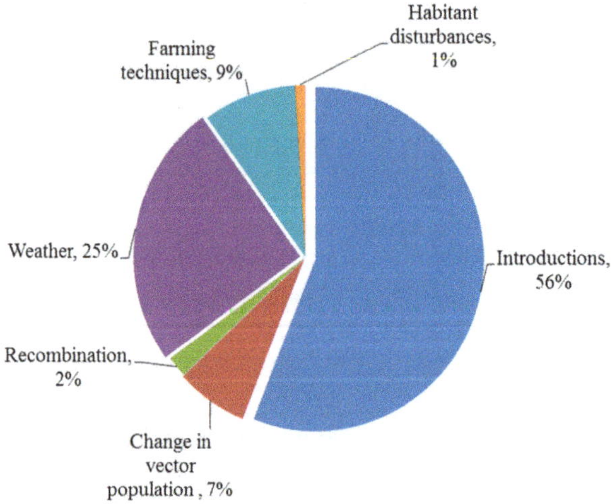

Figure 3 Drivers of disease outbreaks – introductions of new pathogens are the major driver of disease outbreaks (Anderson et al., 2004).

plant viruses might be seed borne in at least one of their hosts (Stace-Smith and Hamilton, 1988; Vurro et al., 2010). It is difficult to inspect every plant or seed for pathogens as they are imported. Introduced biological agents, whether they arrive via natural mechanisms of dispersal or through human activity, might remain present for years below the threshold for detection and perception, but emerge later with seemingly suddenness and dramatic destructive intensity (Crooks, 2005).

Many drivers affect the emergence and increased risks of disease in maize production systems worldwide. These include biophysical factors and institutional arrangements related to the agricultural sector (Bebber et al., 2014; Foresight, 2006). Some of the most important drivers are discussed below.

3.1 Biophysical factors

Agronomic practices

Changes in agronomic practices alter the host–pathogen equilibrium and, depending on the direction to which equilibrium is tipped, may increase or decrease the risk of diseases (King et al., 2006). Changes in crop management techniques, particularly the intensification of cropping, reduction in crop rotations and increase in monocultures, have been shown to enhance disease severity (Vurro et al., 2010). A good example is the great epidemic of Southern corn leaf blight (SCLB) in 1970–1. This was caused by a variant strain of the fungus, *C. heterostrophus*, named race T, which was specifically virulent for maize containing a cytoplasmically inherited gene for male sterility (Tcms). Because of the advantage conferred by the gene in breeding this self-fertile crop, it had been incorporated into about 85% of the American crop by 1970. As a result and aided by favourable climatic conditions (unusually warm, moist weather), an epidemic causing enormous damage developed in a wholly susceptible host with the epicentre in the 'corn belt', (Pring and Lonsdale, 1989; Rosenzweig et al., 2001). The pandemic was halted by the withdrawal of susceptible varieties and the establishment of new hybrids. SCLB is infamous for having shocked the world's feed market and set a record in terms of economic losses produced on a single agricultural crop in a single season. Subsequent to the disaster, the reason for the specificity and high virulence of *C. heterostrophus* race T for Tcms maize was determined to be the production of a host-selective toxin by the fungus. Although genetic uniformity in the corn crop had laid the groundwork for the fungus, favourable meteorological conditions were instrumental to the outbreak and spread of the disease.

Another example is grey leaf spot (GLS) of maize, caused by *Cercospora zeae-maydis*. The disease emerged as a result of increased adoption of zero tillage and retention of residue cover (Nutter and Jenco, 1992; Ward and Nowell, 1998). In August 1995, yield losses from GLS were as high as 50% in some maize fields in the United States (*Wall Street Journal*, 1995). In 1996, Garst Seeds estimated that GLS had damaged as many as 9.3 million hectares of maize in Iowa, Illinois, Indiana, Ohio, Kentucky and Tennessee, and predicted to affect an additional 11.3 million hectares in other areas of the United States (Garst Seeds, 1996). In Africa, yield reductions of 30–60% have been attributed to GLS, depending on germplasm and environmental conditions (Ward et al., 1997). GLS is now recognized as one of the most significant yield-limiting diseases of maize worldwide, with high incidences reported in Nepal, China, Colombia, Mexico, Brazil and several countries in Africa (Shiferaw et al., 2011).

Global policy on poverty reduction and sustainable economic growth is a major driver affecting disease outcomes in many direct and indirect ways. The current global focus is on the reduction of poverty while great emphasis is laid on wealth generation. In many developing countries, agriculture has been placed at the forefront of schemes for eliminating hunger and reducing poverty and food insecurity. Achieving these goals requires greater investment in agriculture, agricultural research and technology, with intensification and commercialization of farming, and greater trade in agricultural commodities and products. Such intensification leads to specialization, loss of crop genetic diversity and uniformity of crop species and varieties, making crops more vulnerable to epidemics of diseases. More trade in agricultural commodities and products carries greater risks of pathogens being introduced and spreading, and increases the possibility of disease (Anderson et al., 2004).

Travel, migration and trade

Human spatial mobility has increased at least 1000-fold in the past 200 years, with more people travelling faster, farther and less expensively than ever (Cliff and Haggett, 2005; Hufnagel et al., 2004). This has increased the likelihood of new pathogens being introduced into new areas and increased the risk of disease outbreaks (Anderson et al., 2004). The increase in international transportation, travel and trade associated with globalization has amplified the frequency of interactions among people, plants and microbes – providing novel opportunities for the rapid emergence and spread of infectious diseases (IOM, 2010). For instance, in 2007 alone, the United States imported more than 48 million tonnes of agricultural products, only 1–2% of which were inspected for possible pathogens and other pests (Becker, 2009). Increased travel, migration and trade promote both the spread of existing diseases to new populations and regions and also facilitate the emergence of new diseases by bringing different populations into contact with one another. Policies for regional integration will facilitate and increase the movement of crops and crop products within regions. Regionalization, for instance, in the creation of a common seed policy across the East African community, will mean the elimination of barriers to seed movement between and among these countries. This will facilitate and increase the movement of crops and crop products within Africa relatively easily and increase the risk of pathogens spreading between the regions if adequate phytosanitary controls are not used. In addition, the porous borders mean that many plant products, for example, seeds will be moved illegally, increasing the risk of disease spread. The recent outbreak and rapid distribution of MLN exemplified the effects of seeds and regional integration.

Change in cropping systems

EPDs are principally driven by increased globalization, socio-economic development and technological advances. Changes in the cropping system will see changes in pest prevalence simply because of changes in the proportions of the crops that are grown. Intensification, diversification and globalization of agriculture act in a complex, interrelated way to drive disease emergence. Intensification when joined with irrigation leads to the expansion of vector populations and a positive feedback effect on plant pathogen populations. Diversification introduces new hosts and globalization brings new pathogens, promoting the spread of alien diseases and vectors. *Maize chlorotic mottle virus* (MCMV) is thought to have been introduced into Africa via sweet and baby corn seeds imported from Asia (Adams et al., 2013). As these products were traditionally considered vegetable seed, they were not regulated and thus did not conform to quarantine import regulations. The high demand for sweet and baby corn in European markets resulted in continuous and intensified production in East Africa, a region with favourable climatic conditions. In addition, large irrigation projects were also established in most of the region, driven by the ready market for green mealies. Irrigation enabled maize to be continuously and extensively cultivated, making the crop a reservoir of MLN pathogens and the insect vectors. This resulted in a build-up of virus and the thrips vectors leading to disease emergence throughout East Africa.

In Nicaragua, corn stunt complex caused yield losses of 100% on 18 900 hectares of maize and partial losses on an additional 8783 hectares in 1986 (Hruska et al., 1996). The total production lost was 26 768 000 kg, or an 11% reduction in national production. The epidemic was a direct result of a policy by the government to promote maize production during the dry season, as part of a contingency plan to ensure there were adequate supplies of basic grain at a time when war in the interior of the country was constraining the production and transport of grain. As a result of this policy, maize production during the dry season in the Pacific region grew from a few hundred hectares in 1983 to 10 276 hectares by 1986 (Hruska et al., 1996). This led to the elimination of a natural barrier that had reduced the pathogen population in the dry season from December to April, and so, in the subsequent rainy season maize crops were exposed to higher percentage of insect vectors, namely *Dalbulus maydis*, entering the fields. The off-season maize produced under irrigation provided the temporal bridge for the pathogens, which were then transmitted to rainy season plantings and caused an epidemic of corn stunt throughout the country. In general, loss of biodiversity will bring about the uniformity of plant species, increasing their vulnerability to pathogens and facilitating the spread of disease. It may also enable weak pathogens or minor diseases to gain importance. Similarly, loss of fertility will result in less vigorous plants that could be highly susceptible, even to weaker pathogens.

Climate change

The distribution, incidence and severity of plant diseases are influenced by many interacting factors, including environmental conditions, of which climate is the most important factor. Climate change will affect disease outbreaks either directly, by changing the distribution of vectors of diseases, or indirectly, by changing patterns of land use and agriculture that will affect the incidence and severity of diseases (King et al., 2006). Climate change is not just a change in the global temperature; it is also a change in weather patterns. While temperatures affect the potential ranges of infectious diseases and their vectors, weather affects the timing, intensity and location of outbreaks. For example, fungi are highly sensitive to changes in weather and climate – particularly temperature, humidity and wind; these directly influence their growth, spread and survival (Harvell et al., 2002). Global warming will operate through new or better-adapted vector or pathogen species, biotypes or strains. The new vector types may become more efficient in disease transmission or vectors may transmit a disease that they had not transmitted before. A large proportion of damaging maize diseases are caused by viruses, and the majority of these are vector borne. *Maize streak virus* (MSV) is transmitted by leaf hoppers *Cicadulina* spp.; corn stunt complex, a disease endemic in Latin America and distributed from Mexico to Argentina, is caused by a complex of pathogens, including the corn stunt *Spiroplasma kunkelii*, the maize bushy stunt phytoplasma and the maize fine stripe (rayado fino) virus, all transmitted by species of *Dalbulus* leafhoppers, with *D. maydis* being the most important. Severe epidemics are associated with the continuous planting of susceptible cultivars and yield losses of 50% have been documented (Bradfute et al., 1981; Vasquez and Mora, 2007). MCMV is transmitted by many insects, with thrips being the major vector, and combines with the aphid-borne *sugarcane mosaic virus* (SCMV) to cause MLN (Jiang et al., 1992; Cabanas et al., 2013; Moritz et al., 2013). Currently, the abundance of each of these disease vectors is associated with higher than average temperatures and rainfall. It might be expected that future increases in temperature would favour greater vector abundance, raising the level of risk from these diseases (Rosenzweig et al., 2001). In addition, new strains of pathogens could become more virulent or may be more easily transmissible by vectors. Climate change will progressively change what crops are grown where, shaped by the dynamic economics of local markets, imports and exports. All these factors will increase disease risks in maize cultivation.

3.2 Institutional factors

Civil strifes will drive outbreaks of disease where any capacity and infrastructure to prevent and control diseases have broken down. Furthermore, refugees fleeing strife-ridden areas will typically carry plant materials, which could introduce disease into new areas (Foresight, 2006). In addition, the movement of people whether because of strife or for economic reasons will also contribute to the movements of crop germplasm and diseases. The increasing movement of food aid in response to civil strife might continue to carry new pathogens and accidental introduction by researchers bringing in new germplasm may also result in new diseases.

Capacity for prevention and control of disease

Lack of capacity for prevention and control will increase the risk of diseases, and where policies and regulatory frameworks for disease detection, identification, monitoring and control are poor, these will also have a substantial effect on the disease situation. Poor governance and the loss of capacity to recognize and respond to problems of infectious disease are a major issue, especially for developing countries. Better capacity for disease detection is vital, but if this is to be effective, new technologies must be embedded within functional national or international surveillance systems (King et al., 2006).

Agricultural policies

Policies on agricultural trade and production will be a major driver affecting disease outcomes in many direct and indirect ways. Policies promoting intensification and extensification of farming systems will result in specialization, loss of crop genetic diversity and the uniformity of crop species and varieties, expansion of crops into new niches and the introduction of new varieties, germplasm or crops. All of these alter agrarian systems and some could increase vulnerability to epidemics of diseases. Increased trade in agricultural commodities and products results in greater risks of the introduction and spread of pathogens and consequently of disease (Anderson et al., 2004).

4 Overview of strategies for mitigating risks from emerging maize diseases

Preventing the introduction of new pathogens is the first line of defence against disease epidemics. However, even the best prevention efforts will not stop all introductions. Efforts at early detection and rapid response (ED&RR) increase the likelihood that these will be successfully addressed while populations are still localized and levels are not beyond those that can be contained and eradicated (Hughes and Burnett, 2015; Foresight, 2006; NISC, 2003). Once pathogens are widely established, all that might be possible is the partial mitigation of negative impacts. In addition, the costs associated with ED&RR efforts are typically far lower than those of long-term management programmes.

The hallmarks of successful rapid response systems include the following: (i) timely identification of potential threats to allow risk-mitigation measures to be taken; (ii) capacity to detect new pathogens in time to make efficient and environmentally sound decisions; (iii) deployment of effective and environmentally sound responses to invasions to prevent the spread and permanent establishment of a pathogen; (iv) timely communication of information about the status of a disease within an area to decision-makers, and public and trading partners; and (v) use of lessons drawn from past experience to guide current and future efforts (Morens et al., 2004).

Detecting and responding to emerging diseases require a complex series of actions that can be grouped into three main categories: early detection, rapid assessment and rapid response (Morens et al., 2004). It involves a crop health system that encompasses regional cooperation and national responsibility, and harnesses global excellence in terms of the knowledge and methods for implementation (Beed, 2014). This system would be able to control crop diseases in a pre-emptive and cost-efficient manner and avoid the scenario of belatedly combating large-scale epidemics. Some of the essential functions and requirements would include the following; determining the relative and potential risks associated with an introduction; targeted surveillance, detection and report of activities such as data collection and management; fit-for-purpose diagnostics; priority setting; sharing resources across jurisdictional boundaries; control intervention packages; extension mechanisms; enabling policy environments; and R&D on disease control. Adequate staffing and funds with effective public communication are essential to support rapid response activities. Components of early detection, rapid assessment and rapid response systems that are important to the containment of emerging diseases are outlined below.

5 Components of an effective rapid response system

5.1 Early detection

Diagnostic networks

Detection is the first line for mounting an effective response to new introductions or emerging pathogens and diseases. This depends on the vigilance of all stakeholders in the maize value chain, from farmers to scientists and policy makers (Bouwmeester et al.,

2012). Vigilance is always required when it comes to protecting key crop commodities from disease as dynamic selection pressures exacerbated by certain agricultural methods such as mono- and continuous cropping, as is practised for the cultivation of maize will favour the emergence of new diseases. It is important to have proper and effective surveillance, monitoring, detection and diagnostic services in place (Bostock et al., 2014; Epstein et al., 2003; Foresight, 2006). Incorrect identification can be costly. Networks and coordination are important tools for an effective and efficient detection system. The standardization of survey protocols allows the adoption of similar methods to produce comparable results (Bouwmeester et al., 2010). A basic requirement for each sampling point is to record global positioning system coordinates, to enable others revisit sites of interest and concern. This in turn allows for geographic information system analysis through the extrapolation of different global positioning system-linked datasets, such as climate or trade routes combined with disease incidence, over time, and can be used to determine factors influencing disease spread (Bouwmeester et al., 2012; Nakato et al., 2012).

Success in detection and monitoring systems depends on appropriate diagnostic methods to identify the causal agent(s) of the disease (Petter et al., 2008). This is of particular importance in some developing countries, for instance in sub-Saharan Africa, where available technical capacity and financial resources are limited, resulting in failures in recognizing or reporting new disease outbreaks (Smith, 2015; Smith et al., 2008; Waage et al., 2009). Any delay in the diagnosis of a pathogen will, in turn, delay efforts to deploy appropriate control interventions and increase the risk of an epidemic developing. Initiatives to strengthen capacity for the rapid and precise diagnoses of plant pathogens are underway such as those initiated by the International Plant Diagnostic Network (Miller et al., 2010). Diagnostic systems should be linked to effective monitoring and reporting/alarm systems, such as ProMED (the Program for Monitoring Emerging Diseases; http://www.promedmail.org) and HealthMap (http://healthmap.org). These have two approaches for detecting and monitoring outbreaks worldwide in plants: first, by the active reporting of disease outbreaks, and secondly, by capturing diverse online data sources (Madoff, 2004). The management of these systems within international networks provides the global outbreak situation in real-time and allows the implementation of joint and coordinated strategies and to prevent or better manage possible pandemics (Miller et al., 2009). For example, the rapid identification and reporting of MLN viruses were instrumental in achieving the rapid response to contain the disease and prevent its spread. This was made possible by the availability of diagnostic methods specific to MCMV and *Sugarcane mosaic virus* (SCMV) that showed that both viruses were present in maize plants manifesting MLN disease (Wangai et al., 2012). Next-generation sequencing was then used on samples of plants infected with MLN to produce nucleic acid sequences for substantial parts of both viral genomes, thus allowing strain characterization and the development of specific diagnostic assays (Adams et al., 2013). What is of particular interest with MLN is that it took focused diagnostics to identify that the disease was caused by mixed infection by two viruses, reliant on different insect vectors (aphids for SCMV and thrips for MCMV). Furthermore, risks to agriculture can come from anywhere in the world as was demonstrated by the combination on MLN of the indigenous SCMV (Louie, 1980) with the exotic MCMV.

Research

Research is needed to understand, monitor and model parameters that may contribute to invasions, such as climatic conditions, ecosystem disturbance patterns and land use changes. Baseline research is needed to determine if a species is 'new' to an area and to detect changes in pathways. Scientifically sound information is essential to targeting detection programmes. Similarly, surveys for some pathogens must consider the likely spread of disease due to the behaviour of their insect vectors or the actions of people moving infected planting material. Research is also needed in developing new analytical models, diagnostic assays and molecular tools to significantly advance our abilities to respond to disease threats (Voyles et al., 2014).

Training

Training is important for people engaged in early detection and the collection, and reporting of suspect species so that their efforts can support subsequent action, to reduce the frequency of inaccurate reports and the excessive reporting of common species. The need to train those engaged in detection and response networks and to develop educational materials, trainers and related resources is an ongoing process.

Participation of a wide range of stakeholders is essential. The understanding and participation of stakeholders can be facilitated by working with professional societies and scientific organizations, agricultural organizations, conservation and outdoor recreation organizations, community groups and others. Farmers can play important roles in reporting new diseases, as they are in the field and constantly in contact with the crop. However, structures need to be in place to support farmers, especially through extension workers. Once reports have been received, authoritative verification of reported introductions is especially important when the initial report is obtained from people with minimal technical training. Verification can authoritatively determine the presence or absence of a pathogen in an area, whether an initial introduction or the unexpected arrival of previously reported species, and other essential information. It is important that reports and data concerning invasions be broadly accessible, easy to use and routinely exchanged among interested parties. Data can be used to substantiate crop health claims, document ecosystem health, highlight emerging issues, quantify successes, identify problem areas and direct efforts at rapid response. Knowledge and skills obtained from past ED&RR efforts can guide and expedite subsequent efforts. As new ED&RR systems are developed, it is important to include current efforts and systems. There should be mechanisms for communicating with other networks to help identify gaps that may exist, establish research needs and to cooperatively work together to correct deficiencies in programmes.

5.2 Preliminary risk assessments

It is important to prepare preliminary risk assessments for high-priority diseases prior to their detection to facilitate rapid responses to invasions. Risk assessments should include input from stakeholders and consider appropriate legal authorities, so that challenges do not unduly delay an effective response. Many maize diseases occur in one region or continent (Table 2); there is real potential for these pathogens to be introduced elsewhere. Understanding the potential risk from accidental introduction can help in rapid response systems. Lessons learned in managing these diseases in the areas that they are currently present can help to prevent future epidemics. For example, MLN was first reported in Peru in 1973, and then in the United States in 1976 (Castillo and Hebert, 1974; Niblett and Claflin, 1978). Had pest risk analysis (PRA) for this disease been reported, the devastations observed in Africa would have not occurred. However, the information obtained from

Table 2 Maize disease with potential to emerge and cause devastating epidemics

Disease	Pathogen	Distribution	Economic impact	Reference
Stewart's wilt	*Pantoea stewartii*	Canada and the United States	Yield losses of 40 to 100%	Pataky and Ikin, 2003
Tar spot complex	*Phyllachora maydis* and *Monographella maydis*	The United States, Mexico, Central and South America and the West Indies	Yield losses of up to 75% have been observed	Bajet et al., 1994; ProMED-mail, 2009
Grey leaf spot (GLS)	*Cercospora zeae-maydis and C. zeina*	Worldwide, in subtropical and temperate, humid areas	Associated with conservation agriculture	Ward et al., 1999
Banded leaf and sheath blight	*Rhizoctonia solani f.* sp. *Sasakii*	Asia, Africa and America		White, 1999
Java downy mildew	*Peronosclerospora maydis*	Australia and Indonesia	Up to 40% yield reduction	White, 1999
Philippine downy mildew	*Peronosclerospora philippinensis*	Philippines, Indonesia, India, Nepal, China, Thailand and Pakistan	15–40% yield reduction	White, 1999
Sugarcane downy mildew	*Peronosclerospora sacchari*	Japan, Nepal, New Guinea, India, Philippines and Thailand	30–60%	White, 1999
Brown stripe downy mildew	*S. rayssiae* var zeae	India, Nepal, Pakistan and Thailand	Yield loss of 63% reported in India	White, 1999
Corn stunt complex	*Spiroplasma kunkelii*, maize bushy stunt phytoplasma *maize* rayado fino virus	The United States, Mexico, Central and South America	Up to 50% yield loss	Bradfute et al., 1996
Maize chlorotic dwarf virus (MCDV)	Maize chlorotic dwarf virus	The United States and Mexico – vectored by leaf hoppers *(Graminella nigrifrons)*	Losses ranging from 9 to 90%	Louie et al., 1990
Maize chlorotic mottle virus (MCMV)	Maize chlorotic mottle virus	The United States, South America, Asia and Africa	Up to 90% yield loss in severe infections	Mahuku et al., 2015a; Niblett and Claflin, 1978
Maize rayado fino virus (MRFV)	Maize rayado fino virus	The United States, Mexico, Central America and South America	40–50% losses of early infected plants	Gámez, 1980; Bustamante et al., 1998; Vasquez Mora, 2007)
High plains diseases	High plains virus	The United States, Mexico, South America (Chile, Brazil, Peru) and Israel	Yield losses exceeding 50% in several corn fields were observed in Idaho	Jensen and Siefers 1996
Mal del Río Cuarto	Mal del Río Cuarto virus	Argentina	Losses of 5–60% in the Río Cuarto area	Lenardon et al., 1982
Maize streak virus	Maize streak virus	Africa, Madagascar and Mauritius	USD 120–480 million lost annually in Africa from this virus	Karavina, 2014

past outbreaks helped in quickly putting together a response strategy. It is important to conduct rapid risk assessments of newly detected pathogens so that decision-making, whether to contain, treat and monitor a population; monitor only; or ignore can be done while populations are still localized. It is important for risk assessment to draw upon information from existing surveys and data sources whenever possible. Consistent data definitions and interoperable formats are essential so that summary statistics and analyses are readily available to support risk assessments. In some cases, it may be necessary to seek information internationally.

5.3 Rapid response

Because the development of rapid response systems can be time-consuming and require a significant commitment of resources, it is important that each country should have a workforce of people tasked with this activity, who can be quickly mobilized to respond to a threat. In developed countries such as the United States, such networks are available (Stack et al., 2006; Fletcher et al., 2010). However, they are lacking in developing countries, especially in Africa. The outbreak of MLN caught people by surprise. However, the collaboration fostered between national and international organizations was quick and the disease was diagnosed in a relatively short time. However, the event also exposed some shortcomings and emphasized the need to have standing teams with compositions that can be adjusted to meet the specific needs and circumstances of an invasion in a timely manner. It is essential that the roles and responsibilities of stakeholders and officials involved be delineated clearly and understood by all. Previous training in eradication/control methods is essential since often there is not sufficient time to train a response team following the detection of an invasion. Rapid response teams should set appropriate schedules for action. Depending on the likelihood of establishment, the rate of spread and other factors, different scenarios will require differing schedules of response. It is important that plans contain schedules that are appropriate for the specific conditions of the invasion.

Stakeholders' input is essential in the development of plans for rapid response since actions will often be led by local, district or national bodies and may involve personnel from the private sector, working with the support of public officials when requested. Effective working relationships can be fostered among groups and individuals by collaborating with them in the development of response plans. Adequate flexible and available funding is essential to mount an effective, timely and sustained response to new, often unpredictable diseases. Response can span jurisdictional boundaries as epidemics expand rapidly and several years may be required.

It is important to facilitate public understanding of response efforts. A coordinated public outreach effort should be a component of efforts for rapid response. Authoritative balanced information should be provided early in the rapid response timeline. Response efforts may have impacts beyond the targeted species. Public understanding of the anticipated economic and ecological impacts of response actions as well as the expected benefits associated with response efforts is important to success.

A crop health care system can operate effectively only if supported by an enabling policy environment, created through recognition of the importance of protecting crops from key diseases. This supportive political environment is dependent on regional cooperation (to share information), national responsibility and international excellence (to harness knowledge and methods for adaptation to local demands). At every opportunity, capacity should be shared to bring standards and understanding up to similar levels; this can be supported by regional research and trade bodies, governments, scientists, extension, growers and the private sector. Predictive models can determine the influence of the information flow between individuals and organizations and the effect of ecological interactions on disease dissemination and the efficacy of control interventions (Moslonka-Lefebvre et al., 2011; Garrett, 2012).

6 Strategies for managing maize disease outbreaks

Success in managing maize disease outbreaks will depend on the availability of effective systems of preventing the introduction of pathogens, conducting surveillance, investigating and reporting the incidence of infectious disease. These play a crucial role both in managing existing diseases and in contributing to a flexible and adaptable approach to those that may be new and emerging. A second underpinning factor should be capacity building to promote the availability of expertise and trained personnel.

6.1 An effective quarantine system and regulatory framework

The first line of defence against incursions between regions and countries are official measures for quarantine (Waage and Mumford, 2008). These are International Standard Phytosanitary Measures, to limit the introduction and spread of plant pathogens, facilitated by the International Plant Protection Convention (IPPC), through the Food and Agriculture Organization (FAO) of the United Nations (IPPC, 2003). At the regional and national scale, the regulation of quarantine pests and diseases according to IPPC guidelines is performed by plant protection organizations, to provide information regarding pest and disease status to trading partners (IPPC, 2008). However, differing capacities of these organizations among countries and regions greatly influence the effectiveness of implementation.

Quarantine systems in developing countries are generally weak due to a lack of skilled personnel and inadequate infrastructure and funding. There is therefore a strong case for reassessing the level of investment in quarantine services, especially to build human and physical capacity to meet future phytosanitary challenges. A regional approach to quarantine services may be more feasible and cost-effective. The African plant protection services including quarantine systems should have better access to the global plant diseases knowledge-base for more effective use of already available resources. Furthermore, there should be an effective regulatory system for managing EPDs in developing countries. Strengthening the global phytosanitary system of international agreements on the inspection and quarantining of plants to prevent the spread of pathogens and disease requires improved information, standardized testing protocols, improved access to seed treatment protocols and revised pathology research priorities (McGee, 1997), but funding for adequate diagnostic facilities and programmes is often lacking.

6.2 Disease monitoring, surveillance and forecasting

It is almost impossible to prevent the introduction of new diseases or pathogens into a country. However, there is a strong case for investing in, and intensifying disease monitoring, surveillance, forecasting and prediction to stop pathogens becoming a major threat. The surveillance should increasingly employ modern robust and sensitive methods of detection and diagnosis (King et al., 2006; Jebara, 2004). In addition, changes in the genetic structure of crops and pathogens and the trends in environmental, policy and socio-economic factors over time need to be studied. Research on epidemiology and the control of plant diseases should receive increased emphasis to provide evidence that can be used for prediction and pre-emptive action. Input from farmers will be important. Accurate and professional risk assessments will be required to guide the appropriate disease control strategies either at regional, continental or international level. The current impact of plant diseases should be used as a template for the design of control strategies that increasingly apply cutting-edge science to reduce potential risks. Research in this area must change from passive cataloguing to analysing and interpreting the factors contributing to disease emergence. Spatial and temporal analyses of anthropogenic drivers of EPDs, together with an understanding of pathogen biology and evolutionary and environmental factors are required to predict and prevent emergence (Anderson, 1994). Advanced data collection and processing could enable emerging diseases to be quickly spotted, providing valuable time for the development of effective management strategies or to stop their spread in other ways. New techniques for identifying and analysing new pathogens could help us to understand quickly how the newly detected diseases will affect populations, and how they could be controlled before widespread damage is caused. And faster, smaller and cheaper diagnostic devices will open up entirely new possibilities for use by professionals and the individual. An example is the lateral flow antibody-based devices for detecting MLN viruses (DSMZ, 2014). This will enable more effective and more efficient targeting of resources: detecting a new emerging disease in its very early stages, for example, by advanced data analysis, could enable resources to be tightly focused and an outbreak to be stopped in its tracks. In addition, global surveillance and reporting of EPDs require an international coalition that as yet does not exist. The FAO mandates countries to report plant pests and diseases to the IPPC (FAO, 1995), but economic consequences often hinder accurate reporting. The extreme disparity in detection capabilities between nations, reflecting extreme inequalities in wealth, seriously hinders our ability to tackle EPD problems quickly. It is important to recognize that surveillance for EPDs has become a collective responsibility and requires a collective investment. Therefore, the international community should be encouraged to make the investment needed to strengthen global surveillance and early warning networks.

6.3 Understanding host–pathogen interaction and the emergence of races

Many plant pathogens exhibit variations as races and biotypes and can mutate to more virulent forms, often owing to the widespread use of single-gene resistance in widely grown crops (Welz and Geiger, 2000). To ensure sustainable crop production, it will be important to predict the durability of deployed resistance genes through assessing the fitness of the corresponding pathogens. Pathogen fitness is a key driver in the evolution and stability of a pathosystem. Improved methodologies are needed to allow enhanced understanding of pathogenesis, genetic variation and the triggers of change and drivers of epidemics in the future development of improved strategies for resistance management. In addition, the ongoing monitoring of changes in the pathogen will be necessary in order to develop strategies to manage pathogen change in the longer term, and especially to prevent devastating epidemics.

6.4 Developing host plant resistance

Farmers have always welcomed improved varieties with genetic resistance. They favour host plant resistance mainly because it is cost-effective and environmentally sound, but also because alternative disease control methods are unaffordable or less effective. Genes conferring resistance to some maize diseases have been transferred using

Table 3 Development of maize inbred lines with resistance to multiple diseases. The genotypes listed were evaluated in different environments for response to multiple pathogens between 2009 and 2013.

			Mean disease rating on a 1–5 scale				
Stock ID	AD	GLS	MSV	TLB	Rust	SCLB	TSC
DTMA-11	86.37	1.74	1.41	1.41	1.26	1.80	1.40
DTMA-13	87.43	1.93	1.60	1.70	1.38	1.16	1.59
DTMA-17	82.93	1.87	1.12	1.80	1.59	1.44	2.60
DTMA-19	79.53	2.20	1.61	1.77	1.23	2.62	1.87
DTMA-21	85.85	2.05	1.87	2.22	2.17	1.71	1.36
DTMA-22	83.03	2.08	1.72	2.43	1.92	1.58	2.29
DTMA-24	87.99	1.64	1.85	2.47	1.67	2.06	1.91
DTMA-25	84.77	1.82	1.90	2.47	1.17	1.89	2.24
DTMA-26	86.07	1.71	1.60	1.76	1.51	1.50	2.40
DTMA-296	87.28	2.38	1.78	2.47	2.12	2.10	1.81
DTMA-3	85.63	1.43	1.12	1.74	1.30	1.55	2.11
DTMA-34	86.43	2.07	2.32	2.13	1.47	1.50	2.21
DTMA-40	88.19	2.01	2.03	1.70	1.52	1.60	1.51
DTMA-6	84.72	2.28	1.12	1.90	1.56	1.45	1.49
DTMA-7	86.89	2.03	1.96	1.54	1.42	1.35	1.91
DTMA-8	85.31	2.22	1.53	1.93	2.26	1.21	1.44
DTMA-9	86.14	1.93	2.32	1.45	1.51	1.49	1.40
DTMA-90	87.46	2.24	2.37	2.50	1.59	1.07	2.27
Ventry	27.24	0.27	0.21	4.87	0.45	0.37	0.14
VGE	15.00	0.45	0.12	0.19	0.31	0.12	0.22
Vres	25.35	0.22	0.80	0.58	0.28	0.27	0.15
heritability	0.85	0.76	0.62	1.00	0.85	0.88	0.77
LSD (5%)	7.00	0.87	2.51	0.82	0.91	–	0.76
Nlocs	5	6	3	17	5	5	6

Disease rating is on a scale of 1–5; where 1 = no disease symptoms and 5 = severe disease symptoms, with more than 80% of foliage blighted.

AD = days to anthesis, GLS = grey leaf spot, MSV = maize streak virus; TLB = northern corn leaf blight; SCLB = southern corn leaf blight; Rust = common rust; TSC = tar spot complex conventional breeding (Pratt and Gordon, 2006). Possibilities are now wider, given the new developments in genomics, genetic mapping and marker-assisted selection (Bernardo and Yu, 2007; Morris et al., 2003). Considerable efforts have been made in developing resistant inbred lines in maize with quantitative resistance. In addition, inbred lines are now being developed with multiple disease resistance (Table 3). Multiple or durable resistance would be preferable as single-gene (R genes) resistance tends to break down due to changes in pathogen populations. Examples of previous effective resistance genes becoming ineffective and resulting in devastating epidemics can be traced for several crops in the world (Pataky et al., 2001; Welz and Geiger, 2000).

6.5 Pre-emptive breeding

Predictions of what diseases are likely to be a future threat must have a stronger role in the structure of maize breeding programmes than is currently the case. It is possible to anticipate that a disease is absent from an area only by virtue of not having been introduced, or that a particular disease under a changed cropping system may emerge as a major threat. Many maize diseases are regionally confined, for example, tar spot complex and corn stunt complex are in the Americas, and MSV is in Africa (Table 2). It is only a matter of time before these diseases also spread as experience with MLN has shown in East Africa. Pre-emptive breeding, as opposed to the current reactive approach on maize diseases, must be seen as a sound investment if the consequences of shock disease episodes are to be avoided. A retrospective look at past epidemics tells us that a 10–15 year time frame will be required for a control practice for a new disease epidemic to be developed and released to farmers. Pre-emptive breeding can significantly reduce the time needed to produce a new variety or hybrid, and minimize losses from shock events. Genes that impose a high penalty on the adapting pathogen and genes transferred from outside the crop gene pool should help in developing more durable crop resistance. Consequently, widespread use of host plant resistance alone will not, in the long run, be adequate to prevent big losses caused by plant diseases. For some diseases, such as MLN, germplasm with resistance to both viruses has so far not been found, despite the screening of over 30 000 maize entries. For such diseases, integrated management strategies will be necessary for controls that are more effective. Transgenic approaches should also be pursued, especially those based on the principle of pathogen-mediated resistance; this has great potential to induce resistance to homologous pathogens.

6.6 Integrated management

Cultural control, including the use of crop rotation, removal of plant residues, manipulating planting dates, vector control, phytosanitation and roguing, are widely applied in disease management. Where host plant resistance has been integrated with these strategies, the efficacy of the approach in controlling the diseases has been greater. Future disease management strategies should therefore be integrated and built around host plant resistance to achieve more effective controls.

6.7 Effective seed systems

For maize, seed-based solutions (resistant varieties) are a significantly easier method of controlling plant diseases and more effective to implement than the knowledge and labour-intensive technologies such as cultural, biological and chemical control. However, dissemination of disease-resistant crop varieties to smallholders can be grossly affected by ineffective seed systems. Appropriate policies and regulatory frameworks that support the production and marketing of quality seeds of improved varieties could usefully be developed and implemented at all levels. Village- and community-level seed production systems, usually supported by NGOs, have proven successful in Malawi, Kenya, Uganda and Tanzania. Similarly, private sector participation is crucial for sustainable seed systems (USAD-FAS, 2014).

6.8 Networking

Plant diseases and their control have become a transnational challenge that can be met only if international cooperation is strengthened in the following areas:

- The surveillance of maize diseases should become a global activity; individual components of which should operate effectively at country level while regional programmes provide coordination. In parallel there should be early warning systems concerned with the potential for the transnational spread of maize diseases.
- National plant disease control systems should be strengthened within national public plant protection sectors. This requires international and bilateral technical cooperation. The bilateral development agencies and the development banks should join others active in this field. Improved coordination and cooperation are needed between sectors of Government and among governments, industry, consumers and NGOs.
- Risk assessment should be undertaken to an increased extent to establish standards for disease surveillance and detection/diagnosis both nationally and internationally. International cooperation is needed to assist developing countries with the integration of risk analysis into national plant disease control programmes. International bodies, such as FAO and CGIAR, should intensify their efforts on the application of risk assessment for the development of standards for EPDs.
- Many regions of the world do not have the laboratory infrastructure, the human resources or the financial resources to support effective disease surveillance programmes. Yet it is increasingly clear that infectious diseases are a global problem and surveillance is an international responsibility. Investment by richer countries in surveillance capacity in poorer countries may be a sensible response to this problem. One route to a better understanding of the risks of future diseases is a better understanding of past and present examples of their emergence and spread. Collation and critical analysis of such data could provide valuable information. Every new outbreak provides valuable data for research, if appropriate efforts are made to capture them and to make them available.

7 The emergence of MLN in Eastern Africa

MLN was first reported in the lower parts of Longisa division of Bomet District of Kenya in September 2011(Wangai et al., 2012), and in all the major maize-growing districts (MDRAT, 2012) in the following year. The disease has since spread to other East and Central Africa

Table 4 Global occurrence and reports of MCMV and MLN

Location	Date reported	Potyvirus[a]	Reference
Peru	1973	NR	Castillo and Herbert, 1974
US Mainland (Kansas and Nebraska)	1976	WSMV	Niblett and Claflin, 1978
Argentina	1982	–	Teyssandier et al., 1981
Thailand	1982	–	Klingkong and Sutabutra, 1982
Mexico	1987	–	Delgadillo and Gaytan, 1987
US Hawaii	1992	MDMV	Jiang et al., 1992
China	2011	NR	Xie et al., 2011
Kenya	2012	SCMV	Wangai et al., 2012
Rwanda	2013	SCMV	Adams et al., 2014
DRC[b]	2013	SCMV	Lukanda et al., 2014
Taiwan	2014	SCMV	Deng et al., 2014
Ethiopia	2014	SCMV	Mahuku et al., 2015b

[a]Complementing potyviruses reported with MCMV. WSMV, *Wheat Streak Mosaic Virus*; MDMV, *Maize Dwarf Mosaic Virus*; SCMV, *Sugarcane Mosaic Virus*; NR, not reported.

[b]DRC, Democratic Republic of Congo.

Figure 4 Maize plants inoculated with MCMV, SCMV and the combination of MCMV and SCMV to give MLN in the greenhouse.

(ECA) countries including Tanzania, Uganda, South Sudan, Ethiopia and Rwanda (IPPC, 2014; Mahuku et al., 2015). MLN is caused by the simultaneous infection of maize plants with MCMV (genus *Machlomovirus*) and with any of the cereal-infecting viruses in the

Table 5 Timelines of initial observation for *Maize chrolotic mottle virus* (MCMV) and MLN in Africa.

Country	Month/year of initial observation
Kenya	September 2011
Tanzania	August 2012
Uganda	October 2012
South Sudan	January 2013
Rwanda	June 2013
Democratic Republic of Congo	March 2014
Ethiopia	June 2014

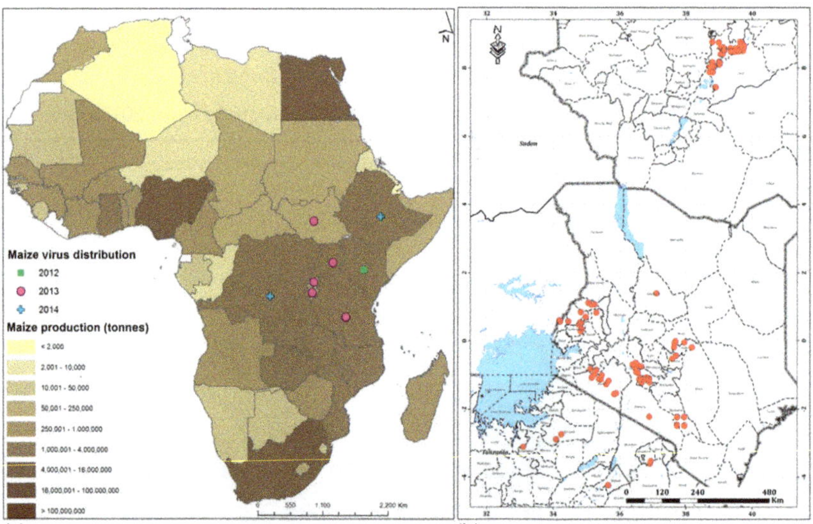

(a) (b)

Figure 5 (a) Production of maize (average of 2006–8; Source: FAOstat) and occurrence of MCMV in Africa; (b) distribution of MLN in Ethiopia, Kenya and Tanzania.

Potyviridae family, such as SCMV (genus *Potyvirus*), *maize dwarf mosaic virus* (MDMV, genus *Potyvirus*) and *wheat streak mosaic virus* (WSMV, genus *Potyvirus*). When infecting maize, MCMV or SCMV alone typically produces milder symptoms (Fig. 4). However, when both viruses infect the plant at the same time, a rapid synergistic reaction is produced that seriously damages or kills infected plants (Mahuku et al., 2015a).

MCMV was first reported in Peru in 1973 (Castillo and Hebert, 1974), in the United States in 1976 (Niblett and Claflin, 1978) and then in China in 2010 (Xie et al., 2011). In Africa, after initially being reported in Kenya in 2011 (Wangai et al., 2012), it was subsequently reported in Tanzania (CIMMYT, 2012), Uganda (IPPC, 2014), Rwanda (Adams et al., 2014), Democratic Republic of Congo (Lukanda et al., 2014) and Ethiopia (Mahuku et al., 2015b).

Occurrence of MCMV coincided with outbreaks of MLN in all these countries. Given the poor phytosanitary systems, porous borders and inadequate regulatory regimes in the ECA region, the disease is likely to spread to all countries (FSNWG, 2012) bordering Kenya (Fig. 5). Both viruses involved in MLN infection are reportedly transmitted by insects, and also through seeds. National and inter-governmental regulatory research and extension organizations, laboratories and the private sector are working together to control the spread of the diseases.

8 Responding to the MLN outbreak and minimizing its impact

Surveys have been conducted to assess the extent of spread of MCMV in Africa (Table 5; Fig. 5). The viral genome of MCMV isolates in East African countries was characterized: this revealed 99% sequence homology among the East African isolates and also with those reported in China (Mahuku et al., 2015a). However, the SCMV isolates from East Africa were found to be diverse with about 1–11% divergence in coat protein sequence. Cheap and robust diagnostics have been developed for the detection of both viruses. IITA developed a cheap in-house ELISA test for MCMV and trained national partners from Tanzania and Kenya on its use. IITA also developed a reverse transcription loop-mediated isothermal amplification (RT-LAMP)-based assay and a multiplex PCR for the detection of MCMV and SCMV in maize. These assays offer sensitive detection of the two causal viruses . A lateral flow diagnostic kit using antisera developed against the African isolates of SCMV and MCMV has been developed by DSMZ (Germany). Operating in much the same as a test kit for the rapid confirmation of pregnancy, these rapid diagnostic tools have brought diagnosis to the field for efficient monitoring of these diseases. ELISA, LAMP and RT-PCR assays have also been developed for the detection of SCMV. Efforts are ongoing to validate the lateral flow diagnostic kits to identify all SCMV isolates found in Africa.

8.1 Prevention and control

The most effective control for MCMV has been achieved through the integration of cultural practices with the use of insecticides and host resistance (Nelson et al., 2011). In the United States, a number of potential sources of resistance to MCMV have been identified; however, a lot still needs to be done to develop commercial varieties with adequate resistance (Nelson et al., 2011). Crop rotation with non-maize crops has been shown to reduce the incidence of MCMV in subsequent years (Phillips et al., 1982; Uyemoto, 1983). One of the factors identified as driving the incidence and severity of MLN in Kenya was the continuous production of maize that can lead to a constantly available reservoir of the virus and its vectors. To mitigate the effects of MLN, farmers were provided with information on management practices promoting good agricultural practices such as crop rotation; roguing and burning of infected plants; weed management to eliminate vector reservoirs; planting only certified seeds and diversifying farm enterprises and avoiding the continuous planting of maize. Furthermore, farmers were advised to plant maize at the onset of the main rainy season and to avoid late planting, as disease incidence and severity/impact were highest in late-planted maize.

Phytosanitary measures and regulations

A number of phytosanitary measures and regulations were put in place to minimize further spread of MLN. Following stakeholders' meetings involving seed companies and researchers, internal quarantine restricting the movement of maize seeds has been set up to prevent seeds being moved from MLN-infected to non-infected areas. Recommendations have been made to produce seeds in areas that are known as disease-free or are outside the endemic region. For example, the southern highlands in Tanzania are free from MLN; as a result, seed companies operating in Tanzania agreed to concentrate seed productions there. Efforts are underway through the FAO to harmonize sanitary and phytosanitary services regulations across the East African region enforce the necessary plant health regulations to address the MLN crisis. These issues include the development of PRA for MLN, standardized protocols for surveys, diagnostics, quarantine measures and movement of seeds across borders and protocols for certifying seeds as free of MLN viruses (Mezzalama et al., 2015).

Information dissemination

A major activity in managing the spread of MLN has been to build awareness along the maize value chain on the impact of the disease. Farmers were informed on appropriate agronomical practices to reduce disease incidence and severity, including crop rotation, a maize-free season, early planting, timely detection and destruction of infected plants, planting of certified maize seeds and crop diversification where possible (CIMMYT, 2013). Meetings were conducted to build awareness and brief farmers about MLN and what they should do. The government of Kenya set up a task force to carry out a survey in all maize-growing areas across the country and recommend the best ways of controlling the disease. Similar efforts have been established in all the countries where MLN has been reported. A number of workshops have been conducted to mobilize maize growers, researchers and policy makers across Africa to be vigilant and on the lookout for MLN. In 2015, a high-level international conference was held in Nairobi to share knowledge on methods of diagnostics and screening for MLN, identify ways of curbing its spread across Africa and to foster international collaboration. Brochures were produced with information in both English and local languages to inform farmers of the MLN crisis and the measures to adopt to minimize its impact.

Since the outbreak of MCMV/MLN in Kenya, substantial progress has been made in understanding disease biology, diversity, epidemiology and control (Mahuku et al., 2015a). Further research should focus on generating information on the diversity and distribution of the viruses causing MLN in the region, and come up with simple and effective detection methods, such as the use of lateral flow devices, that can be adopted by smallholder farmers and extension agents. Advantage can be taken of the availability of smartphones, to build this capacity for farmers to take photos of suspect plants; to be followed by confirmation from trained people. This could help in disease surveillance, and early identification and in shaping rapid response systems. Surveillance activities are ongoing in West and Southern Africa to map the distribution of the disease better.

8.2 Understanding the transmission of MLN viruses

Although aphids spread SCMV, different vectors have been associated with the spread of MCMV. In the United States, MCMV is transmitted by at least six species of chrysomelidae beetles, including the cereal leaf beetle (*Oulema melanopa*), the corn flea beetle (*Chaetocnema pulicaria*) and the western corn rootworm (*Diabrotica virgifera*). In Hawaii, the corn thrips, *Frankliniella williamsi* Hood (Thysanoptera: Thripidae), has been identified as the main vector (Nault et al., 1981; Jiang et al., 1992; Cabanas et al., 2013). In Kenya, high populations of corn thrips were present in infested fields (Moritz et al., 2013), a good indication that thrips are potential vectors in the ECA region. However, no further information on potential vectors of MCMV is available in Africa, where a wide diversity of arthropods has been noticed in maize fields. The role and efficiency of each of the potential vectors are being established. The spread of MCMV through seeds has also been identified as critical. Research is underway to ascertain the level of seed transmission; preliminary results have shown that it is higher than previously reported, with up to 15% of seed transmission recorded in one hybrid (Ben Lockhart, pers. comm.). The rate of disease spread in the region has been extremely rapid, raising the question of the significance of each of these methods in disease dissemination (Fig. 5). There is a need to determine the efficiency of different vectors in virus spread and to monitor the extent to which different seed systems adopted in different countries contribute to the spread of the viruses. The survival of the viruses within a farming system is important in designing a management strategy. Viruses causing MLN may survive in different ways in the absence of a maize crop, including through maize seeds, in vectors, plant debris and in alternative hosts. Some of the recommended ways of managing the disease include crop rotation and the planting of non-maize crops for a period. Although these are good recommendations, the presence of viruses in other sources may have a negative effect on management strategies. There is ongoing research to determine the survival of viruses causing MLN in the maize-farming systems (intercrop, crop rotation or sequential planting, monoculture, relay cropping, zero tillage) and their contribution to disease development.

8.3 Developing host resistance

Identification and development of maize varieties with resistance to MLN are important in the management of the disease. Since September 2013, CIMMYT, in partnership with KARLO, IITA, the Seed Trade Association of Kenya, Association for Strengthening Agricultural Research in ECA (ASARECA) and ICIPE, has led efforts to identify sources of resistance, and replace existing susceptible varieties with those that are resistant. With funding from the Bill and Melinda Gates Foundation (BMGF) and the Syngenta Foundation for Sustainable Agriculture (SFSA), a dedicated screening facility was opened at KARLO's facilities in Naivasha, Kenya, to support the identification of MLN-resistant germplasm (CIMMYT, 2013). More than 30000 genotypes, including elite lines, experimental, pre-commercial and commercial hybrids and landraces from both private and public institutions, have been screened at the facility. Most of the genotypes screened so far have been susceptible to the disease, and very few have shown promise as potential sources of tolerance. Promising inbred lines with MLN resistance have been shared with public and private institutions in sub-Saharan Africa. Intensive efforts are ongoing to rapidly develop and deploy MLN-resistant maize germplasm that can replace the existing MLN-vulnerable varieties grown across East Africa. The initial screening of thousands of experimental and pre-commercial hybrids has resulted in finding a few tolerant lines, but to date, high levels of resistance to MLN/MCMV have not been identified. Although screening continues for germplasm resistant to the disease, there might be a need to seek other sources of genes for resistance (e.g., through transgenics). However, protocols for handling MLN viruses and evaluating for resistance have been developed in collaboration with scientists from the United States, who have had extensive experience working with MCMV. These protocols have come in handy in working with MLN. Networking and collaboration have proved to be a key feature in tackling MLN. Although control measures are yet to be found, the spread will be contained, as pre-emptive measures, have been adopted to manage the disease.

8.4 Overview of research efforts in the region

Multidisciplinary teams of researchers from local and international universities and research institutions, regulatory bodies, extension agents, NGOs, CBOs and communication specialists have been assembled in different countries within the region. Scientific support has been tapped from laboratories all over the world. Different R&D projects aim to manage the threat of MLN by combining state-of-the art science with on-field experience in scaling up technologies and innovations for disease management. Key donors supporting research work in the region include national governments, ASARECA, CGIAR, the World Bank (through KAPAP in Kenya), USAID, BMGF and SFSA. There is ongoing research to understand the epidemiology of the MLN viruses, map distribution, and conduct surveillance studies and screening to identify sources of resistance for use in breeding . Intensive efforts are underway to develop high levels of resistance to MCMV/MLN, including the identification of molecular markers for the selection of promising genotypes to fast-track development of resistant varieties (Gowda et al., 2015).

8.5 Lessons from the MLN outbreak in Africa

Plant diseases do not respect national boundaries. To address plant health issues, a regional approach must be taken that helps harmonize regulatory efforts and ensure effective regulatory implementation. It is futile for one nation to try and eradicate a disease if the neighbouring country does not. Therefore, support for capacity-building at regional and national levels is essential to contain diseases and avert detrimental impacts. A regional umbrella organization with plant health experts must be in place to support regulatory harmonization efforts, help ensure effective implementation and support national-level experts on plant disease issues as they arise.

The national, regional and international response to minimize the devastation of MLN in East Africa is impressive. However, the case of MLN is a good example to illustrate how having a strong and regionally harmonized plant health regulatory system, where each East African nation works collaboratively, would have lessened the severity of MLN's impact. The extent of the distribution of MLN in Kenya suggests that the disease had been present for some time before it was formally identified. A strong plant health regulatory system that works collaboratively with other nations, with effective surveillance systems, would have identified MLN earlier and implemented a regional plan for crisis management. This would have implemented immediate quarantine zones to contain and minimize the disease, shared information across the region and with the IPPC and helped to rally resources for a faster response, including research on the disease and methods of control. A robust plant health system would have minimized the damage, economic loss and threat to food security that quickly developed as MLN spread unabated. It is important to note that plant health systems are designed to address the normal threats to plant health in the region, in addition to organize crisis interventions such as the one required for MLN.

The lack of diagnostic capacity, in both personnel and infrastructure, has meant that samples had to be sent outside Africa for diagnosis, taking valuable time that could have been used to implement mitigation strategies. Because of the lack of networks for ED and RR, the appearance of MLN caught the research community by surprise, and took a while to mobilize resources, and assemble the relevant teams to deploy control measures. There is an urgent need to establish functional structures to prevent another epidemic.

9 Conclusions

Success in mitigating the effects of disease outbreaks depends on the efficacy and effectiveness of systems in reporting disease outbreaks and developing rapid coordinated responses. A disease can be controlled efficiently only if the causal agent is known. The rapid decrease in the number of New Disease Reports over the past 100 years is ample evidence of an increasingly failing capacity to recognize and control current, new and emerging diseases in Africa, even though the number of such reports has rapidly increased in Europe (Waage et al., 2009). A consequence of this has been the rapid emergence of new diseases in several food security crops, including cassava with cassava mosaic disease and cassava brown streak disease (Legg et al., 2014), maize with MLN (Mahuku et al., 2015a), banana with bunchy top disease, *Xanthomonas* wilt and Panama disease (Beed, 2014; Kumar et al., 2015) and soybean with rust (Murithi et al., 2016) as case examples.

The introduction of new pests, and with it the risk of new outbreaks, continues to occur, driven by globalization, increasing human mobility, climate change, and the evolution and adaptation of pathogens or their vectors (Anderson et al., 2004; Käferstein and Abdussalam, 1999). This threat is best addressed through reliance on rapid detection, diagnosis, response and containment. Early disease diagnosis and pathogen identification are essential elements of a rapid response system. Linked to these are the availability of the necessary diagnostic expertise, infrastructure and resources. Simple, less-expensive and fit-for-purpose diagnostic tests are crucial for accurate identification, especially for the majority of developing countries with scarce resources. Emphasis should be placed on developing these low-cost, easy-to-use diagnostic tests that will ensure information required for assessing and combating disease outbreaks is obtained in a timely manner.

The availability of adequate monitoring systems allows the initiation of appropriate measures to prevent the rapid spread of diseases, and also timely interventions to reduce the impact of a pandemic if the disease should spread. A number of diseases, such as MSV, corn stunt complex and tar spot complex, are known to occur in some maize-growing environments. It is only a matter of time before these diseases spread to new maize-growing areas. A pre-emptive measure to minimize outbreaks from these pathogens is through risk analysis; this would be evaluated based on current knowledge and combined with results from predictive models (IPPC, 2003). Predictive models can be used to help quantify the risk of a disease epidemic for any specified location based on known epidemiological principles and environmental parameters and thus help further prioritize localities for surveillance activities (Isabirye and Rwomushana, 2016; Jeger et al., 2007). Models can be used for evaluating the likelihood of a given pathogen being introduced and established, then spreading and emerging; to assess risk and target strategies for surveillance and diagnostic methods; and also to determine the likely effect on disease epidemics of mixed cropping systems, host resistance, landscapes and climate change (Garrett et al., 2009, 2011).

Control of diseases has become a transnational challenge that can be met only if international cooperation is strengthened in the following areas:

- The disease surveillance is taken as a global activity; the individual components of which operate effectively at the country level while regional programmes provide coordination. In parallel, there should be early warning systems to prevent transnational spread of pathogens.
- National plant protection systems in the developing countries are strengthened. This requires international and bilateral technical cooperation, as well as improved coordination and cooperation between different sectors of government, and cooperation among government, industry, consumers and nongovernmental organizations.
- Risk assessment is undertaken as part of the emergency response system. Developing countries need assistance to integrate risk analysis into national plant disease monitoring and control programmes.

10 Where to look for further information

Navas-Castillo, J., Fiallo-Olivé, E. and Sánchez-Campos, S. (2011), 'Emerging Virus Diseases Transmitted by Whiteflies', *Annual Review of Phytopathology* 49, 219–48.

Perrings, C., Burgiel, S., Lonsdale, M., Mooney, H. and Williamson, M. (2010), 'International cooperation is the solution to trade-related invasive species risks', *Annals of the New York Academy of Sciences* 1195, 198–212.

MacLeod, A., Pautasso, M., Jeger, M. J. and Haines-Young, R. (2010), 'Evolution of the international regulation of plant pests and challenges for future plant health', *Food Security* 2, 49–70.

Magarey, R. D., Dolezal, W. E. and Moore, T. J. (2009), 'Worldwide monitoring systems: The need for public and private collaboration'. In U. Gisi, I. Chet and M. L. Gullino (eds), *Plant Pathology in the 21st Century*, vol. 1., Springer, Dordrecht, The Netherlands: Recent Developments in Management of Plant Diseases, pp. 349–55.

11 Acknowledgements

The authors gratefully acknowledge funding support for research work on MLN from the CGIAR Research Program on MAIZE, the Africa RISING (USAID), BMGF and the SFSA.

12 References

Adams, I. P., Miano, D. W., Kinyua, Z. M., Wangai, A., Kimani, E., Phiri, N., Reeder, R., Harju, V., Glover, R., Hany, U., Souza-Richards, R., Deb Nath, P., Nixon, T., Fox, A., Barnes, A., Smith, J., Skelton, A., Thwaites, R., Mumford, R. and Boonham, N. (2013), 'Use of next-generation sequencing for the identification and characterization of *Maize chlorotic mottle virus* and *Sugarcane mosaic virus* causing maize lethal necrosis in Kenya', *Plant Pathology* 62, 741–9.

Adams, I. P., Harju, V. A., Hodges, T., Hany, U., Skelton, A., Rai, S., Deka, M. K., Smith, J., Fox, A., Uzayisenga, B., Ngaboyisonga, C., Uwumukiza, B., Rutikanga, A., Rutherford, M., Ricthis, B., Phiri, N. and Boonham, N. (2014), 'First report of maize lethal necrosis disease in Rwanda', *New Disease Reports*, 29, 22.

Anderson, P. K. (1994), 'A conceptual framework of integrated epidemiology for application to emerging disesaes', *Annals of the New York Academy of Sciences* 740, 439–44.

Anderson, P. K., Cunningham, A. A., Patel, N. G., Morales, F. J., Epstein, P. R. and Daszak, P. (2004), 'Emerging infectious diseases of plants: Pathogen pollution, climate change and agrotechnology drivers', *Trends in Ecology & Evolution* 19(10), 535–44.

Bajet, N. B., Renfro, B. L. and Valdez, C. J. M. (1994), 'Control of tar spot of maize and its effect on yield', *International Journal of Pest Management* 40(2), 121–5.

Bebber, D. P., Holmes, T. and Gurr, S. J. (2014), 'The global spread of crop pests and pathogens', *Global Ecology and Biogeography* 23, 1398–1407.

Beed, F. D. (2014), 'Managing the biological environment to promote and sustain crop productivity and quality', *Food Security* 6, 169–86.

Becker, G. S. (2009), 'U.S. food and agricultural imports: Safeguards and selected issues', CRS Report RL34198.

Bernardo, R. and Yu, J. M. (2007), 'Prospects for genomewide selection for quantitative traits in maize', *Crop Science* 47(3), 1082–90.

Bostock, R. M., Thomas, C. S., Hoenisch, R. W., Golino, D. A. and Vidalakis, G. (2014), 'Plant health: How diagnostic networks and interagency partnerships protect plant systems from pests and pathogens', *California Agriculture* 68, 117–24.

Bouwmeester, H., Abele, S., Manyong, V. M., Legg, C., Mwangi, M., Nakato, V., Coyne, D. and Sonder, K. (2010), 'The potential benefits of GIS techniques in disease and pest control: An example based on a regional project in Central Africa', *Acta Horticulturae* 879, 333–40.

Bouwmeester, H., Heuvelink, G. M. B., Legg, J. P. and Stoorvogel, J. J. (2012), 'Comparison of disease patterns assessed by three independent surveys of cassava mosaic virus disease in Rwanda and Burundi', *Plant Pathology* 61, 399–412.

Bradfute, O. E., Tsai, J. A. and Gordon, D. T. (1981), 'Corn stunt spiroplasma and viruses associated with a maize disease epidemic in southern Florida', *Plant Disease* 65, 837–41.

Bustamente, P. I., Hammong, R. and Ramirez, P. (1998), 'Evaluation of maize germ plasm for resistance to maize rayado fino virus', *Plant Disease* 82, 50–6.

Cabanas, D. W., Higashi, C. H. V. S. and Bressan, A. (2013), 'Dissecting the mode of Maize Chlorotic Mottle Virus transmission (Tombusviridae: Machlomovirus) by *Frankliniella williamsi* (*Thysanoptera: Thripidae*)', *Journal of Economic Entomology* 106, 16–24.

Castillo, J. and Hebert, T. (1974), 'Nueva enfermedad virosa afectando al maiz en el Peru', *Fitopatologia* 9(2), 79–84.

CIMMYT. (2013), 'Maize lethal necrosis (MLN) disease in Kenya and Tanzania: Facts andactions', CIMMYTKARI. http://www.cimmyt.org/en/wherewe-work/africa/item/maize-lethal-necrosismln-disease-in-kenya-and-tanzania-factsand-actions.

CIMMYT. (2012), 'Annual Report. CGIAR Research Program MAIZE', CIMMYT, Mexico, D. F., Mexico.

CIMMYT. (2013), 'Annual Report. CGIAR Research Program MAIZE', CIMMYT, Mexico, D. F., Mexico.

Cliff, A. and Haggett, P. (2004), 'Time, travel and infection', *British Medical Bulletin* 69(1), 87–99.

Condon, M. (1997), 'Implications of plant pathogens to international trading of seeds'. In D. C. McGeee (ed.), *Plant Pathogens and the Worldwide Movement of Seeds*, St. Paul, MN: APS Press, pp. 17–30.

Contreras, S. (2002), 'The International Seed Industry'. In M. McDonald and S. Contreras (eds), *Proceedings International Seed Seminar: Trade, Production and Technology*, Santiago, Chile: Pontificia Universidad Católica de Chile, Facultad de Agronomía e Ingeniería Forestal, 15–16 Octubre 2002, pp. 1–9. http://seedbiology.osu.edu/HCS630_files/March%2027/Global%20 Seed%20Industr,%20text.pdf (accessed 10 January 2016).

Crooks, J. A. (2005), 'Lag times and exotic species: The ecology and management of biological invasions in slow-motion', *Ecoscience* 12, 316–29.

Daszak, P., Cunningham, A. A. and Hyatt, A. D. (2000), 'Emerging infectious diseases of wildlife – threats to biodiversity and human health', *Science* 287(5452), 443–9.

De Groote, H., Oloo, F., Tongruksawattana, S. and Das, B. (2016), 'Community-survey based assessment of the geographic distribution and impact of maize lethal necrosis (MLN) disease in Kenya', *Crop Protection* 82, 30–5.

Delgadillo, S. F. and Gaytán, B. R. (1987), 'Identificación de la enfermedad de la necrosis letal del maíz en Guanajuato', *Rev. Mex. Fitopatologia* 5, 21–6.

Deng, T. C., Chou, C. M., Chen, C. T., Tsai, C. H. and Lin, F. C. (2014), 'First report of *Maize chlorotic mottle virus* on Sweet Corn in Taiwan', *Plant Disease* 98, 1748.

DSMZ. (2014), 'Data Sheet on Maize Lethal Necrosis (MLN) Disease', Leibniz Institut DSMZ GmbH, Plant Virus Department, Inhoffenstraße 7 b, 38124 Braunschweig, Germany.

Epstein, P. R., Chivian, E. and Frith, K. (2003), 'Emerging diseases threaten conservation', *Environmental Health Perspectives* 111(10), A506–7.

FAO (1995), *International Standards for Phytosanitary Measures: Principles of Plant Quarantine as Related to International Trade*, Food and Agriculture Organisation of the United Nations. World Agricultural Information Centre (WAICENT), Publication No 1. http://www.furs.si/law/FAO/ZVR/ENG/ISPM1.pdf (accessed 20 January 2016).

Fisher, M. C., Henk, D. A., Briggs, C. J., Brownstein, J. S., Madoff, L. C., McCraw, S. L. and Gurr, S. J. (2012), 'Emerging fungal threats to animal, plant and ecosystem health', *Nature* 484, 186–94.

Fletcher, J., Luster, D., Bostock, R., Buras, J., Cardwell, K., Gottwald, T., McDaniel, L., Royer, M. and Smith, K. (2010), 'Emerging infectious plant disease'. In W. M. Scheld, M. L. Grayson and J. M. Hughes (eds), *Emerging Infections*, vol. 9, Washington, DC: ASM Press, pp. 337–66.

Foresight. (2006), *Infectious Diseases: Preparing for the Future. Executive Summary*, London: Office of Science and Innovation.

FSNWG. (2012), 'Regional Food Security Situation and Outlook', Food Security and Nutrition Working Group, http://www.disasterriskreduction.net/east-central-africa/fsnwg.

Gámez, R. (1980), 'Rayado fino virus of maize in the American tropics', *Tropical Pest Management* 26, 26–33.

Garst Seeds. (1996), 'Garst establishes a G. L. S. standard', Garst Seeds, online publication, 980318.

Garrett, K. A. (2012), 'Information networks for disease: Commonalities in human management networks and within-host signaling networks', *European Journal of Plant Pathology* 133, 75–88.

Garrett, K. A., Zúñiga, L. N., Roncal, E., Forbes, G. A., Mundt, C. C., Su, Z. and Nelson, R. J. (2009), 'Intraspecific functional diversity in hosts and its effect on disease risk across a climatic gradient', *Ecological Applications* 19, 1868–83.

Garrett, K. A., Forbes, G. A., Savary, S., Skelsey, P., Sparks, A. H., Valdivia, C., van Bruggen, A. H. C., Willocquet, L., Djurle, A., Duveiller, E., Eckersten, H., Pande, S., Vera Cruz, C. and Yuen, J. (2011), 'Complexity in climate-change impacts: an analytical framework for effects mediated by plant disease', *Plant Pathology* 60, 15–30.

Gowda, M., Das, B., Makumbi, D., Babu, R., Semagn, K., Mahuku, G., Olsen, M. S., Jumbo, M. B., Beyene, Y. and Prasanna, B. M. (2015), 'Genome-wide association and genomic prediction of resistance to maize lethal necrosis disease in tropical maize germplasm', *Theoretical and Applied Genetics* 128, 1957–68.

Harvell, C. D., Mitchell, C. E., Ward, J. R., Altizer, S., Dobson, A. P., Ostfeld, R. S. and Samuel, M. D. (2002), 'Climate warming and disease risks for terrestrial and marine biota', *Science* 296(5576), 2158–62.

Hruska, A. J., Gladstone, S. M. and Obado, R. (1996), 'Epidemic roller coaster: Mize Stunt Disease in Nicaragua', *American Entomologist* 42, 248–52.

Hughes, G. and Burnett, F. J. (2015), 'Integrating experience, evidence and expertise in the crop protection decision process', *Plant Disease* 99, 1197–1203.

Hufnagel, L., Brockmann, D. and Geisel, T. (2004), 'Forecast and control of epidemics in a globalized world', *Proceedings of the National Academy of Sciences, USA* 101(42), 15124–9.

IOM (Institute of Medicine). (2008), *Global Climate Change and Extreme Weather Events. Understanding the Contributions to Infectious Disease Emergence: Workshop Summary*, Washington, DC: The National Academies Press.

IOM (Institute of Medicine). (2010), *Infectious Disease Movement in a Borderless World*, Washington, DC: The National Academies Press.

IPPC (2003), 'Identification of risks and management of invasive alien species using the IPPC framework', *Proceedings Workshop on Invasive Alien Species International Plant Protection Convention*, 22–26 September, Germany: Braunschweig.

IPPC (2008), 'Considerations for developing a phytosanitary capacity building strategy for developing countries', *Proceedings Open-ended Working Group on Building National Phytosanitary Capacity*, 8–12 December, Rome, Italy: FAO. https://www.ippc.int/IPP/En/default.jsp (accessed 4 January 2014).

IPPC (2014), 'New pest of maize: maize lethal necrosis in Uganda', IPPC Official Pest Report, No. UGA-01/2, No. UGA-01/2. Rome, Italy: FAO. https://www.ippc.int/.

Isabirye, B. E. and Rwomushana, I. (2016), 'Current and future potential distribution of maize chlorotic mottle virus and risk of maize lethal necrosis disease in Africa', *Journal of Crop Protection* 5(2), 215–28.

Jebara, K. B. (2004), 'Surveillance, detection and response: Managing emerging diseases at national and international levels', *OIE Revue Scientifique et Technique* 23(2), 709–15.

Jeger, M. J., Pautasso, M., Holdenrieder, O. and Shaw, M. W. (2007), 'Modelling disease spread and control in networks: implications for plant sciences', *New Phytologist* 174, 279–97.

Jensen, S. G. and Siefers, D. L. (1996), 'A new disease of maize and wheat in the high plains', *Plant Disease* 80, 1387–90.

Jiang, X. Q. T., Meinke, L. J., Wright, R. J., Wilkinson, D. R. and Campbell, J. E. (1992), 'Maize chlorotic mottle virus in Hawaiian-grown maize–vector relations, host range and associated viruses', *Crop Protection* 11, 248–54.

Jones, R. A. C. (2009), 'Plant virus emergence and evolution: Origins, new encounter scenarios, factors driving emergence, effects of changing world conditions, and prospects for control', *Virus Research* 141, 113–30.

Jones, K. E., Patel, N. G., Levy, M. A., Storeygard, A., Balk, D., Gittleman, J. L. and Daszak, P. (2008), 'Global trends in emerging infectious diseases', *Nature* 451, 990–3.

Josling, T., Roberts, D. and Orden, D. (2003), *Food Regulation and Trade. Towards a Safe and Open Global System*, Washington, DC: Institute for International Economics.

Käferstein, F. and Abdussalam, M. (1999), 'Food safety in the 21st century', *Bulletin World Health Organization* 77(4), 347–51.

Karavina, C. (2014), 'Maize streak virus: A review of pathogen occurrence, biology and management options for smallholder farmers', *African Journal of Agricultural Research* 9(36), 2736–47.

King, D. A., Peckham, C., Waage, J. K., Brownlie, J. and Woolhouse, M. E. J. (2006), 'Infectious diseases: Preparing for the future', *Science* 313, 1392–3.

Klingkong, T. and Sutabutra, T. (1982), 'Anew virus disease of maize in Thailand'. In D. T. Gordon, J. K. Knoke, L. R. Nault and R. M. Ritter (eds), *Proceedings of the International Maize Virus Disease Colloquium and Workshop*, Wooster, OH: OARDC, pp. 191–3.

Kumar, P. L., Selvarajan, R., Iskra-Caruana, M.-L., Chabannes, M. and Hanna, R. (2015), 'Biology, etiology and control of virus diseases of banana and plantain', *Advances in Virus Research* 91, 229–69.

Legg, J., Somado, E. A., Barker, I., Beach, L., Ceballos, H., Cuellar, W., Elkhoury, W., Gerling, D., Helsen, J., Hershey, C., Jarvis, A., Kulakow, P., Kumar, L., Lorenzen, J., Lynam, J., McMahon, M., Maruthi, G., Miano, D., Mtunda, K., Natwuruhunga, P., Okogbenin, E., Pezo, P., Terry, E., Thiele, G., Thresh, M., Wadsworth, J., Walsh, S., Winter, S., Tohme, J. and Fauquet, C. (2014), 'A global alliance declaring war on cassava viruses in Africa', *Food Security* 6, 231–48.

Lenardon, S. L., March, G. J., Nome, S. F. and Ornaghi, J. A. (1998), 'Recent Outbreak of "Mal de Rio Cuarto" Virus on Corn in Argentina', *Plant Disease* 82, 448.

Louie, R. (1980), 'Sugarcane mosaic virus in Kenya', *Plant Disease* 64, 944–7.

Louie, R., Knoke, J. K. and Findley, W. R. (1990), 'Elite Maize Germplasm: Reactions to Maize Dwarf Mosaic and Maize Chlorotic Dwarf Viruses', *Crop Science* 30, 1210–15.

Lukanda, M., Owati, A., Ogunsanya, P., Valimunzigha, K., Katsongo, K., Ndemere, H. and Kumar, L. P. (2014), 'First report of *Maize chlorotic mottle virus* infecting maize in the Democratic Republic of the Congo', *Plant Disease* 98, 1448–9.

Madoff, L. C. (2004), 'ProMED-mail: An early warning system for emerging diseases', *Clinical Infectious Diseases* 39(2), 227–32.

Mahuku, G., Lockhart, B. E., Wanjala, B., Jones, M. W., Kimunye, J. N., Stewart, L. R., Cassone, B. J., Subramanian, S., Nyasani, J., Kusia, E., Kumar, L., Niblett, C. L., Kiggundu, A., Asea, G., Pappu, H., Wangai, A., Prasanna, B. M. and Redinbaugh, P. (2015a), 'Maize lethal necrosis (MLN), an emerging threat to maize-based food security in sub-Saharan Africa', *Phytopathology* 105, 956–65.

Mahuku, G., Wangai, A., Sadessa, K., Teklewold, A., Wegary, D., Ayalneh, D., Adams, I., Smith, J., Bottomley, E., Bryce, S., Braidwood, L., Feyissa, B., Regassa, B., Wanjala, B., Kimunye, J. N., Mugambi, C. , Monjero, K. and Prasanna, B. M. (2015b), 'First Report of *Maize chlorotic mottle virus* and Maize Lethal Necrosis on Maize in Ethiopia', *Plant Disease* 99(12), 1870.

McGee, D. C. (1997), *Plant Pathogens and the Worldwide Movement of Seeds*, St. Paul, MN: APS Press.

MDRAT (2012), *The Status of Maize Lethal Necrosis Disease and General Maize Performance in Kenya*, Kenya: Multi-Disciplinary Rapid Assessment Team, Ministry of Agriculture.

Mezzalama, M., Das, B. and Prasanna, B. M. (2015), *MLN Pathogen Diagnosis, MLN-free Seed Production and Safe Exchange to Non-Endemic Countries*, Mexico, D.F.: CIMMYT Brochure.

Miller, S. A., Beed, F. D. and Harmon, C. L. (2009), 'Plant disease diagnostic capabilities and networks', *Annual Review of Phytopathology* 47, 15–38.

Miller, S. A., Kinyua, Z. M., Beed, F., Harmon, C. L., Xin, J., Vergot, P., Momol, T., Gilbertson, R. and Garcia, L. (2010), 'The International Plant Diagnostic Network (IPDN) in Africa: Improving capacity for diagnosing diseases of banana (Musa spp.) and other African crops', *Acta Horticulturae*, 879, 341–8.

Morens, D. M., Folkers, G. K. and Fauci, A. S. (2004), 'The challenge of emerging and re-emerging infectious diseases', *Nature* 430, 242–9.

Moritz, G., Brandt, S., Triapitsyn, S. and Subramanian, S. (2013), 'Identification and information tools for pest thrips in East Africa. QAAFI Biological Information Technology (QBIT)', Brisbane, Australia: The University of Queensland. ISBN 978-1-74272-0687-8, http://thripsnet.zoologie.uni-halle.de/key-server-neu/.

Morris, M., Dreher, K., Ribaut, J. M. and Khairallah, M. (2003), 'Money matters (II): Costs of maize inbred line conversion schemes at CIMMYT using conventional and marker-assisted selection', *Molecular Breeding* 11(3), 235–47.

Moslonka-Lefebvre, M., M., Finley, A., Dorigatti, I., Dehnen-Schmutz, K., Harwood, T., Jeger, M. J., Xu, X., Holdenrieder, O. and Pautasso, M. (2011), 'Networks in plant epidemiology: From genes to landscapes, countries, and continents', *Phytopathology* 101, 392–403.

Murithi, H. M., Beed, F., Tukamuhabwa, P., Thomma, B. P. H. J. and Joosten, M. H. A. J. (2016), 'Soybean production in eastern and southern Africa and threat of yield loss due to soybean rust caused by *Phakopsora pachyrhizi*', *Plant Pathology* 65 (2), 176–88.

Nakato, G. V., Beed, F. D., Ramathani, I., Kubiriba, J., Rwomushana, I. and Opio, F. (2012), 'Risk of banana Xanthomonas wilt spread through trade', *Journal of Crop Protection* 2(2), 151–61.

Nault, L. R., Gordon, D. T. and Loayza, J. C. (1981), 'Maize virus and mycoplasma diseases in Peru', *International Journal of Pest Management* 27(3), 363–9.

Nelson, S., Brewbaker, J. and Hu, J. (2011), 'Maize chlorotic mottle', PD-79. College of Tropical Agriculture and Human Resources, University of Hawaii at Manoa, pp. 1–6.

Niblett, C. and Clafin, L. (1978), 'Corn lethal necrosis, a new virus disease of corn in Kansas', *Plant Distributional Reports* 62, 15–19.

NISC (National Invasive Species Council). (2003), 'General guidelines for the establishment and evaluation of invasive species early detection and rapid response systems', version 1, 16p.

Nutter Jr., F. W. and Jenco, J. H. (1992), 'Development of critical-point yield loss models to estimate yield losses in corn caused by *Cercospora zeae-maydis*', *Phytopathology* 82, 994.

Oerke, E. C. (2006), 'Crop losses to pests', *Journal of Agricultural Science* 144, 31–43.

Oerke, E. C., Dehne, H. W., Schohnbeck, F. and Weber, A. (1995), *Crop Production and Crop Protection: Estimated Losses in Major Food and Cash Crops*, Amsterdam, The Netherlands and New York: Elsevier.

Pataky, J. and Ikin, R. (2003), 'Pest risk analysis: The risk of introducing *Erwinia stewarti* in maize seed. http://www.worldseed.org/wp-content/uploads/2015/10/Erwinia_stewartii.pdf (Cited 3 March 2016).

Pataky, J. K., Pate, M. C. and Hulbert, S. H. (2001), 'Resistance genes in the rp1 region of maize effective against *Puccinia sorghi* virulent on the Rp1-D gene in North America', *Plant Disease* 85, 165–8.

Petter, F., Roy, A. S. and Smith, I. (2008), 'International standards for the diagnosis of regulated pests', *European Journal of Plant Pathology* 121, 331–7.

Phillips, N. J., Uyemoto, J. K. and Wilson, D. L. (1982), '*Maize chlorotic mottle virus* and crop rotation: Effect of sorghum on virus incidence', *Plant Disease* 66, 376–9.

Pratt, R. C. and Gordon, S. G. (2006), 'Breeding for resistance to maize foliar pathogens', *Plant Breeding Reviews* 26, 119–73.

Pring, D. R. and Lonsdale, D. M. (1989), 'Cytoplasmic male sterility and maternal inheritance of disease susceptibility in maize', *Annual Review of Phytopathology* 27, 483–502.

ProMED-mail (2009), 'PRO/PL> Undiagnosed fungus, maize – Guatemala (02): tar spot', Archive number: 20090420.1491.

Ríos-Herrera, E. N., Partida-Martínez, L. P., Ochoa-Fuentes, Y. M., Cerna-Chávez, E., Hernández-Castillo, F. D., Flores-Olivas, A., Olalde-Portugal, V. and Rodríguez-Guerra, R. (2016), 'First Report on the Presence of *Phyllachora* sp. in Corn Crops at Toluca, Estado de Mexico', *American Journal of Plant Sciences* 7, 733–9.

Rosenzweig, C., Iglesias, A., Yang, X. B., Epstein, P. R. and Chivian, E. (2001), 'Climate change and extreme weather events: Implications for food production, plant diseases, and pests', *Global Change and Human Health* 2, 90–104.

Ruhl, G., Romberg, M. K., Bissonnette, S., Plewa, D., Creswell, T. and Wise, K. A. (2016), 'First Report of Tar Spot on Corn Caused by *Phyllachora maydis* in the United States', *Plant Disease* 100(2), 1496. http://dx.doi.org/10.1094/PDIS-12-15-1506-PD.

Scholthof, K. B. G. (2007), 'The disease triangle: Pathogens, the environment, and society', *Nature Reviews Microbiology* 5, 152–6.

USDA-FAS. (2014), 'Maize Lethal Necrosis – The growing challenge in Eastern Africa', GAIN Report. Global Agricultural Information Network, Washington DC. http://www.fas.usda.gov/data/kenya-maize-lethal-necrosis-growing-challenge-eastern-africa

Security (CCAFS). Available online at: www.ccafs.cgiar.org.

Shiferaw, B., Prasanna, B. M., Hellin, J. and Banziger, M. (2011), 'Crops that feed the world 6. Past success and future challenges to the role played by maize in global food security', *Food Security* 3, 307–27.

Smith, J. (2015), 'Crops, crop pests and climate change – why Africa needs to be better prepared', CCAFS Working. Paper no. 114. Copenhagen, Denmark. CGIAR Research Program on Climate Change, Agriculture and Food (CCAFS). https://cgspace.cgiar.org/handle/10568/66597

Smith, J. J., Waage, J., Woodhall, J. W., Bishop, S. J. and Spence, N. J. (2008), 'The challenge of providing plant pest diagnostic services for Africa', *European Journal of Plant Pathology* 121, 365–75.

Stack, J., Cardwell, K., Hammerschmidt, R., Byrne, J., Loria, R., Snover-Clift, K., Baldwin, W., Wisler, G. Beck, H., Bostock, R., Thomas, C. and Luke, E. (2006), 'The national plant diagnostic network', *Plant Disease* 90(2), 128–36.

Stace-Smith, R. and Hamilton, R. I. (1988), 'Inoculum thresholds of seedborne pathogens: Viruses', *Phytopathology* 78, 875–80.

Teyssandier, E. E., Nome, S. F. and Dalbo, E. (1981), 'Maize virus diseases in Argentina'. In D. T. Gordon, J. K. Knoke, L. R. Nault and R. M. Ritter (eds), *Proceedings of the International Maize Virus*

Disease Colloquium and Workshop, 2–6 August 1982, Wooster, Ohio: Ohio State University, Ohio Agricultural Research and Development Center (OARDC), pp. 93–9.
Ullstrup, A. J. (1972), 'The impacts of the Southern Corn Leaf Blight Epidemics of 1970-1971', *Annual Review of Phytopathology* 10, 37–50.
Uyemoto, J. K. (1983), 'Biology and control of *Maize chlorotic mottle virus'*, *Plant Disease* 67, 7–10.
Vasquez, J. and Mora, E. (2007), 'Incidence of and yield loss caused by Maize rayado fino virus in maize cultivars in Ecuador', *Euphytica* 153, 339–42.
Voyles, J., Kilpatrick, A. M., Collins, J. P., Fisher, M. C., Frick, W. F., McCallun, H., Willis, C. K., Blehert, D. S., Murray, K. A., Puschendorf, R., Rosenblum, E. B., Bolker, B. M., Cheng, T. L., Langwig, K. E., Lindner, D. L., Toothman, M., Wilber, M. Q. and Briggs, C. J. (2014), 'Moving beyond too little, too late: Managing emerging infectious diseases in wild populations requires international policy and partnerships', *EcoHealth*, doi: 10.1007/s10393-014-0980-5.
Vurro, M., Bonciani, B. and Vannacci, G. (2010), 'Emerging infectious diseases of crop plants in developing countries: Impact on agriculture and socio-economic consequences', *Food Security* 2(2), 113–32.
Waage, J. K., Woodhall, J. W., Bishop, S. J., Smith, J. J., Jones, D. J. and Spence, N. J. (2009), 'Patterns of plant pest introductions in Europe and Africa', *Agricultural Systems* 99(1), 1–5.
Waage, J. K. and Mumford, J. D. (2008), 'Agricultural biosecurity', *Philosophical Transactions of the Royal Society B*, 363, 863–76.
Wall Street Journal. (1995), 'Fungal disease is threatening US', 21 August, p. C1.
Wangai, A. W., Redinbaugh, M. G., Kinyua, Z. M., Miano, D. W., Leley, P. K., Kasina, M., Mahuku, G., Scheets, K. and Jeffers, D. (2012), 'First report of *Maize chlorotic mottle virus* and maize lethal necrosis in Kenya', *Plant Disease* 96, 1582.
Ward, J. M. J. and Nowell, D. C. (1998), 'Integrated management for the control of maize gray leaf spot', *Integrated Pest Management Reviews* 3, 1–12.
Ward, J. M. J., Stromberg, E. L., Nowell, D. C. and. Nutter Jr, F. W. (1999), 'Gray leaf spot: A disease of global importance in maize production', *Plant Disease* 83, 884–95.
Ward, J. M. J., Laing, M. D. and Cairns, A. L. P. (1997), 'Management practices to reduce gray leaf spot of maize', *Crop Science* 37, 1257–62.
Welz, H. G. and Geiger, H. H. (2000), 'Genes for resistance to northern corn leaf blight in diverse maize populations', *Plant Breeding* 119, 1–14.
White, G. G (1999), *Compendium of Corn Diseases*, St. Paul, MN: APS Press.
Xie, L., Zhang, J., Wang, Q., Meng, C., Hong, J. and Zhou, X. (2011), 'Characterization of maize chlorotic mottle virus associated with maize lethal necrosis disease in China', *Journal of Phytopathology* 159(3), 191–3.

Chapter 17

Controlling aflatoxins in maize in Africa: strategies, challenges and opportunities for improvement

Amare Ayalew and Martin Kimanya, Partnership for Aflatoxin Control in Africa, Ethiopia; Limbikani Matumba, Lilongwe University of Agriculture and Natural Resources, Malawi; Ranajit Bandyopadhayay and Abebe Menkir, International Institute of Tropical Agriculture (IITA), Nigeria; and Peter Cotty, USDA-ARS, USA

1 Introduction

Maize (*Zea mays*) is the most important food staple in sub-Saharan Africa (SSA), accounting for up to 70% of the total human calorific intake (Byerlee and Hiesey, 1996; Martin et al., 2000). In southern Africa, per capita annual consumption of maize remains well over 100 kg, with countries such as Malawi and South Africa being reported to have reached as high as 181 kg and 195 kg, respectively (CIMMYT, 1999). Trends indicate that the production and consumption of maize in Africa will continue to grow in the next couple of decades (Wu et al., 2011).

Unfortunately, maize is subjected to pre- and post-harvest contamination with aflatoxins, which are acutely toxic, immunosuppressive, mutagenic, teratogenic and carcinogenic. (Williams et al., 2004). Aflatoxins are produced mainly by some strains of the related species, *Aspergillus flavus* and *A. parasiticus*. In east Africa, consumption of such heavily contaminated maize has been the cause of acute outbreaks of aflatoxin poisoning, which

http://dx.doi.org/10.19103/AS.2016.0002.23

caused hundreds of deaths (Probst et al., 2007). Chronic exposure to aflatoxins has been reported to lead to malnutrition and stunted growth in children and to many other disabilities (Gong et al., 2004; Khlangwiset et al., 2011). There are reports showing that aflatoxins increase the rate of progression from HIV infection to AIDS (Jolly et al., 2013; Jolly, 2014). Considering the heavy dietary reliance on maize in Africa, the associated exposure of the population to aflatoxins is high. Aflatoxin contamination of maize also limits trade and value addition through processing by food and feed manufacturers. The best approach is prevention, both in the field and during storage, and strategies for control have been discussed earlier (Bruns, 2003; Chulze, 2010). However, environmental conditions and socio-economics in Africa are complex, requiring a tailored strategy. This chapter discusses options and interventions for aflatoxin control and management in maize in the continent, the associated challenges and opportunities for improvement.

2 Aflatoxin contamination in maize

Maize is one of the most susceptible crops to aflatoxin contamination. High levels of aflatoxin have been reported in maize across Africa. Notably, 55% of 350 maize samples collected across Kenya in 2004 had aflatoxin levels greater than 20 µg/kg, with 7% of the samples registering levels greater than 1000 µg/kg (Lewis et al., 2005, 51). In Nigeria% of 70 maize grain samples collected across the country in 2013 had aflatoxin concentrations above 10 µg/kg (Adutenji et al., 2014). A survey in Malawi in 2014, which involved 90 maize samples, detected aflatoxin in 34% of the samples with a mean aflatoxin content of 164 ± 45 (max 878) µg/kg (Matumba et al., 2014). The monitoring and surveillance of aflatoxin in Africa is intermittent due to resource constraints, and more systematic, multi-season evidence is needed on the magnitude of contamination.

Toxigenic fungi may infect maize and produce aflatoxin while grain is maturing in the field, in mature grains (pre harvest), during harvest, or during storage and processing. The biosynthetic pathway of aflatoxin in *A. flavus* has been studied extensively and is now quite well understood (Bhatnagar et al., 2006; Yu, 2012). However, the signaling processes which turn on aflatoxin biosynthesis during fungal contamination of crops are still not well understood. Nonetheless the role of humidity, temperature, plant nutrition, and disease and pest damage has been extensively studied.

Water activity (a_w) is one of the most important determining factors in the fungal colonisation of maize grain. a_w is a measure of water availability in hygroscopic products such as food grains and is equivalent to the percentage equilibrium relative humidity of the product expressed as a decimal. Minimum a_w levels of 0.82–0.83 and 0.87 have been reported, respectively, for growth and aflatoxin production by *A. flavus* and *A. parasiticus* (Pitt and Miscamble, 1995). The effect of temperature on the growth of *A. flavus* and aflatoxin synthesis has also been widely studied. The growth of *A. flavus* has been reported to occur from as low as 8°C (Diener and Davis, 1970) up to 43°C (Rambo et al., 1975; Sorenson et al., 1967), with the optimum temperature range for aflatoxin production being 24–35°C (Northolt et al., 1976, 1977). Under favourable conditions, *A. flavus* is capable of producing aflatoxin within 24 hours after infestation (Gwinner et al., 1996).

Total preclusion of aflatoxin contamination in maize is impractical because most critical factors for its production are extrinsic to human control. This is particularly the case in tropical Africa where high ambient humidity makes the control of commodity moisture difficult (Chulze, 2010). In some parts of Africa there are additional compounding social factors such as theft of maize still standing in the fields (McCall, 1985) which compels farmers to harvest and store the crop before adequate field drying has taken place, thus increasing the risk of aflatoxin contamination. Nonetheless, it is possible to keep aflatoxin levels low. This is most effectively achieved by following both good agricultural practices (GAP) during the seed management and field production stages, and good manufacturing practices (GMP) during the handling, storage, processing, and distribution of maize for human food and animal feed. A complementary management system is the Hazard Analysis Critical Control Point (HACCP), which has been successfully applied in Southern Africa in commercially produced peanut butter (FAO, 2003). However, small-scale farmers have resource and capacity constraints on the application of measures to address invisible poisons. Most aflatoxin control in the near future will come from specific technologies which are suitable for adoption. The subsequent sections elaborate the means by which some practices and actions could assist in mitigating aflatoxin contamination of maize. The practical challenges of the interventions are also highlighted.

3 Pre-harvest aflatoxin control

Pre-harvest aflatoxin contamination is a complex problem influenced by many biotic and abiotic factors. A multi-pronged approach is therefore necessary for its control. (Cary et al., 2011). Pre-harvest aflatoxin prevention strategies may be broadly classified as: 1) enhancement of the host (crop) fungal resistance, 2) the avoidance of fungal contamination and 3) the reduction of toxigenic fungal population. Specifically, this includes crop rotation, tillage, timely planting of agro-ecologically adapted varieties or disease-resistant varieties, appropriate levels of crop density, irrigation and fertilisation management , weed control, effective disease and pest management, including the use of biocontrol agents.

3.1 Crop rotation

The continuous cultivation of maize on the same land may contribute to the build-up of toxigenic fungal populations in the soil with an increased risk of fungal infection and aflatoxin contamination. Crop rotation potentially reduces the population of aflatoxin-producing fungal communities (Griffin et al., 1981; Jaime-Garcia and Cotty, 2004, 2006, 2010). Studies have shown that fields previously cropped to maize have significantly greater quantities of *A. flavus* than those previously cropped to either cotton or sorghum, with the latter having the lowest populations (Jaime-Garcia and Cotty, 2010). Maize cobs and residues, which are naturally good substrates for fungal proliferation, are left in the field after harvest and are subsequently colonised by toxigenic fungi (Zummo, 1991; Jaime-Garcia and Cotty, 2004). This results in an increase of fungal populations during the next cropping season. The same explanation holds for the high toxigenic fungal populations in fields previously cropped to groundnuts (Ortiz et al., 2011). It is for this reason that crop rotation involving maize and groundnuts does not help in reducing toxigenic fungal populations (Griffin et al., 1981). An aflatoxin-smart crop rotation pattern should involve cultivating crops such as soya, sorghum, millet, cassava, and yams which are not good substrates for fungal growth. Unfortunately, in most parts of Africa the implementation of such an effective crop rotation is greatly constrained by socio-economic factors, including the shrinking size of smallholder farms due to rural population growth and rising land scarcity (Hazell, 2005; Jayne et al., 2012; Thierfelder et al., 2013), and over-reliance on maize by human populations (Byerlee and Hiesey, 1996; Martin et al., 2000).

3.2 Tillage

No-tillage production systems have been shown to provide benefits in long-term soil sustainability, partly through the elimination of soil and carbon losses caused by tillage operations (Teasdale et al., 2007). However, there is sufficient evidence to show that no-till production systems encourage the build-up of aflatoxin-producing fungi (Zablotowicz et al., 2007) and other mycotoxigenic fungi and mycotoxin contaminants (Xu, 2003, 2008; Bernhoft et al., 2010; Blandino et al., 2010). No-tillage practices generally increase soil organic matter in the surface soil (Reeves, 1997; Locke et al., 2005), which consequently supports fungal proliferation. Residues also act as a source of increased fungal inoculum. It should be noted that extensive data suggest comparable soil conservation benefits could be realised if residues are incorporated during tillage (Baggs et al., 2000; Witt et al., 2000; Bakht et al., 2009). However, in areas vulnerable to erosion, the no-till practice might be an option for safeguarding soil, and in this case, a crop other than maize which is less susceptible to aflatoxin contamination would be ideal.

3.3 Host plant resistance

Efforts to reduce aflatoxin accumulation in maize have long been focused on identifying and selecting germplasm with natural host resistance factors (Widstrom et al., 1984; Brown et al., 2001; Warburton et al., 2011). Native resistance to aflatoxin contamination in maize is polygenic and complex and thus advances in the area have been slow (Cary et al., 2011). The greater effect of dominant genes on resistance to aflatoxin accumulation when compared with additive gene action were also reported (Gardner et al., 1987; Campbell et al., 1997; Busboom and White, 2004). The widely accepted tenet is that genetic variation for aflatoxin resistance exists in maize, making host plant resistance a viable option for aflatoxin control. The primary focus of resistance has been on the kernel, both at the pericarp and at the sub-pericarp level (Brown et al., 1993). A number of studies have demonstrated the protective role of pericarp waxes in kernel resistance (Guo et al., 1996; Gembeh et al., 2001; Rajasekaran et al., 2013).

Breeding lines and populations with high resistance under varying environments have been released (Williams and Windham, 2012; Williams et al., 2005; Mayfield et al., 2012). However, the utilisation of maize resistance in practical aflatoxin management is not yet commensurate with progress in identifying sources of resistance. Researchers ascribe this to the complex nature of the inheritance of resistance and to the erratic nature of infection by *A. flavus* within a field and across seasons which limits the transfer of resistance to desirable maize inbred lines (Brooks et al., 2005). As the known resistant maize lines have generally poor agronomic quality, markers would facilitate the incorporation of aflatoxin resistance into commercially acceptable lines. Resistance associated proteins (RAPs) and stress-related proteins have been identified (Chen et al., 2004a,b) but their use in developing commercial maize varieties has yet to be demonstrated.

Researchers have observed the lowest levels of aflatoxin contamination in an aflatoxin resistant genotype that was severely drought stressed. They concluded that resistance in this genotype involved complex mechanisms such as the inhibition of fungal infection and growth or inhibition of aflatoxin biosynthesis after infection (Kebede et al., 2012). Physical traits such as husk openness/tightness and ear orientation after maturity may also influence the pre-harvest aflatoxin contamination. It has been established that kernels of loose-husked maize are more prone to mould infection and aflatoxin contamination than those which are tight-husked (Widstrom et al., 1981; Betran and Isakeit, 2004). Ears which remain upright are likely to be infected by mould especially when exposed to prolonged rainfall (Munkvold, 2003b). Unfortunately, some of these traits are not preferred by farmers. For instance, tight husks are not favoured by most US farmers (Warburton and Williams, 2014).

3.4 Timely planting of agro ecologically adapted varieties

Early planting may shift the period between anthesis and dough development in maize to a time frame in the growing season when drought and heat stress are less likely to be encountered (Bruns, 2003).

However, in some cases, early planting increases the risk of pre-harvest contamination. For instance, early planted maize in Georgia, USA, is at higher risk because the critical grain filling period, which begins 20 days after silking, falls when the seasonal maximum and minimum temperatures are the highest and when net evaporation (moisture stress) is at its peak (Widstrom et al., 1990). It is important for varieties to be well adapted to local conditions. Matching the variety to the agro-ecology allows the crop to mature before moisture stress occurs. However, the crop should not mature too early during the rainfall season, as in most parts of Africa, alternative drying techniques (e.g. forced air dying) are not practically achievable. Rainfall patterns have also become so irregular due to climate change that 'variety-agroecology-time of planting' matching has become almost impossible.

3.5 Water stress management

Drought stress is a major contributory factor in pre-harvest aflatoxin contamination. Maize plants exposed to drought stress are more susceptible to infection by toxigenic fungi than are unstressed plants (Kebede et al., 2012). The decrease in a_w reduces phytoalexin production in crops (Wotton and Strange, 1985; Dorner et al., 1989). Phytoalexins are antimicrobial substances synthesised by plants that accumulate rapidly at areas of pathogen infection. In developing groundnut kernels, for example, sufficient phytoalexins are produced at high a_w (>0.97) to inhibit growth of *A. flavus* and subsequent aflatoxin contamination. Moisture loss in pre-harvest kernels lowers the capacity of phytoalexin production and is eventually lost at a_w < 0.95. In immature groundnuts, aflatoxin does not form until phytoalexin production ceases in drought-stressed plants (Dorner et al., 1989). Although there are no data relating phytoalexin and aflatoxin contamination in pre-harvest maize, correlation between end of season drought stress and aflatoxin contamination is well established. Drought stress during grain filling results in cracks which may enhance *A. flavus* infection and subsequent aflatoxin development (Smart et al., 1990). Periodic drought stress aggravates aflatoxin contamination by eliminating microbial competitors for the toxigenic fungi, resulting in high populations of the aflatoxin producers (Sorenson et al., 1984; Diener et al., 1987).

In sub-Saharan Africa, maize is principally grown under a rainfed system, and unfortunately in recent years, the region has been plagued by both recurrent and severe droughts (Jones and Thornton, 2003). As a mitigation and adaptation measure for climate change, conservation agriculture involving surface residue cover has been widely promoted (Fowler and Rockstrom, 2001). However, this strategy, of which the main aim is the management of soil nutrients and moisture, is to a great extent insensitive to mycotoxin management as it increases the fungal spore load (see Section 3.2). In this context, alternative water harvesting and management techniques such as tied-ridging (Motsi et al., 2004), residue incorporation into the soil (Baggs et al., 2000; Witt et al., 2000; Bakht et al., 2009) and the use of live mulch (Sharma et al., 2010) would reduce aflatoxin risk in addition to their role in improving crop yield and quality. Soils with high organic matter have a greater moisture-holding capacity than those with a lower content and consequently reduce the risk of moisture stress and susceptibility to aflatoxin contamination. Equally important is the maintenance of optimal plant density. The wide spacing of maize plants may exacerbate moisture loss through evaporation during drought stress periods. On the other hand, dense planting leads to competition for nutrient and moisture and results in compromised plant vigour (making plants prone to pest and disease attack, lodging and mycotoxin risk) (Zhang, et al., 2014). Similarly, high crop densities and excessive weed growth may deplete available soil moisture.

3.6 Adequate soil nutrient supply

Adequate soil nutrients promote plant vigour, assisting in resistance to pest and disease pressures. Nitrogen stress can significantly increase aflatoxin contamination in maize, especially where there is drought stress and high temperatures during pollination and seed maturation (Jones 1987; Payne et al., 1989). Nitrogen stress may also result from drought stress due to insufficient uptake of nutrients (Jones, 1987). However studies conducted in Benin have found higher aflatoxin concentrations when the N-fertiliser source was urea with NH_4^+ than when NPK-fertilisers were used (Cotty and Cardwell, 1999; Hell et al., 2003). Contrary effects of urea on aflatoxin contamination were observed in experiments in Italy and in the United States where a negative correlation was found between the N rate and aflatoxin B1 contamination (Tubajika et al., 1999; Blandino et al., 2008). It is logical that application of NPK-fertiliser would lower aflatoxins as phosphorus is necessary for the development of root systems and thus for nutrient and water uptake. Similarly, calcium has an important role in root growth and development in addition to neutralisating low pH which has a deleterious effect on the growing crop. Potassium regulates cellular hydration and stomatal activity. High zinc concentration in soils is associated with germ aflatoxin contamination; however the presence of high levels of phosphate makes zinc unavailable to the *A. flavus* (Lacey and Magan, 1991). Therefore, adequate application of these fertilisers is likely to counter drought stress and subsequently to deter fungal attack and aflatoxin contamination. Unfortunately, fertilisers are beyond the reach of African smallholder farmers (Vanlauwe et al., 2014), in spite of a substantial increase in fertiliser usage in sub-Saharan Africa due to subsidy programmes in recent years (Jayne et al., 2013). Some cultural practices thought to substitute for fertilisers, for example, alley cropping systems, integration of herbaceous legumes, improved legume tree fallows and biomass transfer systems (Sanchez et al., 1997; Carsky et al., 2001) have also been found impractical due to ineffectiveness (poor crop response), high labour intensity and the lack of immediate benefits to farmers (Sidibé, 2005; Hauser et al., 2006).

3.7 Seed vigour, pest and diseases

Seed vigour is related to aflatoxin resistance. Most *Aspergillus* species are opportunistic pathogens and therefore their ability to invade a host chiefly depends on the plant's loss of natural protection, for example, the prevalence of damage, deformity and disease (Leger et al., 2000). High drying and storage temperatures compromise seed vigour (Dickie et al., 1990; Ellis et al., 1991; Ortiz et al., 2016). Poor seeds will probably grow into weaker plants which are more vulnerable to mould, other pest attacks and aflatoxin contamination. Pest attack and diseases may produce plant stress and facilitate the infection of the grain by mycotoxin-producing fungi (Hell et al., 2000; Mukanga et al., 2010; Ni et al., 2014). Of particular importance is the stem-borer beetle, which is one of the most injurious pre-harvest pests of maize in Africa (Khan et al., 2000). Fortunately, management strategies for this pest now exist. These include: utilisation of synthetic sex pheromones, cultural control, inter-cropping and habitat management, management of crop residues, manipulation of sowing dates and densities, fertiliser, host plant resistance and biological control. Kfir et al. (2002) provide a review of these management options.

3.8 Biological control

Biological control is the purposeful utilisation of introduced or resident living organisms, other than disease-resistant host plants, to suppress the activities and/or populations of one or more plant pathogens (Pal and Gardener, 2006). Successful biological control of aflatoxin contamination requires potential populations of aflatoxin-producing members of *Aspergillus* section Flavi, causal agents of crop contamination with aflatoxins, to be sufficiently reduced to lower the aflatoxin content of treated crops. Several fungi and bacteria have been evaluated as agents for the prevention of aflatoxin contamination during the pre-harvest stages of crop production.

Bacteria frequently display excellent activity in inhibiting *Aspergillus* species and aflatoxin biosynthesis in environments with relatively high a_w. This has resulted in early industrial interest in bacteria (e.g. *Bacillus subtilis*) and/or bacterial products (e.g. Iturins, a group of cyclic lipopeptides) as a source of biocontrol products for preventing aflatoxin formation in crops (Kimura and Hirano, 1988; Ono and Kimura, 1991). In more recent work, bacteria including members of the genera *Bacillus*, *Pseudomonas*, *Ralstonia*, and *Burkholderia* isolated from pistachios in California were able to completely inhibit the growth of *A. flavus* under in vitro conditions (Palumbo et al., 2006). Interest in various genotypes of *Bacillus subtilis* and *P. solanacearum* as inhibitors of aflatoxin accumulation has also continued, including bacteria collected from Argentinian soil (Nesci et al., 2005). In India, *Bacillus subtilis* Strain G1 was found to suppress *A. flavus* populations in the soil, *A. flavus* infection and aflatoxin B1 biosynthesis in groundnut kernels and to increase the pod yield in greenhouse and field experiments (Shifa et al., 2016). Similar inhibitory effects of selected antagonistic strains of fluorescent *Pseudomonas*, rhizobacteria and *Bacillus* strains have also been demonstrated in greenhouse and field experiments using groundnut as the test crop (Anjaiah et al., 2006; Navya et al., 2015). There may be human health concerns over some bacteria (Govan and Deretic, 1996; Parke and Gurian-Sherman, 2001) which have potential as biocontrol agents, especially those which are active at the high temperatures in which aflatoxin-producing fungi thrive (i.e. 37°C, human body temperature). Furthermore, most bacteria with potential as active ingredients, when applied at economically feasible rates for the management of aflatoxins, are not adequately competitive in hot, dry field conditions under which the fungi that produce aflatoxins dominate crop environments.

The production of massive amounts of conidia (spores produced by mitosis) and the mycelial habit allow some filamentous fungi to be effective biocontrol agents, even under the dry conditions in which aflatoxin-producing fungi are very competitive. Of the many fungal genera which are of interest as biocontrol agents, *Trichoderma* is perhaps the most frequently studied for the management of plant diseases (Howell, 2003; Mukherjee et. al., 2013). Many genotypes of *Trichoderma* have significant potential as active ingredients for biocontrol products. As a result, the genus *Trichoderma* has a commercial application in controlling a large number of plant diseases (Lorito and Woo, 2015). Inoculation with *Trichoderma* spp. on groundnut resulted in a significant reduction of seed infection by *A. flavus*, and also reduced >50% of the *A. flavus* populations in the geocarposphere of groundnut (Anjaiah et al., 2006). *Trichoderma harzianum* Strain 77, the active ingredient of the registered biopesticide Eco-77® in South Africa, has been evaluated for aflatoxin reduction in sweet corn (Sivparsad and Laing, 2016). This strain is a mycoparasite causing hyphal lysis of *A. flavus*. When sprayed on silk 10–12 days post-mid silk stage, Strain 77 lowered aflatoxins in sweet corn, but spraying silk before or after this stage had little effect. Therefore, the window of application for effective control of aflatoxin using Strain 77 is extremely short. There are practical challenges associated with the short window of application since silking is rarely uniform across a maize field. Because sweet corn is harvested with a high moisture content and consumed fresh, aflatoxin contamination is usually not of commercial concern. The studies mentioned above were primarily carried out under laboratory and greenhouse conditions but these 'proof-of-concept' studies were rarely extended to on-farm field conditions or progressed to the registration stage necessary for the commercial use of microbial biopesticides.

The species of *Aspergillus* which produce aflatoxins are of great genetic, morphological and chemical diversity. Some genotypes of the fungus produce large quantities of aflatoxins with the potential to produce over a million µg/kg of aflatoxins in a single corn kernel. Other genotypes produce no aflatoxins and are known as atoxigenic strains. The two major morphotypes of *A. flavus* are the L and S strains, named for the size of sclerotia produced (S – small, L – large). Genotypes belonging to the S-strain morphotype produce on average more aflatoxins than genotypes belonging to the L-strain. Aflatoxin production within the L-strain is highly variable and nearly all the atoxigenic genotypes found naturally within *A. flavus* belong to the L-morphotype.

Appropriately timed introduction of endemic atoxigenic (non-toxic) vegetative compatibility groups (VCGs) of *A. flavus* into agricultural fields alters the composition of fungal communities, increasing the frequency of the applied atoxigenic *A. flavus* and decreasing both the frequency of aflatoxin producers and the quantity of aflatoxins in crops (Cotty, 1994; Probst et al., 2011). These changes to the average aflatoxin-producing potential of fungal communities can be induced without increasing the overall quantity of *A. flavus* in the environment (Atehnkeng et al., 2014). Competitive exclusion of aflatoxin-producing fungi by endemic atoxigenic VCGs is a proven non-toxic biological control technology which reduces aflatoxins during both crop development and post-harvest storage, as well as throughout the value chain (Cotty et al., 2007; Atehnkeng et al., 2014; Bandyopadhyay et al., 2016). Two biocontrol products containing atoxigenic strain active ingredients – Aspergillus flavus AF36 and Afla-Guard® – are registered for use in the United States, where large acreages of cottonseed, maize, groundnut and pistachios are treated with these biocontrol products annually (Cotty et al., 2008; Dorner, 2009; Weaver et al., 2015).

Atoxigenic-strain-based biocontrol technology using the competitive displacement principle has been successfully adapted for use on maize and groundnut in several African nations. Native, widely distributed and locally adapted atoxigenic VCGs have been identified from several countries and genotypes belonging to these VCGs have been incorporated as active ingredients in biocontrol products bearing the trademark Aflasafe (Bandyopadhyay et al., 2016). Each of the three currently registered Aflasafe products (Aflasafe for Nigeria, Aflasafe KE01 for Kenya, and Aflasafe SN01 for Senegal) contains a different mixture of four atoxigenic *A. flavus* genotypes highly adapted to the target nation's agroecosystem. Consistently high levels of efficacy in reducing aflatoxin contamination in maize and groundnut by country-specific Aflasafe products have been obtained for several years in Burkina Faso, Kenya, Nigeria, Senegal, Gambia and Zambia. The biocontrol programme currently includes countries in East Africa (Burundi, Rwanda, Tanzania and Uganda), West Africa (Ghana) and Southern Africa (Malawi, Mozambique and Zambia), due to the widespread need to address aflatoxin exposure and the efficacy of the biocontrol approach. The active ingredient genotypes used in the biocontrol products carry over both between seasons and throughout the value chain, reducing contamination in storage and transport, even where conditions favour fungal growth.

Annual carry-over of the influences of atoxigenic strain applications result in additional benefits when treatments are made in multiple seasons, and in widespread benefits due to reduced aflatoxins throughout the environment. These include reduced exposure to aflatoxins in crop fragments during harvest or processing and in conidia from high aflatoxin producers. Multi-year influences mean a single application of an atoxigenic strain biocontrol product may benefit not only the treated crop, but also rotation crops and second season crops which miss a treatment (Cotty et al., 2007). As the safety of fungal communities within treated fields improves, so does the safety of fungal communities in areas neighbouring treated fields (Cotty et al., 2008).

In spite of the high efficacy of the atoxigenic strain-based biocontrol approach, several concerns have been raised about the technology. These include: use of sorghum as a carrier in the formulated product, potentially high distribution costs due to the bulky nature of the formulated product, lack of aflatoxin-conscious markets willing to pay a premium for low aflatoxin maize, efficacy of the product during drought, potential risk of allergies and/or aspergillosis, influence of biocontrol products on other mycotoxins and the soil microenvironment, and recombination between atoxigenic and toxigenic genotypes under natural conditions. All these concerns have been addressed by Bandyopadhyay et al. (2016) and Adhikari et al. (2017) and both the risks and costs appear minimal when compared to the health benefits for the African population, the potential gain of high value export and the growing formal domestic markets.

Atoxigenic strain-based biological control has advanced through the stages of product development, registration by regulators and commercialisation and is on the path towards widespread adoption by farmers. Thus, the biocontrol product can be made available to farmers and other end-users; a manufacturing plant has begun to produce Aflasafe in Nigeria and a small-scale modular manufacturing plant is under construction in Kenya. A model for creating sustainable market demand for Aflasafe in the maize value chain is being piloted under the AgResults Aflasafe Initiative (in Nigeria http://agresults.org/en/283/NigeriaAflasafePilot), where farmers have purchased Aflasafe to treat about 30 000 ha (at a cost of $12–20/ha depending on the exchange rate between Naira and the US$ and an application rate of 10 kg/ha). The Kenyan government has procured the Aflasafe product and enabled smallholder farmers to treat nearly 23,000 ha in aflatoxin-prone areas in the interests of public health and to improve the marketability of maize grains. A Senegalese agribusiness firm provided 40 tons of Aflasafe to its contract growers in 2016 to improve the safety and marketability of groundnuts procured from farmers (Bandyopadhyay et al., 2016). In all instances of scaling up mentioned above, biocontrol is used as a component of integrated management in which several other practices, such as rapid drying and good storage, are also promoted. In order to encourage Aflasafe use and commercialisation, licensing mechanisms for manufacturing, marketing and distribution are required. USAID and the Bill & Melinda Gates Foundation have recently funded a technology transfer and commercialisation initiative to scale-up use of Aflasafe on 500 000 ha in 11 African nations through private, public, or public-private partnerships.

4 Preventing post-harvest aflatoxin contamination

4.1 Time of harvest and drying

Adverse weather at harvesting causes slow and inadequate drying of maize which results in fungal growth and possible aflatoxin formation. Thus the timing of harvest can therefore have a major influence on ultimate aflatoxin contamination in the maize. Whereas early harvest followed by effective drying can prevent pre-harvest fungal growth and toxin production, late harvesting may also be preventive because it allows field drying to proceed until a desirable moisture content is reached, particularly where there is little likelihood of pre-harvest infection (Munkvold, 2003a).

It is important to ensure that timely harvested maize is correctly handled to prevent contamination. Immediately, after harvest, it should be dried as quickly as possible to a level of moisture content below 13.5% in order to stop or prevent fungal growth. As fungal growth proceeds at lower a_w in grain already invaded by fungi, it is advisable to maintain the moisture content of stored seed at even lower levels. Drying relies on the sun in Africa , thus taking longer to reach a safe moisture level. If maize is not properly dried, aflatoxin contamination can increase ten-fold in the three days following harvest (Tanboon-ek, 1989). This is particularly the case in regions where humidity and rainfall remain high at harvesting, delaying the harvest and constraining drying. It is also a challenge in regions where the next rainy season starts early when crops are still in the field (Kaaya and Kyamuhangire, 2005).

Farmers are advised to dry maize outside the field and to avoid drying on bare ground. A practical approach in Africa is drying on a raised platform, using a bin for unshelled maize and tarpaulins for shelled maize. It is also necessary to promote the use of other appropriate technologies for the quick drying of maize and other crops to ensure optimum moisture levels. The use of solar or hot air drying and monitoring of moisture content during drying are recommended practices.

4.2 Proper storage

After drying, maize should be stored in dry, sanitary conditions to prevent pest attacks and to minimise aflatoxin contamination. When good storage practices are observed, aflatoxin levels in maize are not believed to increase (Munkvold and Desjardins, 1997). In Africa, however, aflatoxin contamination during storage is common because the storage conditions are favourable for its growth. Storage structures are characterised by excessive heat, high humidity, lack of aeration and insect and storage damage. Generally, maize cobs should be stored in a well-aerated drying bin. The Codex Alimentarius Commission (CAC) recommends the use of dry and well-constructed structures which provide protection from rain, drainage and ground water, the entry of rodents and birds, and which ideally enable minimum temperature fluctuations (CAC, 2002). In addition to possession of good storage facilities, farmers are required to clear the facilities of remains of the previous harvests and to destroy infested crop residues (Munkvold, 2003a). The monitoring of quality and insect infestation status is necessary. If pest infestation is observed, the application of chemicals to prevent further infestation and fungal growth is recommended (Munkvold, 2003b).

Higher levels of aflatoxin-producing fungi were found in maize that had been shelled immediately after harvest when compared to that stored in cob form (Mora end Lacey, 1997). Storing maize in polypropylene bags (a common practice in Africa) should be discouraged because it enhances fungal growth and aflatoxin contamination due to poor aeration, especially where the maize has been improperly dried (Udoh et al., 2000). The hermetic storage of maize offers a promising option for reducing post-harvest losses and minimising aflatoxin increase during storage. Williams et al. (2014) evaluated Purdue Improved Crop Storage (PICS) bags, which are triple layer hermetic bags, and reported that no aflatoxin was detected in maize stored in PICS bags over a period of two months, even at seed moisture levels of 18 to 21%. However, breaking the hermetic seal of the PICS bags – opening the bags for about 30 minutes, in the experiment by Tubbs et al. (2016) – increased fungal growth and aflatoxin contamination. A recent study also confirmed that the storage of maize in PICS bags by rural farmers can prevent accumulation of aflatoxins at seed moisture levels below 14% (Ng'ang'a et al., 2016).

4.3 Control of fungal growth in storage

Juglal et al. (2002) evaluated the effect of nine different oils on the growth of *Aspergillus parasiticus* and *Fusarium verticillioides* in stored food. Commonly occurring mycotoxigenic fungi can be controlled with clove oil (containing eugenol). The researchers attributed the inhibitory effects of spices and herbs at least in part to phenolic compounds such as coumarins and flavonoides. However, the use of plant products in control of mycotoxins is still at an experimental stage. Udoh et al. (2000) pointed out that care must be taken in using plant materials to prevent mycotoxins as some of these materials are natural media for the growth of *Aspergillus flavus*. Ammonium bicarbonate provides an inexpensive, safe, easily applicable and highly effective antimycotic agent which can be used to prevent mycotoxin formation in bulk grain storage systems, especially in resource-poor developing countries. The compound completely inhibited both growth and mycotoxin production by *Aspergillus* isolates investigated at levels up to 1% which still rendered the grains organoleptically acceptable (Samapundo et al., 2007). Where high levels of insect infestation are found during monitoring, maize cobs should be shelled and the grains stored in jute bags. It should, however, be noted that storing maize as grains increases the likelihood of contamination.

5 Removing aflatoxin contamination

5.1 Sorting

Freshly harvested maize may be contaminated with aflatoxins. The extent of contamination may increase when post-harvest operations such as drying, storage, distribution and processing are not well controlled. One of the means of ensuring that maize found to be contaminated, either pre-harvest or post-harvest, is rendered safe for human consumption is to remove the contaminated portion of the kernels or cobs. Removal of whole contaminated kennels can be achieved by sorting or flotation. Johansson et al. (2000) found that about 60% of aflatoxins was found in damaged kernels, broken kernels and foreign material. In a more recent experiment, Pearson et al. (2010) was able to reduce aflatoxin content by up to 82% in samples of yellow maize with initial aflatoxin level higher than 10 µg/kg, and 38% by sorting in samples contaminated with less than 10 µg/kg. The same approach was applied in white maize, reducing 46% of the aflatoxin content in the first sorting and 88% after a re-sorting. The sorting is based on the visible appearance of the kernel to the naked eye. The change in the appearance could be a result of direct fungal infection and growth or of fungi-induced biochemical changes (da Gloria, 2011). Size and density have also been employed to separate fungus-invaded and potentially aflatoxin-contaminated kernels from sound ones. Size sorting (Piedade et al., 2002) is often carried out through the use of sieves, whereas winnowing and flotation rely on the weight of the kernels or their components when subjected to a stream of air or placed in a liquid, respectively. For instance, removal of small and shrivelled (lighter) peanut kernels has been shown to substantially reduce the mean aflatoxin content of the remaining larger, sound kernels (Whitaker et al., 2005; Dorner, 2008). Huff (1980) recorded a 60% reduction in aflatoxin in maize through flotation.

However, sorting and flotation yield two products: a 'clean' or low aflatoxin fraction and a 'dirty' or high aflatoxin content fraction. For instance, Piedade et al. (2002) found that the aflatoxin concentration of particulates segregated on the basis of size was approximately three times that of the original maize. If the high aflatoxin fraction is not correctly discarded, consumption of this fraction will pose a greater danger than the original material. The efficiency of hand sorting and density segregation as aflatoxin-reduction tools depends on the product type, size (quantity), and the way in which the technique is applied. Hand sorting is labour intensive and is therefore unlikely to be adopted by mid-level traders and commercial storage facilities, unless demanded by law or the benefits justify its use. However, this method can be useful and effective at the rural household level where quantities in the range of 10–15 kg are processed for grinding maize into flour for home consumption.

5.2 Dry milling

Dry milling is widely used in the processing of maize. The capacity of the process to yield aflatoxin-safe products is highly dependent on the initial level of toxin in the original material. Studies have shown that during the process of dry milling, aflatoxin is primarily segregated in the bran and fines fraction (Njapau et al., 1998; Mutungi et al., 2008), and less than 12% of the aflatoxin in the original material remains in the main product (endosperm/grits and low fat flour) (Scott, 1984). Siwela et al. (2005) showed that aflatoxin concentration in maize meal was reduced by approximately 92% after dehulling maize grains. The distribution of aflatoxins in dry-milled maize fractions was also evaluated by Castells et al. (2008), who found higher levels of aflatoxins in the outer layers of the kernels, while processed products from the inner parts of the grain, such as maize meal and flaking grits, had decreased mycotoxin levels. Pietri et al. (2009) found reductions of 8.0% in aflatoxin levels after cleaning (from a 5 µg/kg contaminated maize sample) and 57.0% (from a 120 µg/kg sample). The subsequent removal of bran and germ led to a further decrease in contamination levels in products destined for human consumption. However, the components usually intended for animal feed production were the most contaminated. Relying on this method for aflatoxin management in Africa is a challenge because many producers cannot afford drying operations which separate bran and germ from the grains, nor can they comply with requirements or guidance that may give rise to food insecurity. Methods or technologies that prevent contamination, pre and post-harvest are more practical and therefore recommended.

5.3 Wet milling

Wet milling is also widely used in the processing of maize. In the wet milling process, aflatoxin, and fibre and germ are segregated primarily (40% and 30 – 42%, respectively) in the water used for steeping (Njapau et al., 1998). This is because aflatoxins have relatively high solubility in the water fraction. When the wet milling process is used for the production of starch, 1% or less of the original quantity of aflatoxin ends up in the starch (Yahl et al., 1971). Despite its potential for reduction of aflatoxin contamination in food, wet milling is impractical under current African conditions as most people rely on home grown maize which is dry milled and often consumed as whole meal.

5.4 Solvent extraction

Solvent extraction relies on the solubility of aflatoxin in various organic solvents such as ethanol, isopropanol and methanol, and can be an effective means of removing aflatoxin from a food matrix. Aflatoxin extraction from meals using organic solvents is routine and is the basis for most laboratory analysis methods. However, the process may also remove vital nutrients and alter or diminish desirable sensory characteristics of the remaining food material. It is not practical in a rural setting due to the potential hazard of handling the solvents and the associated costs (Shapira, 2004).

6 Detoxification

Food processing technologies have been demonstrated to reduce aflatoxin levels in contaminated food to varying extents. These include heating and certain chemical processes such as nixitamilisation (de Arriola, 1987), and ammoniation (Park et al., 1988). Nixitamilisation is applied to human food whereas ammoniation is used on commodities intended for animal feed.

6.1 Heating

Heating operations such as baking, cooking, frying, and roasting show limited effectiveness in reducing aflatoxin contamination as aflatoxins only decompose at temperatures as high as 269°C (Quadri et al., 2013) and the effect of heat is dependent on the matrix and its moisture content (Njapau et al.,1998; Shapira, 2004). In general, boiling in water destroys less than 50% of the aflatoxin content of a commodity (Njapau et al., 1998). During stiff porridge cooking, which is widely practised in Africa, the aflatoxin content of the flour is minimally reduced by about 10–18% (Njapau et al., 1998; Mutungi et al., 2008). Similar reductions were observed inextrusion cooking in which aflatoxin B1 reduction in maize flour ranged from 10.0% to 25.0% (Cazzaniga et al., 2001). Heating cannot therefore be promoted as a technology for preventing exposure to aflatoxins.

6.2 Nixtamalization

Nixtamalization is the technique for preparing foods such as tortillas and masa from maize which involves cooking and steeping maize kernels in alkaline water (NaOH or Ca(OH)2) (de Arriola et al., 1987). It is a simple and low-cost aflatoxin decontamination process which is reported as eliminating 68–90% of aflatoxin in maize (de Arriola et al., 1987; Perez-Flores et al., 2011). The effectiveness of nixtamalisation may be increased by modifications such as the use of calcium sulphate (Maya-Cortes et al., 2010), and microwave heating (Perez-Flores et al., 2011). The latter showed that microwave heating or modification of the tortilla-making process caused a decrease of 68.0–84.0% in aflatoxin content, and after an extract acidification (as occurs during digestion), there was an increase of up to 3.0% as compared to the 33% reported by Mendez-Albores (2003) when using the traditional method of tortilla production.

Reduction of aflatoxin levels resulting from food processing does not necessarily mean decreased toxicity of the compounds as the toxin may not have been destroyed but may be bound to the food matrix or have been changed to an unknown degradation product (Park and Kim, 2006). It is therefore essential to carry out tests to determine the toxicity and biological activity of the remaining compounds as well as to develop analytical methodologies capable of detecting those bound and changed products. Vazquez-Duran et al. (2014) reported cytotoxic and genotoxic activity in tortilla extracts, supporting the observations that there are acid-induced reversions of the altered aflatoxin molecule to the original toxic chemical structure. Generally, nixtamalisation is a simple and effective aflatoxin decontamination tool which can be applied to maize for human consumption. However, consideration should be given to determination of the lowest strength of the alkaline solution, the duration of exposure to the alkali and the process temperature during the exposure.

6.3 Ammoniation

Ammoniation, like nixtamalisation, is an alkali-based decontamination process which may be applied to maize, groundnuts, and cottonseed, using gaseous or liquid ammonia. In practice, the process uses ammonium hydroxide or gaseous ammonia for the detoxification of mycotoxins in maize, cotton and groundnut meals (Kabak et al., 2006). The ammonia can be applied to whole kernels or meals using two processes which differ in operating pressure, temperature, and duration. One process uses high pressure (45–55 psi) or high temperature (80–120°C) for a short period (20–60 minutes), while the other employs ambient pressure or ambient temperature and requires 14–28 days. Ammoniation of grains is considered to be the most acceptable and practical industrial scale method (Moss, 1998) for the detoxification of mycotoxins in food. However, kernel discoloration and the strong odour that occurs after ammoniation are major limitations to the adoption of the processing technique for human food (Samapundo et al., 2007).

7 Role of policy and public awareness in aflatoxin control

The complex nature of aflatoxin contamination requires a holistic approach for effective mitigation of contamination in maize value chains as well as in other commodities. The Partnership for Aflatoxin Control in Africa (PACA) works to drive systemic change by leveraging the convening power of the African Union Commission (AUC) and the wide sphere of influence of country governments to forge partnerships and work with a range of stakeholders. Governments play a key role in the regulation of aflatoxin levels and enforcement of standards. These are essential drivers in creating incentives for producers and traders to apply aflatoxin mitigation measures as well as to achieve market distinction between compliant and non-compliant produce.

At the country level, PACA coordinates the strategic activities necessary to create systemic changes by: supporting policy development based on credible evidence, clarifying and supporting aflatoxin standards, creating locally appropriate alternative use frameworks, establishing surveillance capacity, promoting user adoption of aflatoxin mitigation technologies and practices, and increasing general public awareness of the problem and mitigation options. At the continental level, PACA convenes African leaders and stakeholders to identify and drive strategic actions, gather and disseminate knowledge, and build awareness of the problem and mitigation options. The role of information campaigns in raising awareness of aflatoxin and in creating behaviour change, such as an increase in the number of consumers sorting and discarding affected grains, was reported by James et al. (2007).

To date, PACA and the governments of six pilot countries in Africa (Gambia, Malawi, Nigeria, Senegal, Tanzania and Uganda) have generated empirical evidence on the nature and impact of aflatoxin contamination in maize and other value chains as well as on public health and national economies, and have developed comprehensive Aflatoxin Control National Plans. PACA has also launched a unique one-stop portal known as Africa Aflatoxin Information Management System (AfricaAIMS). This provides comprehensive information ranging from aflatoxin levels in key crops (mainly maize and groundnuts) in different countries, to testing methods and data on trade volumes, losses and rejections, and on aflatoxin-related health impacts. This unique resource will support policy and interventions for more effective aflatoxin control in agriculture and food systems in Africa. Testing and scaling of pre-harvest maize (biocontrol) and post-harvest (storage and drying technologies) is underway in Africa by PACA partners: among others IITA and AflaSTOP project, respectively.

8 Conclusion and future trends

As with most food safety issues, aflatoxin contamination in maize is complex but preventable and can be managed to acceptable levels. There is an urgent need for integrated use of GAP including crop rotation, timely planting, use of agro-ecologically adapted varieties, effective management of nutrients, disease and pests, including the use of biocontrol agents, timely harvesting, efficient moisture control, cleaning, sorting, improved storage and processing. However, performing only one or a few of these steps in the production chain may not adequately lower aflatoxin contamination.

Notwithstanding the fact that the majority of African smallholders struggle with the implementation of GAP due to limited knowledge and lack of resources, governments should strive to change the landscape. GAP are key to the increased agricultural production and reduced post-harvest losses, necessary for achieving agricultural sustainability in the continent. The demonstrated efficacy of biocontrol technology is a tool which forms an integral part of aflatoxin management in maize production systems. Efforts at identifying the atoxigenic strains of *A. flavus* which are widespread across regions offer the potential to develop biocontrol products for use in multiple countries with economies of scale. The progress made so far in identifying available genetic variations for resistance to aflatoxin in maize, and the improvements in evaluation methods suggest resistant maize lines could become integral components of managing aflatoxin contamination. RNAi gene silencing for characterisation of resistance factors and other recent genetic tools will be instrumental in achieving a breakthrough in the development of maize lines resistant to aflatoxin contamination and/or *A. flavus* invasion and which possess commercially desirable traits. Efficient and effective drying methods suitable for adoption by value chain actors in Africa are expected to find their way into the market in the near future. The promotion of hermetic storage technologies aimed at reducing post-harvest losses will also benefit users.

The implementation of aflatoxin control and management interventions necessitates increased public awareness of the health risk associated with consuming aflatoxin-contaminated food, so that there is market distinction as intermediate and end-users become willing to pay a higher price for better quality. With this in mind, governments should prioritise the raising of public awareness on the subject.

9 Where to look for further information

The following online resources provide additional information on the topic of aflatoxin/mycotoxin control. This list is not exhaustive.

1 The Partnership for Afltoxins Control in Africa (PACA) publishes on its website -http://www.aflatoxinpartnersip.org – diverse information about aflatoxin control activities in Africa.
2 In 2015 the International Institute of Tropical Agriculture published information covering various aspects of aflatoxins in 11 Technical Policy Papers. These papers are available in a folder called East African Community Policy Briefs, accessible through: http://www.aflasafe.com/policy-briefs
3 General information about aflatoxins is accessible through the following links:
   a) https://www.fao.org/docrep/X5036E/x5036E0M.HTM
   b) https://www.efsa.europa.eu/en/topics/topic/aflatoxins-food
   c) https://www.iita.org/iita.../aflatoxin-policy-and-program-for-the-east-africa-region-appear
   d) http://www.mycored.eu
   e) https://www.mycotox-society.org
   f) https://www.mytoolbox.eu
   g) https://www.nal.usda.gov/fsrio/aflatoxins
4 The health impacts of aflatoxins are documented in the following monograph: https://monographs.iarc.fr/ENG/Monographs/vol100F/mono100F-23.pdf
5 Information about use of atoxigenic fungi to prevent growth of toxigenic fungi is available through the following links:
   a) https://cals.arizona.edu/research/cottylab/)
   b) https://www.aspergillusflavus.org/
6 There are various links for information about analytical methods and commercial products for aflatoxin analysis. (*Disclaimer*:The inclusion of links for commercial

products should not be construed as an endorsement of any product over others, by the authors or the publisher. The links below offer examples of rapid methods for aflatoxin analysis that are used in quite a number of publications from Africa):

a) https://www.beaconkits.com/
b) https://www.envirologix.com/mycotoxin-testing/aflatoxin-testing
c) http://helica.com/food-safety/downloadablebrochures/.
d) https://www.r-biopharm.com/products/food-feed-analysis/mycotoxins/aflatoxin
e) https://www.romerlabs.com/en/analytes/mycotoxins/aflatoxin-testing
f) https://www.vicam.com/aflatoxin-test-kits
g) https://www.ifpri.org/publication/improving-diagnostics-aflatoxin-detection
h) https://*www.zapmeta.ws/Aflatoxin+Testing*

7 Maximum limits for aflatoxins in food recommneded by Codex Alimentarius Commission are accessible through the following:

a) Food and Agriculture Oragnisation of the United Naions: https://www.fao.org/input/download/standards/17/CXS_193e_2015.pdf
b) Codex Alimentarius Commission: http://www.fao.org/fao-who-codexalimentarius/standards/en/

8 The USA and European Union publish maximum limits for aflatoxins in food as regulated in the respecitve regions. The standards can be accessed through the links below:

a) US Food and Drugs Administration: https://www.fda.gov/food/guidanceregulation/.../ucm077969.htm
b) European union: http://eur-lex.europa.eu/LexUriServ/LexUriServ.do?uri=CONSLEG:2002L0032:20100302:EN:PDF

10 References

Adhikari, B. N., Bandyopadhyay, R. and Cotty, P. J. (2017), Degeneration of aflatoxin gene clusters in *Aspergillus flavus* from Africa and North America, *AMB Express*, 6(1), 62. doi: 10.1186/s13568-016-0228-6.

Anjaiah, V., Thakur, R. P. and Koedam, N. (2006), Evaluation of bacteria and *Trichoderma* for biocontrol of pre-harvest seed infection by *Aspergillus flavus* in groundnut, *Biocontrol Sci. Techn.*, 16, 431–6.

Atehnkeng, J., Ojiambo, P. S., Cotty, P. J. and Bandyopadhyay, R. (2014), Field efficacy of a mixture of atoxigenic Aspergillus flavus Link:Fr vegetative compatibility groups in preventing aflatoxin contamination in maize (Zea mays L.), *Biol. Control*, 72, 62–70.

Baggs, E. M., Watson, C. A. and Rees, R. M. (2000), The fate of nitrogen from incorporated cover crop and green manure residues, *Nutr. Cycl. Agroecosys.*, 56(2), 153–63.

Bakht, J., Shafi, M., Jan, M. T. and Shah, Z. (2009), Influence of crop residue management, cropping system and N fertilizer on soil N and C dynamics and sustainable wheat (Triticumaestivum L.) production, *Soil Till. Res.*, 104(2), 233–40.

Bandyopadhyay, R., Ortega-Beltran, A., Akande, A., Atehnkeng, J., Mutegi, C. K., Kaptoge, L., Senghor, A. L., Adhikari, B. N. and Cotty, P. J. (2016), Biological control of aflatoxins in Africa: current status and potential challenges in the face of climate change, *World Mycotoxin J.*, 9, 771–89.

Bernhoft, A., Clasen, P. E., Kristoffersen, A. B. and Torp, M. (2010), Less *Fusarium* infestation and mycotoxin contamination in organic than in conventional cereals, *Food Addit. Contam.*, 27(6), 842–52.

Betran, F. J. and Isakeit, T. (2004), Aflatoxin accumulation in maize hybrids of different maturities, *Agron. J.*, 96(2), 565–70.

Bhatnagar, D., Cary, J. W., Ehrlich, K., Yu, J. and Cleveland, T. E. (2006), Understanding the genetics of regulation of aflatoxin production and Aspergillus flavus development, *Mycopathologia*, 162(3), 155–66.

Blandino, M., Pilati, A., Reyneri, A. and Scudellari, D. (2010), Effect of maize crop residue density on Fusarium head blight and on deoxynivalenol contamination of common wheat grains, *Cereal Res. Commun.*, 38(4), 550–9.

Blandino, M., Reyneri, A. and Vanara, F. (2008), Influence of nitrogen fertilization on mycotoxin contamination of maize kernels, *Crop Prot.*, 27(2), 222–30.

Brown, R. L., Chen, Z. Y., Menkir, A., Cleveland, T. E., Cardwell, K., Kling, J. and White, D. G. (2001), Resistance to aflatoxin accumulation in kernels of maize inbreds selected for ear rot resistance in West and Central Africa, *J. Food Prot.*, 64(3), 396–400.

Brown, R. L., Cotty, P. J., Cleveland, T. E. and Widstrom, N. W. (1993), Living maize embryo influences accumulation of aflatoxin in maize kernels, *J. Food Prot.*, 56(11), 967–71.

Bruns, H. A. (2003), Controlling aflatoxin and fumonisin in maize by crop management, *Toxin Rev.*, 22, 153–73.

Byerlee, D. and Heisey, P. W. (1996), Past and potential impacts of maize research in Sub Saharan Africa: a critical assessment, *Food Policy*, 21, 255–77.

CAC (2002), *Proposed Draft Code of Practice for the Prevention (reduction) of Mycotoxin Contamination in Cereals, Including Annexes on Ochratoxin A, Zearalenone, Fumonisins and Trichothecenes*, Joint FAO/WHO Food Standards Programme: Rotterdam, The Netherlands.

Carsky, R. J., Becker, M. and Hauser, S. (2001), Mucuna cover crop fallow systems: potential and limitations, pp. 111–36. In Tian, G., Ishida, F. and Keatinge, J. D. H. (Eds), *Sustaining Soil Fertility in West-Africa*, SSSA Special Publication Number 58, Madison, USA.

Cary, J. W., Rajasekaran, K., Brown, R. L., Luo, M., Chen, Z. Y. and Bhatnagar, D. (2011), Developing resistance to aflatoxin in maize and cottonseed, *Toxins*, 3(6), 678–96.

Castells, M., Marin, S., Sanchis, V. and Ramos, A. J. (2008), Distribution of fumonisins and aflatoxins in corn fractions during industrial cornflake processing, *Int. J. Food Microbiol.*, 123(1–2), 81–7.

Cazzaniga, D., Basilico, J. C., Gonzalez, R. J., Torres, R. L. and de Greef, D. M. (2001), Mycotoxins inactivation by extrusion cooking of corn flour, *Lett. Appl. Microbiol.*, 33, 144–7.

Chulze, S. N. (2010), Strategies to reduce mycotoxin levels in maize during storage: a review, *Food Addit. Contam. Part A Chem. Anal. Control Expo. Risk Assess*, 27, 651–7.

CIMMYT. (1999), 1997/98 CIMMYT World Maize Facts and Trends; Maize Production in Drought-Stressed Environments: Technical Options and Research Resource Allocation, CIMMYT: Mexico, D.F.

Cotty, P. J. and Cardwell, K. F. (1999), Divergence of West African and North American Communities of *Aspergillus* Section Flavi, *Appl. Environ. Microbiol.*, 65(5), 2264–6.

Cotty, P. J. (1994), Influence of field application of an atoxigenic strain of *Aspergillus flavus* on the population of *A. flavus* infecting cotton bolls and on the aflatoxin content of cottonseed, *Phytopathologia*, 84, 1270–7.

Cotty, P. J., Antilla, L. and Wakelyn, P. J. (2007), Competittive exclusion of aflatoxin producers: Farmer driven research and development, pp. 242–53. In Vincent, C., Goettel, N. and Lazarovits, G. (Eds), *Biological Control: A Global Perspective*. CAB International: Oxfordshire, UK.

Cotty, P. J., Bayman, P., Egel, D. S. and Elias, K. S. (1994), Agriculture, aflatoxins and *Aspergillus*, pp. 1–27. In Powell, K. A., Renwick, A. and Peberdy, J. F. (Eds), *The Genus Aspergillus*. Plenum Press, New York.

Cotty, P. J., Probst, C. and Jaime-Garcia, R. (2008), Etiology and management of aflatoxin contamination, pp. 287–99. In Leslie, J. F., Bandyopadhyay, R. and Visconti, A. (Eds), *Mycotoxins: Detection Methods, Management, Public Health and Agricultural Trade*. CABI Publishing: Wallingford, UK.

da Gloria, E. M. (2011), Aflatoxin contamination distribution among grains and nuts, aflatoxins, pp. 1–17. In Torres-Pacheco, I. (Ed.), *Detection, Measurement and Control*. In Tech: Europe, Rijeka, Croatia.

Dickie, J. B., Ellis, R. H., Kraak, H. L., Ryder, K. and Tompsett, P. B. (1990), Temperature and seed storage longevity, *Ann. Bot.*, 65(2), 197–204.

Diener, U. L. and Davis, N. D. (1970), Limiting temperature and relative humidity for aflatoxin production by *Aspergillus flavus* in stored peanuts, *J. Am. Oil Chem. Soc.*, 47(9), 347–51.

Diener, U. L., Cole, R. J., Sanders, T. H., Payne, A., Lee, L. S. and Klich. M. A. (1987), Epidemiology of aflatoxin formation by *Aspergillus flavus. Ann. Rev. Phytopathol.*, 25, 249–70.

Dorner, J. (2009), Development of biocontrol technology to manage aflatoxin contamination in peanut, *Peanut Sci.*, 36, 60–7.

Dorner, J. W. (2008), Management and prevention of mycotoxins in peanuts. *Food Addit. Contam.*, 25(2), 203–8.

Dorner, J. W., Cole, R. J., Sanders, T. H. and Blankenship, P. D. (1989), Interrelationship of kernel water activity, soil temperature, maturity, and phytoalexin production in preharvest aflatoxin contamination of drought-stressed peanuts, *Mycopathologia*, 105(2), 117–28.

Ellis, R. H., Hong, T. D. and Roberts, E. H. (1991), Effect of storage temperature and moisture on the germination of papaya seeds, *Seed Sci. Res.*, 1(01), 69–72.

FAO (Food and Agriculture Organization of the United Nations) (2003), Manual on the application of the HACCP system in mycotoxin prevention and control, Chapter 3 Illustrative examples of applications of HACCP to mycotoxin control, pp. 82–90.

Fowler, R. and Rockstrom, J. (2001), Conservation tillage for sustainable agriculture: an agrarian revolution gathers momentum in Africa, *Soil Till. Res.*, 61(1), 93–108.

Gembeh, S. V., Brown, R. L., Grimm, C. and Cleveland, T. E. (2001), Identification of chemical components of corn kernel pericarp wax associated with resistance to *Aspergillus flavus* infection and aflatoxin production, *J. Agric. Food Chem.*, 49(10), 4635–41.

Gong, Y., Hounsa, A., Egal, S., Turner, P. C., Sutcliffe, A. E., Hall, A. J. and Wild, C. P. (2004), Postweaning exposure to aflatoxin results in impaired child growth: a longitudinal study in Benin, West Africa, *Environ. Health Perspect.*, 112(13), 1334–8.

Govan, J. R. and Deretic, V. (1996), Microbial pathogenesis in cystic fibrosis: mucoid *Pseudomonas aeruginosa* and *Burkholderia cepacia*, *Microbiol. Rev.*, 60, 539–74.

Griffin, G. J., Garren, K. H. and Taylor, J. D. (1981), Influence of crop rotation and minimum tillage on the population of *Aspergillus flavus* group in peanut field soil, *Plant Dis.*, 65, 898–900.

Guo, B. Z., Russin, J. S., Cleveland, T. E., Brown, R. L. and Damann, K. E. (1996), Evidence for cutinase production by *Aspergillus flavus* and its possible role in infection of corn kernels, *Phytopathologia*, 86(8), 824–9.

Gwinner, J., Harnisch, R. and Muck, O. (1996), *Manual of the Prevention of Post-Harvest Grain Losses*, GTZ: Eschborn, Germany, p. 330.

Hauser, S., Nolte, C. and Carsky, R. J. (2006), What role can planted fallows play in the humid and sub-humid zone of West and Central Africa?, *Nutr. Cycl. Agroecosys.*, 76, 297–318.

Hazell, P. B. R. (2005), Is there a future for small farms?, *Agr. Econ.*, 32, 93–101. http://doi.org/10.1111/j.0169-5150.2004.00016.x.

Hell, K., Cardwell, K. F. and Poehling H. M. (2003), Relationship between management practices, fungal infection and aflatoxin for stored maize in Benin, *J. Phytopathol.*, 151(11–12), 690–8.

Hell, K., Cardwell, K. F., Setamou, M. and Schulthess, F. (2000), Influence of insect infestation on aflatoxin contamination of stored maize in four agroecological regions in Benin, *Afr. Entomol.*, 8(2), 169–77.

Howell, C. R. (2003), Mechanisms employed by *Trichoderma* species in the biological control of plant diseases: The history and evolution of current concepts, *Plant Dis.*, 87, 4–10.

Huff, W. E. (1980), A physical method for the segregation of aflatoxin-contaminated maize, *Cereal Chem.*, 57(4), 236–8.

Jaime-Garcia, R. and Cotty, P. J. (2004), *Aspergillus flavus* in soils and corncobs in south Texas, Implications for management of aflatoxins in corn-cotton rotations, *Plant Dis.*, 88, 1366–71.

Jaime-Garcia, R. and Cotty, P. J. (2006), Spatial relationships of soil texture and crop rotation to Aspergillus flavus community structure in south Texas, *Phytopathologia*, 96, 599–607.

Jaime-Garcia, R. and Cotty, P. J. (2010), Crop rotation and soil temperature influence the community structure of Aspergillus flavus in soil, *Soil Biol. Biochem.*, 42, 1842–7.

James, B., Adda, C., Cardwell, K., Annang, D., Hell, K., Korie, S., Edorth, M., Gbeassonr, F., Nagatey, K. and Houenou, G. (2007), Public information campaign on aflatoxin contamination of maize grains in market stores in Benning, Ghana and Togo, *Food Addit. Contam.*, 24(11), 1–9.

Jayne, T. S., Chamberlin, J. and Muyanga, M. (2012), Emerging land issues in African agriculture: implications for food security and poverty reduction strategies, *Stanford University Global Food Policy and Food Security Symposium Series*. Center for Food Security and the Environment and the Freeman Spogli Institute for International Studies: Stanford, CA.

Jayne, T. S., Mather, D., Mason, N. and Ricker-Gilbert, J. (2013), How do fertilizer subsidy programs affect total fertilizer use in sub-Saharan Africa? Crowding out, diversion, and benefit/cost assessments, *Agric. Econ.*, 44(6), 687–703.

Johansson, A. S., Whitaker, T. B., Hagler, W. M., Giesbrecht, F. G., Young, J. H. and Bowman, D. T. (2000), Testing shelled maize for aflatoxin, part I: estimation of variance components, *J. AOAC Int.*, 83(5), 1264–9.

Jones, P. G. and Thornton, P. K. (2003), The potential impacts of climate change on maize production in Africa and Latin America in 2055, *Global Environ. Change*, 13(1), 51–9.

Jones, R. K. (1987), The influence of cultural practices on minimizing the development of aflatoxin in field maize, pp. 136–144. In Zuber, M. S., Lillehoj, E. B. and Renfro, B. L. (Eds.), *Aflatoxin in Maize: Proc. Workshop*. International Maize and Wheat Improvement Center: Mexico, D.F.

Juglal, S., Govinden, R. and Odhav, B. (2002), Spice oils for the control of co-occurring mycotoxin-producing fungi, *J. Food Protect.*, 65, 683–7.

Kaaya, A. N. and Kyamuhangire, W. (2005), The effect of storage time and agroecological zone on mould incidence and aflatoxin contamination of maize from traders in Uganda, *Int. J. Food Microbiol.*, 110, 217–23.

Kabak, B., Dobson, A. D. W. and Var, I. (2006), Strategies to prevent mycotoxin contamination of food and animal feed, A Review. *Crit. Rev. Food Sci. Nutr.*, 46, 593–619

Kebede, H., Abbas, H. K., Fisher, D. K. and Bellaloui, N. (2012), Relationship between aflatoxin contamination and physiological responses of corn plants under drought and heat stress, *Toxins*, 4(11), 1385–403.

Kfir, R., Overholt, W. A., Khan, Z. R. and Polaszek, A. (2002), Biology and management of economically important lepidopteran cereal stem borers in Africa, *Ann. Rev. Entomol.*, 47(1), 701–31.

Khan, Z. R., Pickett, J. A., Berg, J. V. D., Wadhams, L. J. and Woodcock, C. M. (2000), Exploiting chemical ecology and species diversity: stem borer and striga control for maize and sorghum in Africa, *Pest Manage Sci.*, 56(11), 957–62.

Khlangwiset, P., Shephard, G. S. and Wu, F. (2011), Aflatoxins and growth impairment: a review, *Crit. Rev. Toxicol.*, 41(9), 740–55.

Kimura, N. and Hirano, S. (1988), Inhibitory strains of Bacillus subtilis for growth and aflatoxin production of aflatoxigenic fungi, *Agric. Biol. Chem.*, 52, 1173–9.

Lacey, J. and Magan, N. (1991), Fungi in cereal grains: their occurrence and water and temperature relationships, pp. 77–118. In Chelkowski, J. (Ed.), *Cereal Grain. Mycotoxin and Quality in Drying and torage*. Elsevier: Amsterdam.

Leger, R. J. S., Screen, S. E. and Shams-Pirzadeh, B. (2000), Lack of host specialization in *Aspergillus flavus, Appl. Environ. Microb.*, 66(1), 320–4.

Lewis, L., Onsongo, M., Njapau, H., Schurz-Rogers, H., Luber, G., Kieszak, S., Nyamongo, J., Backer, L., Dahiye, A. M., Misore, A., DeCock, K. and Rubin, C. (2005), Aflatoxin contamination of commercial maize products during an outbreak of acute aflatoxicosis in eastern and central Kenya, *Environ. Health Perspect.*, 113(12), 1763–7.

Locke, M. A., Zablotowicz, R. M. and Weaver, M. A. (2005), Herbicide fate under conservation tillage, cover crop, and edge-of-field management practices, pp, 373–92. In Singh, H. P., Batish, D. R. and Kohli, R. K. (Eds), *Handbook of Sustainable Weed Management*. Haworth: New York, NY.

Lorito, M. and Woo, S. L. (2015), *Trichoderma*: A multi-purpose tool for integrated pest management, pp. 345–53. In Lugtenberg, B. (Ed.), *Principles of Plant-Microbe Interactions: Microbes for Sustainable Agriculture*. Springer International Publishing: Cham, Switzerland.

Martin, R. V., Washington, R. and Downing, T. E. (2000), Seasonal maize forecasting for South Africa and Zimbabwe derived from an agroclimatological model. *J. Appl. Meteorol.*, 39, 1473–9.

Matumba, L., Sulyok, M., Monjerezi, M., Biswick, T. and Krska, R. (2014), Fungal metabolites diversity in maize and associated human dietary exposures relate to micro-climatic patterns in Malawi. *Worl Mycotoxin J.*, 8(3), 269–82.

Maya-Cortes, D. C., Cardenas, J. D. F., Garnica-Romo, M. G., Cuevas-Villanueva, R. A., Cortes-Martınez, R., Veles-Medina, J. J. and Martınez-Flores, H. E.(2010), Whole-grain maize tortilla prepared using an ecological nixitamilisation process and its impact on the nutritional value, *Int. J. Food Sci. Tech.*, 45, 23–8

Mayfield, K., Betrán, F. J., Isakeit, T., Odvody, G., Murray, S. C., Rooney, W. L. and Landivar, J. C. (2012), Registration of maize germplasm lines Tx736, Tx739, and Tx740 for reducing preharvest aflatoxin accumulation, *J. Plant Regist.*, 6(1), 88–94.

McCall, M. K. (1985), The significance of distance constraints in peasant framing with special references to Sub-Sahran Africa, *Appl. Geogr.*, 5, 325–45.

Mora, M. and Lacey, J. (1997), Handling and aflatoxin contamination of white maize in Costa Rica, *Mycopathologia*, 138, 77–89.

Moss, M. O. (1998), Recent studies of mycotoxins, *J. Appl. Microb. Symp. Suppl.*, 84, 62S–76S.

Motsi, K. E., Chuma, E. and Mukamuri, B. B. (2004), Rainwater harvesting for sustainable agriculture in communal lands of Zimbabwe, *Phys. Chem. Earth, Pt. A/B/C*, 29(15), 1069–73.

Mukanga, M., Derera, J., Tongoona, P. and Laing, M. D. (2010), A survey of pre-harvest ear rot diseases of maize and associated mycotoxins in south and central Zambia, *Int. J. Food Microb.*, 141(3), 213–21.

Mukherjee, P. K., Horwitz, B. A., Herrera-Estrella, A., Schmoll, M., and Kenerley, C. M. (2013), *Trichoderma* research in the genome era, *Annu. Rev. Phytopathol.*, 51, 105–29.

Munkvold, G. P. (2003a), Cultural and genetic approaches to managing mycotoxins in maize, *Annu. Rev. Phytopathol.*, 41, 99–116.

Munkvold, G. P. (2003b), Epidemiology of *Fusarium* diseases and their mycotoxins in maize ears, *Eur. J. Plant Pathol.*, 109, 705–713.

Munkvold, G. P. and Dejardins, A. E. (1997), Fumonisins in maize; Can we reduce their occurrence?, *Plant Dis.*, 18, 556–65.

Mutungi, C., Lamuka, P., Arimi, S., Gathumbi, J. and Onyango, C. (2008), The fate of aflatoxins during processing of maize into muthokoi – A traditional Kenyan food, *Food Control*, 19(7), 714–21.

Navya, H. M., Naveen, J., Hariprasad, P. and Niranjana, S. R. (2015), Beneficial rhizospheric microorganisms mediated plant growth promotion and suppression of aflatoxigenic fungal and aflatoxin contamination in groundnut seeds, *Ann. Appl. Biol.*, 167, 225–35.

Nesci, A. V, Bluma, R. V. and Etcheverry, M. G. (2005), In vitro selection of maize rhizobacteria to study potential biological control of *Aspergillus* section Flavi and aflatoxin production, *Eur. J. Plant Pathol.*, 113, 159–71.

Ng'ang'a, J., Mutegi, C., Imathiu, S. and Affognon, H. (2016), Effect of triple-layer hermetic bagging on mould infection and aflatoxin contamination of maize during multi-month on-farm storage in Kenya. *J. Stored Produce Res.*, 69, 119–28.

Ni, X., Wilson, J. P., Toews, M. D., Buntin, G. D., Lee, R. D., Li, X. and Huffaker, A. (2014), Evaluation of spatial and temporal patterns of insect damage and aflatoxin level in the pre-harvest corn fields to improve management tactics, *Insect Sci.*, 21(5), 572–83.

Njapau, H., Muzungaile, E. and Changa, R. (1998), Effect of village processing techniques on the content of aflatoxins in maize and peanuts in Zambia, *J. Sci. Food Agric.*, 76, 450–6.

Northolt, M. D., Van Egmond, H. P. and Paulsch, W. E., (1977), Differences between *Aspergillus flavus* strains in growth and aflatoxin B1 production in relation to water activity and temperature, *J. Food Protect.*, 40, 778–81.

Northolt, M. D., Verhulsdonk, C. A. H., Soentoro, P. S. S. and Paulsch, W. E. (1976), Effect of water activity and temperature on aflatoxin production by *Aspergillus parasiticus, J. Milk Food Technol.*, 39, 170–4.

Ono, M. and Kimura, N. (1991), Antifungal peptides produced by *Bacillus subtilis* for the biological control of aflatoxin contamination, *Proc. Jpn. Assoc. Mycotoxicol.*, 34, 23–8.

Ortiz, M. P., Barros, G. G., Reynoso, M. M., Torres, A. M., Chulze, S. N. and Ramirez, M. L. (2011), Soil populations of *Aspergillus* section Flavi from the main and new peanut growing areas in Argentina, ISM Conference 2011 'Strategies to reduce the impact of mycotoxins in Latin America in a global context'. Abstract Book.

Ortiz, T. A., Nicoletti, M. A. and Takahashi, L. S. A. (2016), Effect of processing stages in the physiological quality of maize seeds, *Austr. J. Crop Sci.*, 10(6), 819.

Pal, K. K. and Gardener, B. M. (2006), Biological Control of Plant Pathogens, *The Plant Health Instructor*. doi: 10.1094/PHI-A-2006-1117-02.

Palumbo, J. D., Baker, J. L. and Mahoney, N. E. (2006), Isolation of bacterial antagonists of *Aspergillus flavus* from almonds, *Microbial Ecol.*, 52, 45–52.

Park, D. L., Lee, L. S., Price, R. L. and Pohland, A. E. (1988), Review of the decontamination of aflatoxins by ammoniation: Current status and regulation, *J. Assoc. Off. Anal. Chem. J. AOAC*, 71(4), 685–703.

Park, J. W. and Kim, Y. B. (2006), Effect of pressure cooking on aflatoxin B1 in rice, *J. Agric. Food Chem.*, 54, 2431–5.

Parke, J. L. and Gurian-Sherman, D. (2001), Diversity of the Burkholderia cepacia complex and implications for risk assessment of biological control strains, *Annu. Rev. Phytopathol.*, 39, 225–58.

Payne, G. A., Kamprath, E. J. and Adkins, C. R. (1989), Increased aflatoxin contamination in nitrogen-stressed corn, *Plant Dis.*, 73(7), 556–9.

Pearson, T. C., Wicklow, D. T. and Brabec, D. L., (2010), Characteristics and sorting of white food corn contaminated with mycotoxins, *Appl. Eng. Agric.*, 26, 109–13.

Perez-Flores, G. C., Moreno-Martınez, E. and Mendez-Albores, A. (2011), Effect of Microwave heating during alkaline-cooking of aflatoxin-contaminated maize, *J. Food Sci.*, 76(2), T48–52.

Piedade, F. S., Fonseca, H., da Glória, E. M., Calori-Domingues, M. A., Piedade, S. M. S. and Barbin, D. (2002), Distribution of aflatoxins in contaminated maize fractions segregated by size, *Braz. J. Microbiol.*, 33, 12–16.

Pitt, J. I. and Miscamble, B. F. (1995), Water relations of *Aspergillus flavus* and closely related species, *J. Food Protect.*, 58, 86–90.

Probst, C., Bandyopadhyay, R., Price, L. E. and Cotty, P. J. (2011), Identification of atoxigenic *Aspergillus flavus* isolates to reduce aflatoxin contamination of maize in Kenya, *Plant Dis.*, 95, 212–18.

Quadri, S. H. M., Niranjan, M. S., Chaluvaraju, K. C., Shantaram, U. and Zaranappa, E. H. (2013), An Overview on chemistry, toxicity, analysis and control of aflatoxins, *Int. J. Chem. Life Sci.*, 2(1), 1071–8.

Rajasekaran, K., Sickler, C. M., Brown, R. L., Cary, J. W. and Bhatnagar, D. (2013), Evaluation of resistance to aflatoxin contamination in kernels of maize genotypes using a GFP-expressing *Aspergillus flavus* strain, *World Mycotoxin J.*, 6(2), 151–8.

Rambo, G., Tuite J. and Zachariah, G. L. (1975), Fluorescence associated with corn infected with *Aspergillus flavus* and *A. parasiticus* in storage, *Cereal Chem.*, 52(7), 757–64.

Reeves, D. W. (1997), The role of soil organic matter in maintaining soil quality in continuous cropping systems, *Soil Till. Res.*, 43(1), 131–67.

Samapundo, S., Devlieghere, F., De Meulenaer, B., Lamboni, Y., Osei-Nimoh, D. and Debevere, J. (2007), Interaction of water activity and bicarbonate salts in the inhibition of growth and mycotoxin production by *Fusarium* and *Aspergillus* species of importance to corn, *Int. J. Food Microb.*, 116, 266–74.

Sanchez, P. A., Buresh, R. J. and Leakey, R. R. (1997), Trees, soils, and food security, *Philos. Trans. R. Soc. Lond. B Biol. Sci.*, 352(1356), 949–61.

Schindler, A. F., Palmer, J. G. and Eisenberg, W. V. (1967), Aflatoxin production by *Aspergillus flavus* as related to various temperatures, *Appl. Microbiol.*, 15, 1006–9.
Scott, P. M. (1984), Effects of food processing on mycotoxins, *J. Food Protect.*, 47(6), 489–99.
Shapira, R. (2004), Control of mycotoxins in storage and techniques for their decontamination, pp. 190–223. In Magan, N. and Olsen, M. (Eds.), *Mycotoxins in Food Detection and Control*. Woodhead Publishing: Cambridge, England.
Sharma, A. R., Singh, R., Dhyani, S. K. and Dube, R. K. (2010), Moisture conservation and nitrogen recycling through legume mulching in rainfed maize (*Zea mays*)–wheat (*Triticum aestivum*) cropping system, *Nutr. Cycl. Agroecosys.*, 87(2), 187–97.
Shifa, H., Tasneem, S., Gopalakrishnan, C. and Velazhahan, R. (2016), Biological control of pre-harvest aflatoxin contamination in groundnut (*Arachis hypogaea* L.) with *Bacillus subtilis* G1, *Arch. Phytopathol. Plant Protect.*, 49, 1–12.
Sidibé, A. (2005), Farm-level adoption of soil and water conservation techniques in northern Burkina Faso, *Agr. Water Manage*, 71(3), 211–24.
Sivparsad, B. J. and Laing, M. D. (2016), Pre-harvest silk treatment with Trichoderma harzianum reduces aflatoxin contamination in sweetcorn, *J. Plant Dis. Protect.*, 123, 285–93.
Siwela, A. H., Siwela, M., Matindi, G., Dube, S. and Nziramasanga, N. (2005), Decontamination of aflatoxin-contaminated maize by dehulling, *J. Sci. Food Agric.*, 85, 2535–8.
Smart, M. G., Wicklow, D. T. and Caldwell, R. W. (1990), Pathogenesis of Aspergillus ear rot of maize; Light microscopy of fungal spread from wounds, *Phytopathology*, 80, 1287–94.
Sorenson, W. G., Hesseltine, C. W. and Shotwell, O. L. (1967), Effect of temperature on production of aflatoxin on rice by *Aspergillus flavus, Mycopathol. Mycol. Appl.*, 33, 49–55.
Sorenson, W. G., Jones, W., Simpson J. and Davidson J. I. (1984), Aflatoxin in respirable airborne peanut dust, *J. Toxicol. Env. Health*, 14, 525–33.
Tanboon-ek, P. (1989), Control of aflatoxin in maize, pp. 166–72. In Semple, R. L., Frio, A. S., Hicks, P. A. and Lozare, J. V. (Eds.), *Mycotoxin Prevention and Control in Food Grains*. FAO: Rome, Italy.
Teasdale, J. R., Coffman, C. B. and Mangum, R. W. (2007), Potential long-term benefits of no-tillage and organic cropping systems for grain production and soil improvement, *Agron. J.*, 99(5), 1297–305.
Thierfelder, C., Cheesman, S. and Rusinamhodzi, L. (2013), Benefits and challenges of crop rotations in maize-based conservation agriculture (CA) cropping systems of southern Africa, *Int. J. Agric. Sustain.*, 11(2), 108–24.
Trenk, H. L. and Hartman, P. A. (1970). Effects of moisture content and temperature on aflatoxin production, *Appl. Microbiol.*, 19, 781–4.
Tubajika, K. M., Mascagni, H. J., Damann, K. E. and Russin, J. S. (1999), Nitrogen fertilizer influence on aflatoxin contamination of corn in Louisiana, *J. Agric. Food Chem.*, 47(12), 5257–60.
Tubbs, T., Baribusta, D. and Woloshuk, C. (2016), Impact of opening hermetic storage bags on grain quality, fungal growth and aflatoxin accumulation, *J. Stored Prod. Res.*, 69, 276–81.
Udoh, J. M., Cardwel, K. F. and Ikotun, T. (2000), Storage structures and aflatoxin content of maize in five agro-ecological zones of Nigeria, *J. Stored Prod. Res.*, 36, 187–201.
Vanlauwe, B., Wendt, J., Giller, K. E., Corbeels, M., Gerard, B. and Nolte, C. (2014), A fourth principle is required to define conservation agriculture in sub-Saharan Africa: the appropriate use of fertilizer to enhance crop productivity, *Field Crops Res.*, 155, 10–13.
Warburton, M. L. and Williams, W. P. (2014), Aflatoxin resistance in maize: what have we learned lately?, *Adv. Bot.*, 2014, 10.
Warburton, M. L., Brooks, T. D., Windham, G. L. and Williams, W. P. (2011), Identification of novel QTL contributing resistance to aflatoxin accumulation in maize, *Mol. Breed.*, 27(4), 491–9.
Weaver, M. A., Abbas, H. K., Falconer, L. L., Allen, T. W., Pringle, H. L. and Sciumbato, G. L. (2015), Biological control of aflatoxin is effective and economical in Mississippi field trials, *Crop Prot.*, 69, 52–5.
Whitaker, T. B., Slate, A. B. and Johansson, A. S. (2005), Sampling feeds for mycotoxin analysis, pp. 1–23. In: Diaz, D. (Ed.), *The Mycotoxin Blue Book*. Nottingham University Press: Nottingham, United Kingdom.

Widstrom, N. W., McMillian, W. W., Beaver, R. W. and Wilson, D. M. (1990), Weather-associated changes in aflatoxin contamination of preharvest maize, *J. Prod. Agr.*, 3(2), 196–9.

Widstrom, N. W., Wilson, D. M. and McMillian, W. W. (1981), Aflatoxin contamination of preharvest corn as influenced by timing and method of inoculation, *Appl. Environ. Microbiol.*, 42(2), 249–51.

Widstrom, N. W., Wilson, D. M. and McMillian, W. W. (1984), Ear resistance of maize inbreds to field aflatoxin contamination, *Crop sci.*, 24(6), 1155–7.

Williams, J. H., Phillips, T. D., Jolly, P. E., Stiles, J. K., Jolly, C. M. and Aggarwal, D. (2004), Human aflatoxicosis in developing countries: a review of toxicology, exposure, potential health consequences, and interventions, *Am. J. Clin. Nutr.*, 80, 1106–22.

Williams, S. B., Baribusta, D. and Woloshuk, C. (2014), Assessing Purdue Improved Crop Storage (PICS) bags to mitigate fungal growth and aflatoxin contamination, *J. Stored Prod. Res.*, 59, 190–6.

Williams, W. P. and Windham, G. L. (2012), Registration of Mp718 and Mp719 germplasm lines of maize, *J. Plant Regist.*, 6(2), 200–2.

Williams, W. P., Windham, G. L. and Buckley, P. M. (2005), Enhancing maize with resistance to *Aspergillus flavus* infection and aflatoxin accumulation, pp. 379–94. In Abbas, H. K. (Ed.), *Aflatoxin and food Food safety Safety.* CRC: New York.

Wilson, D. M, and Payne G. A., (1994), Factors affecting *Aspergillus flavus* group infection and aflatoxin contamination in crops, pp. 309–25. In Eaton, D. L. and Groopman, J. D. (Eds.), *Toxicology of Aflatoxins: Human Health, Veterinary, and Agricultural Significance.* Academic Press: San Diego, California.

Witt, C., Cassman, K. G., Olk, D. C., Biker, U., Liboon, S. P., Samson, M. I. and Ottow, J. C. G. (2000), Crop rotation and residue management effects on carbon sequestration, nitrogen cycling and productivity of irrigated rice systems, *Plant Soil*, 225(1–2), 263–78.

Wotton, H. R. and strange, R. N. (1985), Circumstantial evidence for phytoalexin involvement in the resistance of peanuts to *Aspergillus flavus, Microbiology*, 131(3), 487–94.

Wu, F., Narrod, C., Tiongco, M. and Liu, Y. (2011), *The Health Economics of Aflatoxin: Global Burden of Disease.* Aflacontrol Working Paper 4. International Food Policy Research Institute: Washington, DC.

Xu, X. (2003), Effects of environmental conditions on the development of *Fusarium* ear blight, *Eur. J. Plant Pathol.*, 109(7), 683–9.

Yahl, K. R., Watson, S. A., Smith, R. J. and Barabolok, R. (1971), Laboratory wet-milling of maize containing high levels of aflatoxin and a survey of commercial wet milling products, *Cereal Chem.*, 48, 385–91.

Yu, J. (2012), Current understanding on aflatoxin biosynthesis and future perspective in reducing aflatoxin contamination, *Toxins*, 4(11), 1024–57.

Zablotowicz, R. M., Abbas, H. K. and Locke, M. A. (2007), Population ecology of *Aspergillus flavus* associated with Mississippi Delta soils, *Food Addit. Contam.*, 24(10), 1102–8

Zhang, Q., Zhang, L., Evers, J., van der Werf, W., Zhang, W. and Duan, L. (2014), Maize yield and quality in response to plant density and application of a novel plant growth regulator, *Field Crops Res.*, 164, 82–9.

Zummo, N. (1991), Concurrent infection of individual corn kernels with white and green isolates of *Aspergillus flavus, Plant Dis.*, 75, 910–13.

Chapter 18

Integrated weed management in maize cultivation: an overview

Khawar Jabran, Duzce University, Turkey; Mubshar Hussain, Bahauddin Zakariya University, Pakistan; and Bhagirath Singh Chauhan, The University of Queensland, Australia

1 Introduction

Maize (*Zea mays* L., also known as corn) is among the top three cereals of the world and is grown across the globe under very diverse environmental conditions (Yamasaki et al. 2005). It has a wide range of uses (Jabran et al. 2007). According to the FAO, world production in 2012 was over 870 million tonnes, grown on 158 million hectares of land (FAO 2013). Sources such as the FAO's Agricultural Market Information System and The International Grains Council have forecasted production increasing to as high as 990 million tonnes in 2014–15 grown on almost 200 million hectares. According to the OECD-FAO (2014), by 2023, maize will account for the greatest share (34%) of the total area harvested.

Over 80% of maize production by volume comes from the Americas (53%) and Asia (28%), followed by Europe (15%). Key areas of cultivation include the US maize belt, north-eastern China and Eastern Europe. Major producers in 2012 included the United States (over 270 million tonnes), China (over 200 million tonnes) and Brazil (71 million tonnes), followed by Mexico, Argentina, India, Ukraine, Indonesia, France and Canada. A major part of the total acreage for maize cultivation is in the developing world where it plays a key role in the food security and livelihoods of millions of poor farmers. Nevertheless,

http://dx.doi.org/10.19103/AS.2016.0002.21

Table 1 Maize area, production and yield over the world and in major maize-producing regions in 2013

Region	Area (m ha)	Production (m tonnes)	Yield (kg/ha)
World (total)	184.2	1016.7	5520
Africa	35.02	71.6	2045
America	70.7	522.6	7392
Asia	59.3	304.3	5124
Europe	18.97	117.5	6190

Source: http://faostat.fao.org/site/567/DesktopDefault.aspx?PageID=567#ancor.

a significant contribution of maize grains comes from developed countries compared to the developing world (see Table 1). High-yielding genotypes, use of modern machinery and better plant protection techniques are the factors that contribute to higher yields in developed countries compared to developing countries.

2 Weeds and their impact on maize cultivation

Seventy-nine weed species have been found to infest maize in different areas of the world. Table 2 summarizes the weed flora reported in different maize-growing regions of the world. Maize is heavily infested by broadleaved and grass weeds with more occasional problems from sedges (Khan et al. 2013). In some maize-growing areas, parasitic weeds have also been reported to heavily infest maize crops (Kiwia et al. 2009; Hooper et al. 2010). Globally, the most important weeds are

- *Amaranthus retroflexus* L.
- *Chenopodium album* L.
- *Convolvulus arvensis* L.
- *Digitaria sanguinalis* (L.) Scop.
- *Echinochloa crus-galli* (L.) P. Beauv.
- *Sorghum halepense* (L.) Pers.
- *Capsella bursa-pastoris* (L.) Medik.
- *Cirsium arvense* (L.) Scop.
- *Datura stramonium* L.
- *Portulaca oleracea* L.
- *Striga hermonthica* (Delile) Benth.
- *Xanthium strumarium* L.

Dewar (2009) reviewed the weed flora of maize in European countries:

- *Chenopodium* spp. and *E. crus-galli* were common in Spain, Hungary, France and Czech Republic.
- *Amaranthus* spp. were common in maize fields of Spain, France and Czech Republic.
- *Digitaria* spp. were common in maize fields of Germany and France.

Figure 1 A maize field infested with *Echinochloa* spp. (Photo by Khawar Jabran).

To take one country as an example, common maize weeds in Romania are *Cirsium arvense* (L.) Scop., *Setaria* spp., *Sorghum halepense* (L.) Pers., *Cynodon dactylon* (L.) Pers. and *Xanthium* spp. (Dewar, 2009). In contrast, *E. crus-galli*, *Setaria* spp. and *S. halepense* were the most common weeds in Hungary (Zalai et al. 2013; Andújar et al. 2011).

In Brazil, 14 weed species were noted in organically and conventionally sown maize crops (Favarato et al. 2014). Out of these, *Galinsoga quadriradiata* Ruiz & Pav., *Oxalis* spp. and *Stachys arvensis* (L.) L. were the most prevalent weeds in both systems. Similarly, in a study conducted in China, *Eleusine indica* (L.) Gaertn., *A. retroflexus*, *Setaria faberi* Herrm. and *Portulaca oleracea* L. were the most common among the weed flora noted in maize (Zhang et al. 2013). Parasitic weeds such as *S. hermonthica* and *Striga hermonthica* (Delile) Benth. pose the most serious threat in the maize-growing areas of Africa (Kiwia et al. 2009; Hooper et al. 2010; Avedi et al. 2014).

Weeds have been reported to reduce grain yields in maize by 20–98%, with 40% losses often recorded (see Table 3) (Oerke 2006). Weed density of 1 plants m^{-2} for grasses, 6 plants m^{-2} for broadleaves or 0.1 plants m^{-2} for perennials may cause a significant decrease in maize productivity (Gutjahr et al. 2012). A mixture of grasses, sedges and broadleaved weeds has been shown to significantly reduce maize grain weight and yield (Khan et al. 2013). *E. crus-galli* can cause 24–34% yield losses in hybrid maize, and *Echinochloa colona* (L.) Link and *T. portulacastrum* were noted to cause 35% and 37%

Table 2 Weed flora of maize reported in different regions

Weeds	Country	References
Abutilon theophrasti Medik.	Serbia	(Simić et al. 2012)
	Hungary	(Zalai et al. 2013)
	China	(Zhang et al. 2013)
Acalypha australis L.	China	(Zhang et al. 2013)
Ageratum conyzoides L.	Cameroon	(Fontem and Chikoye 2012)
Amaranthus retroflexus L.	Czech Republic, France, Spain	(Dewar 2009)
	United States	(Cordill and Grift 2011)
	Greece	(Efthimiadou et al. 2012)
	Slovenia	(Lešnik et al. 2012)
	Serbia	(Simić et al. 2012)
	Hungary	(Zalai et al. 2013)
	China	(Zhang et al. 2013)
	Czech Republic	(Andr et al. 2014)
	Italy	(Pannacci and Tie 2014)
Amaranthus spp.	Brazil	(Favarato et al. 2014)
Amaranthus viridis L.	Pakistan	(Khan et al. 2013)
Ambrosia artemisiifolia L.	Slovenia	(Lešnik et al. 2012)
	Hungary	(Zalai et al. 2013)
Ammi majus L.	Italy	(Pannacci and Tie 2014)
Artemisia absinthium L.	Brazil	(Favarato et al. 2014)
Atriplex patula L.	Czech Republic, Germany	(Dewar 2009)
Bidens pilosa L.	Mexico	(Ramos-Zapata et al. 2012)
	Brazil	(Favarato et al. 2014)
Brachiaria brizantha (A.Rich.) Stapf	Canada	(Marin and Weiner 2014)
Capsella bursa-pastoris (L.) Medik.	Germany	(Gutjahr et al. 2012)
	Lithuania	(Romaneckas et al. 2012)
	Hungary	(Zalai et al. 2013)
	Denmark	(Ali et al. 2015)
Calystegia hederacea Wall.	China	(Zhang et al. 2013)
Cenchrus echinatus L.	Brazil	(Dan et al. 2011)
Chenopodium album L.	Czech Republic, France, Hungary, Spain	(Dewar 2009)
	United States	(Cordill and Grift 2011)
	Germany	(Gutjahr et al. 2012)
	Serbia	(Simić et al. 2012)

Weeds	Country	References
	Lithuania	(Romaneckas et al. 2012)
	Slovenia	(Lešnik et al. 2012)
	China	(Zhang et al. 2013)
	Czech Republic	(Andr et al. 2014)
	Italy	(Pannacci and Tie 2014)
	New Zealand	(Rahman et al. 2014)
	Denmark	(Ali et al. 2015)
Chenopodium hybridum L.	Serbia	(Simić et al. 2012)
Cichorium intybus L.	Hungary	(Zalai et al. 2013)
Cirsium arvense (L.) Scop.	Romania	(Dewar 2009)
	Germany	(Gutjahr et al. 2012)
	Hungary	(Zalai et al. 2013)
	Denmark	(Ali et al. 2015)
Commelina benghalensis L.	Brazil	(Favarato et al. 2014)
Convolvulus arvensis L.	France	(Dewar 2009)
	Pakistan	(Khalil et al. 2010)
	Slovenia	(Lešnik et al. 2012)
	Pakistan	(Khan et al. 2013)
	Hungary	(Zalai et al. 2013)
Cynodon dactylon L.	Pakistan	(Khan et al. 2013)
	Brazil	(Favarato et al. 2014)
Cyperus esculentus L.	South Africa	(Bezuıdenhout et al. 2012)
Cyperus rotundus L.	Pakistan	(Khalil et al. 2010)
	Pakistan	(Khan et al. 2013)
	Brazil	(Favarato et al. 2014)
Cyperus spp.	Cameroon	(Fontem and Chikoye 2012)
Datura stramonium L.	Hungary	(Dewar 2009)
	Iran	(Hassan et al. 2010)
	Iran	(Karimmojeni et al. 2010)
	Serbia	(Simić et al. 2012)
Digera muricata (L.) Mart.	Pakistan	(Khan et al. 2013)
Digitaria insularis (L.) Fedde	Mexico	(Ramos-Zapata et al. 2012)
Digitaria sanguinalis (L.) Scop	France, Germany	(Dewar 2009)
	USA	(Cordill and Grift 2011)

(*Continued*)

Table 2 (*Continued*)

Weeds	Country	References
	Pakistan	(Khan et al. 2013)
	Hungary	(Zalai et al. 2013)
	China	(Zhang et al. 2013)
	Brazil	(Favarato et al. 2014)
Echinochloa crus-galli (L.) P.Beauv.	Czech Republic, France, Hungary, Spain	(Dewar 2009)
	Romania	(Rusu et al. 2010)
	Greece	(Travlos et al. 2011)
	Belgium	(Cauwer et al. 2012)
	Germany	(Gutjahr et al. 2012)
	Hungary	(Kukorelli et al. 2012)
	Slovenia	(Lešnik et al. 2012)
	Pakistan	(Khan et al. 2013)
	Hungary	(Zalai et al. 2013)
	China	(Zhang et al. 2013)
	Czech Republic	(Andr et al. 2014)
	Brazil	(Galon et al. 2014)
	Italy	(Pannacci and Tie 2014)
Echinochloa muricata (P.Beauv.) Fernald	Belgium	(Cauwer et al. 2012)
Echinochloa colona (L.) Link	Pakistan	(Fahad et al. 2014)
Eleusine indica (L.) Gaertn.	China	(Zhang et al. 2013)
	Brazil	(Favarato et al. 2014)
Elytrigia repens (L.) Nevski	Slovenia	(Lešnik et al. 2012)
Euphorbia heterophylla L.	Brazil	(Favarato et al. 2014)
Fallopia convolvulus (L.) Á.Löve	Slovenia	(Lešnik et al. 2012)
	Hungary	(Zalai et al. 2013)
	Italy	(Pannacci and Tie 2014)
Galium aparine L.	Germany	(Dewar 2009)
Galinsoga quadriradiata Ruiz & Pav.	Brazil	(Favarato et al. 2014)
Lamium amplexicaule L.	Denmark	(Ali et al. 2015)
Leptochloa spp.	Pakistan	(Khan et al. 2013)
Matricaria spp.	Czech Republic, Germany	(Dewar 2009)
Mercurialis annua L.	Czech Republic	(Andr et al. 2014)

Weeds	Country	References
Oryza sativa L. (weedy rice)	Brazil	(Galon et al. 2014)
Oxalis spp.	Brazil	(Favarato et al. 2014)
Panicum maximum Jacq.	Cameroon	(Fontem and Chikoye 2012)
Panicum miliaceum L.	Hungary	(Kukorelli et al. 2012)
Persicaria lapathifolia (L.) Delarbre	Hungary	(Zalai et al. 2013)
Physalis angulata L.	China	(Zhang et al. 2013)
Plantago major L.	Lithuania	(Romaneckas et al. 2012)
Parthenium hysterophorus L.	Pakistan	(Safdar et al. 2015)
Poa annua L.	Lithuania	(Romaneckas et al. 2012)
	Denmark	(Ali et al. 2015)
Polygonum aviculare L.	Greece	(Efthimiadou et al. 2012)
	Italy	(Pannacci and Tie 2014)
	Denmark	(Ali et al. 2015)
Polygonum lapathifolium L.	Lithuania	(Romaneckas et al. 2012)
	Slovenia	(Lešnik et al. 2012)
Polygonum persicaria L.	Italy	(Pannacci and Tie 2014)
	Denmark	(Ali et al. 2015)
Portulaca oleracea L.	Greece	(Efthimiadou et al. 2012)
	Pakistan	(Khan et al. 2013)
	China	(Zhang et al. 2013)
	Italy	(Pannacci and Tie 2014)
Priva lappulacea (L.) Pers.	Mexico	(Ramos-Zapata et al. 2012)
Rumex obtusifolius L.	Brazil	(Favarato et al. 2014)
Sanvitalia procumbens Lam.	Mexico	(Ramos-Zapata et al. 2012)
Setaria pumila (Poir.) Roem. & Schult.	Hungary	(Zalai et al. 2013)
Setaria faberi Herrm.	United States	(Cordill and Grift 2011)
	China	(Zhang et al. 2013)
Setaria glauca (L.) Beauv.	Romania	(Rusu et al. 2010)
Setaria viridis (L.) Beauv.	France, Hungary, Romania	(Dewar 2009)
	Hungary	(Zalai et al. 2013)
Setaria verticillata (L.) P.Beauv.	Hungary	(Kukorelli et al. 2012)
Sida acuta Burm.	Mexico	(Ramos-Zapata et al. 2012).
Sinapis arvensis L.	Lithuania	(Romaneckas et al. 2012)

(Continued)

Table 2 (*Continued*)

Weeds	Country	References
Solanum nigrum L.	Serbia	(Simić et al. 2012)
	China	(Zhang et al. 2013)
Sonchus arvensis L.	Lithuania	(Romaneckas et al. 2012)
Sonchus oleraceus L.	Brazil	(Favarato et al. 2014)
Sorghum halepense (L.) Pers.	Hungary, Romania, Spain	(Dewar 2009)
	Spain	(Andújar et al. 2011)
	Spain	(Andújar et al. 2013)
	Pakistan	(Khan et al. 2013)
Stachys arvensis (L.) L.	Brazil	(Favarato et al. 2014)
Stellaria media (L.) Vill.	Lithuania	(Romaneckas et al. 2012)
	Denmark	(Ali et al. 2015)
Striga asiatica (L.) Kuntze	Angola	(Dovala and Monteiro 2014)
Striga hermonthica (Delile) Benth.	Kenya	(Kiwia et al. 2009)
	Kenya	(Hooper et al. 2010)
	Kenya	(Avedi et al. 2014)
Taraxacum officinale F.H.Wigg.	Lithuania	(Romaneckas et al. 2012)
Trianthema portulacastrum L.	Pakistan	(Khalil et al. 2010)
	Pakistan	(Khan et al. 2012)
	Pakistan	(Fahad et al. 2014)
Tripleurospermum inodorum (L.) Sch.Bip.	Denmark	(Ali et al. 2015)
Veronica persica Poir.	Denmark	(Ali et al. 2015)
Xanthium strumarium L.	Iran	(Hassan et al. 2010)
	Iran	(Karimmojeni et al. 2010)
	Serbia	(Simić et al. 2012)
	Hungary	(Zalai et al. 2013)

yield losses, respectively, in maize (Fahad et al. 2014). The ear numbers of maize were negatively correlated with the density of *Striga asiatica* (L.) Kuntze (Dovala and Monteiro 2014). *Parthenium hysterophorus* L. is an invasive weed species, which has invaded many parts of the world. Recently, this weed was found to infest maize, absorb nutrients in high quantities (N = 78 and K = 64 kg ha^{-1}) and decrease maize grain yields by 21–50% (Safdar et al. 2015). A density of 1.3 plants m^{-2} of this weed was sufficient to cause yield losses in maize (Safdar et al. 2015).

As can be seen from the effects of weeds on yields, maize plants are highly sensitive to the presence of weeds. Maize plants can compete with and suppress weeds if the weed densities are low. However, at high densities, weeds quickly develop more leaf area and

Table 3 Yield losses caused by weeds in maize crop in different regions

Weeds	Yield losses (%)	Country	References
E. crus-galli	24–34	Greece	(Travlos et al. 2011)
A. conyzoides, *P. maximum*, *Cyperus* spp.	62	Cameroon	(Fontem and Chikoye 2012)
A mixture of broadleaved weeds, grasses and sedges	98	Pakistan	(Khan et al. 2013)
T. portulacastrum	37	Pakistan	(Fahad et al. 2014)
E. colona	35	Pakistan	(Fahad et al. 2014)
Oryza sativa L. (weedy rice)	43	Brazil	(Galon et al. 2014)
Grasses and broadleaved	84	Italy	(Pannacci and Tie, 2014)
C. album	24	New Zealand	(Rahman et al. 2014)
T. portulacastrum (3–18 plants m^{-2})	4–30	Pakistan	(Saeed et al. 2010)
Abutilon theophrasti Medik., *A. retroflexus*, *C. album*, *C. hybridum*, *D. stramonium*, *S. nigrum*, *X. strumarium*	33	Serbia	(Simić et al. 2012)
Parthenium hysterophorus L. (5–20 plants m^{-2})	21–50	Pakistan	(Safdar et al. 2015)

a stronger canopy than maize, and take up more resources (Hassan et al. 2010). As an example, one *E. crus-galli* plant growing in maize and reaching maturity can produce more than 1000 seeds, effectively outpacing maize plants in the competition for resources (Travlos et al. 2011). As an illustration of the effects of weed density, ear numbers of maize have been negatively correlated with the density of *Striga asiatica* (L.) Kuntze (Dovala and Monteiro 2014). Exposure of maize plants to a high density of weeds has been shown to reduce the number of maize leaf tips, leaf appearance rate, biomass of leaf, root and stem, leaf area and seedling biomass (Page et al. 2009). Maize plants facing a competition from weeds such as *E. crus-galli* and *Setaria glauca* (L.) Beauv., for example, had very weak shoots and stunted growth (Rusu et al. 2010).

Weeds both compete for resources and interfere with the growth activities of maize plants. Weed infestation, for example, can reduce the nitrogen available to maize plants by 50% (Lindquist et al. 2010). It has been found that overall nutrient uptake by maize plants is significantly decreased when growing under competition with weeds (Cury et al. 2012). The results of a study from Pakistan indicated that weeds might benefit equally from applied fertilizers as maize plants (Khan et al. 2013). Weeds can also damage maize by their allelopathic activity (the production of biochemicals by one plant, which can affect the growth and survival of others). Konstantinović (2014) reported that allelopathic extracts from *A. retroflexus* inhibited seedling growth in maize. Similarly, the root exudates of *Ambrosia artimisiifolia* L. significantly reduced the emergence and plant biomass of maize (Vidotto et al. 2014). *X. strumarium* was found to interfere with the growth activities of maize throughout its growing period (Karimmojeni et al. 2010).

3 The use of herbicides

There are a number of ways of applying herbicides, including pre-plant, pre-emergence, early and late post-emergence applications of herbicides to deal with different stages of weed growth, using mixtures of herbicides or rotating different herbicides to suppress weeds (Norsworthy et al. 2012). Jhala et al. (2014) have discussed in detail the use of herbicides to control weeds in maize.

Both pre-emergence and post-emergence herbicides can be used for weed control in maize (Dewar 2009; Peña-Asin et al. 2013; Zhang et al. 2013; Rahman et al. 2014). Pre-emergence herbicides include *S*-metolachlor, pendimethalin, acetochlor, mesotrione, metosulam, dimethenamid-P, flufenacet, aclonifen, isoxaflutole and aclonifen. Post-emergence herbicides include nicosulfuron, 2,4-D, bromoxynil, dicamba, mesotrione, foramsulfuron, fluroxypyr, pyridate, topramezone, terbuthylazine, rimsulfuron, bentazone and sulcotrione.

The pre-emergence herbicides isoxaflutole plus acetochlor were effective in completely suppressing grass and broadleaved weeds in maize, and increasing maize grain yield by 48% (Simić et al. 2012). In Czech Republic, the pre-emergence herbicide dimenthamid-P plus pendimethalin, isoxaflutole plus thiencarbazone and metolachlor plus terbuthylazine effectively controlled four important maize weeds (Andr et al. 2015). Dimenthamid-P plus pendimethalin was effective against *A. retroflexus* and *C. album* (>80%); isoxaflutole plus thiencarbazone was effective against *E. crus-galli, A. retroflexus* and *C. album* (88–96%), and metolachlor plus terbuthylazine was effective against *E. crus-galli, A. retroflexus, C. album* and *Mercurialis annua* L. (65–92%) (Andr et al. 2014).

The post-emergence herbicide nicosulfuron was effective in suppressing maize weeds such as *Panicum maximum* Jacq., *Cyperus* spp. and *Ageratum conyzoides* L. (Fontem and Chikoye 2012). This herbicide applied at 14 days after sowing decreased weed biomass by more than 90% and increased the maize grain yield by more than 60% (Fontem and Chikoye 2012). The post-emergence application of cycloxydim (150 g a.i. ha^{-1}) in cycloxydim-tolerant maize was effective in controlling the annual maize weeds [*Setaria verticillata* (L.) P. Beauv., *E. crus-galli, Panicum miliaceum* L.] (Kukorelli et al. 2012).

A study from Germany reported the use of the pre- and post-emergence herbicides mesotrione plus bromoxynil for the control of broadleaved, nicosulfuron for the control of grasses, and dicamba for the control of perennial weeds in maize crop using site-specific weed management tools (Gutjahr et al. 2012). In China, nicosulfuron, nicosulfuron plus mesotrione and topramizone were effective in controlling narrow and broadleaved weeds of maize while mesotrione was not effective against grass weeds (Zhang et al. 2013).

Cauwer et al. (2012) evaluated the sensitivities of two noxious maize weeds, that is, *E. crus-galli* and *Echinochloa muricata* (P. Beauv.) Fernald, to post-emergence herbicides tembotrione, nicosulphuron, sulcotrione, topramezone and mesotrione. Both the weeds were equally sensitive to nicosulphuron, tembotrione and mesotrione. However, much higher (5 to 14 times) doses of sulcotrione and topramezone were needed to achieve a good control of *E. muricata* compared with *E. crus-galli*. Herbicide application at a proper time was an important factor in achieving a satisfactory control of *E. muricata*. Dan et al. (2011) evaluated the efficacy of different rates of atrazine for the control of *Cenchrus echinatus* L., an important weed of maize in Brazil. Atrazine dose of 3.5 kg a.i. ha^{-1} or higher was helpful in effectively controlling this weed. The herbicide could be applied at the two-leaf growth stage in order to obtain effective control.

Weeds are not evenly distributed in agricultural fields, which makes effective weed control a difficult task (Gutjahr et al. 2012). The heterogeneous distribution of weeds in the crop fields can lead to overuse of herbicides in patches with few or no weeds. Site-specific weed control has been suggested for such cases (Ali et al. 2015). Site-specific herbicide applications can result in reduced herbicide use as well as economic and environmental benefits. In a study from Germany, a geographical positioning system (GPS) was used to develop maps of weed distribution in maize fields in order to apply site-specific weed control (Gutjahr et al. 2012). Only 53% of the cropped area of maize field required herbicides when site-specific spray of herbicide was made using a GPS-controlled multiple tank sprayer (Gutjahr et al. 2012). Nevertheless, the price of site-specific weed recognition and related herbicide application technology is still higher than its economic benefits (through herbicide saving) (Andújar et al. 2013). Further research work is desired to make this technology cheaper and accessible to farmers (Andújar et al. 2013). Gutjahr et al. (2012) have reported problems such as slow working of the imaging system for real-time weed identification and poor performance of the existing injection systems for site-specific weed control.

Although they have been very successful in controlling weed populations and improving maize yields, continuous long-term application of the same herbicides in the same area has led to increasing herbicide resistance amongst some weeds (Powles and Yu 2010). There are now 61 herbicide-resistant weed species documented in maize (Heap 2015). At the same time, new herbicides have been difficult to develop (Duke 2012). New Zealand provides an example of the problem. *C. album* is among the most dominant maize weeds in New Zealand. Farmers had used atrazine herbicide extensively to control this weed, which has led to herbicide resistance evolution in *C. album* against atrazine (James et al. 2000). This led the farming community to shift to the use of dicamba for post-emergence control of *C. album*. Recently, this weed was reported to have evolved resistance against dicamba herbicide too (Rahman et al. 2014). In laboratory experiments, *C. album* could survive a fourfold application (of normal rate) of dicamba while under field conditions, this weed could tolerate dicamba application by eightfold of the normal rate. Other herbicides such as bromoxynil, pyridate, nicosulfuron and mesotrione were effective in controlling the dicamba-resistant populations of *C. album* but these herbicides too could suffer from the evolution of resistance over time (Rahman et al. 2014).

Environmental contamination resulting from the overuse of pesticides is also an important problem in modern agriculture. As an example, the maize herbicide terbuthylazine and its metabolite desethylterbuthylazine were found to have polluted lowland natural springs in Italy (Laini et al. 2012). There have been a range of studies on the environmental effects of herbicides such as Hayes et al. (2003) and Ying et al. (2005). Microbes (such as *Pseudomonas* spp.) can be used to degrade herbicide residues; for example, Fernández et al. (2013) isolated three strains of *Arthrobacter* spp., which were effective in degrading atrazine herbicide under field conditions. However, there is widespread recognition that it is better to avoid these residues in the first place rather than have to find ways of removing them. Problems such as herbicide resistance and potential contamination mean that there is a need to reduce herbicide use wherever possible by combining it with other means of suppressing weeds (Harker 2013).

4 Integrated weed management

A simple definition of integrated weed management (IWM) is 'the use of more than one weed management tactic (biological, chemical, cultural, and/or physical)' (Harker and O'Donovan 2013). A good introduction to IWM is provided by Liebman et al. (2007). Examples of the range of IWM techniques include

- Chemical, for example, use of herbicides
- Cultural, for example, use of stale seed beds
- Physical, for example, mechanical weeding

IWM is not about excluding one technique in favour of another, but developing an effective combination of techniques to ensure the sustainable control of weed populations.

Employing a range of techniques to control weeds can prevent the emergence of one or a few dominant weed species that can compete effectively for resources with crops (Derkson et al. 2002). Whilst many weed species have advantages such as higher relative growth rates compared to many crops, they can have weaknesses such as a higher susceptibility to shade (Fenner 1978; Mohler 2015). One objective of IWM is to exploit these weaknesses by providing conditions that inhibit weed growth whilst favouring the growth of crops, for example supporting the development of crop canopy cover, which maximizes crop health whilst suppressing weed growth. The use of a range of methods for control also makes it possible to use herbicides more selectively, slowing the development of herbicide resistance amongst weed populations.

5 IWM techniques: land preparation, cultivars, planting, cultivation and allelopathy

A key concept for weed control is the critical period for weed control (CPWC), defined as the period during which a crop must be kept weed free to prevent losses of more than 5% (Hall et al. 1992). The CPWC for maize ranges from 1 to 8 weeks after crop emergence until canopy closure occurs (Page et al. 2012). Research suggests that weeds emerging four weeks after maize have limited impact on yields (Knezevic et al. 2002). This makes early season weed control particularly important in achieving a good yield.

Young maize seedlings are poor competitors against weeds. Stale seedbed management involves delaying seeding until after tillage for weed control (Rao 2000). Providing maize with a weed-free environment at seedling growth stage can help to avoid weed–crop competition during the initial growth period. A stale seedbed can be applied with either a weed harrow or seedbed cultivator (ENDURE 2008). Weeds are allowed to germinate and then eradicated by hoeing or the use of herbicides (Jhala et al. 2014). Increasing the density of planting, for example, by narrowing rows between crops, creates a denser crop canopy, which can help to suppress weed growth, though this has to be offset against potentially making weeding more difficult (Mohler 2001b; Weiner et al. 2010; Colbach et al. 2014). Targeted spraying of rows (known as band spraying) combined with inter-row cultivation (e.g. using hoes or rolling cultivators) helps to manage weeds with reduced herbicide use (ENDURE 2008).

Zero-till systems can be beneficial for weed management by improving soil health and water availability to crops, allowing them to compete more effectively with weeds, as well as protecting insects such as ground beetles which may eat weed seeds (Alignier et al. 2015; Kulkarni et al. 2015). However, some studies have shown that minimizing soil disturbance can allow stronger weed growth compared to conventional tillage (Gul et al. 2009).

Chemicals exuded from plants in the environment can be used to control weeds, a phenomenon known as allelopathy (Farooq et al. 2011; Jabran and Farooq 2013; Jabran et al. 2015). One way of implementing allelopathic control is to use allelopathic mulches. Straw mulch can significantly reduce weed infestation in maize (Ngwira et al. 2014). In addition to allelopathic effects, mulch also suppresses weeds by reducing access to sunlight. The granules of *Aglaia odorata* Lour., possessing allelopathic properties, were found to effectively suppress weeds in maize (Laosinwattana et al. 2012). Granules of this plant applied as mulch significantly decreased the emergence of weeds such as *Amaranthus gracilis* Desf. (56%), *Digitaria adscendens* (Kunth) Henrard (97%) and *T. portulacastrum* (47%). Maize grain yield with the application of allelopathic granules was similar with that of obtained by atrazine application (Laosinwattana et al. 2012). The results of a study from Kenya indicated that mulches were also effective in suppressing *S. hermonthica*, which is a notorious weed infesting maize fields in many regions of Africa (Midega et al. 2013). The use of allelopathic water extracts is another method to control weeds in maize. Herbage from plants with established allelopathic potential is soaked in water in order to let the allelochemicals leached in water (Jabran et al. 2010a,b; Farooq et al. 2011). These liquids can then be sprayed to control weeds in maize (Khan et al. 2012).

6 IWM techniques: herbicide tolerant cultivars, rotations, allelopathy, intercropping and cover crops

Genetically modified herbicide-tolerant maize cultivars are now widely used in the United States and many maize-producing countries (Lundmark 2007; Green 2012). Examples of herbicide-tolerant maize cultivars include T25 (tolerant against glufosinate), and NK603 (tolerant against glyphosate) (Meissle et al. 2011). By extending the potential range and strength of herbicide use, herbicide-tolerant maize provides more flexibility in weed management (Green and Castle 2010).

Crop rotation has been shown to help manage weeds in cereals and reduce the need for herbicides (Anderson 2005; Anderson et al. 2007; Koocheki et al. 2009; Beres et al. 2010; Davies et al. 2012). An example is a two-year maize–soya bean rotation, particularly if a different tillage system is used for each crop (Jhala et al. 2014). Rotations introduce varying patterns of resource competition, allelopathic interference, soil disturbance and mechanical damage that disrupt weed proliferation. It is important to note that maize has allelopathic properties which can affect the growth of plants such as soya bean (Ahmad and Bano 2013; Ayeni and Kayode 2013).

Intercropping of allelopathic crops in the main crop is an important way of weed control (Jabran et al. 2015). *S. hermonthica* has been successfully controlled in many African countries by intercropping an allelopathic crop such as *Desmodium uncinatum* (Jacq.) DC. (Midega et al. 2013). The root exudates from *D. uncinatum* have been known to cause abortive germination in *S. hermonthica* (Khan et al. 2011). Hooper et al. (2010) found isoschaftoside to be the biologically active compound responsible for *S. hermonthica* suppression. In addition to providing effective control of *S. hermonthica*, *D. uncinatum* intercropping also improved maize growth, functioning and grain yield (Midega et al. 2013). Some other allelopathic crops are also available for intercropping in maize in order to control weeds. For instance, studies from Pakistan indicated that the intercropping of allelopathic sorghum (*Sorghum bicolor* L.) plants in maize helped to effectively control weeds such as *Cyperus rotundus* L., *T. portulacastrum* and *Convolvulus arvensis* L. (Khalil et al. 2010; Mahmood et al. 2013).

Cover crops can also be used for weed suppression. In addition to weed control, cover crops can also provide additional benefits, such as enhanced soil health and crop nutrition (Ramos-Zapata et al. 2012). For example, a cover crop of *Mucuna deeringiana* (Bort) Merr. reduced maize weeds by over 50% and also improved the growth of beneficial arbuscular mycorrhizal fungi (Ramos-Zapata et al. 2012). Legumes can make potentially valuable cover crops because of their ability to fix nitrogen. In one study, the leguminous cover crops were evaluated for weed control in maize (intercropped in cassava), only two [*Stylosanthes guianensis* (Aubl.) Sw. and *Mucuna pruriens* (L.) Dc] were found to be effective in suppressing weeds (Olorunmaiye 2010).

It is also important to note that the allelopathic properties of some cover crops, whilst suppressing weeds, can also inhibit maize plant growth, making it important to select the cover crop carefully. As an example, whilst providing effective weed control of *Cyperus esculentus* L., cover crops such as annual ryegrass and stooling rye were found to delay the emergence and growth of maize plants (Bezuidenhout et al. 2012). Romaneckas et al. (2012) evaluated several cover crops for weed control in maize, but found that, although they were effective in suppressing weed growth, they also negatively affected the maize development. The crops that effectively suppressed weeds and had least negative effect on maize productivity were spring barley (*Hordeum vulgare* L.), spring oilseed rape (*Brassica napus* L.) and white mustard (*Sinapsis alba* L.) (Romaneckas et al. 2012).

7 IWM techniques: mechanical control and flame weeding

Mechanical means of weed control in maize remain important, including traditional methods such as hand hoeing which remain widely used by smallholders (Peña-Asin et al. 2013). Rotary hoes are effective for small-seeded broadleaved weeds and grasses (Jhala et al. 2014). More than one mechanical tool may be combined to improve mechanical weed control (Pannacci and Tie 2014). Combined applications of harrowing (accomplished with a 3-m wide powered rotary hoe) and hoeing-ridging (hoeing with a powered hoe following by ridge making in order to bury weeds in soil) helped to lower the weed cover in maize fields by nearly 90% compared with the non-treated fields (Pannacci and Tie 2014). The split-hoe is a tool, which leaves 10-cm of unworked soil in the maize rows and has a cultivation depth of 3–4 cm. This tool could help to decrease weed seed production (>60%), weed ground cover, biomass and density by 70–90% in the maize fields (Pannacci and Tie 2014). The finger-weeder is another tool for intra-row weeding in maize that helps in suppressing weeds at an early growth stage (Pannacci and Tie 2014). Cordill and Grift (2011) have designed an intra-row mechanical weeding machine with tines cultivating a

5-cm space in 15-cm wide maize rows. This mechanical weeding machine caused 9–26% damage to maize plants, illustrating the challenges in designing effective mechanical weeders.

The use of flame weeding is gaining attention owing to its potential for weed control, for example in organic maize production, although there remains a risk of damaging maize plants (Bong and Grundy 2001; Lague et al. 2001; Ascard et al. 2007). In a study from the United States, the five-leaf stage was found most suitable for flame application to avoid injury to maize plants; application of propane at 85 kg ha^{-1} at the five-leaf stage did not have a negative effect on maize yield (Ulloa et al. 2011).

8 Conclusions and future trends

It has been shown that combining a range of weed techniques can be effective in providing effective, economical and potentially more sustainable weed control (Anderson 2003; Norsworthy and Frederick 2005). As an example, atrazine application rates to control *T. portulacastrum* in maize fields could be lowered by 30–50% if applied in combinations with allelopathic water extracts of sorghum (*Sorghum bicolor* L.) and sunflower (*Helianthus annuus* L.) (Khan et al. 2012). This combination of herbicide (at a lower rate application) plus allelopathic water extract application was found to improve maize yield by 31–36%, only a little below the yield achieved with a standard herbicide dose (Khan et al. 2012). Similarly, combining herbicides with mechanical control helps achieve maize fields clean from weeds (Pannacci and Tie 2014). In a recent study conducted in three important maize-growing regions of Europe, combining an early post-emergence herbicide application with mechanical hoeing helped reduce herbicide use, and provided satisfactory maize grain yields and economical weed control (Vasileiadis et al. 2015). The potential economic benefits of using IWM in Africa have also been explored (Chikoye et al. 2004; Olorunmaiye 2011).

9 Where to look for further information

A good introduction to IWM is Mohler et al. (2007). A good overview of the application of IWM in maize is Jhala et al. (2014). Two useful websites are

http://www.iwss.ino
http://weedsmart.org.au

10 References

Ahmad, N. and Bano, A. (2013). Impact of allelopathic potential of maize (*Zea Mays* L.) on physiology and growth of soybean (*Glycine Max.* (L.) Merr.). *Pakistan Journal of Botany* 45(4), 1187–92.

Ali, A., Streibig, J. C., Christensen, S. and Andreasen, C. (2015). Image-based thresholds for weeds in maize fields. *Weed Research* 55, 26–33.

Alignier, A., Meiss, H., Petit, S. and Reboud, X. (2008). Variation of post-dispersal weed seed predation according to weed species, space and time. *Journal of Plant Diseases and Protection* 221–6 (supplement).

Anderson, R. L. (2003). An ecological approach to strengthen weed management in the semiarid great plains. *Advances in Agronomy* 80, 33–62.

Anderson, R. L. (2005), A multi-tactic approach to manage weed population dynamics in crop rotations. *Agronomy Journal* 97, 1579–83.

Anderson, R. L., Stymiest, C. E., Swan, B. A. and Rickertsen, J. R. (2007). Weed community response to crop rotations in western South Dakota. *Weed Technology* 21, 131–5.

Andr, J., Hejnak, V., Jursik, M. and Fendrychova, V. (2014). Effects of application terms of three soil active herbicides on herbicide efficacy and reproductive ability for weeds in maize. *Plant Soil and Environment* 60, 452–8.

Andrea Fernandez, L., Valverde, C. and Anahi Gomez, M. (2013). Isolation and characterization of atrazine-degrading Arthrobacter sp strains from Argentine agricultural soils. *Annals of Microbiology* 63, 207–14.

Andujar, D., Ribeiro, A., Fernandez-Quintanilla, C. and Dorado, J. (2013). Herbicide savings and economic benefits of several strategies to control *Sorghum halepense* in maize crops. *Crop Protection* 50, 17–23.

Andujar, D., Ruiz, D., Ribeiro, A., Fernandez-Quintanilla, C. and Dorado, J. (2011). Spatial distribution patterns of Johnsongrass (*Sorghum halepense*) in corn fields in Spain. *Weed Science* 59, 82–9.

Ascard, J., Hatcher, P., Melander, P. and Upadhyaya, M. (2007). Thermal weed control. In M. Upadhyaya and E. Blackshaw (eds), *Non-chemical Weed Management*. Wallingford, UK: CABI International.

Avedi, E. K., Ochieno, D. M. W., Ajanga, S., Wanyama, C., Wainwright, H., Elzein, A. and Beed, F. (2014). *Fusarium oxysporum* f. sp strigae strain Foxy 2 did not achieve biological control of *Striga hermonthica* parasitizing maize in Western Kenya. *Biological Control* 77, 7–14.

Ayeni, M. and Kayode, J. (2013). Allelopathic effects of extracts from maize roots and rise husk residues on the germination and growth of *Bidens pilosa L. Journal of Agricultural Science* 5(4), 22–30.

Beres, B. L., Harker, K. N., Clayton, G. W., Bremer, E., Blackshaw, R. E. and Graf, R. J. (2010). Weed-competitive ability of spring and winter cereals in the Northern Great Plains. *Weed Technology* 24, 108–16.

Bezuidenhout, S. R., Reinhardt, C. F. and Whitwell, M. I. (2012). Cover crops of oats, stooling rye and three annual ryegrass cultivars influence maize and *Cyperus esculentus* growth. *Weed Research* 52, 153–60.

Bond, W. and Grundy, A. (2001). Non-chemical wed management in organic farming systems. *Weed Research* 41, 383–405.

Chikoye, D., Schulz, S. and Ekeleme, F. (2004). Evaluation of integrated weed management practices for maize in the northern Guinea savanna of Nigeria. *Crop Protection* 23(11), 895–900.

Colbach, N., Collard, A., Guyot, S. H. M., Meziere, D. and Munier-Jolain, N. (2014). Assessing innovative sowing patterns for integrated weed management with a 3D crop: Weed competition model. *European Journal of Agronomy* 53, 74–89.

Cordill, C. and Grift, T. E. (2011). Design and testing of an intra-row mechanical weeding machine for corn. *Biosystems Engineering* 110, 247–52.

Cury, J. P., Santos, J. B., Silva, E. B., Byrro, E. C. M., Braga, R. R., Carvalho, F. P. and Valadao Silva, D. (2012). Accumulation and partitioning of dry matter and nutrients in maize cultivars in competition with weed. *Planta Daninha* 30, 287–96.

Dan, H. A., Dan, L. G. M., Barroso, A. L. L., Oliveira Jr, R. S., Alonso, D. G. and Finotti, T. R. (2011). Effect of the growth stage of *Cenchrus echinatus* on weed suppression imposed by atrazine. *Planta Daninha* 29, 179–84.

Davis, A. S., Hill, J. D., Chase, C. A., Johanns, A. M. and Liebman, M. (2012). Increasing cropping system diversity balances productivity, profitability and environmental health. *PLoS ONE* 7, e47149. doi:10.1371/journal.pone.0047149.

De Cauwer, B., Rombaut, R., Bulcke, R. and Reheul, D. (2012). Differential sensitivity of *Echinochloa muricata* and *Echinochloa crus-galli* to 4-hydroxyphenyl pyruvate dioxygenase- and acetolactate synthase-inhibiting herbicides in maize. *Weed Research* 52, 500–9.

Derksen, D. A., Anderson, R. L., Blackshaw, R. E. and Maxwell, B. (2002). Weed dynamics and management strategies for cropping systems in the Northern Great Plains. *Agronomy Journal* 94, 174–85.

Dewar, A. M. (2009). Weed control in glyphosate-tolerant maize in Europe. *Pest Management Science* 65, 1047–58.

Dovala, A. C. and Monteiro, A. (2014). Influência do azoto amoniacal na emergência de *Striga asiatica* em milho (Planalto Central de Angola) of Angola). *Revista de Ciências Agrárias* 37, 89–99.

Duke, S. O. (2012). Why have no new herbicide modes of action appeared in recent years? *Pest Management Science* 68, 505–12.

ENDURE (2008). Integrated Weed Management Case Study – Guide Number 1: Maize Cropping with Less Herbicide, European Network for the Durable Exploitation of Crop Protection Strategies, Denmark (www.edure-network.eu).

Efthimiadou, A., Froud-Williams, R. J., Eleftherohorinos, I., Karkanis, A. and Bilalis, D. J. (2012). Effects of organic and inorganic amendments on weed management in sweet maize. *International Journal of Plant Production* 6, 291–307.

Fahad, S., Hussain, S., Saud, S., Hassan, S., Muhammad, H., Shan, D., Chen, C., Wu, C., Xiong, D., Khan, S. B., Jan, A., Cui, K. and Huang, J. (2014). Consequences of narrow crop row spacing and delayed *Echinochloa colona* and *Trianthema portulacastrum* emergence for weed growth and crop yield loss in maize. *Weed Research* 54, 475–83.

FAO (2013). Food and Agriculture Organization. Available online: http://faostat3.fao.org/browse/rankings/countries_by_commodity/E (Accessed 30 September 2015).

Farooq, M., Jabran, K., Cheema, Z. A., Wahid, A. and Siddique, K. H. M. (2011). The role of allelopathy in agricultural pest management. *Pest Management Science* 67, 493–506.

Favarato, L. F., Galvao, J. C. C., Souza, J. L., Guarconi, R. C., Souza, C. and Cunha, D. N. (2014). Population density and weed infestation in organic no-tillage corn cropping system under different soil covers. *Planta Daninha* 32, 739–46.

Fenner, M. (1978). Susceptibility to shade in seedlings of colonizing and closed turf species. *New Phytologist* 81, 739–44.

Fontem, L. A. and Chikoye, D. (2012). Efficacy of herbicide formulations for weed control in maize in a humid tropical environment. *Journal of Food Agriculture & Environment* 10, 1572–74.

Galon, L., Concenco, G., Oliveira Pinto, J. J., Marques, R. F. and Andres, A. (2014). Imidazolinone-tolerant maize as a tool for weed control in flooded rice production systems. *Maydica* 59, 130–7.

Green, J. M. and Castle, L. A. (2010). Transitioning from single to multiple herbicide-resistant crops. In V. K. Nandula (ed.), *Glyphosate Resistance in Crops and Weeds*, New York: John Wiley & Sons, Inc, pp. 67–91.

Green, J. M. (2012). The benefits of herbicide-resistant crops. *Pest Management Science* 68, 1323–31.

Gul, B., Marwat, K. B., Hassan, G., Khan, A., Hashim, S. and Khan, I. A. (2009). Impact of tillage, plant population and mulches on biological yield of maize. *Pakistan Journal of Botany* 41, 2243–9.

Gutjahr, C., Soekefeld, M. and Gerhards, R. (2012). Evaluation of two patch spraying systems in winter wheat and maize. *Weed Research* 52, 510–19.

Hall, M., Swanton, C. and Anderson, G. (1992). The critical period of weed control in grain corn. *Weed Science* 40, 441–7.

Harker, K. N. (2013). Slowing weed evolution with integrated weed management. *Canadian Journal of Plant Science* 93, 759–64.

Harker, K. N. and O'Donovan, J. T. (2013). Recent weed control, weed management and integrated weed management. *Weed Technology* 27, 1–11.

Hayes, T., Haston, K., Tsui, M., Hoang, A., Haeffele, C. and Vonk, A. (2003). Atrazine-induced hermaphrodisom in American leopard frogs. *Environmental Health Perspectives* 111, 568–75.

Heap, I. M. (2015). The international survey of herbicide resistant weeds. http://www.weedscience.org. (Accessed 14 September 2015).

Hooper, A. M., Tsanuo, M. K., Chamberlain, K., Tittcomb, K., Scholes, J., Hassanali, A., Khan, Z. R. and Pickett, J. A. (2010). Isoschaftoside, a C-glycosylflavonoid from *Desmodium uncinatum* root exudate, is an allelochemical against the development of *Striga*. *Phytochemistry* 71, 904–8.

Jabran, K., Cheema, Z. A., Farooq, M. and Hussain, M. (2010a). Lower doses of pendimethalin mixed with allelopathic crop water extracts for weed management in canola (*Brassica napus*). *International Journal of Agriculture & Biology* 12, 335–40.

Jabran, K., Farooq, M., Hussain, M. and Ali, M. (2010b). Wild oat (*Avena fatua* L.) and canary grass (*Phalaris minor* Retz.) management through allelopathy. *Journal of Plant Protection Research* 50, 41–4.

Jabran, K., Mahajan, G., Sardana, V. and Chauhan, B. S. (2015). Allelopathy for weed control in agricultural systems. *Crop Protection* 72, 57–65.

Jabran, K. and Farooq, M. (2013). Implications of potential allelopathic crops in agricultural systems. In Z. A. Cheema et al. (eds), *Allelopathy: Current Trends and Future Applications*, 349–85. New York, The Netherlands: Springer. doi: 10.1007/978-3-642-30595-5_15.

Jabran, K., Ata, Z. and Farooq, M. (2007). *Maize: Cereal with Variety of Uses*. Daily Dawn, Lahore, 12 March 2007.

James, T. K., Rahman, A. and Mellsop, J. (2000). Weed competition in maize crop under different timings for post-emergence weed control. *New Zealand Plant Protection* 53, 269–72.

Jhala, A., Knezecic, S., Ganie, Z. and Singh, M. (2014). Integrated weed management in maize. In B. Chauhan and G. Mahajan (2014). *Recent Advances in Weed Management*, 177–196. New York, USA: Springer Science.

Karimmojeni, H., Mashhadi, H. R., Alizadeh, H. M., Cousens, R. D. and Mesgaran, M. B. (2010a). Interference between maize and *Xanthium strumarium* or *Datura stramonium*. *Weed Research* 50, 253–61.

Karimmojeni, H., Mashhadi, H. R., Shahbazi, S., Taab, A. and Alizadeh, H. M. (2010b). Competitive interaction between maize, *Xanthium strumarium* and *Datura stramonium* affecting some canopy characteristics. *Australian Journal of Crop Science* 4, 684–91.

Khalil, S. K., Mehmood, T., Rehman, A., Wahab, S., Khan, A. Z., Zubair, M., Mohammad, F., Khan, N. U. and Khalil, A. I. H. (2010). Utilization of allelopathy and planting geometry for weed management and dry matter production Of maize. *Pakistan Journal of Botany* 42, 791–803.

Khan, M. A., Kakar, S., Marwat, K. B. and Khan, I. A. (2013). Differential response of *Zea mays* L. in relation to weed control and different macronutrient combinations. *Sains Malaysiana* 42, 1395–1401.

Khan, M. B., Ahmad, M., Hussain, M., Jabran, K., Farooq, S. and Waqas-Ul-Haq, M. (2012). Allelopathic plant water extracts tank mixed with reduced doses of atrazine efficiently control *Trianthema portulacastrum* L. in *Zea mays* L. *Journal of Animal and Plant Sciences* 22, 339–46.

Khan, Z., Midega, C., Pittchar, J., Pickett, J. and Bruce, T. (2011). Push-pull technology: A conservation agriculture approach for integrated management of insect pests, weeds and soil health in Africa UK government's Foresight Food and Farming Futures project. *International Journal of Agricultural Sustainability* 9, 162–70.

Kiwia, A., Imo, M., Jama, B. and Okalebo, J. R. (2009). Coppicing improved fallows are profitable for maize production in striga infested soils of western Kenya. *Agroforestry Systems* 76, 455–65.

Knezevic, S., Evans, S., Blankenship, E., Van Acker, R. and Lindquist, H. (2002). Critical period of weed control: The concept and data analysis. *Weed Science* 50, 773–86.

Konstantinovic, B., Blagojevic, M., Konstantinovic, B. and Samardzic, N. (2014). Allelopathic effect of weed species *Amaranthus retroflexus* L. on maize seed germination. *Romanian Agricultural Research* 31, 315–21.

Koocheki, A., Nassiri, M., Alimoradi, L. and Ghorbani, R. (2009). Effect of cropping systems and crop rotations on weeds. *Agronomy for Sustainable Development* 29, 401–8.

Kukorelli, G., Reisinger, P. and Kazinczi, G. (2012). Results of the study of cross-resistance and effect of herbicide on crops in the production of cycloxydim-tolerant maize (*Zea mays* L). *Maydica* 57, 188–93.

Kulkarni, S. S., Dosdall, L. M. and Willenborg, C. J. (2015). The role of ground beetles (Coleoptera: Carabidae) in weed seed consumption: A review. *Weed Science* 63, 355–76.

Lague, C., Gill, J. and Peloquin, G. (2001). Thermal control in plant protection. In C. Vincent, B. Panneton and F. Fleurat-Lessard (eds), *Physical Control Methods in Plant Protection*, 35–46. Berlin, Germany: Springer.

Laini, A., Bartoli, M., Lamastra, L., Capri, E., Balderacchi, M. and Trevisan, M. (2012). Herbicide contamination and dispersion pattern in lowland springs. *Science of the Total Environment* 438, 312–18.

Laosinwattana, C., Teerarak, M. and Charoenying, P. (2012). Effects of *Aglaia odorata* granules on the seedling growth of major maize weeds and the influence of soil type on the granule residue's efficacy. *Weed Biology and Management* 12, 117–22.

Lesnik, M., Kramberger, B. and Vajs, S. (2012). The effects of drift-reducing nozzles on herbicide efficacy and maize (*Zea mays* L.) yield. *Zemdirbyste-Agriculture* 99, 371–8.

Liebman, M., Mohler, C. and Staver, C. (2007). *Ecological Management of Agricultural Weeds*. Cambridge, UK: Cambridge University Press.

Lindquist, J. L., Evans, S. P., Shapiro, C. A. and Knezevlc, S. Z. (2010). Effect of nitrogen addition and weed interference on soil nitrogen and corn nitrogen nutrition. *Weed Technology* 24, 50–8.

Lundmark, C. (2007). Genetically modified maize. *BioScience* 57, 996.

Mahmood, A., Cheema, Z. A., Mushtaq, M. N. and Farooq, M. (2013). Maize-sorghum intercropping systems for purple nutsedge management. *Archives of Agronomy and Soil Science* 59, 1279–88.

Marin, C. and Weiner, J. (2014). Effects of density and sowing pattern on weed suppression and grain yield in three varieties of maize under high weed pressure. *Weed Research* 54, 467–74.

Meissle, M., Romeis, J. and Bigler, F. (2011). Bt maize and integrated pest management-a European perspective. *Pest Management Science* 67, 1049–58.

Midega, C. A. O., Pittchar, J., Salifu, D., Pickett, J. A. and Khan, Z. R. (2013). Effects of mulching, N-fertilization and intercropping with *Desmodium uncinatum* on *Striga hermonthica* infestation in maize. *Crop Protection* 44, 44–9.

Mohler, C. L. (2001a). Weed life history: Identifying vulnerabilities. In M. Liebman, C. L. Mohler and C. P. Staver (eds), *Ecological Management of Agricultural Weeds*, 40–98. Cambridge, UK: Cambridge University Press, 532p.

Mohler, C. L. (2001b). Enhancing the competitive ability of crops. In M. Liebman, C. L. Mohler and C. P. Staver (eds), *Ecological Management of Agricultural Weeds*, 269–321. Cambridge, UK: Cambridge University Press, 532p.

Ngwira, A. R., Aune, J. B. and Thierfelder, C. (2014). On-farm evaluation of the effects of the principles and components of conservation agriculture on maize yield and weed biomass in Malawi. *Experimental Agriculture* 50, 591–610.

Norsworthy, J. K. and Frederick, J. R. (2005). Integrated weed management strategies for maize production on the southeastern coastal plains of North America. *Crop Protection* 24(2), 119–26.

Norsworthy, J. K., Ward, S. M., Shaw, D. R., Llewellyn, R. S., Nichols, R. L., Webster, T. M., Bradley, K. W., Frisvold, G., Powles, S. B., Burgos, N. R., Witt, W. W. and Barrett, M. (2012). Reducing the risks of herbicide resistance: Best management practices and recommendations. *Weed Science Special Issue*, 31–62.

Oerke, E. C. (2006). Crop losses to pests. *Journal of Agricultural Science* 144, 31–43.

Olorunmaiye, P. M. (2010). Weed control potential of five legume cover crops in maize/cassava intercrop in a Southern Guinea savanna ecosystem of Nigeria. *Australian Journal of Crop Science* 4, 324–9.

Olorunmaiye, P. M. (2011), Economic viability of integrated weed management in maize/cassava intercrop in the Guinea savanna ecology of Nigeria. *Agriculture and Biology Journal of North America* 2(3), 522–8.

Page, E., Cerrudo, D., Westra, P., Louis, M., Smith, K., Foresman, C., Wright, H. and Swanton, C. (2012). Why early season weed control is important in maize. *Weed Science*, 60, 423–30.

Page, E. R., Tollenaar, M., Lee, E. A., Lukens, L. and Swanton, C. J. (2009). Does the shade avoidance response contribute to the critical period for weed control in maize (*Zea mays*)? *Weed Research* 49, 563–71.

Pannacci, E. and Tei, F. (2014). Effects of mechanical and chemical methods on weed control, weed seed rain and crop yield in maize, sunflower and soyabean. *Crop Protection* 64, 51–9.

Pena-Asin, J., Costar, A. and Alvarez, A. (2013). Effect of weeding management on the performance of local maize populations. *Spanish Journal of Agricultural Research* 11, 1078–84.

Powles, S. B. and Yu, Q. (2010). Evolution in action: Plant resistance to herbicides. *Annual Review of Plant Physiology* 61, 317–47.

Rahman, A., James, T. and Trolove, M. (2014). Characteristics and control of dicamba-resistant common lambsquarters (*Chenopodium album*). *Weed Biology and Management* 14, 88–98.

Ramos-Zapata, J. A., Marrufo-Zapata, D., Guadarrama, P., Carrillo-Sanchez, L., Hernandez-Cuevas, L. and Caamal-Maldonado, A. (2012). Impact of weed control on arbuscular mycorrhizal fungi in a tropical agroecosystem: A long-term experiment. *Mycorrhiza* 22, 653–61.

Rao, V. S. (2000). *Principles of Weed Science*. London, UK: Taylor and Francis.

Romaneckas, K., Adamaviciene, A., Pilipavicius, V., Sarauskis, E., Avizienyte, D. and Buragiene, S. (2012). Interaction of maize and living mulch. Crop weediness and productivity. *Zemdirbyste-Agriculture* 99, 23–30.

Rusu, T., Gus, P., Bogdan, I., Moraru, P. I., Pop, A. I., Sopterean, M. L. and Pop, L. I. (2010). Influence of infestation with *Echinochloa crus-galli* species on crop production in corn. *Journal of Food Agriculture & Environment* 8, 760–63.

Saeed, M., Marwat, K. B., Hassan, G., Khan, A. and Khan, I. A. (2010). Interference of horse purslane (*Trianthema portulacastrum* L.) with maize (*Zea mays* L.) at different densities. *Pakistan Journal of Botany* 42, 173–9.

Safdar, M. E., Tanveer, A., Khaliq, A. and Riaz, M. A. (2015). Yield losses in maize (*Zea mays*) infested with parthenium weed (*Parthenium hysterophorus* L.). *Crop Protection* 70, 77–82.

Simic, M., Dolijanovic, Z., Maletic, R., Stefanovic, L. and Filipovic, M. (2012). Weed suppression and crop productivity by different arrangement patterns of maize. *Plant Soil and Environment* 58, 148–53.

Travlos, I. S., Economou, G. and Kanatas, P. J. (2011). Corn and barnyardgrass competition as influenced by relative time of weed emergence and corn hybrid. *Agronomy Journal* 103, 1–6.

Ulloa, S. M., Datta, A., Bruening, C., Neilson, B., Miller, J., Gogos, G. and Knezevic, S. Z. (2011). Maize response to broadcast flaming at different growth stages: Effects on growth, yield and yield components. *European Journal of Agronomy* 34, 10–19.

Vasileiadis, V. P., Otto, S., van Dijk, W., Urek, G., Leskovsek, R., Verschwele, A., Furlan, L. and Sattin, M. (2015). On-farm evaluation of integrated weed management tools for maize production in three different agro-environments in Europe: Agronomic efficacy, herbicide use reduction, and economic sustainability. *European Journal of Agronomy* 63, 71–8.

Vidotto, F., Tesio, F. and Ferrero, A. (2013). Allelopathic effects of *Ambrosia artemisiifolia* L. in the invasive process. *Crop Protection* 54, 161–7.

Weiner, J., Andersen, S. B., Wibke, K.-M., Griepentrog, H. W. and Olsen J. M. (2010). Evolutionary agroecology: The potential for cooperative, high density, weed-suppressing cereals. *Evolutionary Applications* 3, 473–9.

Yamasaki, M., Tenaillon, M. I., Bi, I. V., Schroeder, S. G., Sanchez-Villeda, H., Doebley, J. F., Gaut, B. S. and McMullen, M. D. (2005). A large-scale screen for artificial selection in maize identifies candidate agronomic loci for domestication and crop improvement. *The Plant Cell* 17, 2859–72.

Ying, G., Kookana, R. and Mallavarpi, M. (2005). Release and behavior of trazine residues in stabalised contaminated soils. *Environmental Pollution* 134, 71–7.

Zalai, M., Dorner, Z. and Keresztes, Z. (2014). Seasonal weed structure of maize in the light of farming systems. *Applied Ecology and Environmental Research* 12, 765–76.

Zhang, J., Zheng, L., Jaeck, O., Yan, D., Zhang, Z., Gerhards, R. and Ni, H. (2013). Efficacy of four post-emergence herbicides applied at reduced doses on weeds in summer maize (*Zea mays* L.) fields in North China Plain. *Crop Protection* 52, 26–32.

Chapter 19

Weed management of maize grown under temperate conditions: the case of Europe and the United States

Vasileios P. Vasileiadis and Maurizio Sattin, National Research Council (CNR), Institute of Agro-Environmental and Forest Biology, Italy; and Per Kudsk, Aarhus University, Department of Agroecology, Denmark

1 Introduction

Maize (*Zea mays* L.) is one of the most important crops in the world covering an area of approximately 183 million hectares for grain maize production globally and reaching a production of around 1000 Mt in 2014 (FAOSTAT, 2016). The United States was the number one maize-producing country in 2014 with a 35% share of the global production, followed by China (21%), Brazil (7.8%), Argentina (3.2%) and Ukraine (2.8%), with 7.5% of the global production of grain maize within the 28-member states of the European Union (EU; EUROSTAT, 2016). In Europe, the area with green maize grown for silage or energy is almost the same size as the area with grain maize (EUROSTAT, 2016).

Maize is very sensitive to early weed competition due to a slow early growth rate and wide row spacing (Cerrudo et al., 2012) and is often infested by weeds that are highly competitive for water, nutrients and light causing significant yield losses (Sattin et al., 1992). The extent of yield losses caused by weed competition depends on several factors such as weed density, weed species growth rate, weed size, weed canopy shape, uniformity of weed distribution, weed emergence date relative to the crop, crop

http://dx.doi.org/10.19103/AS.2016.0002.24

competitiveness and density, soil type and crop fertilization (Eleftherohorinos, 2014). When weeds and crop co-exist in the early growth stages of the crop, there is no direct resource competition as plants are small and widely spaced with little requirements for nutrients and water. However, experiments demonstrated that plant competition is triggered initially by the red to far-red light ratio (R:FR) reflected by neighbouring plants well before direct competition starts (Sattin et al., 1994). Once a plant is physiologically triggered into a shade avoidance response, these plants do so at a physiological cost, which may constrain plant development and possibly reduce reproductive fitness (Liu et al., 2009). At some point, depending on weed species, direct interference between the maize crop and weeds begins for both above and below ground plant organs, resulting in resource limitations (Aldrich and Kremer, 1997; Rajcan and Swanton, 2001). The critical period for weed control (CPWC) is defined as the period in the crop growth cycle during which weeds must be controlled to prevent crop yield losses (Zimdahl, 1988). Several studies have determined the effects of weed competition on yield loss and the critical period during which maize is mostly affected by competition with weeds (Zanin et al., 1994; Norsworthy and Oliveira, 2004; Otto et al., 2009; Cerrudo et al., 2012; Page et al., 2012; Knezevic and Datta, 2015). For instance, Page et al. (2012) reported that grain yield loss increased as weed control was delayed reaching a maximum of 65%, and that the beginning of the CPWC occurred, on average, between the third- and fifth-leaf stages and ended at the ten-leaf stage.

For more than 50 years, herbicides have been the primary tool for weed control in all major crops, greatly reducing yield losses and stabilizing potential weed infestations at acceptable levels in most situations (Chikowo et al., 2009). Besides their high efficacy and reliability, the rapid adoption of herbicides has also been driven by their relative low cost, compared to non-chemical weed control practices (Fernandez-Cornejo et al., 2014). The introduction and the rapid diffusion of genetically modified herbicide-tolerant (GMHT) crops, with maize varieties covering the largest acreage, have led to a complete revamping of weed management practices (Livingston et al., 2015). Together with the environmental concerns associated with the use of herbicides, which has become a key societal issue particularly in Europe, other concerns of agronomic nature have also arisen in the last decades. The repeated and intensive use of herbicides can rapidly cause shifts in the weed population towards more tolerant, difficult-to-control weeds and also result in the evolution of herbicide-resistant (HR) weed populations, especially when herbicides with the same mode of action are being used or weed control techniques are not diversified (Vencill et al., 2012). For example, in countries where GMHT maize that is tolerant to glyphosate has been adopted, allowing for multiple applications of glyphosate in the same crop year, the overreliance on this technology has led to many cases of glyphosate-resistant (GR) weeds. The Weed Science Society of America (WSSA) listed 33 GR weed species of which many were found in glyphosate-tolerant crops, including maize (Heap, 2016).

Maize covers the second largest production area globally after wheat (FAOSTAT, 2016) and more than 85% of the total maize area in the United States, the top maize-producing country, was cultivated with GMHT maize in 2013 (Livingston et al., 2015). This emphasizes the urgent need for more integrated and diverse approaches to weed management, with a more sustainable herbicide use and an overall reduction in the reliance on herbicides to reduce the selection pressure on weeds and minimize any adverse impacts on the environment and human health.

2 Maize cropping systems and weed flora in the United States and Europe

Maize in the United States and Europe is cultivated as grain for food, feed and processing, as well as green maize for silage or ethanol and biogas production, thus having an impact on key sectors such as livestock and energy production (USDA, 2011; Vasileiadis et al., 2011). In the United States, maize is grown in most States, but production is concentrated in the Midwest, with Iowa and Illinois being the two top maize-producing States accounting for about one-third of the production in the country (USDA, 2011). In 2014, the total maize area harvested was 33.6 million hectares of grain maize and 2.6 million hectares of silage maize, with a total production of 361 million tonnes of grain (FAOSTAT, 2016).

Weed seedbank emergence studies for 22 sites across the maize-growing area, from Ohio in the east to Colorado in the west and from Minnesota in the north to Missouri in the south, indicated 15 important weed species – *Abutilon theophrasti* Medik., *Amaranthus* spp., *Ambrosia artemisiifolia* L., *Chenopodium album* L., *Helianthus annuus* L., *Kochia scoparia* (L.) Schrad., *Panicum miliaceum* L., *Polygonum aviculare* L., *Polygonum convolvulus* L., *Polygonum pensylvanicum* L., *Portulaca oleracea* L., *Setaria faberi* Herrm., *Setaria pumila* (Poir.) Roemer and J. A. Schultes, *Setaria viridis* (L.) Beauv. and *Solanum physalifolium* Rusby (Forcella et al., 1997). Another survey conducted across Indiana indicated that *Ambrosia trifida* L., *Cirsium arvense* (L.) Scop., *C. album*, *Xanthium strumarium* L. and *Abutilon theophrasti* were the most problematic summer-annual and perennial weeds, whereas *Stellaria media* (L.) Vill., *Conyza canadensis* (L.) Cronq., *Taraxacum officinale* G. H. Weber ex Wiggers and *Lamium amplexicaule* L. were the most problematic winter-annual weeds (Gibson et al., 2005). Since their introduction in 1996, US farmers have progressively adopted GMHT maize and mainly glyphosate-tolerant varieties. GMHT maize was grown on 89% of the total maize area in 2015 (USDA, 2016). Types of maize-based cropping systems (MBCS) in the United States are mainly continuous GMHT maize (i.e. monoculture) or GMHT maize in rotation with GMHT soya bean (Benbrook, 2012). The introduction of GMHT crops in the United States contributed to a slowdown of the evolution of weeds resistant to the acetolactate synthase (ALS)- and protoporphyrinogen oxidase (PPO)-inhibiting herbicides (Norsworthy et al., 2012). In the last decade, however, the repeated and extensive use of glyphosate has resulted in a steadily increasing number of cases of GR weeds, including *Amaranthus palmeri* S. Wats., *Amaranthus tuberculatus* (Moq.) Sauer, *A. artemisiifolia*, *A. trifida*, *Conyza bonariensis* (L.) Cronq., *C. canadensis*, *Echinochloa colona* (L.) Link, *K. scoparia* and *Lolium perenne* ssp. *multiflorum* (Lam.) Husnot (Heap, 2016).

It is in southern Europe that grain maize production is concentrated, while green maize is typically grown in northern Europe. The cropping area within the 28-member states of the EU touched 9.6 million hectares in 2014 for grain maize and 6.1 million hectares for green maize. Green maize for energy is mainly cultivated in Germany. The total production was 76.8 million tonnes of grain and 245.8 million tonnes of green maize. The largest maize producers are France, Germany, Romania, Hungary, Poland and Italy, where maize is grown on more than 1 million hectares, with France having the highest grain maize production (18.5 million tonnes) and Germany the highest green maize production (99.2 million tonnes) (EUROSTAT, 2016). In a recent expert-based survey in four European regions, the differences that characterize MBCS in Europe were reported (Vasileiadis et al., 2011). The authors classified MBCS according to the type of maize production (silage or grain), cropping sequence (crop rotation or continuous maize) and irrigation (irrigated or not), resulting in eight MBCS combinations. According to this survey, different MBCS predominated in the four EU regions reflecting the diversity in climatic and soil conditions across Europe. Moving from northern to southern Europe, the types and number of MBCS increased, with differences in the choice of rotational crops. In northern Europe (i.e. The Netherlands and Denmark), silage maize is grown either continuously or rotated and always without irrigation. Rotational crops in the Netherlands include grass, cereals, potato (*Solanum tuberosum* L.) and sugar beet (*Beta vulgaris* L.), while in Denmark, it is mostly cereals with undersown grass or a grass and clover (*Trifolium* spp.) mixture. In the central-eastern part of Europe, represented in the survey by two counties in Hungary, grain maize is mainly cultivated continuously without irrigation (Tolna region), while in Békés, 85% of the total maize area is grain maize grown in rotation with winter wheat (*Triticum aestivum* L.), sunflower (*Helianthus annuus* L.) and oilseed rape (*Brassica napus* L.) without irrigation. In south-western Europe (Ebro Valley, Spain), three cropping systems are found with grain and silage maize in rotation and continuous grain maize all with irrigation. The rotated maize systems cover 83% of the total maize acreage, with winter wheat being the most common rotational crop, while the remaining 17% is cultivated with continuous grain maize. In the Po Valley (Italy) in southern Europe, five MBCS were identified with grain and silage maize in rotation and irrigated covering about 50% of the total maize area, and continuous grain and silage maize with irrigation covering another 30% of the area. The remaining 20% is cultivated with grain maize in rotation and without irrigation. Winter wheat or barley (*Hordeum vulgare* L.) is the typical crop rotated with maize in the Po Valley; however, rotations with soya bean (*Glycine max* (L.) Merr.) or sugar beet, grass or alfalfa do also exist in this region.

The diversity in climatic conditions and soil characteristics across Europe, together with different cultivation practices, influences the weed flora in the MBCS. A survey of weed species distribution and incidence in maize production in 11 European countries was recently reported by Jensen et al. (2011). In this survey, a total of 203 different weed species were reported, of which 61 annual and perennial weed species were reported as being 'very common' in at least one country. *Chenopodium album* was the only weed species that was assessed as being 'very common' in all 11 countries. Similarly, another expert-based weed survey conducted by Meissle et al. (2010) covering 11 European maize-growing regions identified *C. album* as the most important dicotyledonous species, and *Echinochloa crus-galli* (L.) Beauv. and *S. viridis* as the most important monocotyledonous weeds. The most important weeds in European maize production, as identified by Meissle et al. (2010) according to their significance and occurrence in 11 European regions, can be seen in Fig. 1.

3 Weed management in maize cropping systems in the United States and Europe

Maize in the United States and Europe is either grown continuously or in rotation with other crops, and weed management is mainly herbicide based, with 98% of the total maize crop area in the United States (Fernandez-Cornejo et al., 2014) and more than 90% in 11 European regions being treated with herbicides at least once in a season (Meissle et al., 2010). In the United States, the rapid adoption and use of GMHT maize that is tolerant to glyphosate, and its rotation with GMHT soya bean, also tolerant to glyphosate, allowed

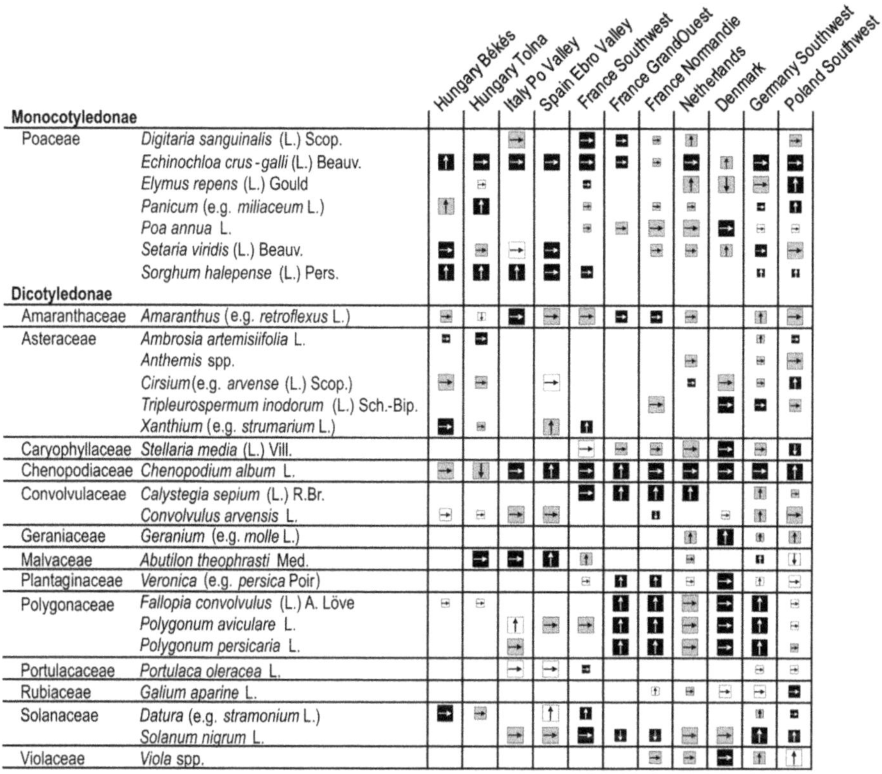

Figure 1 Most important weeds in European maize production. Significance is represented by symbol colour: black = high, grey = medium, white =low. Occurrence is represented by symbol size: large = widespread and regularly, medium = widespread and occasionally, small = regionally and rare. The 5-year population development is represented by arrows: up = increasing, horizontal = stable, down = decreasing. Reprinted from Meissle et al. (2010).

farmers to rely on just one herbicide to achieve effective weed management, improved yield and profitability for more than a decade. However, in recent years, several key weeds have evolved resistance, and using glyphosate alone has proved unsustainable (Green and Owen, 2011). Nowadays, growers need to rethink their weed management practices and use glyphosate with cultural, mechanical and other herbicide options in integrated systems to reduce weed shifts to difficult-to-control and HR weeds (Green, 2012).

Reduction of pesticide use, as well as the risks associated with pesticides, is a key issue on the EU's agricultural political agenda (Vasileiadis et al., 2011). Directive 2009/128/EC on the sustainable use of pesticides aims to reduce the reliance on, as well as the risks and adverse impacts of, pesticides on human health and the environment. A key instrument to reach this goal is to promote the implementation of integrated pest management (IPM), which was made to adopt compulsorily by all professional users of pesticides since 1 January 2014 (Directive 2009/128/EC; European Parliament, 2009). Integrated weed management (IWM) is an important component of IPM that aims at preserving crop yield and grower's profit while minimizing the impact on the environment and human health through the concerted use of preventive tactics, scientific knowledge, management skills, monitoring procedures and efficient use of weed control practices (Buhler, 2002). IWM has the potential to suppress weed densities to manageable levels, to reduce the environmental impact of individual weed management practices, to increase cropping system sustainability and to reduce selection pressure for weed resistance to herbicides (Harker and O'Donovan, 2013). An IWM programme is developed by combining preventive, cultural, mechanical, biological and chemical control methods making use of available decision-aid methods like decision support systems (DSS) and/or predictive models of weed emergence, which are discussed hereafter. Like IPM, IWM is a dynamic approach that is based on general principles that need to be adapted to local conditions (Barzman et al., 2015)

3.1 Preventive and cultural methods

Preventing the introduction of a weed into an area or preventing its spread within a field is a vital part of an IWM programme (Vencill et al., 2012). Methods include using clean crop seeds not contaminated with weed seeds, cleaning field machinery before moving into a new field, managing weeds at field margins and preventing the spread of weed propagules by transport of livestock, manures or compost, or through irrigation or drainage waters (Walker, 1995). Effective scouting to check efficacy of treatments and keeping an inventory of weed species, location, density and size is another preventive method and a vital component of any successful weed management programme, including those aiming at mitigating or managing HR weeds (Norsworthy et al., 2012).

Crop rotation with the diverse practices and tools that characterize weed management in different crops is the primary IWM tool for disrupting the growing cycle of difficult-to-control or HR weeds. Cropping system diversification through the use of longer rotations of dissimilar crop species, for example, alternating winter and summer crops, can increase the range of stress and mortality factors that regulate weed population dynamics, maintaining effective weed control while reducing the use of herbicides (Liebman et al., 2014). Within a rotation, the choice of crops determines both the current and the potential future weed problems that a grower will face (Bond and Grundy, 2001). For herbicide resistance management, crop rotation increases diversity within cropping systems and provides more opportunities for rotating herbicides with different modes of action (Vencill et al., 2012). When a severe weed infestation occurs, it could be reduced by growing winter-annual and summer-annual crops in rotation, permitting weed control with tillage and/or herbicide treatments in between crops, or rotating dicotyledonous and monocotyledonous crops to permit the use of a wider range of herbicides (Unger et al., 1999). Research on MBCS has proven the positive impact of crop rotation on reducing the weed seedbank, as well as on weed population dynamics, in comparison with continuous maize systems. For example, Schreiber (1992) reported that soya bean–maize or soya bean–wheat–maize rotation resulted in a reduction of *S. faberi* seeds in the soil, compared to continuous maize, regardless of tillage system or herbicide applications. Bàrberi et al. (1997) determined significant differences in total weed density and weed species composition in maize, soya bean, sugar beet and sunflower rotation with wheat under different input levels in a 4-year experiment. Doucet et al. (1999) reported that crop rotation had a greater effect on weed density when combined with herbicide application, especially for wheat and maize, which had higher weed densities in monocultures than in rotations. Buhler et al. (2001) reported that seed densities of *Amaranthus rudis* Sauer declined each year in a maize–soya bean–maize rotation, whereas *Setaria* spp. declined by almost 90% in the first year of maize with no further changes in the following years of soya bean and maize production. Liebman et al. (2014) reported 71% lower seedling emergence of *A. theophrasti* and 83% greater seed mortality in a 3-year maize–soya bean–triticale (*Triticosecale* Wittmack) + red clover (*Trifolium pratense* L.) and a 4-year maize–soya bean–triticale + lucerne (*Medicago sativa* L.)–lucerne systems than in a 2-year maize–soya bean rotation.

Another management practice that has received increased attention by researchers is weed suppression by inclusion of cover crops in the rotation. The inclusion of cover crops (legume or grasses) in the rotation improves soil fertility and prevents erosion, whereas the decomposing cover crop residues may release allelochemicals that can prevent or slowdown germination of weed seeds and weed growth (Liebman and Davis, 2000). Grundy et al. (1999) reported that allelopathic properties of cover crops may also reduce weed development, but that weed suppression was primarily caused by competition for growth factors. Research has mostly focused on cover crops in simplified systems based on simple rotations, no-tillage and GMHT crops (Melander et al., 2013). Recently, the integration of cover crops with herbicide mixtures that incorporate multiple sites of action in GMHT maize has been proposed to mitigate the selection of GR *A. palmeri* biotypes (Wiggins et al., 2015). In an organic maize system, Gieske et al. (2015) observed that the inclusion of fall-seeded radish as a cover crop did not affect maize yield and reduced the densities of *Thlaspi arvense* L. from 9 to <1 plant m^{-2} and of *C. canadensis* from 6 to 2 plants m^{-2}, as well as the cover of summer-annual weeds from 4 to 0% preventing these weeds from setting seed. Among the few studies conducted in maize in Europe, Moonen and Bàrberi (2004) observed a 25% reduction in the total number of seeds in the weed seedbank after 7 years of a rye cover crop in a conventional tillage maize system when compared to the non-cover-cropped system, whereas in the no-tillage maize system, cover cropping with subterranean clover (*Trifolium subterraneum* L.) resulted in an average reduction of 22% of the weed seedbank.

Enhancing crop competitiveness is another cultural control method that can be achieved by providing the best possible environment for crop growth combined with tactics that will reduce the density and vigour of weeds (Buhler, 2002), especially at early growth stages. Practices that will directly enhance crop competitive ability against weeds include competitive crop cultivars, appropriate sowing date, narrow row spacing, increased plant density and fertilizer management (Buhler and Gunsolus, 1996; Bàrberi, 2002). Competitive crop cultivars possess traits conferring a higher competitive ability against weeds (Bàrberi, 2002). Usually, these traits are related to rapid germination and root development, rapid early vegetative growth and vigour, rapid canopy closure and high leaf area index, profuse tillering or branching, increased leaf duration, greater plant height and allelopathy (Pester et al., 1999).

The modification of crop sowing date, density and sowing pattern may also reduce weed emergence and increase crop competitive ability (Mohler, 1996). Late sowing allows for the control of early germinating weeds by tillage or pre-sowing herbicide treatments, while early sowing increases the importance of late winter- and early spring-emerging weeds (Otto et al., 2009). Narrower crop row spacing and increased crop plant density have also been proven to promote crop competitiveness over weeds by allowing the crop to form a continuous and dense canopy more quickly and intercept more light relative to the associated weeds (Teasdale, 1995; Murphy et al., 1996; Norsworthy and Oliveira, 2004; Vencill et al., 2012). In this regard, Teasdale (1995) reported that growing maize in narrower row spacing, 38-cm-wide compared to 76-cm-wide rows, with increased density resulted in more effective weed control and reduced herbicide requirements. Similarly, Murphy et al. (1996) demonstrated that narrower row spacing and high maize density significantly reduces biomass of late emerging weeds.

Fertilizer management can also alter the competitive balance between crops and weeds (Bàrberi, 2002). The greatest competition among plants is usually for nitrogen (N), which is the major nutrient input that farmers use in order to increase crop yield (Raun and Johnson, 1999). Weed interference may be reduced by management strategies that maximize nutrient uptake by crops and minimize nutrient availability to weeds (DiTomaso, 1995). The application timing and method of placement of N fertilizer are two major factors influencing crop–weed competition (Blackshaw et al., 2004). Appropriate timing of N mineral fertilization has been proposed in integrated cropping systems as a means to unbalance crop–weed competition for nutrient to the benefit of the crop (Bàrberi, 2002). Banded application of N over the crop row or placement at an appropriate soil depth might also allow the timely supply of N to crops while delaying N uptake from weeds (DiTomaso, 1995). Scharf et al. (2002) suggested that delaying N application until late spring may not affect maize grain yield when compared to pre-emergence application. In contrast, Johnson et al. (2007) reported that N accumulation of *A. trifida* was higher when N was applied late in the season in maize, compared to the early season application, whereas maize plants treated with N at planting accumulated up to 27% more dry weight and 26% more leaf area. Nevertheless, Harbur and Owen (2004) stated that N management may alter weed competitiveness and weed seed production in maize, but fertilization recommendations may vary with the weed species that are present.

3.2 Tillage and mechanical weed control

The effect of primary tillage on summer-annual weed population dynamics is complex and is affected by several factors, with weed seed distribution in the soil being the most important (Buhler et al., 1997). Inversion tillage through mouldboard ploughing reduces weed populations by burying a large proportion of freshly shed seeds deep in the soil profile, whereas chisel ploughing retains them nearer the surface facilitating germination and emergence, thus increasing the need for weed control (Ball, 1992; Vasileiadis et al., 2007). In general, reduced and especially no-tillage systems are more dependent on herbicides and more frequently infested by HR weeds (Vencill et al., 2012). Reduced tillage systems tend to increase densities of annual weed species due to increased seed accumulation at the soil surface and the growth potential of perennial weeds as the subterranean plant parts are not disturbed and herbicides applied to control annual weeds are not effective on established perennial plants (Buhler et al., 1994; Bàrberi et al., 1998). Reduced tillage also influences the timing of weed emergence, compared to inversion tillage (Bullied et al., 2003). No-tillage systems, in contrast to reduced tillage systems, increase the growth potential of certain weed species, such as annual grasses and Asteraceae that have been reported to emerge rapidly under no-tillage conditions (Ball and Miller, 1990; Oryokot et al., 1997). Previous studies have also documented weed population shifts when changing from conventional- to conservation-tillage systems. Buhler and Daniel (1988) reported that *S. faberi* density was more than eight times greater in no-tillage than in conventional tillage, in contrast to *A. theophrasti* density, which was about six times greater in conventional tillage than in no-tillage. Populations of several annual grasses, including *S. faberi* (Buhler and Mester, 1991), *S. viridis* and *S. glauca* (Forcella and Lindstrom, 1988), were found to increase faster over years in reduced tillage

Figure 2 Inter-row cultivation (hoeing) in maize in Italy (Copyright: V. P. Vasileiadis).

systems than in conventional tillage systems. Tuesca et al. (2001) also determined that in the absence of tillage, annual grass populations increased in maize–soya bean rotation.

In the United States, Argentina and Brazil, the adoption of GMHT maize allowed producers to adopt conservation-tillage production, including no-tillage (Duke and Powles, 2009; Givens et al., 2009; Vencill et al., 2012). Because glyphosate is a broad-spectrum herbicide that controls most weeds present in maize fields it was predominantly used alone which led to a strong selection pressure on weeds in GMHT maize and the evolution of glyphosate resistance (Duke and Powles, 2009). Therefore, it is even more important for farmers growing GMHT maize to adopt integrated management practices so that conservation tillage will be a sustainable production practice (Vencill et al., 2012).

Tillage for seedbed preparation can also result in the reduction of annual weed populations, especially when a false seedbed technique is adopted, that is, weed seed germination is triggered prior to final seedbed preparation (Buhler and Gunsolus, 1996; Bond and Grundy, 2001). On the other hand, inter-row cultivation (Fig. 2) after crop establishment is effective in inhibiting perennial species development and removing annual weeds; thus, reducing seed production and influencing the seedbank (Ball and Miller, 1990; Buhler, 2002; Melander et al., 2005). Ground-driven and rotating weeding devices, such as 'spiders' (curved teeth), finger weeders mounted on inter-row cultivators and disc hillers can be used for both inter-row and intra-row weed control and can operate successfully in maize cropping (van der Weide et al., 2008; Melander et al., 2013).

3.3 Biological control

Biological control of weeds is defined as the use of an agent, that is, arthropods (insects and mites), plant pathogens (fungi, bacteria, viruses and nematodes), fish, birds and other animals, a complex of agents or biological processes to bring about weed suppression (WSSA, 2016). Classical biological control, which is the control of invasive weeds with natural enemies originating from the native range of the weed, has proven to be a viable strategy for managing weeds in areas subjected to low-intensity management, such as rangelands, forests, preserved natural areas and some waterways (Vencill et al., 2012). No commercial biological control products are available for weed control in maize yet, and this is mainly due to the substantial investment that is needed to develop and realize such projects. Nevertheless, there is good potential for biological control agents against some important weed species found in maize crop both in the United States and in Europe. For example, Ghorbani et al. (2000) determined that a spore concentration of 10^7 spores ml^{-2} of *Alternaria alternata* (Fr.) Keissler, a common saprophytic fungus, in a rapeseed oil emulsion and given a 24-h dew period caused 100% mortality of *Amaranthus retroflexus* L. plants when applied at the four-leaf stage, with the highest levels of plant mortality at post-inoculum temperatures between 20 and 30°C. The authors concluded that *A. alternata* has a great potential as a mycoherbicide under specific environmental conditions. In a greenhouse study, Sciegienka et al. (2011) investigated the single and combined effects of two biological control agents, the stem-mining weevil *Hadroplontus litura* and the pathogen *Pseudomonas syringae* pv. *tagetis* and a reduced or full application of glyphosate on the growth of *C. arvense*. They reported that individual control methods impacted the growth of *C. arvense*, with *H. litura* decreasing root biomass by 18%, *P. syringae* by 33% (late application) and 69% (early application), whereas the reduced and full rates of glyphosate decreased root biomass by 45 and 86%, respectively. No significant interactions were found between *H. litura*, *P. syringae* and glyphosate on *C. arvense* root biomass, indicating that the combined effects of control methods were additive rather than antagonistic or synergistic.

3.4 Chemical control

Herbicides have been very efficient in controlling annual and perennial weed species in maize cropping systems worldwide. Herbicides reduce weed densities, even surviving weed plants are often less vigorous, and hence reduce the number of weed seeds that are produced and enter the seedbank (Bàrberi et al., 1998). They can be applied pre-sowing, pre-emergence or post-emergence to the maize crop depending on herbicide, soil and climatic conditions, weed composition and infestation levels, and farmers' perception of chemical weed management. Pre-sowing herbicides are usually soil-incorporated in order to be effective, and they are applied from a few days to several months before crop sowing, depending on their soil persistence and tolerance of the crop (Vencill et al., 2012). Pre-emergence herbicides are applied at sowing or before maize emergence and as the pre-sowing herbicides include active ingredients that provide residual activity keeping the field clean for maize to emerge and establish during the first growth stages. Sometimes pre-emergence herbicides can be applied after maize emergence as an early post-emergence at 2–3 maize leaf stage to extend the residual control and cover the CPWC. The performance of pre-emergence herbicides is influenced by precipitation around application in order to incorporate and activate them. Post-emergence herbicides, applied after maize emergence, include those that are foliar-applied, and they are referred to either as contact herbicides, if the herbicide enters the plants but stay close to the site of uptake, or as systemic, if the herbicide enters the plant and moves within it to the site of herbicide action (Vencill et al., 2012). Pre-emergence and post-emergence herbicides are mainly applied as broadcast, but sometimes also as a band, in the maize row (Fig. 3), for example, when combined with inter-row cultivation (Vasileiadis et al., 2015).

A major problem related to chemical weed control is the evolution of HR populations driven by the selection pressure exerted through intense herbicide use and favoured by

Figure 3 Pre-emergence herbicide in band application at maize sowing in Italy (Copyright: V. P. Vasileiadis).

over-simplification and intensification of cropping systems. Maize was first affected globally by widespread weed biotypes resistant to PSII inhibitors (Heap, 2016). However, nowadays they do not represent an agronomic problem due to the availability of herbicides with alternative modes of action and to the lower fitness of the resistant biotypes, compared to the susceptible ones. As pointed out previously, the main HR issue in maize crops is now related to glyphosate resistance detected in many weeds infesting GMHT maize varieties, especially in no-tillage conditions (Heap, 2016; Vencill, 2012). In several locations multiple resistances to glyphosate and to herbicides with different modes of actions has evolved (e.g. Bell et al., 2013). In Europe, where GMHT varieties are not cultivated, a few populations of *E. crus-galli* and *Sorghum halepense* resistant to ALS inhibitors have been recently reported in maize crops in southern Europe (Heap, 2016; Panozzo et al., 2015). Mitigating the evolution of herbicide resistance depends on the implementation of IWM (Barzman et al., 2015), that is, reducing selection through diversification of weed control techniques, minimizing the spread of resistance genes and genotypes via pollen or propagule dispersal and eliminating additions of weed seed to the soil seedbank (Norsworthy et al., 2012).

4 Decision-support tools and bottlenecks hindering IWM implementation

4.1 Decision-aid methods

DSS can assist farmers in making decisions about 'if' and 'how' to treat, and field surveys can provide the information needed to run the DSS (Berti et al., 2003). Most of the existing DSS for weed control in maize and other crops are aiming at optimizing herbicide use (Swinton et al., 2002; Bennett et al., 2003; Berti et al., 2003; Sønderskov et al., 2014, 2015); nevertheless, few models have been developed to aid in a range of management practice decisions within the cropping system (Monjardino et al., 2003; Benjamin et al., 2009). DSS for optimizing herbicide use are estimating the need for weed control, where the necessary efficacy depends on field-specific weed population with each weed species causing a certain magnitude of yield loss (Sønderskov et al., 2015), with recommendations often being optimized on the basis of net return (Bennett et al., 2003; Berti et al., 2003). However, DSS are still not broadly adopted and used by farmers (Berti et al., 2003; Swanton et al., 2008), even though some DSS could reduce herbicide inputs significantly (Rydahl et al., 2009). Reasons for this non-use are, for example the relatively low cost of routine herbicide treatments (Rydahl et al., 2009), the additional costs for weed monitoring and DSS being unable to give information about 'when to treat' as this can be provided only if the daily weed density is known (Masin et al., 2011). The prediction of weed density dynamics could be achieved by integrating a predictive model of weed emergence that estimates the seedling emergence dynamics throughout the crop cycle using the actual weed density, that is, the density measured on the field at a given date (Masin et al., 2011).

Numerous predictive models of weed emergence have been developed and some tested in maize crops (Forcella et al., 2000; Grundy, 2003; Dorado et al., 2009; Masin et al., 2012, 2014), with the most recent ones being based on the thermal and hydrothermal concept, with the latter being more reliable under water-limiting conditions for weed emergence (Vasileiadis et al., 2011). The output of such models is the emerged weeds percentage out of the total number of plants that may potentially emerge during the crop cycle. This information can assist in identifying the most efficient timing for either chemical or mechanical weed control, which usually coincides when the predicted percentage of emergence is high (more than 80%, if the weed phenological stage allows it) (Masin et al., 2011). This can allow for reductions in herbicide inputs as a second treatment may be avoided or by replacing herbicides by mechanical weeding, that is, with a positive impact on the environment, net profit and society's acceptance (Vasileiadis et al., 2011).

4.2 Lack of IWM implementation

Limited on-farm implementation and incorporation of IWM components into a fully functional cropping system is caused by gaps and barriers to IWM adoption (Swanton et al., 2008; Brewer and Goodell, 2012; Barzman et al., 2014). Despite decades of research and the availability of several IWM tools, farmers are still reluctant to adopt alternative weed control measures as they are not willing to face increased cropping system complexity and higher perceived agronomic and economic risks when compared to chemical weed control (Jones and Medd, 2005; Deytieux et al., 2012). Based on recent risk analyses in spring wheat, those perceptions are unfounded as IWM strategies were found to be less variable (i.e. less risky), depleted the weed seedbank more quickly and had more economic benefits, compared to post-emergence herbicide-only systems (Jones and Medd, 2005; Jones et al., 2006). Nevertheless, most growers are unfortunately unable to realize the economic benefits that can derive after the adoption of IWM strategies (Swanton et al., 2008). In order to achieve IWM adoption and promote its implementation, robust evidence on the agronomic, economic and environmental sustainability of IWM strategies, by assessing and validating them at real farm scale (i.e. using farm equipment, and under diverse climatic and soil conditions), is needed and the results should be disseminated properly to raise awareness and motivate stakeholders to adopt IWM (Vasileiadis et al., 2015). In this context, the approach followed by Vasileiadis et al. (2015) who studied the agronomic performance and environmental and economic sustainability of three IWM tools tested in three European regions serves as an example and will be presented in the following paragraph.

5 Case study: IWM tools as evaluated in three European maize production regions

The EU Project PURE (Pesticide Use-and-Risk reduction in European farming systems with IPM, http://www.pure-ipm.eu) that ran during 2011–15 provided IPM solutions and a practical toolbox for their implementation in key European farming systems (annual arable and vegetable, perennial and protected crops) (Lescourret, 2014). Within PURE, a work package was devoted to on-farm experimentation in maize where various IWM tools were tested on large plots in real field conditions and under diverse European grain maize-producing conditions. IWM tools were selected on the basis of the outcomes of the EU projects ENDURE (European Network for Durable Exploitation of Crop Protection Strategies; http://www.endure-network.eu) (Meissle et al., 2010; Vasileiadis et al., 2011, 2013), and after consultations with local stakeholders (e.g. extension services, agrochemical companies, farmers, academic and research institutions). As a case study, we will discuss results obtained on IWM tools tested in three maize-growing regions during 2011–12 and published by Vasileiadis et al. (2015). The IWM tools tested in that study can be found in Table 1.

The objective of the study was to evaluate IWM tools for direct weed control in maize in on-farm experiments (i.e. real field conditions on commercial farms, with natural weed flora) that can be as efficient and as profitable as the conventional approach (broadcast herbicide application) in the region, while exhibiting a lower impact on the environment and human health. Weed flora analyses showed that *C. album* was the only weed species detected in all plots, countries and years, confirming its reputation of being commonly

Table 1 Integrated weed management tools tested in maize during the two experimental periods of the EU project PURE as tested by Vasileiadis et al. (2015)

Countries/region	Conventional weed management	Integrated weed management tools
Southern Germany (2 farms)	Broadcast post-emergence herbicide	Early post-emergence herbicide in band application combined with hoeing followed by a second hoeing
Northern Italy (5 farms)	Broadcast pre- and post-emergence herbicide followed by hoeing	Early post-emergence herbicide broadcast application when indicated by a predictive model of weed emergence, followed by hoeing
Slovenia (2 farms)	Broadcast post-emergence herbicide	Tine harrowing at 2–3rd leaf stage of maize and low dose of post-emergence herbicide

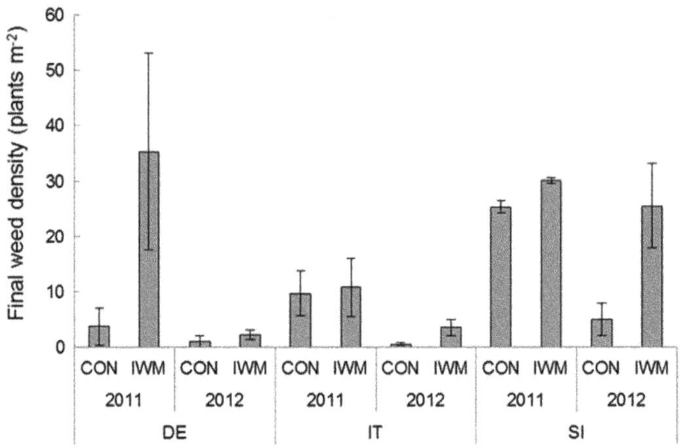

Figure 4 Final weed density (weeds m^{-2}; columns indicate the mean and bars the standard error) as affected by the integrated weed management (IWM) and conventional management (CON) tested in the different countries for each growing season (from Vasileiadis et al., 2015). DE: Germany; IT: Italy; SI: Slovenia.

found and of major importance in maize in Europe (Meissle et al., 2010; Jensen et al., 2011). Diversity analyses showed little evidence of truly dominant species across plots, years and countries before treatments, that is, there was a uniform weed distribution across the experimental sites. Results showed that the performance of the early post-emergence herbicide broadcast application recommended by the predictive model of weed emergence ALERTINF (Masin et al., 2012) and scouting, followed by hoeing in Italy, was as efficient as the conventional approach in both years (Fig. 4).

Indeed, the use of scouting and the predictive model for decision-aid indicated that no post-emergence herbicide was necessary in 2011 at any of the five farms in Italy, mainly due to dry weather conditions hampering weed emergence, that is, the only weed control measure was an inter-row cultivation, whereas in 2012, a post-emergence herbicide application was recommended in four out of five farms, followed by inter-row cultivation in all cases. This resulted in a 64% reduction of the treatment frequency index (TFI) over two years, compared to the conventional weed management (Fig. 5). Moreover, total costs (i.e. herbicide plus application costs) were also greatly reduced (−82 € ha^{-1}) over the two years, whereas no significant differences were determined in maize grain yield and gross margins between IWM and conventional management (Fig. 6). Thus, two years of study has revealed that the use of scouting and the predictive model provided efficient weed control decisions, reduced costs, maintained yields and gross margin, and reduced environmental impacts.

In contrast, for the IWM tools tested in Germany and Slovenia, IWM had significantly lower weed control in the years 2011 and 2012, respectively. The authors stated that in Germany in 2011, *C. album* and *C. polyspermum* were only partially controlled by inter-row cultivation due to a late second treatment caused by soil conditions not being suitable for mechanical weed control (up to 100 mm rainfall in June). Similarly, in Slovenia in 2012, early germinated established plants of *A. retroflexus*, *C. polyspermum* and *E. crus-galli* escaped the tine harrowing and a new flush of weeds shortly after applying reduced rates

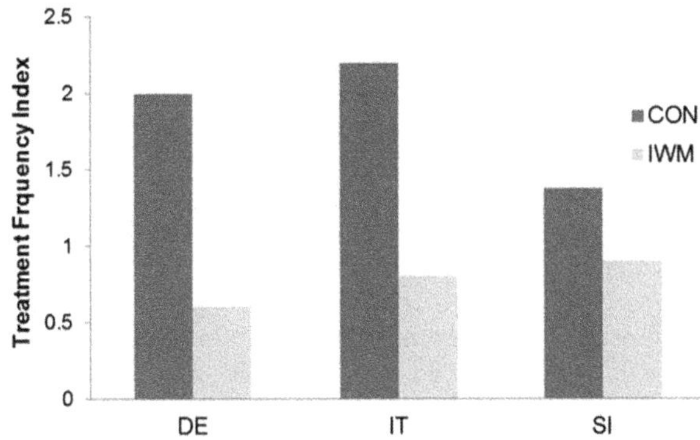

Figure 5 Treatment frequency index (TFI) under conventional (CON) and integrated weed management (IWM) in the three European regions-experimental sites over the two years of study (from Vasileiadis et al., 2015). Treatment frequency index (TFI): the number of full rate treatments: $TFI = 1/n\sum_{t=1}^{t=T} D_t / DAp_t$ with n: number of years in the crop sequence, T: total number of pesticide treatments, D: applied rate of commercial product, DAp: approved/registered rate for the commercial product (Gravesen, 2003). DE: Germany; IT: Italy; SI: Slovenia.

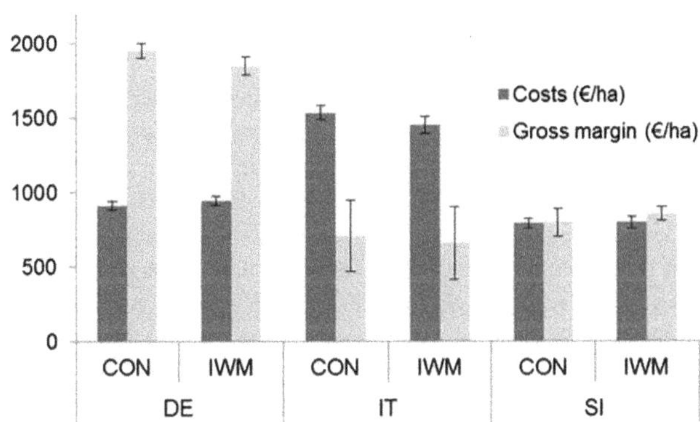

Figure 6 Total costs (€ ha^{-1}) and gross margin (€ ha^{-1}) under conventional (CON) and integrated weed management (IWM) in the three European regions-experimental sites over the two years of the study (from Vasileiadis et al., 2015). DE: Germany; IT: Italy; SI: Slovenia.

of herbicides led to a low control level in IWM. This result stresses the importance of timeliness for post-emergence weed control operations (Parker et al., 2006). Adverse weather conditions or the time needed to complete field operations on large acreages of cropland can reduce the efficacy of weed control eventually affecting yields and impairing farmer's incentives to adopt IWM. Nevertheless, in both countries over the two-year period no significant difference was determined in grain yield and gross margin. In Germany, the use of band application and hoeing gave a 70% reduction in TFI and in Slovenia the harrowing and the reduced doses resulted in a 36% reduction in TFI, compared to conventional weed management across farms and years, respectively.

This two-year study suggested that IWM in Europe is not about a unique weed control strategy but is based on general principles that must be adapted to address locally adapted agro-environmental and social conditions. In this regard, the authors proposed that policy- and decision-makers should closely involve the regional advisory services for the general implementation of IPM (e.g. organize open field visits for dissemination of successful IWM strategies, establish farmer training programmes, encourage the formation of farmers groups to exchange experiences of IWM implementation), and enhancing knowledge on the sustainable use of herbicides. It was also shown that there is always the risk of timeliness for post-emergence weed control operations that can affect the efficacy of an IWM tool and enabling some weeds to survive the treatment, as it was seen in Germany and Slovenia. However, it was reasoned that even if weed escapes could happen and seeds will be produced and shed, in an IWM context, this seedbank increase could be buffered by crop rotation and a higher level of diversity in the system (i.e. winter and spring or summer crops, cover crops), stressing the importance of diversified crop rotations. Overall, the authors concluded that the IWM tools tested in the three European regions provided sufficient weed control, did not cause any significant yield reduction, greatly reduced the reliance on herbicides for weed control in maize and were economically sustainable as no differences were observed in gross margin when compared to the conventional approach.

6 Summary and future trends

Weed management in maize production is mainly dependent on herbicide use since it is perceived by farmers worldwide to be the most economic and effective means for weed control. Besides the fact that herbicide use has resulted in contamination of surface and ground waters in Europe (Haarstad and Ludvigsen, 2007) as well as in the United States (Holman et al., 2000), the repeated and intensive use of herbicides with the same modes of action has also led to the weed flora shifts where tolerant, difficult-to-control weeds dominate the weed flora and has promoted the evolution of HR weeds, especially in the United States and South America. A more integrated and diverse approach to weed management in maize is essential to prevent these adverse effects and increase the sustainability of MBCS. IWM in maize combining preventive, cultural, mechanical and chemical control methods, with the possible integration of decision-aid methods like DSS and/or predictive models of weed emergence, has the potential to suppress weed densities to manageable levels, reduce the environmental impact of individual weed management practices, increase cropping system sustainability and reduce selection pressure for weed resistance to herbicides. Nevertheless, IWM implementation in both the United States and Europe is limited. Reasons for this are that the potential economic benefits of IWM in most cases cannot be realized within a few years but will typically materialize over a longer time span. Furthermore, many farmers perceive IWM to be a higher risk approach and, as seen in the PURE case study, this can sometimes be the case. Targeting farm subsidies to farmers adopting IWM or setting up a private insurance system may overcome farmer's reluctance to take IWM to heart. Also, robust evidence on the benefits of IWM should be disseminated to farmers to raise awareness and provide motive for its adoption. Indeed, research on IWM tools and methods is vast worldwide as discussed in this chapter and agronomic benefits of single weed control techniques and of IWM have been reported. According to a review by Harker and O'Donovan (2013), covering a period from 1995 to 2012, the number of IWM research articles published (*viz.* 697) is very low, compared to the number of research papers on single weed control tools (*viz.* 9964), and this trend has to change in the coming years.

There is a need to change the focus of weed management research from individual crops to the cropping system studying, for example the impact of crop rotation and of combining diverse IWM tools on weed population and seedbank dynamics. In maize-growing areas in Europe, an expert-based survey (interviews with advisors, academics and stakeholders from agrochemical industries) conducted in four European regions recommended to focus on crop rotation/cropping sequence, early detection methods, precision/patch spraying using GPS spray maps, DSS, predictive weed emergence models, innovative mechanical weed control methods and cover crops (Vasileiadis et al., 2011). These innovative tools were suggested, as they could (1) be efficient to control weeds (agronomic impact), (2) reduce the use of or dependence on pesticides against them (environmental impact), (3) result in a net profit of the systems within a time frame of 3–4 years (economic impact) and (4) be accepted by society in terms of their environmental and health impact, and safety of end products (social impact). The survey highlights the will of European stakeholders to change the current weed control practices and should guide future research efforts.

7 Where to look for further information

The following EU research projects provide a lot of information on IPM in maize cropping systems:

EU Network of Excellence ENDURE – European Network for Durable Exploitation of Crop Protection Strategies. Available at: http://www.endure-network.eu

EU Project PURE – Pesticide Use-and-Risk reduction in European farming systems with Integrated Pest Management. Available at: http://www.pure-ipm.eu

More information about weed species, weed management, HR weeds in maize crop and research activities of weed science societies on this subject can be found below:

European Weed Research Society (EWRS). Available at: http://www.ewrs.org/

Heap, I. (2016), *International Survey of Herbicide Resistant Weeds*. Online. Internet. Friday, 8 January 2016. Available at: http://www.weedscience.org/In.asp

International Weed Science Society (IWSS). Available at: http://www.iwss.info/

Weed Science Society of America (WSSA). Available at: http://wssa.net/

Official journal of EWRS – 'Weed Research'. Available at: http://onlinelibrary.wiley.com/journal/10.1111/%28ISSN%291365-3180

Gruppo Italiano di Lavoro sulla Resistenza agli Erbicidi (GIRE, Italian herbicide resistance working group). Database of herbicide resistance in Italy. Available at: http://www.resistenzaerbicidi.it

Official journals of WSSA – 'Weed Science', 'Weed Technology' and 'Invasive Plant Science and Management'. Available at: http://wssajournals.org/

8 References

Aldrich, R. J. and Kremer, R. J. (1997), *Principles in Weed Management*. 2nd Edition, Iowa State University Press/Ames, p. 455.

Ball, D. A. and Miller, S. D. (1990), Weed seed population response to tillage and herbicide use in three irrigated cropping sequences. *Weed Sci.* 38, 511–17.
Ball, D. A. (1992), Weed seedbank response to tillage, herbicides and crop rotation sequence. *Weed Sci.* 40, 654–9.
Bàrberi, P., Silvestri, N. and Bonari, E. (1997), Weed communities of winter wheat as influenced by input level and rotation. *Weed Res.* 37, 301–13.
Bàrberi, P., Cozzani, A., Macchia, M. and Bonari, E. (1998), Size and composition of the weed seedbank under different management systems for continuous maize cropping. *Weed Res.* 38, 319–34.
Bàrberi, P. (2002), Weed management in organic agriculture: are we addressing the right issues? *Weed Res.* 42, 177–93.
Barzman, M. S., Bertschinger, L., Dachbrodt-Saaydeh, S., Graf, B., Jensen, J. E., Joergensen, L. N., Kudsk, P., Messéan, A., Moonen, A. C., Ratnadass, A., Sarah, J. L. and Sattin, M. (2014), Integrated Pest Management policy, research and implementation: European initiatives, in R. Peshin and D. Pimentel (eds), *Integrated Pest Management: Experiences with Implementation*, Global Overview, Springer, pp. 415–28.
Barzman, M., Bàrberi, P., Nicholas, A., Birch, E., Boonekamp, P., DachbrodtSaaydeh, S., Graf, B., Hommel, B., Jensen, J. E., Kiss, J., Kudsk, P., Lamichhane, J. R., Messéan, A., Moonen, A. C., Ratnadass, A., Ricci, P., Sarah, J. L. and Sattin, M. (2015), Eight principles of Integrated Pest Management. *Agron. for Sust. Develop.* doi: 10.1007/s13593-015-0327-9.
Bell, M. S., Hager, A. G. and Tranel, P. J. (2013), Multiple resistance to herbicides from four site-of-action groups in waterhemp (*Amaranthus tuberculatus*). *Weed Sci.* 61, 460–8.
Benbrook, C. M. (2012), Impacts of genetically engineered crops on pesticide use in the U.S. – the first sixteen years. *Environmental Sciences Europe* 24(24), 2190–4715.
Benjamin, L. R., Milne, A. E., Parsons, D. J., Cussans, J. and Lutman, P. J. W. (2009), Using stochastic dynamic programming to support weed management decisions over a rotation. *Weed Res.* 49, 207–16.
Bennett, A. C., Price, A. J., Sturgill, M. C., Buol, G. S. and Wilkerson, G. G. (2003), HADSS (TM), pocket HERB (TM), and WebHADSS (TM): decision aids for field crops. *Weed Technol.* 17, 412–20.
Berti, A. and Zanin, G. (1997), GESTINF: a decision model for post-emergence weed management in soybean (*Glycine max* (L) Merr). *Crop Prot.* 16, 109–16.
Berti, A., Bravin, F. and Zanin, G. (2003), Application of decision-support software for postemergence weed control. *Weed Sci.* 51, 618–27.
Blackshaw, R. E., Molnar, L. J. and Janzen, H. H. (2004), Nitrogen fertilizer timing and application method affect weed growth and competition with spring wheat. *Weed Sci.* 52, 614–22.
Bond, W. and Grundy, A. C. (2001), Non-chemical weed management in organic farming systems. *Weed Res.* 41, 383–405.
Brewer, M. J. and Goodell, P. B. (2012), Approaches and incentives to implement integrated pest management that addresses regional and environmental issues. *Annu. Rev. Entomol.* 57, 41–59.
Buhler, D. D. and Daniel, T. C. (1988), Influence of tillage systems on giant foxtail, Setaria faberi, and velvetleaf, *Abutilon theophrasti*, density and control in corn, *Zea mays*. *Weed Sci.* 36, 642–7.
Buhler, D. D. and Mester, T. C. (1991), Effect of tillage system on the emergence depth of giant (*Setaria faberi*) and green foxtail (*Setaria viridis*). *Weed Sci.* 39, 200–3.
Buhler, D. D., Stoltenberg, D. E., Becker, R. L. and Gunsolus, J. L. (1994), Perennial weed populations after 14 years of variable tillage and cropping practices. *Weed Sci.* 42, 205–9.
Buhler, D. D. and Gunsolus, J. L. (1996), Effect of date of preplant tillage and planting on weed populations and mechanical weed control in soybean (*Glycine max*). *Weed Sci.* 44, 373–9.
Buhler, D. D., Hartzler, R. G. and Forcella, F. (1997), Implications of weed seedbank dynamics to weed management. *Weed Sci.* 45, 329–36.
Buhler, D. D., Kohler, K. A. and Thompson, R. L. (2001), Weed seedbank dynamics during a five-year crop rotation. *Weed Technol.* 15, 170–6.
Buhler, D. D. (2002), Challenges and opportunities for integrated weed management. *Weed Sci.* 50, 273–80.

Bullied, W. J., Marginet, A. M. and Van Acker, R. C. (2003), Conventional and conservation tillage systems influence emergence periodicity of annual weed species in canola. *Weed Sci.* 51, 886–97.

Cerrudo, D., Page, E. R., Tollenaar, M., Stewart, G. and Swanton, C. J. (2012), Mechanisms of yield loss in maize caused by weed competition. *Weed Sci.* 60, 225–32.

Chikowo, R., Faloya, V., Petit, S. and Munier-Jolain, N. M. (2009), Integrated Weed Management systems allow reduced reliance on herbicides and long-term weed control. *Agric. Ecosyst. Environ.* 132, 237–42.

Claerhout, S., Reheul, D. and De Cauwer, B. (2015), Sensitivity of *Echinochloa crus-galli populations* to maize herbicides: a comparison between cropping systems. *Weed Res.* 55, 470–81.

Deytieux, V., Nemecek, T., Freiermuth Knuchel, R., Gaillard, G. and Munier-Jolain, N. M. (2012), Is Integrated Weed Management efficient for reducing environmental impacts of cropping systems? A case study based on life cycle assessment. *Eur. J. Agronomy* 36, 55–65.

DiTomaso, J. M. (1995), Approaches for improving crop competitiveness through the manipulation of fertilization strategies. *Weed Sci.* 43, 491–7.

Dorado, J., Sousa, E., Calha, I. M., Gonzalez-Andujar, J. L. and Fernandez-Quintanilla, C. (2009), Predicting weed emergence in maize crops under two contrasting climatic conditions. *Weed Res.* 49, 251–60.

Doucet, C., Weaver, S. E., Hamill, A. S. and Zhang, J. (1999), Separating the effects of crop rotation from weed management on weed density and diversity. *Weed Sci.* 47, 729–35.

Duke, S. O. and Powles, S. B. (2009), Glyphosate-resistant crops and weeds: now and in the future. *Agbioforum* 12, 346–57.

Eleftherohorinos, I. G. (2014), Ζιζανιολογία: Ζιζάνια, Ζιζανιοκτόνα, Περιβαλλον, Αρχές και μέθοδοι διαχείρισης, 4η Έκδοση (*Weed Science: Weeds, Herbicides, Environment, Principles and Methods of Weed Management, 4nd Edition)*, Agrotypos Press.

European Parliament (2009), Directive 2009/128/EC of the European parliament and of the council. http://eur-lex.europa.eu/LexUriServ/LexUriServ.do?uri=OJ:L:2009:309:0071:0086:EN:PDF

EUROSTAT (2016), Statistical office of the European Commission. Available from: http://ec.europa.eu/eurostat/data/database (accessed 3 January 2016).

FAOSTAT (2016), Food and Agriculture Organization of the United Nations, Statistics Division, Rome, Italy. Available from: http://faostat3.fao.org/faostat-gateway/go/to/download/Q/QC/E (accessed 3 January 2016).

Fernandez-Cornejo, J., Nehring, R., Osteen, C., Wechsler, S., Martin, A. and Vialou, A. (2014), Pesticide Use in U. S. Agriculture: 21 Selected Crops, 1960–2008. Economic Information Bulletin No.124. Available from: http://www.ers.usda.gov/publications/eib-economic-information-bulletin/eib124.aspx (accessed 4 January 2016).

Forcella, F. and Lindstrom, M. J. (1988), Weed seed populations in ridge and conventional tillage. *Weed Sci.* 36, 500–2.

Forcella, F., Wilson, R. G., Dekker, J., Kremer, R. J., Cardina, J., Anderson, R. L., Alm, D., Renner, K. A., Harvey, R. G., Clay, S. and Buhler, D. D. (1997), Weed seed bank emergence across the Corn Belt. *Weed Sci.* 40, 67–76.

Forcella, F., Benech-Arnold, R. L., Sanchez, R. and Ghersa, C. M. (2000), Modeling seedling emergence. *Field Crop Res.* 67, 123–39.

Gibson, K. D., Johnson, W. G. and Hillger, D. E. (2005), Farmer perceptions of problematic corn and soybean weeds in Indiana. *Weed Technol.* 19(4), 1065–70.

Gieske, M. F., Wyse, D. L. and Durgan, B. R. (2015), Spring and fall-seeded radish cover crop effects on weed management in corn. *Weed Technol.* (In press).

Givens, W. A., Shaw, D. R., Kruger, G. R., Johnson, W. G., Weller, S. C., Young, B. G., Wilson, R. G., Owen, M. D. K. and Jordan, D. (2009), Survey of tillage trends following the adoption of glyphosate-resistant crops. *Weed Technol.* 23, 150–6.

Ghorbani, R., Seel, W., Litterick, A. and Leifert, C. (2000), Evaluation of Alternaria alternata for biological control of *Amaranthus retroflexus*. *Weed Sci.* 48(4), 474–80.

Gravesen, L. (2003), The treatment frequency index – an indicator for pesticide use and dependency as well as overall load on the environment. In *Reducing Pesticide Dependency in Europe to Protect Health, Environment and Biodiversity, Copenhagen, Pesticides Action Network Europe (PAN)*, Pure Conference.

Green, J. M. (2012), The benefits of herbicide-resistant crops. *Pest Manag. Sci.* 68(10), 1323–31.

Green, J. M. and Owen, M. D. K. (2011), Herbicide-resistant crops: utilities and limitations for herbicide-resistant weed management. *J. Agric. Food Chem.* 59(11), 5819–29.

Grundy, A. C. (2003), Predicting weed emergence: a review of approaches and future challenges. *Weed Res.* 43, 1–11.

Grundy A. C., Bond, W. Burston, S. and Jackson, L. (1999), Weed suppression by crops. In *Proceedings 1999 Brighton Conference – Weeds*, Brighton, UK, pp. 957–62.

Haarstad, K., Ludvigsen, G. H. (2007), Ten years of pesticide monitoring in Norwegian ground water. *Ground Water Monit.* R 27, 75–89.

Harbur, M. M. and Owen, M. D. K. (2004), Response of three annual weeds to corn population density and nitrogen fertilization timing. *Weed Sci.* 52(5), 845–53.

Harker, K. N. and O'Donovan, J. T. (2013), Recent weed control, weed management and Integrated Weed Management. *Weed Technol.* 27, 1–11.

Heap, I. (2016), The International Survey of Herbicide Resistant Weeds. Online. Internet. Wednesday, 21 March 2016. Available www.weedscience.org.

Holman, R. E., Leidy, R. B. and Walker, A. E. (2000), Evaluation of selected pesticides in North Carolina surface water supply: intake study. *J. Am. Water Resour. Assoc.* 36, 75–85.

Jensen, P. K., Bibard, V., Czembor, E., Dumitru, S., Foucart, G., Froud-Williams, R. J., Jensen, J. E., Saavedra, M., Sattin, M., Soukup, J., Palou, A. T., Thibord, J.-B., Voegler, W. and Kudsk, P. (2011), Survey of weeds in maize crops in Europe. *DJF Report Agricultural Science* No. 149, p. 44.

Johnson, W. G., Ott, E. J., Gibson, K. D., Nielsen, R. L. and Bauman, T. T. (2007), Influence of nitrogen application timing on low density giant ragweed (*Ambrosia Trifida*) interference in corn. *Weed Technol.* 21(3), 763–7.

Jones, R. E. and Medd, R. W. (2005), A methodology for evaluating risk and efficacy of weed management technologies. *Weed Sci.* 53, 505–14.

Jones, R., Cacho, O. and Sinden, J. (2006), The importance of seasonal variability and tactical responses to risk on estimating the economic benefits of integrated weed management. *Agric. Econ.* 35, 245–56.

Knezevic, S. Z. and Datta, A. (2015), The Critical Period for Weed Control: Revisiting Data Analysis. *Weed Sci. Special Issue*, 63, 188–202.

Lescourret, F. (2014), IPM and the integration of innovative application techniques in IPM. *Asp. Appl. Biol.* 122, 153–7.

Liebman, M. and Davis, A. S. (2000), Integration of soil, crop and weed management in low-external-input farming systems. *Weed Res.* 40, 27–47.

Liebman, M., Miller, Z. J., Williams, C. L., Westerman, P. R., Dixon, P. M., Heggenstaller, A., Davis, A. S., Menalled, F. D. and Sundberg, D. N. (2014), Fates of *Setaria faberi* and *Abutilon theophrasti* seeds in three crop rotation systems. *Weed Res.* 54, 293–306.

Liu, J. G., Mahoney, K. J., Sikkema, P. H. and Swanton, C. J. (2009), The importance of light quality in crop–weed competition. *Weed Res.* 49, 217–24.

Livingston, M., Fernandez-Cornejo, J., Unger, J., Osteen, C., Schimmelpfennig, D., Park, T. and Lambert, D. (2015), The Economics of Glyphosate Resistance Management in Corn and Soybean Production. Economic Information Bulletin No.184. Available from: http://www.ers.usda.gov/publications/err-economic-research-report/err184.

Masin, R., Vasileiadis, V. P., Loddo, D., Otto, S. and Zanin, G. (2011), A single-time survey method to predict the daily weed density for weed control decision-making. *Weed Sci.* 59(2), 270–5.

Masin, R., Loddo, D., Benvenuti, S., Otto, S. and Zanin, G. (2012), Modeling weed emergence in Italian maize fields. *Weed Sci.* 60(2), 254–9.

Masin, R., Loddo, D., Gasparini, V., Otto, S. and Zanin, G. (2014), Evaluation of weed emergence model alertInf for maize in soybean. *Weed Sci.* 62(2), 360–9.
Meissle, M., Mouron, P., MUS, T., Bigler, F., Pons, X., Vasileiadis, V. P., Otto, S., Antichi, D., Kiss, J., Pálinkás, Z., Dorner, Z., van der Weide, R., Groten, J., Czembor, E., Adamczyk, J., Thibord, J.-B., Melander, B., Cordsen Nielsen, G., Poulsen, R. T., Zimmermann, O., Verschwele, A. and Oldenburg, E. (2010), Pests, pesticide use and alternative options in European maize production: current status and future prospects. *J. Appl. Entomol.* 134, 357–75.
Melander, B., Rasmussen, I. A. and Barberi, P. (2005), Integrating physical and cultural methods of weed control – examples from European research. *Weed Sci.* 53, 369–81.
Melander, B., Munier-Jolain, N., Charles, R., Wirth, J., Schwarz, J., van der Weide, R., Bonin, L., Jensen, P. K. and Kudsk, P. (2013), European perspectives on the adoption of nonchemical weed management in reduced-tillage systems for arable crops. *Weed Technol.* 27(1), 231–40.
Mohler, C. L. (1996), Ecological bases for the cultural control of annual weeds. *J. Production Agriculture* 9, 468–74.
Monjardino, M., Pannell, D. J. and Powles, S. B. (2003), Multispecies resistance and integrated management: a bioeconomic model for integrated management of rigid ryegrass (*Lolium rigidum*) and wild radish (*Raphanus raphanistrum*). *Weed Sci.* 51, 798–809.
Moonen, A. C. and Bàrberi, P. (2004), Size and composition of the weed seedbank after 7 years of different cover-crop-maize management systems. *Weed Res.* 44, 163–77.
Murphy, S. D., Yukubu, Y., Weise, S. W. and Swanton, C. J. (1996), Effect of planting patterns and inter-row cultivation on competition between corn and late emerging weeds. *Weed Sci.* 44, 856–70.
Norsworthy, J. K. and Oliveira, M. J. (2004), Comparison of the critical period for weed control in wide- and narrow-row corn. *Weed Sci.* 52, 802–7.
Norsworthy, J. K., Ward, S. M., Shaw, D. R., Llewellyn, R. S., Nichols, R. L., Webster, T. M., Bradley, K. W., Frisvold, G., Powles, S. B., Burgos, N. R., Witt, W. W. and Barrett, M. (2012), Reducing the risks of herbicide resistance: Best management practices and recommendations. *Weed Sci. Special Issue*, 31–62.
Oryokot, J. O. E., Murphy, S. D. and Swanton, C. J. (1997), Effect of tillage and corn on pigweed (*Amaranthus* spp.) seedling emergence and density. *Weed Sci.* 45, 120–6.
Otto, S., Masin, R., Casari, G. and Zanin, G. (2009), Weed–corn competition parameters in late-winter sowing in Northern Italy. *Weed Sci.* 57(2), 194–201.
Page, E. R., Cerrudo, D., Westra, P., Loux, M., Smith, K., Foresman, C., Wright, H. and Swanton, C. J. (2012), Why early season weed control is important in Maize. *Weed Sci.* 60, 423–30.
Panozzo, S., Scarabel, L., Balogh, Á., Heini, J., Dancza, I. and Sattin, M. (2015), First European cases of *Sorghum halepense* (L.) resistant to ALS inhibitors: resistance patterns and mechanisms. In *Proceedings of 17th European Weed Research Society Symposium, Weed Management in Changing Environments*, 23–6 June 2015, Montpellier, France, p. 97.
Parker R. G., York, A. C. and Jordan D. L. (2006), Weed control in glyphosate-resistant corn as affected by preemergence herbicide and timing of postemergence herbicide application. *Weed Technol.* 20, 564–70.
Rajcan, I. and Swanton, C. J. (2001), Understanding maize–weed competition: resource competition, light quality and the whole plant. *Field Crops Res.* 71, 139–50.
Raun, W. R. and Johnson, G. V. (1999), Improving nitrogen use efficiency for cereal production. *Agronomy Journal* 9, 357–63.
Rydahl, P., Been, T., Berti, A., Evans, N., Gouache, D., Gutsche, V., Jensen, J. E., Kapsa, J., Levay, N., Munier-Jolain, N., Nibouche, S. and Raynal, M. (2009), Review of New Technologies Critical to Effective Implementation of Decision Support Systems and Farm Management Systems. http://www.endure-network.eu/about_endure/all_the_news/dss_helping_farmers_make_smart_decisions.
Pester, T. A., Burnside, O. C. and Orf, J. H. (1999), Increasing crop competitiveness to weeds through crop breeding. *J. Crop Prod.* 2, 31–58.
Sattin, M., Zanin, G. and Berti, A. (1992), Case history for weed competition/population ecology: velvetleaf (*Abutilon theophrasti*) in corn (*Zea mays*). *Weed Technol.* 6, 213–19.

Sattin, M., Zuin, M. C. and Sartorato, I. (1994), Light quality beneath field-grown maize, soybean and wheat canopies – red:far red variations. *Physiol. Plantarum* 91, 322–8,

Sciegienka, J. K, Keren, E. N. and Menalled, F. D. (2011), Interactions between two biological control agents and an herbicide for Canada thistle (*Cirsium arvense*) Suppression. *Invasive Plant Sci. and Manag.* 4,151–8.

Scharf, P. C., Wiebold, W. J. and Lory, J. A. (2002), Corn yield response to nitrogen fertilizer timing and deficiency level. *Agron. J.* 94, 435–41.

Schreiber, M. M. (1992), Influence of tillage, crop rotation, and weed management on giant foxtail (*Setaria faberi*) population dynamics and corn yield. *Weed Sci.* 40, 645–53.

Sønderskov, M., Kudsk, P., Mathiassen, S., Bøjer, O. M. and Rydahl, P. (2014), Decision support system for optimized herbicide dose in spring barley. *Weed Technol.* 28, 19–27.

Sønderskov, M., Fritzsche, R., de Mol, F., Gerowitt, B., Goltermann, S., Kierzek, R., Krawczyk, R., Bøjer, O. M. and Rydahl, P. (2015), DSSHerbicide: Weed control in winter wheat with a decision support system in three South Baltic regions e Field experimental results. *Crop Prot.* 76, 15–23.

Swanton, C. J., Mahoney, K. J., Chandler, K. and Gulden, R. H. (2008), Integrated weed management: knowledge-based weed management systems. *Weed Sci.* 56, 168–72.

Swinton, S. M., Renner, K. A. and Kells, J. J. (2002), On-Farm Comparison of Three Postemergence Weed Management Decision Aids in Michigan. *Weed Technol.* 16(3), 691–8.

Teasdale, J. R. (1995), Influence of narrow row/high population corn (*Zea mays*) on weed control and light transmission. *Weed Technol.* 9, 113–18.

Tuesca, D., Puricelli, E. and Papa, J. C. (2001), A long term study of weed flora shifts in different tillage systems. *Weed Res.* 41, 369–82.

Unger, P. W., Miller, S. D. and Jones, O. R. (1999), Weed seeds in long-term dryland tillage and cropping system plots. *Weed Res.* 39, 213–23.

United States Department of Agriculture (USDA) – Economic Research Service (2011), Corn. Available from: http://www.ers.usda.gov/topics/crops/corn.aspx (accessed 8 January 2016).

United States Department of Agriculture (USDA) – National Agricultural Statistics Service (2016), Available from: http://quickstats.nass.usda.gov/ (accessed 8 January 2016).

Van der Weide, R. Y., Bleeker, P. O., Achten, V. T. J. M., Lotz, L. A. P., Fogelberg, F. and Melander, B. (2008), Innovation in mechanical weed control in crop rows. *Weed Res.* 48, 215–24.

Vasileiadis, V. P., Froud-Williams, R. J. and Eleftherohorinos I. G. (2007), Vertical distribution, size and composition of the weed seedbank under various tillage and herbicide treatments in a sequence of industrial crops. *Weed Res.* 47, 222–30.

Vasileiadis, V. P., Sattin, M., Otto, S., Veres, A., Pálinkás, Z., Ban, R., Pons, X., Kudsk, P., van der Weide, R., Czembor, E., Moonen, A. C. and Kiss, J. (2011), Crop protection in European maize-based cropping systems: current practices and recommendations for innovative Integrated Pest Management. *Agr. Syst.* 104, 533–40.

Vasileiadis, V. P., Moonen, A. C., Sattin, M., Otto, S., Pons, X., Kudsk, P., Veres, A., Dorner, Z., van der Weide, R., Marraccini, E., Pelzer, E., Angevin, F. and Kiss, J. (2013), Sustainability of European maize-based cropping systems: Economic, environmental and social assessment of current and proposed innovative IPM-based systems. *Eur. J. Agron.* 48, 1–11.

Vasileiadis, V. P., Otto, S., van Dijk, W., Urek, G., Leskovšek, R., Verschwele, A., Furlan, L. and Sattin, M. (2015), On-farm evaluation of integrated weed management tools for maize production in three different agro-environments in Europe: agronomic efficacy, herbicide use reduction and economic sustainability. *Eur. J. Agron.* 63, 71–8.

Vencill, W. K., Nichols, R. L., Webster, T. M., Soteres, J. K., Mallory-Smith, C., Burgos, N. R., Johnson, W. G. and McClelland, M. R. (2012), Herbicide Resistance: Toward an understanding of resistance development and the impact of herbicide-resistant crops. *Weed Sci. Special Issue*, 2–30.

Walker, R. H. (1995), Preventative weed management. in A. E. Smith, ed. Handbook of Weed Management Systems. New York: Marcel Dekker, Pages 35–50.

Wiggins, M. S., McClure, M. A., Hayes, R. M. and Steckel, L. E. (2015), Integrating cover crops and POST herbicides for glyphosate-resistant palmer amaranth (*Amaranthus palmeri*) control in corn. *Weed Technol.* 29(3), 412–18.

Zanin, G., Berti, A. and Sattin, M. (1994), Estimation of economic thresholds for weed control in maize in northern Italy. In F. Tei and Onofri Perugia, *Proceedings of the 5th EWRS Mediterranean Symposium: Weed Control in Sustainable Agriculture in the Mediterranean Area*, Italy, 6–8 June 1994, pp. 51–8.

Zimdahl R. L. (1988), The concept and application of the critical weed-free period. In M. A. Altieri and M. Liebmann (eds), *Weed Management in Agroecosystems: Ecological Approaches*, CRC Press, Boca Raton, pp. 145–55.

Index

Lightning Source UK Ltd.
Milton Keynes UK
UKHW051509181118
332453UK00003BA/15/P